Springer Texts in Statistics

Springer Texts in Statistics (STS) includes advanced textbooks from 3rd- to 4th-year undergraduate courses to 1st- to 2nd-year graduate courses. Exercise sets should be included. The series editors are currently Genevera I. Allen, Richard D. De Veaux, and Rebecca Nugent. Stephen Fienberg, George Casella, and Ingram Olkin were editors of the series for many years.

More information about this series at http://www.springer.com/series/417

Ronald Christensen

Plane Answers to Complex Questions

The Theory of Linear Models

Fifth Edition

Springer

Ronald Christensen
Department of Mathematics and Statistics
University of New Mexico
Albuquerque, NM, USA

ISSN 1431-875X ISSN 2197-4136 (electronic)
Springer Texts in Statistics
ISBN 978-3-030-32099-7 ISBN 978-3-030-32097-3 (eBook)
https://doi.org/10.1007/978-3-030-32097-3

This Springer imprint is published by the registered company Springer Nature Switzerland AG
The registered company address is: Gewerbestrasse 11, 6330 Cham, Switzerland

To Dad, Mom, Sharon, Fletch, and Don

Prefaces

Critical assessment of data is the essential task of the educated mind.
Professor Garrett G. Fagan, Pennsylvania State University.
The last words in his audio course *The Emperors of Rome*, The Teaching Company.

Preface to the Fifth Edition

I prepared the fifth edition of *Plane Answers (PA-V)* in conjunction with a new edition of *Advanced Linear Modeling (ALM-III)* (Christensen, 2019). The emphasis in both revisions was to include more material on *Statistical Learning*. *ALM-III* has far more changes in it than *PA-V*. (*ALM-III* is about 50% longer than *ALM-II*.) In *ALM-III*, all but the first three chapters (and Chapter 13) are devoted to dependent data. I regretfully concluded that almost all of the mixed models chapter in *PA* needed to go into *ALM-III*. The one exception is that I moved the discussion of BLUP into Chapter 6 of *PA-V*.

The biggest changes in *PA-V* are listed below:

- Section 1.3 has been restructured to isolate the more difficult parts.
- Section 2.9 is a new section on biased estimation and the variance-bias tradeoff.
- Subsection 3.2.1 is a short new subsection that introduces the importance of small F statistics.
- A new Exercise 3.7b helps establish Fieller's method prior to its application in Exercises 6.9.1, 2, 3.
- Section 4.1 contains some cleaner notation for one-way ANOVA computations.
- Section 5.1 is a new section containing my overall view of common multiple comparison procedures.
- Subsubsection 6.3.3.2 discusses best predictors for loss functions other than squared error.
- Section 6.6 now contains the discussion of BLUP.
- The section on polynomial regression and one-way ANOVA now contains a table of polynomial contrasts.
- Subsection 7.5.3 contains new material on characterizing the interaction space in an unbalanced two-way ANOVA.

- Subsection 9.1.1 introduces ACOVA ideas for models with dependent or heteroscedastic data.
- I thought about just deleting Section 9.3 but opted for attempting to make it more relevant.
- Section 11.2 has some new results on checking whether models qualify as generalized split plot models.
- New Subsection 11.2.3 addresses the analysis of (generalized) split plot designs when there is missing data in the subplots.
- As mentioned, mixed models got moved to *ALM-III* because it fits naturally into *ALM-III*'s emphasis on dependent data.
- Subsection 12.4.1 includes additional discussion of testing for heteroscedasticity.
- Subsection 12.4.2 introduces the Huber–White sandwich estimator.
- I changed the order of the chapters on variable selection and collinearity from previous versions of *PA* in order to smooth the presentation with *ALM-III*.
- The collinearity chapter contains a new Section 13.2 on variance inflation factors.
- The singular value decomposition of a matrix X, Theorem 13.3.1, has been generalized and the relationship between ridge regression and principal component regression explicated.
- Section 13.6 is a new section on different approaches to estimation. While this stands on its own, it also motivates material in *ALM-III*.
- Section 14.1 now discusses information criteria for model selection as well as cost complexity pruning.
- Section 14.2 has examples illustrating issues with larger data sets.
- Section 14.3 contains more discussion of variable selection.
- Section 14.4 introduces *boosting*, *bagging* and the random part of *random forests*. The application of these subjects is closely related to nonparametric regression as discussed in Chapter 1 of *ALM-III*.
- Appendix B has a number of refinements in the results. I also decided to rename "orthogonal matrices" as "orthonormal matrices" because it is a clearly better name.
- Appendix D has a new section on identifiability.

A big part of the effort in producing *PA-V* was just cleaning the text. After 30 years you would think, by now, I would be happy with it.

While *PA* is a book on Linear Model Theory, Christensen (2015) illustrates the use of most of the theory presented in this book. There are a number of related topics discussed on my website `http://www.stat.unm.edu/~fletcher/` in various places. These include computer code for the applications book as well as for *ALM-III*, cf. `http://www.stat.unm.edu/~fletcher/Rcode.pdf` and `http://www.stat.unm.edu/~fletcher/R-ALMIII.pdf`.

I have quite assiduously avoided doing asymptotic theory in *PA*, and that remains true in *PA-V*. There are many sources that discuss asymptotics for linear model theory. The appendix to Christensen and Lin (2015) uses a number of the most important results.

I would like to thank Fletcher Christensen and Joe Cavanaugh both of whom I have used as a sounding board for years on linear model issues. I thank Mohammad Hattab for numerous suggestions. Since the last edition of the book, Steve Fienberg and Ingram Olkin have both died. They were the subject matter editors for Springer when *PA* was first published in 1987. I sent the book to a lot of publishers and Steve was the only person who took seriously the efforts of an assistant professor from Montana State University. (Steve had been one of my professors at the University of Minnesota.) He recommended it to Ingram who both liked it a lot and gave me a large number of suggestions (virtually all of which remain in the book). I owe both of them a great debt!

Please note that while most of this book's examples, exercises, and figures draw from real data and are cited accordingly, a few of them are based on simulated data.

Some people think that Plane Answers is an example of the old maxim, "If all you have is a hammer, everything looks like a nail." I prefer to think that if you have a good enough hammer, almost everything actually is a nail.

Ronald Christensen
Albuquerque, New Mexico, 2018

Preface to the Fourth Edition

As with the prefaces to the second and third editions, this focuses on changes to the previous edition. The preface to the first edition discusses the core of the book.

Two substantial changes have occurred in Chapter 3. Subsection 3.3.2 uses a simplified method of finding the reduced model and includes some additional discussion of applications. In testing the generalized least squares models of Section 3.8, even though the data may not be independent or homoscedastic, there are conditions under which the standard F statistic (based on those assumptions) still has the standard F distribution under the reduced model. Section 3.8 contains a new subsection examining such conditions.

The major change in the fourth edition has been a more extensive discussion of best prediction and associated ideas of R^2 in Sections 6.3 and 6.4. It also includes a nice result that justifies traditional uses of residual plots. One portion of the new material is viewing best predictors (best linear predictors) as perpendicular projections of the dependent random variable y into the space of random variables that are (linear) functions of the predictor variables x. A new subsection on inner products and perpendicular projections for more general spaces facilitates the discussion. While these ideas were not new to me, their inclusion here was inspired by deLaubenfels (2006).

Section 9.1 has an improved discussion of least squares estimation in ACOVA models. A new Section 9.5 examines Milliken and Graybill's generalization of Tukey's one degree of freedom for nonadditivity test.

A new Section 10.5 considers estimable parameters that can be known with certainty when $C(X) \not\subset C(V)$ in a general Gauss–Markov model. It also contains a relatively simple way to estimate estimable parameters that are not known with certainty. The nastier parts in Sections 10.1–10.4 are those that provide sufficient generality to allow $C(X) \not\subset C(V)$. The approach of Section 10.5 seems more appealing.

In Sections 12.4 and 12.6, the point is now made that ML and REML methods can also be viewed as method of moments or estimating equations procedures.

The biggest change in Chapter 13 is a new title. The plots have been improved and extended. At the end of Section 13.6, some additional references are given on case deletions for correlated data as well as an efficient way of computing case deletion diagnostics for correlated data.

The old Chapter 14 has been divided into two chapters, the first on variable selection and the second on collinearity and alternatives to least squares estimation. Chapter 15 includes a new section on penalized estimation that discusses both ridge and lasso estimation and their relation to Bayesian inference. There is also a new section on orthogonal distance regression that finds a regression line by minimizing orthogonal distances, as opposed to least squares, which minimizes vertical distances.

Appendix D now contains a short proof of the claim: If the random vectors x and y are independent, then any vector-valued functions of them, say $g(x)$ and $h(y)$, are also independent.

Another significant change is that I wanted to focus on Fisherian inference, rather than the previous blend of Fisherian and Neyman–Pearson inference. In the interests of continuity and conformity, the differences are soft-pedaled in most of the book. They arise notably in new comments made after presenting the traditional (one-sided) F test in Section 3.2 and in a new Subsection 5.6.1 on multiple comparisons. The Fisherian viewpoint is expanded in Appendix F, which is where it primarily occurred in the previous edition. But the change is most obvious in Appendix E. In all previous editions, Appendix E existed just in case readers did not already know the material. While I still expect most readers to know the "how to" of Appendix E, I no longer expect most to be familiar with the "why" presented there.

Other minor changes are too numerous to mention and, of course, I have corrected all of the typographic errors that have come to my attention. Comments by Jarrett Barber led me to clean up Definition 2.1.1 on identifiability.

My thanks to Fletcher Christensen for general advice and for constructing Figures 10.1 and 10.2. (Little enough to do for putting a roof over his head all those years. :-)

Ronald Christensen
Albuquerque, New Mexico, 2010

Preface to the Third Edition

The third edition of *Plane Answers* includes fundamental changes in how some aspects of the theory are handled. Chapter 1 includes a new section that introduces generalized linear models. Primarily, this provides a definition so as to allow comments on how aspects of linear model theory extend to generalized linear models.

For years, I have been unhappy with the concept of estimability. Just because you cannot get a linear unbiased estimate of something does not mean you cannot estimate it. For example, it is obvious how to estimate the ratio of two contrasts in an ANOVA, just estimate each one and take their ratio. The real issue is that if the model matrix X is not of full rank, the parameters are not identifiable. Section 2.1 now introduces the concept of identifiability and treats estimability as a special case of identifiability. This change also resulted in some minor changes in Section 2.2.

In the second edition, Appendix F presented an alternative approach to dealing with linear parametric constraints. In this edition, I have used the new approach in Section 3.3. I think that both the new approach and the old approach have virtues, so I have left a fair amount of the old approach intact.

Chapter 8 contains a new section with a theoretical discussion of models for factorial treatment structures and the introduction of special models for homologous factors. This is closely related to the changes in Section 3.3.

In Chapter 9, reliance on the normal equations has been eliminated from the discussion of estimation in ACOVA models—something I should have done years ago! In the previous editions, Exercise 9.3 has indicated that Section 9.1 should be done with projection operators, not normal equations. I have finally changed it. (Now Exercise 9.3 is to redo Section 9.1 with normal equations.)

Appendix F now discusses the meaning of small F statistics. These can occur because of model lack of fit that exists in an unsuspected location. They can also occur when the mean structure of the model is fine but the covariance structure has been misspecified.

In addition, there are various smaller changes including the correction of typographical errors. Among these are very brief introductions to nonparametric regression and generalized additive models, as well as Bayesian justifications for the mixed model equations and classical ridge regression. I will let you discover the other changes for yourself.

Ronald Christensen
Albuquerque, New Mexico, 2001

Preface to the Second Edition

The second edition of *Plane Answers* has many additions and a couple of deletions. New material includes additional illustrative examples in Appendices A and B and Chapters 2 and 3, as well as discussions of Bayesian estimation, near replicate lack of fit tests, testing the independence assumption, testing variance components, the interblock analysis for balanced incomplete block designs, nonestimable constraints, analysis of unreplicated experiments using normal plots, tensors, and properties of Kronecker products and Vec operators. The book contains an improved discussion of the relation between ANOVA and regression, and an improved presentation of general Gauss–Markov models. The primary material that has been deleted are the discussions of weighted means and of log-linear models. The material on log-linear models was included in Christensen (1997), so it became redundant here. Generally, I have tried to clean up the presentation of ideas wherever it seemed obscure to me.

Much of the work on the second edition was done while on sabbatical at the University of Canterbury in Christchurch, New Zealand. I would particularly like to thank John Deely for arranging my sabbatical. Through their comments and criticisms, four people were particularly helpful in constructing this new edition. I would like to thank Wes Johnson, Snehalata Huzurbazar, Ron Butler, and Vance Berger.

Ronald Christensen
Albuquerque, New Mexico, 1996

Preface to the First Edition

This book was written to rigorously illustrate the practical application of the projective approach to linear models. To some, this may seem contradictory. I contend that it is possible to be both rigorous and illustrative, and that it is possible to use the projective approach in practical applications. Therefore, unlike many other books on linear models, the use of projections and subspaces does not stop after the general theory. They are used wherever I could figure out how to do it. Solving normal equations and using calculus (outside of maximum likelihood theory) are anathema to me. This is because I do not believe that they contribute to the understanding of linear models. I have similar feelings about the use of side conditions. Such topics are mentioned when appropriate and thenceforward avoided like the plague.

On the other side of the coin, I just as strenuously reject teaching linear models with a coordinate free approach. Although Joe Eaton assures me that the issues in complicated problems frequently become clearer when considered free of coordinate systems, my experience is that too many people never make the jump from coordinate free theory back to practical applications. I think that coordinate free theory is better tackled after mastering linear models from some other approach. In particular, I think it would be very easy to pick up the coordinate free approach after learning the material in this book. See Eaton (1983) for an excellent exposition of the coordinate free approach.

By now it should be obvious to the reader that I am not very opinionated on the subject of linear models. In spite of that fact, I have made an effort to identify sections of the book where I express my personal opinions.

Although in recent revisions I have made an effort to cite more of the literature, the book contains comparatively few references. The references are adequate to the needs of the book, but no attempt has been made to survey the literature. This was done for two reasons. First, the book was begun about 10 years ago, right after I finished my Masters degree at the University of Minnesota. At that time, I was not aware of much of the literature. The second reason is that this book emphasizes a particular point of view. A survey of the literature would best be done on the literature's own terms. In writing this, I ended up reinventing a lot of wheels. My apologies to anyone whose work I have overlooked.

Using the Book

This book has been extensively revised, and the last five chapters were written at Montana State University. At Montana State, we require a year of Linear Models for all of our statistics graduate students. In our three-quarter course, I usually end the first quarter with Chapter 4 or in the middle of Chapter 5. At the end of winter quarter, I have finished Chapter 9. I consider the first nine chapters to be the core material of the book. I go quite slowly because all of our Masters students are required to take the course. For Ph.D. students, I think a one-semester course might be the first nine chapters, and a two-quarter course might have time to add some topics from the remainder of the book.

I view the chapters after 9 as a series of important special topics from which instructors can choose material but which students should have access to even if their course omits them. In our third quarter, I typically cover (at some level) Chapters 11 to 14. The idea behind the special topics is not to provide an exhaustive discussion but rather to give a basic introduction that will also enable readers to move on to more detailed works such as Cook and Weisberg (1982) and Haberman (1974).

Appendices A–E provide required background material. My experience is that the student's greatest stumbling block is linear algebra. I would not dream of teaching out of this book without a thorough review of Appendices A and B.

The main prerequisite for reading this book is a good background in linear algebra. The book also assumes knowledge of mathematical statistics at the level of, say, Lindgren or Hogg and Craig. Although I think a mathematically sophisticated reader could handle this book without having had a course in statistical methods, I think that readers who have had a methods course will get much more out of it.

The exercises in this book are presented in two ways. In the original manuscript, the exercises were incorporated into the text. The original exercises have not been relocated. It has been my practice to assign virtually all of these exercises. At a later date, the editors from Springer-Verlag and I agreed that other instructors might like more options in choosing problems. As a result, a section of additional exercises was added to the end of the first nine chapters and some additional exercises were added to other chapters and appendices. I continue to recommend requiring nearly all of the exercises incorporated in the text. In addition, I think there is much to be learned about linear models by doing, or at least reading, the additional exercises.

Many of the exercises are provided with hints. These are primarily designed so that I can quickly remember how to do them. If they help anyone other than me, so much the better.

Acknowledgements

I am a great believer in books. The vast majority of my knowledge about statistics has been obtained by starting at the beginning of a book and reading until I covered what I had set out to learn. I feel both obligated and privileged to thank the authors of the books from which I first learned about linear models: Daniel and Wood, Draper and Smith, Scheffé, and Searle.

In addition, there are a number of people who have substantially influenced particular parts of this book. Their contributions are too diverse to specify, but I should mention that, in several cases, their influence has been entirely by means of their written work. (Moreover, I suspect that in at least one case, the person in question will be loath to find that his writings have come to such an end as this.) I would like to acknowledge Kit Bingham, Carol Bittinger, Larry Blackwood, Dennis Cook, Somesh Das Gupta, Seymour Geisser, Susan Groshen, Shelby Haberman, David Harville, Cindy Hertzler, Steve Kachman, Kinley Larntz, Dick Lund, Ingram Olkin, S. R. Searle, Anne Torbeyns, Sandy Weisberg, George Zyskind, and all of my students. Three people deserve special recognition for their pains in advising me on the manuscript: Robert Boik, Steve Fienberg, and Wes Johnson.

The typing of the first draft of the manuscript was done by Laura Cranmer and Donna Stickney.

I would like to thank my family: Sharon, Fletch, George, Doris, Gene, and Jim, for their love and support. I would also like to thank my friends from graduate school who helped make those some of the best years of my life.

Finally, there are two people without whom this book would not exist: Frank Martin and Don Berry. Frank because I learned how to think about linear models in a course he taught. This entire book is just an extension of the point of view that I developed in Frank's class. And Don because he was always there ready to help—from teaching my first statistics course to being my thesis adviser and everywhere in between.

Since I have never even met some of these people, it would be most unfair to blame anyone but me for what is contained in the book. (Of course, I will be more than happy to accept any and all praise.) Now that I think about it, there may be one exception to the caveat on blame. If you don't like the diatribe on prediction in Chapter 6, you might save just a smidgen of blame for Seymour (even though he did not see it before publication).

Ronald Christensen
Bozeman, Montana, 1987

References

Christensen, R. (1997). *Log-linear models and logistic regression* (2nd ed.). New York: Springer.

Christensen, R. (2015). *Analysis of variance, design, and regression: Linear modeling for unbalanced data* (2nd ed.). Boca Raton: Chapman and Hall/CRC Press.

Christensen, R. (2019). *Advanced linear modeling III: Statistical learning and dependent data* (3rd ed.). Springer-Verlag, New York.

Christensen, R. & Lin, Y. (2015). Lack-of-fit tests based on partial sums of residuals. *Communications in Statistics, Theory and Methods, 44* 2862–2880.

Cook, R. D., & Weisberg, S. (1982). *Residuals and influence in regression*. New York: Chapman and Hall.

deLaubenfels, R. (2006). The victory of least squares and orthogonality in statistics. *The American Statistician, 60,* 315–321.

Eaton, M. L. (1983). *Multivariate statistics: a vector space approach*. New York: Wiley. Reprinted in 2007 by IMS Lecture Notes–Monograph Series.

Haberman, S. J. (1974). The Analysis of Frequency Data. University of Chicago Press, Chicago.

Contents

Chapter 1
Introduction

Abstract This chapter introduces the general linear model, illustrating how it subsumes a variety of standard applied models. It also introduces random vectors and matrices and the distributions that will be used with them. Finally, it introduces generalized linear models, which are different from general linear models. It is important to be familiar with the background in Appendices A through E before getting into the book.

This book is about the theory of linear models. A typical model considered is

$$Y = X\beta + e,$$

where Y is an $n \times 1$ vector of random observations, X is an $n \times p$ matrix of known constants called the *model* (or *design*) *matrix*, β is a $p \times 1$ vector of unobservable fixed parameters, and e is an $n \times 1$ vector of unobservable random errors. Both Y and e are random vectors.

Linear models specify the expected value of the observed data Y as a linear function of the parameter vector β. To be a linear model the errors must have mean zero, i.e., $\mathrm{E}(e) = 0$. In a *standard linear model* we assume that the errors have a common unknown variance and are uncorrelated, i.e., $\mathrm{Cov}(e) = \sigma^2 I$ where σ^2 is some unknown parameter. (The operations $\mathrm{E}(\cdot)$ and $\mathrm{Cov}(\cdot)$ will be defined formally in the next section.)

Although our primary object is to explore models that can be used to predict future observable events, much of our effort will be devoted to drawing inferences, in the form of point estimates, tests, and confidence regions, about the parameters β and σ^2. In order to get tests and confidence regions for a standard linear model, we will assume that the errors are independent with normal distributions, i.e., e has an n-dimensional normal distribution with mean vector 0 and covariance matrix $\sigma^2 I$, written $e \sim N(0, \sigma^2 I)$.

R. Christensen, *Plane Answers to Complex Questions*, Springer Texts in Statistics,
https://doi.org/10.1007/978-3-030-32097-3_1

The essence of linear model theory is to decompose the observations Y into $Y = \hat{Y} + \hat{e}$. Here $\hat{Y}$ is a vector of *fitted values* that contains all the information for estimating the unknown parameter vector β. It is somewhere in the vector space spanned by the columns of the model matrix X, i.e., $\hat{Y} \in C(X)$. With any good statistical procedure, it is necessary to investigate whether the assumptions that have been made are reasonable. $\hat{e}$ is a vector of *residuals* that contains the information used for checking the assumptions built into the standard linear model and for estimating parameters associated with the errors. In standard linear model theory, the residuals are used to estimate the unknown variance parameter σ^2. Methods for evaluating the validity of the assumptions will consist of both formal statistical tests and the informal examination of residuals. We will also consider the issue of how to select a model when several alternative models seem plausible.

Applications of linear models often fall into two special cases: *Regression Analysis* and *Analysis of Variance*. Regression Analysis refers to models in which the matrix $X'X$ is nonsingular. Analysis of Variance (*ANOVA*) models are models in which the model matrix consists entirely of zeros and ones. ANOVA models are sometimes called classification models but in recent years that name has been co-opted for models in which the components of Y are binary.

Example 1.0.1 *Simple Linear Regression.*
Consider the model

$$y_i = \beta_0 + \beta_1 x_i + e_i,$$

$i = 1, \ldots, 6$, $(x_1, x_2, x_3, x_4, x_5, x_6) = (1, 2, 3, 4, 5, 6)$, where the e_is are independent $N(0, \sigma^2)$. In matrix notation we can write this as

$$\begin{bmatrix} y_1 \\ y_2 \\ y_3 \\ y_4 \\ y_5 \\ y_6 \end{bmatrix} = \begin{bmatrix} 1 & 1 \\ 1 & 2 \\ 1 & 3 \\ 1 & 4 \\ 1 & 5 \\ 1 & 6 \end{bmatrix} \begin{bmatrix} \beta_0 \\ \beta_1 \end{bmatrix} + \begin{bmatrix} e_1 \\ e_2 \\ e_3 \\ e_4 \\ e_5 \\ e_6 \end{bmatrix}$$

$$Y \quad = \quad X \quad \beta \quad + \quad e.$$

Example 1.0.2 *One-Way Analysis of Variance.*
The model

$$y_{ij} = \mu + \alpha_i + e_{ij},$$

$i = 1, \ldots, 3$, $j = 1, \ldots, N_i$, $(N_1, N_2, N_3) = (3, 1, 2)$, where the e_{ij}s are independent $N(0, \sigma^2)$, can be written as

$$\begin{bmatrix} y_{11} \\ y_{12} \\ y_{13} \\ y_{21} \\ y_{31} \\ y_{32} \end{bmatrix} = \begin{bmatrix} 1 & 1 & 0 & 0 \\ 1 & 1 & 0 & 0 \\ 1 & 1 & 0 & 0 \\ 1 & 0 & 1 & 0 \\ 1 & 0 & 0 & 1 \\ 1 & 0 & 0 & 1 \end{bmatrix} \begin{bmatrix} \mu \\ \alpha_1 \\ \alpha_2 \\ \alpha_3 \end{bmatrix} + \begin{bmatrix} e_{11} \\ e_{12} \\ e_{13} \\ e_{21} \\ e_{31} \\ e_{32} \end{bmatrix}$$

$$Y \quad = \quad X \quad \beta \quad + \quad e.$$

Examples 1.0.1 and 1.0.2 will be used to illustrate concepts in Chapters 2 and 3.

There are a couple of useful alternative methods for writing $Y = X\beta + e$. Write X in terms of its columns and its rows as

$$X_{n\times p} = [X_1, \ldots, X_p] = \begin{bmatrix} x_1' \\ \vdots \\ x_n' \end{bmatrix}.$$

This leads to

$$Y = \sum_{i=1}^{p} \beta_j X_j + e$$

and also, listing the elements of Y and e in the obvious way,

$$y_i = x_i'\beta + e_i, \quad i - 1, \ldots, n.$$

The approach taken here emphasizes the use of vector spaces, subspaces, orthogonality, and projections. These and other topics in linear algebra are reviewed in Appendices A and B. *It is absolutely vital that the reader be familiar with the material presented in the first two appendices.* Appendix C contains the definitions of some commonly used distributions. Much of the notation used in the book is set in Appendices A, B, and C. To develop the distribution theory necessary for tests and confidence regions, it is necessary to study properties of the multivariate normal distribution and properties of quadratic forms. We begin with a discussion of random vectors and matrices in Section 1.

Exercise 1.1 Write the following models in matrix notation:

(a) Multiple regression

$$y_i = \beta_0 + \beta_1 x_{i1} + \beta_2 x_{i2} + \beta_3 x_{i3} + e_i,$$

$i = 1, \ldots, 6$.

(b) Two-way ANOVA with interaction

$$y_{ijk} = \mu + \alpha_i + \beta_j + \gamma_{ij} + e_{ijk},$$

$i = 1, 2, 3, j = 1, 2, k = 1, 2.$

(c) Two-way analysis of covariance (ACOVA) with no interaction

$$y_{ijk} = \mu + \alpha_i + \beta_j + \gamma x_{ijk} + e_{ijk},$$

$i = 1, 2, 3, j = 1, 2, k = 1, 2.$

(d) Multiple polynomial regression

$$y_i = \beta_{00} + \beta_{10}x_{i1} + \beta_{01}x_{i2} + \beta_{20}x_{i1}^2 + \beta_{02}x_{i2}^2 + \beta_{11}x_{i1}x_{i2} + e_i,$$

$i = 1, \ldots, 6.$

1.1 Random Matrices and Vectors

A random matrix is just a matrix that consists of random variables. In general take $W = [w_{ij}]$ where each w_{ij}, $i = 1, \ldots, r$, $j = 1, \ldots, c$, is a random variable. The expected value of W is taken elementwise, i.e.,

$$\mathrm{E}(W) \equiv [\mathrm{E}(w_{ij})].$$

Clearly, if W_1 and W_2 are random matrices of the same dimensions,

$$\mathrm{E}(W_1 + W_2) = \mathrm{E}(W_1) + \mathrm{E}(W_2).$$

Theorem 1.1.1 *If W is an $r \times c$ random matrix and $A = [a_{ij}]$, $B = [b_{ij}]$ and $C = [c_{ij}]$ are fixed matrices of conformable sizes then*

$$\mathrm{E}(AWB + C) = A\mathrm{E}(W)B + C.$$

Proof We show that every element of the matrix on the left equals every element of the matrix on the right. Take $Q \equiv AWB + C$. It is not too difficult to see that the matrix Q consists of random variables

$$q_{ij} = c_{ij} + \sum_{h=1}^{r} \sum_{k=1}^{c} a_{ih} w_{hk} b_{kj}.$$

By the linearity of expectations for random variables,

$$\mathrm{E}(q_{ij}) = c_{ij} + \sum_{h=1}^{r} \sum_{k=1}^{c} a_{ih} \mathrm{E}(w_{hk}) b_{kj}.$$

The right-hand side is the ij element of $A\mathrm{E}(W)B + C$. □

Let $y_1, \ldots, y_n$ be random variables with $\mathrm{E}(y_i) = \mu_i$, $\mathrm{Var}(y_i) = \sigma_{ii}$, and $\mathrm{Cov}(y_i, y_j) = \sigma_{ij} \equiv \sigma_{ji}$. Writing the random variables as an n-dimensional vector Y, the expected value of Y is

$$\mathrm{E}(Y) = \mathrm{E}\begin{bmatrix} y_1 \\ y_2 \\ \vdots \\ y_n \end{bmatrix} = \begin{bmatrix} \mathrm{E}(y_1) \\ \mathrm{E}(y_2) \\ \vdots \\ \mathrm{E}(y_n) \end{bmatrix} = \begin{bmatrix} \mu_1 \\ \mu_2 \\ \vdots \\ \mu_n \end{bmatrix} \equiv \mu .$$

The definition of the *covariance matrix* of Y is

$$\mathrm{Cov}(Y) \equiv \mathrm{E}\big[(Y-\mu)(Y-\mu)'\big] = \begin{bmatrix} \sigma_{11} & \sigma_{12} & \cdots & \sigma_{1n} \\ \sigma_{21} & \sigma_{22} & \cdots & \sigma_{2n} \\ \vdots & \vdots & \ddots & \vdots \\ \sigma_{n1} & \sigma_{n2} & \cdots & \sigma_{nn} \end{bmatrix} .$$

A random vector is referred to as singular or nonsingular depending on whether its covariance matrix is singular or nonsingular. Sometimes the covariance matrix is called the *variance-covariance matrix* or the *dispersion matrix*.

From Theorem 1.1.1, if Y is an n-dimensional random vector, A is a fixed $r \times n$ matrix, and b is a fixed vector in $\mathbf{R}^r$, then

$$\mathrm{E}(AY + b) = A\mathrm{E}(Y) + b$$

and

$$\mathrm{Cov}(AY + b) = A\mathrm{Cov}(Y)A' .$$

This last equality can be used to show that for any random vector Y, $\mathrm{Cov}(Y)$ is nonnegative definite. It follows that Y is nonsingular if and only if $\mathrm{Cov}(Y)$ is positive definite.

Exercise 1.2 Show that $\mathrm{Cov}(AY + b) = A\mathrm{Cov}(Y)A'$.

Exercise 1.3 Show that $\mathrm{Cov}(Y)$ is nonnegative definite for any random vector Y. Hint: Consider the variance of fixed linear combinations of Y, say $a'Y$.

The covariance of two random vectors with possibly different dimensions can be defined. If $W_{r\times 1}$ and $Y_{s\times 1}$ are random vectors with $\mathrm{E}(W) = \gamma$ and $\mathrm{E}(Y) = \mu$, then the covariance of W and Y is the $r \times s$ matrix

$$\mathrm{Cov}(W, Y) \equiv \mathrm{E}[(W-\gamma)(Y-\mu)'] .$$

In particular, $\mathrm{Cov}(Y, Y) = \mathrm{Cov}(Y)$. If A and B are fixed matrices, and a and b are fixed vectors of conformable sizes, Theorem 1.1.1 quickly yields

$$\mathrm{Cov}(AW + a, BY + b) = A\mathrm{Cov}(W, Y)B'.$$

Another simple consequence of the definition is:

Theorem 1.1.2 *If A and B are fixed matrices and W and Y are random vectors, and if AW and BY are both vectors in $\mathbf{R}^n$, then, assuming that the expectations exist,*

$$\mathrm{Cov}(AW + BY) = A\mathrm{Cov}(W)A' + B\mathrm{Cov}(Y)B' + A\mathrm{Cov}(W, Y)B' + B\mathrm{Cov}(Y, W)A'.$$

Proof Without loss of generality we can assume that $\mathrm{E}(W) = 0$ and $\mathrm{E}(Y) = 0$:

$$\begin{aligned}\mathrm{Cov}(AW + BY) &= \mathrm{E}[(AW + BY)(AW + BY)'] \\ &= A\mathrm{E}[WW']A' + B\mathrm{E}[YY']B' + A\mathrm{E}[WY']B' + B\mathrm{E}[YW']A' \\ &= A\mathrm{Cov}(W)A' + B\mathrm{Cov}(Y)B' + A\mathrm{Cov}(W, Y)B' + B\mathrm{Cov}(Y, W)A'.\end{aligned}$$

□

Exercise 1.4 Let M be the perpendicular projection operator onto $C(X)$. Show that $(I - M)$ is the perpendicular projection operator onto $C(X)^{\perp}$. Find $\mathrm{tr}(I - M)$ in terms of $r(X)$. (This problem has become even easier in this edition of the book.)

Exercise 1.5 For a linear model $Y = X\beta + e$; $\mathrm{E}(e) = 0$, show that $\mathrm{E}(Y) = X\beta$. For a *standard linear model* $Y = X\beta + e$; $\mathrm{E}(e) = 0$; $\mathrm{Cov}(e) = \sigma^2 I$, show that $\mathrm{Cov}(Y) = \sigma^2 I$.

Exercise 1.6 For a linear model $Y = X\beta + e$, $\mathrm{E}(e) = 0$, $\mathrm{Cov}(e) = \sigma^2 I$, the residuals are

$$\hat{e} = Y - X\hat{\beta} = (I - M)Y,$$

where M is the perpendicular projection operator onto $C(X)$. Find

(a) $\mathrm{E}(\hat{e})$.

(b) $\mathrm{Cov}(\hat{e})$.

(c) $\mathrm{Cov}(\hat{e}, MY)$.

(d) $\mathrm{E}(\hat{e}'\hat{e})$.

(e) Show that $\hat{e}'\hat{e} = Y'Y - (Y'M)Y$.

[Note: In Chapter 2 we will show that for a least squares estimate of β, say $\hat{\beta}$, we have $MY = X\hat{\beta}$.]

(f) Rewrite (c) and (e) in terms of $\hat{\beta}$.

1.2 Multivariate Normal Distributions

It is assumed that the reader is familiar with the basic ideas of multivariate distributions. A summary of these ideas is contained in Appendix D.

Let $Z = [z_1, \ldots, z_n]'$ be a random vector with $z_1, \ldots, z_n$ *independent identically distributed (i.i.d.)* $N(0, 1)$ random variables. Note that $\mathrm{E}(Z) = 0$ and $\mathrm{Cov}(Z) = I$. Now consider the transformation $AZ + \mu$, for some fixed $r \times n$ matrix A, and some fixed r vector μ.

Definition 1.2.1 Y has an r-dimensional *multivariate normal distribution* if Y has the same distribution as $AZ + \mu$, i.e., $Y \sim AZ + \mu$, for some n, some fixed $r \times n$ matrix A, and some fixed r vector μ. $\mathrm{E}(Y) = \mu$ and $\mathrm{Cov}(Y) = AA' \equiv V$. We indicate the multivariate normal distribution of Y by writing $Y \sim N(\mu, V)$.

It is not clear that the notation $Y \sim N(\mu, V)$ is well defined, i.e., that a multivariate normal distribution depends only on its mean vector and covariance matrix. We can pick A and B so that $V = AA' = BB'$ where $A \neq B$. In that case, we do not know whether to take $Y \sim AZ + \mu$ or $Y \sim BZ + \mu$. In fact, the number of columns in A and B need not even be the same, so the length of the vector Z could change between $Y \sim AZ + \mu$ and $Y \sim BZ + \mu$. We need to show that it does not matter which characterization is used. We now give such an argument based on characteristic functions. The argument is based on the fact that any two random vectors with the same characteristic function have the same distribution. Appendix D contains the definition of the characteristic function of a random vector.

Theorem 1.2.2 *If $Y \sim N(\mu, V)$ and $W \sim N(\mu, V)$, then Y and W have the same distribution.*

Proof Observe that

$$\varphi_Z(t) \equiv \mathrm{E}[\exp(it'Z)] = \prod_{j=1}^{n} \mathrm{E}[\exp(it_j z_j)] = \prod_{j=1}^{n} \exp(-t_j^2/2) = \exp(-t't/2).$$

Define $Y \sim AZ + \mu$, where $AA' = V$. The characteristic function of Y is

$$\begin{aligned}\varphi_Y(t) &= \mathrm{E}[\exp(it'Y)] = \mathrm{E}[\exp(it'[AZ + \mu])] \\ &= \exp(it'\mu)\varphi_Z(A't) \\ &= \exp(it'\mu)\exp(-t'AA't/2) \\ &= \exp(it'\mu - t'Vt/2).\end{aligned}$$

Similarly,

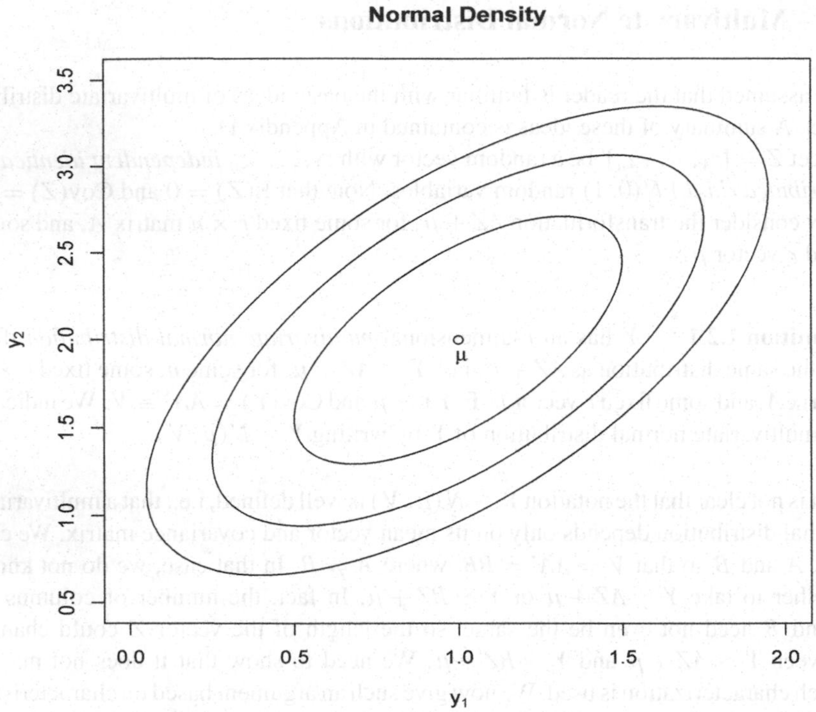

Fig. 1.1 Two-dimensional normal density isobars: $\mu = (1, 2)'$, $v_{11} = 1.0$, $v_{12} = 0.9$, $v_{22} = 2.0$

$$\varphi_W(t) = \exp(it'\mu - t'Vt/2).$$

Since the characteristic functions are the same, $Y \sim W$. □

In other words, if Y has a multivariate normal distribution, the distribution is entirely determined by the mean vector μ and the covariance matrix V because those are the only parameters in the characteristic function.

Suppose that Y is nonsingular and that $Y \sim N(\mu, V)$; then Y has a density. By definition, Y nonsingular means precisely that V is positive definite. By Corollary B.23, we can write $V = AA'$, with A nonsingular. Since $Y \sim AZ + \mu$ involves a nonsingular transformation of the random vector Z, which has a known density, it is quite easy to find the density of Y. The density is

$$f(y) = (2\pi)^{-n/2}[\det(V)]^{-1/2} \exp[-(y - \mu)'V^{-1}(y - \mu)/2],$$

where $\det(V)$ is the determinant of V. Figure 1.1 shows isobars of a normal density.

Exercise 1.7 Show that the function $f(y)$ given above is the density of Y when $Y \sim N(\mu, V)$ and V is nonsingular. Hint: If Z has density $f_Z(z)$ and $Y = G(Z)$, the density of Y is

$$f_Y(y) = f_Z[G^{-1}(y)]|\det(\mathbf{d}G^{-1})|,$$

where $\mathbf{d}G^{-1}$ is the derivative (matrix of partial derivatives) of G^{-1} evaluated at y.

An important and useful result is that for random vectors having a joint multivariate normal distribution, the condition of having zero covariance is equivalent to the condition of independence.

Theorem 1.2.3 *If* $Y \sim N(\mu, V)$ *and* $Y = \begin{bmatrix} Y_1 \\ Y_2 \end{bmatrix}$, *then* $\mathrm{Cov}(Y_1, Y_2) = 0$ *if and only if* Y_1 *and* Y_2 *are independent.*

Proof Partition V and μ to conform with Y, giving $V = \begin{bmatrix} V_{11} & V_{12} \\ V_{21} & V_{22} \end{bmatrix}$ and $\mu = \begin{bmatrix} \mu_1 \\ \mu_2 \end{bmatrix}$. Note that $V_{12} = V_{21}' = \mathrm{Cov}(Y_1, Y_2)$.

$\Leftarrow$ If Y_1 and Y_2 are independent,

$$V_{12} = \mathrm{E}[(Y_1 - \mu_1)(Y_2 - \mu_2)'] = \mathrm{E}(Y_1 - \mu_1)\mathrm{E}(Y_2 - \mu_2)' = 0.$$

$\Rightarrow$ Suppose $\mathrm{Cov}(Y_1, Y_2) = 0$, so that $V_{12} = V_{21}' = 0$. Using the definition of multivariate normality, we will construct a version of Y in which it is clear that Y_1 and Y_2 are independent. Given the uniqueness established in Theorem 1.2.2, this is sufficient to establish independence of Y_1 and Y_2.

Since Y is multivariate normal, by definition we can write $Y \sim AZ + \mu$, where A is an $r \times n$ matrix. Partition A in conformance with $\begin{bmatrix} Y_1 \\ Y_2 \end{bmatrix}$ as $A = \begin{bmatrix} A_1 \\ A_2 \end{bmatrix}$ so that

$$V = \begin{bmatrix} V_{11} & V_{12} \\ V_{21} & V_{22} \end{bmatrix} = \begin{bmatrix} A_1 A_1' & A_1 A_2' \\ A_2 A_1' & A_2 A_2' \end{bmatrix}.$$

Because $V_{12} = 0$, we have $A_1 A_2' = 0$ and

$$V = \begin{bmatrix} A_1 A_1' & 0 \\ 0 & A_2 A_2' \end{bmatrix}.$$

Now let $z_1, z_2, \ldots, z_{2n}$ be i.i.d. $N(0, 1)$. Define the random vectors $Z_1 = [z_1, \ldots, z_n]'$, $Z_2 = [z_{n+1}, \ldots, z_{2n}]'$, and

$$Z_0 = \begin{bmatrix} Z_1 \\ Z_2 \end{bmatrix}.$$

Note that Z_1 and Z_2 are independent. Now consider the random vector

$$W = \begin{bmatrix} A_1 & 0 \\ 0 & A_2 \end{bmatrix} Z_0 + \mu.$$

By definition, W is multivariate normal with $\mathrm{E}(W) = \mu$ and

$$\begin{aligned} \mathrm{Cov}(W) &= \begin{bmatrix} A_1 & 0 \\ 0 & A_2 \end{bmatrix} \begin{bmatrix} A_1 & 0 \\ 0 & A_2 \end{bmatrix}' \\ &= \begin{bmatrix} A_1 A_1' & 0 \\ 0 & A_2 A_2' \end{bmatrix} \\ &= V. \end{aligned}$$

We have shown that $W \sim N(\mu, V)$ and by assumption $Y \sim N(\mu, V)$. By Theorem 1.2.2, W and Y have exactly the same distribution; thus

$$Y \sim \begin{bmatrix} A_1 & 0 \\ 0 & A_2 \end{bmatrix} Z_0 + \mu.$$

It follows that $Y_1 \sim [A_1, 0]Z_0 + \mu_1 = A_1 Z_1 + \mu_1$ and $Y_2 \sim [0, A_2]Z_0 + \mu_2 = A_2 Z_2 + \mu_2$. The joint distribution of (Y_1, Y_2) is the same as the joint distribution of $(A_1 Z_1 + \mu_1, A_2 Z_2 + \mu_2)$. However, Z_1 and Z_2 are independent; thus $A_1 Z_1 + \mu_1$ and $A_2 Z_2 + \mu_2$ are independent, and it follows that Y_1 and Y_2 are independent. □

Exercise 1.8 Show that if Y is an r-dimensional random vector with $Y \sim N(\mu, V)$ and if B is a fixed $n \times r$ matrix, then $BY \sim N(B\mu, BVB')$.

In linear model theory, Theorem 1.2.3 is often applied to establish independence of two linear transformations of the data vector Y.

Corollary 1.2.4 *If $Y \sim N(\mu, \sigma^2 I)$ and if $AB' = 0$, then AY and BY are independent. The result also holds if $Y \sim N(\mu, V)$ and $AVB' = 0$.*

Proof Consider the distribution of $\begin{bmatrix} A \\ B \end{bmatrix} Y$. By Exercise 1.8, the joint distribution of AY and BY is multivariate normal. Since $\mathrm{Cov}(AY, BY) = \sigma^2 AIB' = \sigma^2 AB' = 0$, Theorem 1.2.3 implies that AY and BY are independent. □

Exercise 1.9 For a standard linear model $Y = X\beta + e$, with $e \sim N(0, \sigma^2 I)$, show that $Y \sim N(X\beta, \sigma^2 I)$.

1.3 Distributions of Quadratic Forms

In this section, quadratic forms are defined, the expectation of a quadratic form is found, and a series of results on independence and chi-squared distributions are given.

Definition 1.3.1 Let Y be an n-dimensional random vector and let A be an $n \times n$ matrix. A *quadratic form* is a random variable defined by $Y'AY$ for some Y and A.

Note that since $Y'AY$ is a scalar, $Y'AY = Y'A'Y = Y'(A + A')Y/2$. Since $(A + A')/2$ is always a symmetric matrix, we can, without loss of generality, *restrict ourselves to quadratic forms where A is symmetric*.

The next result is extremely useful.

Theorem 1.3.2 *If* $\mathrm{E}(Y) = \mu$ *and* $\mathrm{Cov}(Y) = V$, *then* $\mathrm{E}(Y'AY) = \mathrm{tr}(AV) + \mu' A\mu$.

Proof

$$\begin{aligned}(Y-\mu)'A(Y-\mu) &= Y'AY - \mu'AY - Y'A\mu + \mu'A\mu,\\ \mathrm{E}[(Y-\mu)'A(Y-\mu)] &= \mathrm{E}[Y'AY] - \mu'A\mu - \mu'A\mu + \mu'A\mu,\end{aligned}$$

so $\mathrm{E}[Y'AY] = \mathrm{E}[(Y-\mu)'A(Y-\mu)] + \mu'A\mu$.

It is easily seen that for any random square matrix W, $\mathrm{E}(\mathrm{tr}(W)) = \mathrm{tr}(\mathrm{E}(W))$. Thus

$$\begin{aligned}\mathrm{E}[(Y-\mu)'A(Y-\mu)] &= \mathrm{E}(\mathrm{tr}[(Y-\mu)'A(Y-\mu)])\\ &= \mathrm{E}(\mathrm{tr}[A(Y-\mu)(Y-\mu)'])\\ &= \mathrm{tr}(\mathrm{E}[A(Y-\mu)(Y-\mu)'])\\ &= \mathrm{tr}(A\mathrm{E}[(Y-\mu)(Y-\mu)'])\\ &= \mathrm{tr}(AV).\end{aligned}$$

Substitution gives

$$\mathrm{E}(Y'AY) = \mathrm{tr}(AV) + \mu'A\mu.$$

□

Exercise 1.10a If W and Y are random vectors with $\mathrm{E}(W) = \gamma$ and $\mathrm{E}(Y) = \mu$, show that

$$\mathrm{E}(Y'AW) = \mathrm{tr}\,[A\mathrm{Cov}(W, Y)] + \mu'A\gamma.$$

We now proceed to give results on chi-squared distributions and independence of quadratic forms. Note that by Definition C.1 and Theorem 1.2.3, if Z is an n-dimensional random vector and $Z \sim N(\mu, I)$, then $Z'Z \sim \chi^2(n, \mu'\mu/2)$.

Theorem 1.3.3 *If Y is a random vector with $Y \sim N(\mu, I)$ and if M is any perpendicular projection matrix, then $Y'MY \sim \chi^2[r(M), \mu'M\mu/2]$.*

Proof Let $r(M) = r$ and let $o_1, \ldots, o_r$ be an orthonormal basis for $C(M)$. Let $O = [o_1, \ldots, o_r]$ so that $M = OO'$. We now have $Y'MY = Y'OO'Y = (O'Y)'(O'Y)$, where $O'Y \sim N(O'\mu, O'IO)$. The columns of O are orthonormal, so $O'O$ is an $r \times r$ identity matrix, and by definition $(O'Y)'(O'Y) \sim \chi^2(r, \mu'OO'\mu/2)$ where $\mu'OO'\mu = \mu'M\mu$. □

Observe that if $Y \sim N\left(\mu, \sigma^2 I\right)$, then $[1/\sigma]Y \sim N\left([1/\sigma]\mu, I\right)$ and $Y'MY/\sigma^2 \sim \chi^2\left[r(M), \mu'M\mu/2\sigma^2\right]$.

Theorem 1.3.6 provides a generalization of Theorem 1.3.3 that is valid for an arbitrary covariance matrix. The next two lemmas are used in the proof of Theorem 1.3.6. The first is also useful in establishing the asymptotic distribution of Pearson's chi-squared statistic for multinomial data. The second is a useful fact whenever dealing with singular covariance matrices.

Lemma 1.3.4 *If $Y \sim N(\mu, M)$ with $\mu \in C(M)$ and if M is a perpendicular projection matrix, then $Y'Y \sim \chi^2[r(M), \mu'\mu/2]$.*

Proof Let O have r orthonormal columns with $M = OO'$. Since $\mu \in C(M)$, $\mu = Ob$. Let $W \sim N(b, I)$, then $Y \sim OW$. Since $O'O = I_r$ is also a perpendicular projection matrix, the previous theorem gives $Y'Y \sim W'O'OW \sim \chi^2(r, b'O'Ob/2)$. The proof is completed by observing that $r = r(M)$ and $b'O'Ob = \mu'\mu$. □

The following lemma establishes that, if Y is a singular random variable, then there exists a proper subset of $\mathbf{R}^n$ that contains Y with probability 1.

Lemma 1.3.5 *If* $\mathrm{E}(Y) = \mu$ *and* $\mathrm{Cov}(Y) = V$, *then* $\Pr[(Y - \mu) \in C(V)] = 1$.

Proof Without loss of generality, assume $\mu = 0$. Let M_V be the perpendicular projection operator onto $C(V)$; then $Y = M_V Y + (I - M_V)Y$. Clearly, $\mathrm{E}[(I - M_V)Y] = 0$ and $\mathrm{Cov}[(I - M_V)Y] = (I - M_V)V(I - M_V)=0$. Thus, $\Pr[(I - M_V) Y = 0] = 1$ and $\Pr[Y = M_V Y] = 1$. Since $M_V Y \in C(V)$, we are done. □

A property that holds with probability one is said to hold *almost surely*. Thus it is almost sure that $(Y - \mu) \in C(V)$, or we might write $(Y - \mu) \in C(V)$ a.s.

Exercise 1.10b Show that if Y is a random vector and if $\mathrm{E}(Y) = 0$ and $\mathrm{Cov}(Y) = 0$, then $\Pr[Y = 0] = 1$. Hint: For a random variable w with $\Pr[w \geq 0] = 1$ and $k > 0$, show that $\Pr[w \geq k] \leq \mathrm{E}(w)/k$. Apply this result to $Y'Y$.

Theorem 1.3.6 *If $Y \sim N(\mu, V)$, then $Y'AY \sim \chi^2[\text{tr}(AV), \mu' A\mu/2]$ provided that (1) $VAVAV = VAV$, (2) $\mu' AVA\mu = \mu' A\mu$, and (3) $VAVA\mu = VA\mu$.*

Proof *The general proof is given in the next subsection.* Here we give a proof for V positive definite.

For V nonsingular, condition (1) of the theorem holds if and only if $AVA = A$, and $AVA = A$ implies conditions (2) and (3). Also, write $V = QQ'$ for Q nonsingular. $AVA = A$ implies that $Q'AQQ'AQ = Q'AVAQ = Q'AQ$, so $Q'AQ$ is a ppo. Also, $AVA = A$ implies that $[Q'A]'[Q'A] = A$. Now consider $Q'AY \sim N(Q'A\mu, Q'AVAQ) = N(Q'A\mu, Q'AQ)$. Since $Q'AQ$ is a ppo, Lemma 1.3.4 applies if $Q'A\mu \in C(Q'AQ)$. But $Q'A\mu = [Q'AQ]Q^{-1}\mu \in C(Q'AQ)$. Applying Lemma 1.3.4,

$$Y'AY = (Q'AY)'(Q'AY) \sim \chi^2\left[r(Q'AQ), (Q'A\mu)'(Q'A\mu)/2\right].$$

However,

$$r(Q'AQ) = \text{tr}(Q'AQ) = \text{tr}(AQQ') = \text{tr}(AV)$$

and

$$(Q'A\mu)'(Q'A\mu) = \mu' AQ'QA\mu = \mu' AVA\mu = \mu' A\mu,$$

so we are done. □

Even if the three conditions of the theorem are not satisfied, it is possible to compute the distribution of the quadratic form, cf. the R package `CompQuadForm`, the references cited in its documentation, and Exercise 1.5.7.

One particularly useful result of the theorem follows.

Corollary 1.3.6a *If $Y \sim N(\mu, V)$ and $\mu \in C(V)$, then*

$$Y'V^-Y \sim \chi^2[r(V), \mu' V^-\mu/2].$$

Proof The three conditions in Theorem 1.3.6 are satisfied so $Y'V^-Y \sim \chi^2[\text{tr}(V^-V), \mu'V^-\mu/2]$. Because VV^- is idempotent, $\text{tr}(V^-V) = \text{tr}(VV^-) = r(VV^-)$. To see that $r(VV^-) = r(V)$, note that the two matrices have the same column space:

$$C(V) = C(VV^-V) \subset C(VV^-) \subset C(V).$$ □

The next three theorems establish conditions under which quadratic forms are independent. Theorem 1.3.7 examines the important special case in which the covariance matrix is a multiple of the identity matrix. In addition to considering independence of quadratic forms, the theorem also examines independence between quadratic forms and linear transformations of the random vector.

Theorem 1.3.7 *If A is symmetric and in (2) B is symmetric, $Y \sim N(\mu, \sigma^2 I)$, and $BA = 0$, then*

(1) $Y'AY$ and BY are independent,
(2) $Y'AY$ and $Y'BY$ are independent,

The result also holds if $Y \sim N(\mu, V)$ and $BVA = 0$.

Proof By Corollary 1.2.4, if $BA = 0$, BY and AY are independent or, more generally, if $BVA = 0$, BY and AY are independent. In addition, as discussed near the end of Appendix D, any function of AY is independent of any function of BY. Since $Y'AY = Y'AA^-AY$ and $Y'BY = Y'BB^-BY$ are functions of AY and BY, the theorem holds.

□

1.3.1 Results for General Covariance Matrices

Proof of Theorem 1.3.6. By Lemma 1.3.5, for the purpose of finding the distribution of $Y'AY$, we can assume that $Y = \mu + e$, where $e \in C(V)$. Using conditions (1), (2), and (3) of the theorem and the fact that $e = Vb$ for some b,

$$\begin{aligned} Y'AY &= \mu'A\mu + \mu'Ae + e'A\mu + e'Ae \\ &= \mu'AVA\mu + \mu'AVAe + e'AVA\mu + e'AVAe \\ &= Y'(AVA)Y. \end{aligned}$$

Write $V = QQ'$ so that $Y'AY = (Q'AY)'(Q'AY)$, where $Q'AY \sim N(Q'A\mu, Q'AVAQ)$. If we can show that $Q'AVAQ$ is a perpendicular projection matrix and that $Q'A\mu \in C(Q'AVAQ)$, then $Y'AY$ will have a chi-squared distribution by Lemma 1.3.4.

Since V is nonnegative definite, we can write $Q = Q_1Q_2$, where Q_1 has orthonormal columns and Q_2 is nonsingular. It follows that

$$Q_2^{-1}Q_1'V = Q_2^{-1}Q_1'[Q_1Q_2Q'] = Q'.$$

Applying this result, $VAVAV = VAV$ implies that $Q'AVAQ = Q'AQ$. Now $Q'AVAQ = (Q'AQ)(Q'AQ)$, so $Q'AQ$ is idempotent and symmetric and $Q'AQ = Q'AVAQ$ so $Q'AVAQ$ is a perpendicular projection operator.

From the preceding paragraph, to see that $Q'A\mu \in C(Q'AVAQ)$ it suffices to show that $Q'AQQ'A\mu = Q'A\mu$. Note that $VAVA\mu = VA\mu$ implies that $Q'AVA\mu = Q'A\mu$. However, since $Q'AVA\mu = Q'AQQ'A\mu$, we are done.

The noncentrality parameter is one-half of

$$(Q'A\mu)'(Q'A\mu) = \mu'AVA\mu = \mu'A\mu.$$

The degrees of freedom are

$$r(Q'AVAQ) = r(Q'AQ) = \mathrm{tr}(Q'AQ) = \mathrm{tr}(AQQ') = \mathrm{tr}(AV). \qquad \square$$

The final two theorems provide conditions for independence of quadratic forms under general covariance matrices. It is possible for the conditions in these theorems to hold with $AVB \neq 0$.

Theorem 1.3.8 *If $Y \sim N(\mu, V)$, A and B are nonnegative definite, and $VAV BV = 0$, then $Y'AY$ and $Y'BY$ are independent.*

Proof Since A and B are nonnegative definite, we can write $A = RR'$ and $B = SS'$. We can also write $V = QQ'$.

$Y'AY = (R'Y)'(R'Y)$ and $Y'BY = (S'Y)'(S'Y)$ are independent

$$\begin{aligned}
&\text{if } R'Y \text{ and } S'Y \text{ are independent}\\
&\text{iff } \mathrm{Cov}(R'Y, S'Y) = 0\\
&\text{iff } R'VS = 0\\
&\text{iff } R'QQ'S = 0\\
&\text{iff } C(Q'S) \perp C(Q'R).
\end{aligned}$$

Since $C(AA') = C(A)$ for any A, we have

$$\begin{aligned}
C(Q'S) \perp C(Q'R) \ &\text{iff } C(Q'SS'Q) \perp C(Q'RR'Q)\\
&\text{iff } [Q'SS'Q][Q'RR'Q] = 0\\
&\text{iff } Q'BVAQ = 0\\
&\text{iff } C(Q) \perp C(BVAQ)\\
&\text{iff } C(QQ') \perp C(BVAQ)\\
&\text{iff } QQ'BVAQ = 0\\
&\text{iff } VBVAQ = 0.
\end{aligned}$$

Repeating similar arguments for the right side gives $VBVAQ = 0$ iff $VBVAV = 0$. $\square$

Theorem 1.3.9 *If $Y \sim N(\mu, V)$ and (1) $VAVBV = 0$, (2) $VAVB\mu = 0$, (3) $VBVA\mu = 0$, (4) $\mu'AVB\mu = 0$, and conditions (1), (2), and (3) from Theorem 1.3.6 hold for both $Y'AY$ and $Y'BY$, then $Y'AY$ and $Y'BY$ are independent.*

Exercise 1.11 Prove Theorem 1.3.9.

Hints: Let $V = QQ'$ and write $Y = \mu + QZ$, where $Z \sim N(0, I)$. Using $\perp\!\!\!\perp$ to indicate independence, show that

$$\begin{bmatrix} Q'AQZ \\ \mu'AQZ \end{bmatrix} \perp\!\!\!\perp \begin{bmatrix} Q'BQZ \\ \mu'BQZ \end{bmatrix}$$

and that, say, $Y'AY$ is a function $Q'AQZ$ and $\mu'AQZ$.

Note that Theorem 1.3.8 applies immediately if AY and BY are independent, i.e., if $AVB = 0$. In something of a converse, if V is nonsingular, the condition $VAVBV = 0$ is equivalent to $AVB = 0$; so the theorem applies only when AY and BY are independent. However, if V is singular, the conditions of Theorems 1.3.8 and 1.3.9 can be satisfied even when AY and BY are not independent.

1.4 Generalized Linear Models

We now give a brief introduction to *generalized linear models*. On occasion through the rest of the book, reference will be made to various properties of linear models that extend easily to generalized linear models. See McCullagh and Nelder (1989) or Christensen (1997) for more extensive discussions of generalized linear models and their applications. First it must be noted that a *general linear model* is a linear model but a *generalized* linear model is a generalization of the concept of a linear model. Generalized linear models include linear models as a special case but also include logistic regression, exponential regression, and gamma regression as special cases. Additionally, log-linear models for multinomial data are closely related to generalized linear models.

Consider a random vector Y with $\mathrm{E}(Y) = \mu$. Let h be an arbitrary function on the real numbers and, for a vector $v = (v_1, \ldots, v_n)'$, define the vector function

$$h(v) \equiv \begin{bmatrix} h(v_1) \\ \vdots \\ h(v_n) \end{bmatrix}.$$

The primary idea of a generalized linear model is specifying that

$$\mu = h(X\beta),$$

where h is a known invertible function and X and β are defined as for linear models. The inverse of h is called the *link* function. In particular, linear models use the identity function $h(v) = v$, logistic regression uses the logistic transform $h(v) = e^v/(1+e^v)$, and both exponential regression and log-linear models use the exponential transform $h(v) = e^v$. Their link functions are, respectively, the identity, logit, and log transforms. Because the linear structure $X\beta$ is used in generalized linear models, many of the analysis techniques used for linear models can be easily extended to generalized linear models.

Typically, in a generalized linear model it is assumed that the y_is are independent and each follows a distribution having density or mass function of the form

$$f(y_i|\theta_i, \phi; w_i) = \exp\left\{\frac{w_i}{\phi}[\theta_i y_i - r(\theta_i)]\right\} g(y_i, \phi, w_i), \tag{1}$$

where $r(\cdot)$ and $g(\cdot, \cdot, \cdot)$ are known functions and θ_i, ϕ, and w_i are scalars. By assumption, w_i is a fixed known number. Typically, it is a known weight that indicates knowledge about a pattern in the variabilities of the y_is. ϕ is either known or is an unknown parameter, but for some purposes is always treated like it is known. It is related to the variance of y_i. The parameter θ_i is related to the mean of y_i. For standard linear models, the assumption is that the y_is are independent $N(\theta_i, \phi/w_i)$, with $\phi \equiv \sigma^2$ and $w_i \equiv 1$. The standard assumption of logistic regression is that the $N_i y_i$s are distributed as independent binomials with N_i trials, success probability

$$\mathrm{E}(y_i) \equiv \mu_i \equiv p_i = e^{\theta_i}/[1 + e^{\theta_i}],$$

$w_i = N_i$, and $\phi = 1$. Log-linear models fit into this framework when one assumes that the y_is are independent Poisson with mean $\mu_i = e^{\theta_i}$, $w_i = 1$, and $\phi = 1$. Note that in these cases the mean is some function of θ_i and that ϕ is merely related to the variance. Note also that in the three examples, the h function has already appeared, even though these distributions have not yet incorporated the linear structure of the generalized linear model.

To investigate the relationship between the θ_i parameters and the linear structure $x_i'\beta$, where x_i' is the ith row of X, let $\dot{r}(\theta_i)$ be the derivative $dr(\theta_i)/d\theta_i$. It can be shown that

$$\mathrm{E}(y_i) \equiv \mu_i = \dot{r}(\theta_i).$$

Thus, another way to think about the modeling process is that

$$\mu_i = h(x_i'\beta) = \dot{r}(\theta_i),$$

where both h and $\dot{r}$ are invertible. In matrix form, write $\theta = (\theta_1, \ldots, \theta_n)'$ so that

$$X\beta = h^{-1}(\mu) = h^{-1}[\dot{r}(\theta)] \quad \text{and} \quad \dot{r}^{-1}[h(X\beta)] = \dot{r}^{-1}(\mu) = \theta.$$

The special case of $h(\cdot) = \dot{r}(\cdot)$ gives $X\beta = \theta$. This is known as a *canonical generalized linear model*, or as using a *canonical link function*. The three examples given earlier are all examples of canonical generalized linear models. Linear models with normally distributed data are canonical generalized linear models. Logistic regression is the canonical model having $N_i y_i$ distributed Binomial(N_i, μ_i) for known N_i with $h^{-1}(\mu_i) \equiv \log(\mu_i/[1 - \mu_i])$. Another canonical generalized linear model has y_i distributed Poisson(μ_i) with $h^{-1}(\mu_i) \equiv \log(\mu_i)$.

1.5 Additional Exercises

Exercise 1.5.1 Let $Y = (y_1, y_2, y_3)'$ be a random vector. Suppose that $\mathrm{E}(Y) \in \mathcal{M}$, where $\mathcal{M}$ is defined by

$$\mathcal{M} = \{(a, a - b, 2b)' | a, b \in \mathbf{R}\}.$$

(a) Show that $\mathcal{M}$ is a vector space.
(b) Find a basis for $\mathcal{M}$.
(c) Write a linear model for this problem (i.e., find X such that $Y = X\beta + e$, $\mathrm{E}(e) = 0$).
(d) If $\beta = (\beta_1, \beta_2)'$ in part (c), find two vectors $r = (r_1, r_2, r_3)'$ and $s = (s_1, s_2, s_3)'$ such that $\mathrm{E}(r'Y) = r'X\beta = \beta_1$ and $\mathrm{E}(s'Y) = \beta_2$. Find another vector $t = (t_1, t_2, t_3)'$ with $r \neq t$ but $\mathrm{E}(t'Y) = \beta_1$.

Exercise 1.5.2 Let $Y = (y_1, y_2, y_3)'$ with $Y \sim N(\mu, V)$, where

$$\mu = (5, 6, 7)'$$

and

$$V = \begin{bmatrix} 2 & 0 & 1 \\ 0 & 3 & 2 \\ 1 & 2 & 4 \end{bmatrix}.$$

Find

(a) the marginal distribution of y_1,
(b) the joint distribution of y_1 and y_2,
(c) the conditional distribution of y_3 given $y_1 = u_1$ and $y_2 = u_2$,
(d) the conditional distribution of y_3 given $y_1 = u_1$,
(e) the conditional distribution of y_1 and y_2 given $y_3 = u_3$,
(f) the correlations $\rho_{12}, \rho_{13}, \rho_{23}$,
(g) the distribution of

$$Z = \begin{bmatrix} 2 & 1 & 0 \\ 1 & 1 & 1 \end{bmatrix} Y + \begin{bmatrix} -15 \\ -18 \end{bmatrix},$$

(h) the characteristic functions of Y and Z.

Exercise 1.5.3 The density of $Y = (y_1, y_2, y_3)'$ is

$$(2\pi)^{-3/2} |V|^{-1/2} e^{-Q/2},$$

where

$$Q = 2y_1^2 + y_2^2 + y_3^2 + 2y_1 y_2 - 8y_1 - 4y_2 + 8.$$

Find V^{-1} and μ.

Exercise 1.5.4 Let $Y \sim N(J\mu, \sigma^2 I)$ and let $O = \left[n^{-1/2}J, O_1\right]$ be an orthonormal matrix.

(a) Find the distribution of $O'Y$.

(b) Show that $\bar{y}_\cdot = (1/n)J'Y$ and that $s^2 = Y'O_1O_1'Y/(n-1)$.

(c) Show that $\bar{y}_\cdot$ and s^2 are independent.

Hint: Show that $Y'Y = Y'OO'Y = Y'(1/n)JJ'Y + Y'O_1O_1'Y$.

Exercise 1.5.5 Let $Y = (y_1, y_2)'$ have a $N(0, I)$ distribution. Show that if

$$A = \begin{bmatrix} 1 & a \\ a & 1 \end{bmatrix} \qquad B = \begin{bmatrix} 1 & b \\ b & 1 \end{bmatrix},$$

then the conditions of Theorem 1.3.7 implying independence of $Y'AY$ and $Y'BY$ are satisfied only if $|a| = 1/|b|$ and $a = -b$. What are the possible choices for a and b?

Exercise 1.5.6 Let $Y = (y_1, y_2, y_3)'$ have a $N(\mu, \sigma^2 I)$ distribution. Consider the quadratic forms defined by the matrices M_1, M_2, and M_3 given below.

(a) Find the distribution of each $Y'M_iY$.

(b) Show that the quadratic forms are pairwise independent.

(c) Show that the quadratic forms are mutually independent.

$$M_1 = \frac{1}{3}J_3^3, \qquad M_2 = \frac{1}{14}\begin{bmatrix} 9 & -3 & -6 \\ -3 & 1 & 2 \\ -6 & 2 & 4 \end{bmatrix},$$

$$M_3 = \frac{1}{42}\begin{bmatrix} 1 & -5 & 4 \\ -5 & 25 & -20 \\ 4 & -20 & 16 \end{bmatrix}.$$

Exercise 1.5.7 Let A be symmetric, $Y \sim N(0, V)$, and $w_1, \ldots, w_s$ be independent $\chi^2(1)$ random variables. Show that for some value of s and some numbers λ_i, $Y'AY \sim \sum_{i=1}^{s} \lambda_i w_i$. Hint: $Y \sim QZ$ so $Y'AY \sim Z'Q'AQZ$. Write $Q'AQ = PD(\lambda_i)P'$.

Exercise 1.5.8 Show that

(a) for Example 1.0.1 the perpendicular projection operator onto $C(X)$ is

$$M = \frac{1}{6}J_6^6 + \frac{1}{70}\begin{bmatrix} 25 & 15 & 5 & -5 & -15 & -25 \\ 15 & 9 & 3 & -3 & -9 & -15 \\ 5 & 3 & 1 & -1 & -3 & -5 \\ -5 & -3 & -1 & 1 & 3 & 5 \\ -15 & -9 & -3 & 3 & 9 & 15 \\ -25 & -15 & -5 & 5 & 15 & 25 \end{bmatrix};$$

(b) for Example 1.0.2 the perpendicular projection operator onto $C(X)$ is

$$M = \begin{bmatrix} 1/3 & 1/3 & 1/3 & 0 & 0 & 0 \\ 1/3 & 1/3 & 1/3 & 0 & 0 & 0 \\ 1/3 & 1/3 & 1/3 & 0 & 0 & 0 \\ 0 & 0 & 0 & 1 & 0 & 0 \\ 0 & 0 & 0 & 0 & 1/2 & 1/2 \\ 0 & 0 & 0 & 0 & 1/2 & 1/2 \end{bmatrix}.$$

References

Christensen, R. (1997). *Log-linear models and logistic regression* (2nd ed.). New York: Springer.
McCullagh, P., & Nelder, J. A. (1989). *Generalized linear models* (2nd ed.). London: Chapman and Hall.

Chapter 2
Estimation

Abstract This chapter focuses on the theory of least squares estimation. It begins with a discussion of identifiability and estimability. It includes discussion of generalized least squares estimation and the possible advantages of biased estimation including Bayesian estimation.

In this chapter, properties of least squares estimates are examined for the model

$$Y = X\beta + e, \quad \mathrm{E}(e) = 0, \quad \mathrm{Cov}(e) = \sigma^2 I.$$

The chapter begins with a discussion of the concepts of identifiability and estimability in linear models. Section 2 characterizes least squares estimates. Sections 3–5 establish that least squares estimates are best linear unbiased estimates, maximum likelihood estimates, and minimum variance unbiased estimates. The last two of these properties require the additional assumption $e \sim N(0, \sigma^2 I)$. Section 6 also assumes that the errors are normally distributed and presents the distributions of various estimates. From these distributions various tests and confidence intervals are easily obtained. Section 7 examines the model

$$Y = X\beta + e, \quad \mathrm{E}(e) = 0, \quad \mathrm{Cov}(e) = \sigma^2 V,$$

where V is a known positive definite matrix. Section 7 introduces generalized least squares estimates and presents properties of those estimates. Section 8 presents the normal equations and establishes their relationship to least squares and generalized least squares estimation. Section 9 introduces the variance-bias tradeoff and illustrates that a little bias in an estimate can sometimes more than pay for itself by decreasing the variance. Section 10 discusses Bayesian estimation, which often exploits the variance-bias tradeoff.

The history of least squares estimation goes back at least to 1805 when Legendre first published the idea. Gauss made important early contributions (and claimed to have invented the method prior to 1805).

R. Christensen, *Plane Answers to Complex Questions*, Springer Texts in Statistics,
https://doi.org/10.1007/978-3-030-32097-3_2

There is a huge body of literature available on estimation and testing in linear models. A few books dealing with the subject are Arnold (1981), Eaton (1983), Graybill (1976), Harville (2018), Monahan (2008), Rao (1973), Ravishanker and Dey (2002), Rencher and Schaalje (2008), Scheffé (1959), Searle (1971), Seber (1966, 1977, 2015), and Wichura (2006).

2.1 Identifiability and Estimability

A key issue in linear model theory is figuring out which parameters can be estimated and which cannot. We will see that what can be estimated are functions of the parameters that are *identifiable*. Linear functions of the parameters that are identifiable are called *estimable* and have linear unbiased estimators. These concepts also have natural applications to generalized linear models. The definitions used here are tailored to (generalized) linear models but our definition of an identifiable parameterization coincides with more common definitions of identifiability; cf. Appendix D.1. The distribution of Y should completely determine $\mathrm{E}(Y)$ along with some covariance parameter(s) [typically σ^2] but the question is whether β is uniquely determined by $\mathrm{E}(Y)$.

Consider the general linear model

$$Y = X\beta + e, \quad \mathrm{E}(e) = 0,$$

where again Y is an $n \times 1$ vector of observations, X is an $n \times p$ matrix of known constants, β is a $p \times 1$ vector of unobservable parameters, and e is an $n \times 1$ vector of unobservable random errors whose distribution does not depend on β. We can only learn about β through $X\beta$. If x_i' is the ith row of X, $x_i'\beta$ is the ith row of $X\beta$ and we can only learn about β through the $x_i'\beta$s. $X\beta$ can be thought of as a vector of inner products between β and a spanning set for $C(X')$. Thus, we can learn about inner products between β and $C(X')$. In particular, when λ is a $p \times 1$ vector of known constants, we can learn about functions $\lambda'\beta$ where $\lambda \in C(X')$, i.e., where $\lambda = X'\rho$ for some vector ρ. These are precisely the estimable functions of β. We now give more formal arguments leading us to focus on functions $\lambda'\beta$ where $\lambda' = \rho'X$ or, more generally, vectors $\Lambda'\beta$ where $\Lambda' = P'X$.

In general, a *parameterization* for the $n \times 1$ mean vector $\mathrm{E}(Y)$ consists of writing $\mathrm{E}(Y)$ as a function of some parameters β, say,

$$\mathrm{E}(Y) = f(\beta).$$

A general linear model is a parameterization

$$\mathrm{E}(Y) = X\beta$$

because $\mathrm{E}(Y) = \mathrm{E}(X\beta + e) = X\beta + \mathrm{E}(e) = X\beta$. A parameterization is identifiable if knowing $\mathrm{E}(Y)$ tells you the parameter vector β.

Definition 2.1.1 The parameter β is *identifiable* if for any β_1 and β_2, $f(\beta_1) = f(\beta_2)$ implies $\beta_1 = \beta_2$. If β is identifiable, we say that the parameterization $f(\beta)$ is identifiable. Moreover, a vector-valued function $g(\beta)$ is identifiable if $f(\beta_1) = f(\beta_2)$ implies $g(\beta_1) = g(\beta_2)$. If the parameterization is not identifiable but nontrivial identifiable functions exist, then the parameterization is said to be *partially identifiable*.

The key point is that if β or a function $g(\beta)$ is not identifiable, it is simply impossible for one to know what it is based on knowing $\mathrm{E}(Y)$. From a statistical perspective, we are considering models for the mean vector with the idea of collecting data that will allow us to estimate $\mathrm{E}(Y)$. If actually knowing $\mathrm{E}(Y)$ is not sufficient to tell us the value of β or $g(\beta)$, no amount of data is ever going to let us estimate them.

In regression models, i.e., models for which $r(X) = p$, the parameters are identifiable. In this case, $X'X$ is nonsingular, so if $X\beta_1 = X\beta_2$, then

$$\beta_1 = (X'X)^{-1}X'X\beta_1 = (X'X)^{-1}X'X\beta_2 = \beta_2$$

and identifiability holds.

For models in which $r(X) = r < p$, there exist $\beta_1 \neq \beta_2$ but $X\beta_1 = X\beta_2$, so the parameters are not identifiable.

For general linear models, the only functions of the parameters that are identifiable are functions of $X\beta$. This follows from the next result.

Theorem 2.1.2 *A function $g(\beta)$ is identifiable if and only if $g(\beta)$ is a function of $f(\beta)$.*

Proof $g(\beta)$ being a function of $f(\beta)$ means that for some function g_*, $g(\beta) = g_*[f(\beta)]$ for all β; or, equivalently, it means that for any $\beta_1 \neq \beta_2$ such that $f(\beta_1) = f(\beta_2)$, $g(\beta_1) = g(\beta_2)$.

Clearly, if $g(\beta) = g_*[f(\beta)]$ and $f(\beta_1) = f(\beta_2)$, then $g(\beta_1) = g_*[f(\beta_1)] = g_*[f(\beta_2)] = g(\beta_2)$, so $g(\beta)$ is identifiable.

Conversely, if $g(\beta)$ is not a function of $f(\beta)$, there exists $\beta_1 \neq \beta_2$ such that $f(\beta_1) = f(\beta_2)$ but $g(\beta_1) \neq g(\beta_2)$. Hence, $g(\beta)$ is not identifiable. □

It is reasonable to estimate any identifiable function. Thus, in a linear model it is reasonable to estimate any function of $X\beta$. It is not reasonable to estimate nonidentifiable functions, because you simply do not know what you are estimating.

The traditional idea of estimability in linear models can now be presented. Estimable functions are linear functions of β that are identifiable.

Definition 2.1.3 A vector-valued linear function of β, say, $\Lambda'\beta$, is *estimable* if $\Lambda'\beta = P'X\beta$ for some matrix P.

Actually, an identifiable linear function of β is a function $g_*(X\beta)$, but since the composite function is linear and $X\beta$ is linear, the function g_* must be linear, and we can write it as a matrix P'.

Clearly, if $\Lambda'\beta$ is estimable, it is identifiable and therefore it is a reasonable thing to estimate. However, estimable functions are not the only functions of β that are reasonable to estimate. For example, the ratio of two estimable functions is not estimable, but it is identifiable, so the ratio is reasonable to estimate. You *can* estimate many functions that are not "estimable." What you cannot do is estimate nonidentifiable functions.

Unfortunately, the term "nonestimable" is often used to mean something other than "not being estimable." You can be "not estimable" by being either not linear or not identifiable. In particular, a linear function that is "not estimable" is automatically nonidentifiable. However, nonestimable is often *taken to mean* a linear function that is not identifiable. In other words, some authors (perhaps, on occasion, even this one) presume that nonestimable functions are linear, so that nonestimability and nonidentifiability become equivalent.

It should be noted that the concepts of identifiability and estimability are based entirely on the assumption that $\mathrm{E}(Y) = X\beta$. Identifiability and estimability do not depend on $\mathrm{Cov}(Y) = \mathrm{Cov}(e)$ (as long as the covariance matrix is not also a function of β).

An important property of estimable functions $\Lambda'\beta = P'X\beta$ is that although P need not be unique, its perpendicular projection (columnwise) onto $C(X)$ is unique. Let P_1 and P_2 be matrices with $\Lambda' = P_1'X = P_2'X$, then

$$MP_1 = X(X'X)^- X'P_1 = X(X'X)^- \Lambda = X(X'X)^- X'P_2 = MP_2.$$

Example 2.1.4 In the simple linear regression model of Example 1.0.1, β_1 is estimable because

$$\frac{1}{35}(-5, -3, -1, 1, 3, 5)\begin{bmatrix} 1 & 1 \\ 1 & 2 \\ 1 & 3 \\ 1 & 4 \\ 1 & 5 \\ 1 & 6 \end{bmatrix}\begin{bmatrix} \beta_0 \\ \beta_1 \end{bmatrix} = (0, 1)\begin{bmatrix} \beta_0 \\ \beta_1 \end{bmatrix} = \beta_1 .$$

β_0 is also estimable. Note that

$$\frac{1}{6}(1,1,1,1,1,1)\begin{bmatrix}1&1\\1&2\\1&3\\1&4\\1&5\\1&6\end{bmatrix}\begin{bmatrix}\beta_0\\\beta_1\end{bmatrix}=\beta_0+\frac{7}{2}\beta_1,$$

so

$$\begin{aligned}\beta_0&=\left(\beta_0+\frac{7}{2}\beta_1\right)-\frac{7}{2}\beta_1\\&=\left[\frac{1}{6}\begin{pmatrix}1\\1\\1\\1\\1\\1\end{pmatrix}-\frac{7}{2}\left(\frac{1}{35}\right)\begin{pmatrix}-5\\-3\\-1\\1\\3\\5\end{pmatrix}\right]'\begin{bmatrix}1&1\\1&2\\1&3\\1&4\\1&5\\1&6\end{bmatrix}\begin{bmatrix}\beta_0\\\beta_1\end{bmatrix}\\&=\frac{1}{30}(20,14,8,2,-4,-10)\begin{bmatrix}1&1\\1&2\\1&3\\1&4\\1&5\\1&6\end{bmatrix}\begin{bmatrix}\beta_0\\\beta_1\end{bmatrix}.\end{aligned}$$

For any fixed number x, $\beta_0+\beta_1 x$ is estimable because it is a linear combination of estimable functions.

Example 2.1.5 In the one-way ANOVA model of Example 1.0.2, we can estimate parameters like $\mu+\alpha_1$, $\alpha_1-\alpha_3$, and $\alpha_1+\alpha_2-2\alpha_3$. Observe that

$$(1,0,0,0,0,0)\begin{bmatrix}1&1&0&0\\1&1&0&0\\1&1&0&0\\1&0&1&0\\1&0&0&1\\1&0&0&1\end{bmatrix}\begin{bmatrix}\mu\\\alpha_1\\\alpha_2\\\alpha_3\end{bmatrix}=\mu+\alpha_1,$$

$$(1,0,0,0,-1,0)\begin{bmatrix}1&1&0&0\\1&1&0&0\\1&1&0&0\\1&0&1&0\\1&0&0&1\\1&0&0&1\end{bmatrix}\begin{bmatrix}\mu\\\alpha_1\\\alpha_2\\\alpha_3\end{bmatrix}=\alpha_1-\alpha_3,$$

but also

$$\left(\frac{1}{3}, \frac{1}{3}, \frac{1}{3}, 0, \frac{-1}{2}, \frac{-1}{2}\right)\begin{bmatrix}1 & 1 & 0 & 0\\1 & 1 & 0 & 0\\1 & 1 & 0 & 0\\1 & 0 & 1 & 0\\1 & 0 & 0 & 1\\1 & 0 & 0 & 1\end{bmatrix}\begin{bmatrix}\mu\\\alpha_1\\\alpha_2\\\alpha_3\end{bmatrix} = \alpha_1 - \alpha_3,$$

and

$$(1, 0, 0, 1, -2, 0)\begin{bmatrix}1 & 1 & 0 & 0\\1 & 1 & 0 & 0\\1 & 1 & 0 & 0\\1 & 0 & 1 & 0\\1 & 0 & 0 & 1\\1 & 0 & 0 & 1\end{bmatrix}\begin{bmatrix}\mu\\\alpha_1\\\alpha_2\\\alpha_3\end{bmatrix} = \alpha_1 + \alpha_2 - 2\alpha_3.$$

We have given two vectors ρ_1 and ρ_2 with $\rho_i' X\beta = \alpha_1 - \alpha_3$. Using M given in Exercise 1.5.8b, the reader can verify that $M\rho_1 = M\rho_2$.

In the one-way analysis of covariance model,

$$y_{ij} = \mu + \alpha_i + \gamma x_{ij} + e_{ij}, \quad \mathrm{E}(e_{ij}) = 0,$$

$i = 1, \ldots, a$, $j = 1, \ldots, N_i$, x_{ij} is a known predictor variable and γ is its unknown coefficient. γ is generally identifiable but μ and the α_is are not. The following results allow one to tell whether or not an individual parameter is identifiable.

Proposition 2.1.6a *For a linear model, write $X\beta = \sum_{k=1}^{p} X_k\beta_k$ where the X_ks are the columns of X. An individual parameter β_j is not identifiable if and only if there exist scalars α_k such that $X_j = \sum_{k \neq j} X_k\alpha_k$.*

Proof To show that the condition on X implies nonidentifiability, it is enough to show that there exist β and β_* with $X\beta = X\beta_*$ but $\beta_j \neq \beta_{*j}$. The condition $X_j = \sum_{k \neq j} X_k\alpha_k$ is equivalent to there existing a vector α with $\alpha_j \neq 0$ and $X\alpha = 0$. Let $\beta_* = \beta + \alpha$ and the proof is complete.

Rather than showing that when β_j is not identifiable, the condition on X holds, we show the contrapositive, i.e., that when the condition on X does not hold, β_j is identifiable. If there do not exist such α_ks, then whenever $X\alpha = 0$, we must have $\alpha_j = 0$. In particular, if $X\beta = X\beta_*$, then $X(\beta - \beta_*) = 0$, so $(\beta_j - \beta_{*j}) = 0$ and β_j is identifiable. □

Proposition 2.1.6b *For a linear model, write $X\beta = \sum_{k=1}^{p} X_k\beta_k$ where the X_ks are the columns of X. An individual parameter β_j is identifiable if and only if for any scalars α_k such that $0 = \sum_k X_k\alpha_k$, $\alpha_j = 0$.*

Proof The condition $0 = \sum_k X_k \alpha_k$ with $\alpha_j = 0$ is equivalent to there existing a vector α with $X\alpha = 0$ and $\alpha_j = 0$. To show that this condition on X implies identifiability, assume β and β_* have $X\beta = X\beta_*$. Then $X(\beta - \beta_*) = 0$, so the jth component of $\beta - \beta_*$ is 0 and $\beta_j = \beta_{*j}$.

Let α be any vector with $X\alpha = 0$. For any β define $\beta_* \equiv \beta + \alpha$. Clearly, $X\beta = X\beta_*$ so by identifiability $\beta_j = \beta_{*j} = \beta_j + \alpha_j$ which implies $\alpha_j = 0$. □

The concepts of identifiability and estimability apply with little change to generalized linear models. In generalized linear models, the distribution of Y is either completely determined by $\mathrm{E}(Y)$ or it is determined by $\mathrm{E}(Y)$ along with another parameter ϕ that is unrelated to the parameterization of $\mathrm{E}(Y)$. A generalized linear model has $\mathrm{E}(Y) = h(X\beta)$. By Theorem 2.1.2, a function $g(\beta)$ is identifiable if and only if it is a function of $h(X\beta)$. However, the function $h(\cdot)$ is assumed to be invertible, so $g(\beta)$ is identifiable if and only if it is a function of $X\beta$. A vector-valued linear function of β, say, $\Lambda'\beta$ is identifiable if $\Lambda'\beta = P'X\beta$ for some matrix P, hence Definition 2.1.3 applies as well to define estimability for generalized linear models as it does for linear models. Propositions 2.1.6a, 2.1.6b also applies without change.

Finally, the concept of estimability in linear models can be related to the existence of linear unbiased estimators. A linear function of the parameter vector β, say $\lambda'\beta$, is estimable if and only if it admits a linear unbiased estimate.

Definition 2.1.7 An estimate $f(Y)$ of $g(\beta)$ is *unbiased* if $\mathrm{E}[f(Y)] = g(\beta)$ for any β.

Definition 2.1.8 $f(Y)$ is a *linear estimate* of $\lambda'\beta$ if $f(Y) = a_0 + a'Y$ for some scalar a_0 and vector a.

Technically, $a_0 + a'Y$ is an *affine* transformation of Y and $a'Y$ is linear. We now see that unbiased linear estimates have to be linear transformations.

Proposition 2.1.9 *An estimate $a_0 + a'Y$ is unbiased for $\lambda'\beta$ if and only if $a_0 = 0$ and $a'X = \lambda'$.*

Proof $\Leftarrow$ If $a_0 = 0$ and $a'X = \lambda'$, then $\mathrm{E}(a_0 + a'Y) = 0 + a'X\beta = \lambda'\beta$.

$\Rightarrow$ If $a_0 + a'Y$ is unbiased, $\lambda'\beta = \mathrm{E}(a_0 + a'Y) = a_0 + a'X\beta$, for any β. Subtracting $a'X\beta$ from both sides gives

$$(\lambda' - a'X)\beta = a_0$$

for any β. If $\beta = 0$, then $a_0 = 0$. Thus the vector $\lambda - X'a$ is orthogonal to any vector β. This can only occur if $\lambda - X'a = 0$; so $\lambda' = a'X$. □

Corollary 2.1.10 *$\lambda'\beta$ is estimable if and only if there exists ρ such that* $\mathrm{E}(\rho'Y) = \lambda'\beta$ *for any β.*

2.2 Estimation: Least Squares

Consider the linear model

$$Y = X\beta + e, \quad \mathrm{E}(e) = 0.$$

Suppose we want to estimate $\mathrm{E}(Y)$. We know that $\mathrm{E}(Y) = X\beta$, but β is unknown; so all we really know is that

$$\mathrm{E}(Y) \in C(X) \equiv \{v|v = X\beta \text{ for some } \beta \in \mathbf{R}^p\}.$$

To estimate $\mathrm{E}(Y)$, we might take the vector in $C(X)$ that is closest to Y. By definition then, an estimate $\hat{\beta}$ is a least squares estimate if $X\hat{\beta}$ is the vector in $C(X)$ that is closest to Y. In other words, $\hat{\beta}$ is a *least squares estimate (LSE)* of β if

$$(Y - X\hat{\beta})'(Y - X\hat{\beta}) = \min_{\beta}(Y - X\beta)'(Y - X\beta).$$

For a vector $\Lambda'\beta$, a least squares estimate is defined as $\Lambda'\hat{\beta}$ for any least squares estimate $\hat{\beta}$.

In this section, least squares estimates are characterized and uniqueness and unbiasedness properties of least squares estimates are given. For standard linear models an unbiased estimate of σ^2 is presented. Finally, at the end of the section, the geometry associated with least squares estimation and unbiased estimation of σ^2 is discussed. The geometry provides good intuition for n-dimensional problems but the geometry can only be visualized in three dimensions. In other words, although it is a fine pedagogical tool, the geometry can only be spelled out for three or fewer data points. The fundamental goal of this book is to build the theory of linear models on vector space generalizations of these fundamentally geometric concepts. We now establish the *fundamental theorem of least squares estimation*, that the vector in $C(X)$ that is closest to Y is the perpendicular projection of Y onto $C(X)$.

Theorem 2.2.1 *$\hat{\beta}$ is a least squares estimate of β if and only if $X\hat{\beta} = MY$, where M is the perpendicular projection operator onto $C(X)$.*

Proof We will show that

$$(Y - X\beta)'(Y - X\beta) = (Y - MY)'(Y - MY) + (MY - X\beta)'(MY - X\beta).$$

Both terms on the righthand side are nonnegative, and the first term does not depend on β. $(Y - X\beta)'(Y - X\beta)$ is minimized by minimizing $(MY - X\beta)'(MY - X\beta)$. This is the squared distance between MY and $X\beta$. The distance is zero if and only if $MY = X\beta$, which proves the theorem. We now establish the equation.

$$\begin{aligned}(Y - X\beta)'(Y - X\beta) &= (Y - MY + MY - X\beta)'(Y - MY + MY - X\beta)\\ &= (Y - MY)'(Y - MY) + (Y - MY)'(MY - X\beta)\\ &\quad + (MY - X\beta)'(Y - MY) + (MY - X\beta)'(MY - X\beta).\end{aligned}$$

However, $(Y - MY)'(MY - X\beta) = Y'(I - M)MY - Y'(I - M)X\beta = 0$ because $(I - M)M = 0$ and $(I - M)X = 0$. Similarly, $(MY - X\beta)'(Y - MY) = 0$. □

Corollary 2.2.2 *$(X'X)^{-}X'Y$ is a least squares estimate of β.*

In Example 1.0.2, with M given in Exercise 1.5.8b, it is not difficult to see that

$$MY = \begin{bmatrix}\bar{y}_{1\cdot}\\ \bar{y}_{1\cdot}\\ \bar{y}_{1\cdot}\\ y_{21}\\ \bar{y}_{3\cdot}\\ \bar{y}_{3\cdot}\end{bmatrix} = \begin{bmatrix}1&1&0&0\\1&1&0&0\\1&1&0&0\\1&0&1&0\\1&0&0&1\\1&0&0&1\end{bmatrix}\begin{bmatrix}0\\ \bar{y}_{1\cdot}\\ y_{21}\\ \bar{y}_{3\cdot}\end{bmatrix} = \begin{bmatrix}1&1&0&0\\1&1&0&0\\1&1&0&0\\1&0&1&0\\1&0&0&1\\1&0&0&1\end{bmatrix}\begin{bmatrix}\bar{y}_{1\cdot}\\ 0\\ y_{21} - \bar{y}_{1\cdot}\\ \bar{y}_{3\cdot} - \bar{y}_{1\cdot}\end{bmatrix}.$$

Thus, both $\hat{\beta}_1 = (0, \bar{y}_{1\cdot}, y_{21}, \bar{y}_{3\cdot})'$ and $\hat{\beta}_2 = (\bar{y}_{1\cdot}, 0, y_{21} - \bar{y}_{1\cdot}, \bar{y}_{3\cdot} - \bar{y}_{1\cdot})'$ are least squares estimates of β. From Example 2.1.5,

$$\alpha_1 - \alpha_3 = (1, 0, 0, 0, -1, 0)X\beta = (1/3, 1/3, 1/3, 0, -1/2, -1/2)X\beta.$$

The least squares estimates are

$$(1, 0, 0, 0, -1, 0)X\hat{\beta} = (1, 0, 0, 0, -1, 0)MY = \bar{y}_{1\cdot} - \bar{y}_{3\cdot},$$

but also

$$\begin{aligned}(1/3, 1/3, 1/3, 0, -1/2, -1/2)X\hat{\beta} &= (1/3, 1/3, 1/3, 0, -1/2, -1/2)MY\\ &= \bar{y}_{1\cdot} - \bar{y}_{3\cdot}.\end{aligned}$$

Moreover, these estimates do not depend on the choice of least squares estimates. Either $\hat{\beta}_1$ or $\hat{\beta}_2$ gives this result.

In Example 1.0.1, $X'X$ has a true inverse, so the unique least squares estimate of β is

$$\hat{\beta} = \left(X'X\right)^{-1}X'Y = \begin{bmatrix}6 & 21\\ 21 & 91\end{bmatrix}^{-1}\begin{bmatrix}1&1&1&1&1&1\\1&2&3&4&5&6\end{bmatrix}Y.$$

An immediate result of Theorem 2.2.1, the uniqueness of perpendicular projection operators (Proposition B.34), and Theorem 2.1.2 as applied to linear models, is that the least squares estimate of any identifiable function is unique. Any least squares estimates $\hat{\beta}_1$ and $\hat{\beta}_2$ have $X\hat{\beta}_1 = X\hat{\beta}_2$, so if $g(\beta)$ is identifiable, $g(\hat{\beta}_1) = g(\hat{\beta}_2)$. In particular, least squares estimates of estimable functions are unique.

Corollary 2.2.3 *The unique least squares estimate of $\rho'X\beta$ is $\rho'MY$.*

Recall that a vector-valued linear function of the parameters, say $\Lambda'\beta$, is estimable if and only if $\Lambda' = P'X$ for some matrix P. The unique least squares estimate of $\Lambda'\beta$ is then $P'MY = \Lambda'\hat{\beta}$.

We now show that the least squares estimate of $\lambda'\beta$ is unique only if $\lambda'\beta$ is estimable.

Theorem 2.2.4 $\lambda' = \rho'X$ *if* $\lambda'\hat{\beta}_1 = \lambda'\hat{\beta}_2$ *for any* $\hat{\beta}_1, \hat{\beta}_2$ *that satisfy* $X\hat{\beta}_1 = X\hat{\beta}_2 = MY$.

Proof Decompose λ into vectors in $C(X')$ and its orthogonal complement. Let N be the perpendicular projection operator onto $C(X')$; then we can write $\lambda = X'\rho_1 + (I - N)\rho_2$. We want to show that $(I - N)\rho_2 = 0$. This will be done by showing that $(I - N)\rho_2$ is orthogonal to every vector in $\mathbf{R}^p$.

By assumption, $\lambda'(\hat{\beta}_1 - \hat{\beta}_2) = 0$, and we know that $\rho_1'X(\hat{\beta}_1 - \hat{\beta}_2) = 0$; so we must have $\rho_2'(I - N)(\hat{\beta}_1 - \hat{\beta}_2) = 0$, and this holds for any least squares estimates $\hat{\beta}_1, \hat{\beta}_2$.

Let $\hat{\beta}_1$ be any least squares estimate and take v such that $v \perp C(X')$, then $\hat{\beta}_2 = \hat{\beta}_1 - v$ is a least squares estimate. This follows because $X\hat{\beta}_2 = X\hat{\beta}_1 - Xv = X\hat{\beta}_1 = MY$. Substituting above gives $0 = \rho_2'(I - N)(\hat{\beta}_1 - \hat{\beta}_2) = \rho_2'(I - N)v$ for any $v \perp C(X')$. Moreover, by definition of N for any $v \in C(X')$, $(I - N)v = 0$. It follows that $\rho_2'(I - N)v = 0$ for any $v \in \mathbf{R}^p$ and thus $(I - N)\rho_2 = 0$. □

When β is not identifiable, sometimes *side conditions* are arbitrarily imposed on the parameters to allow "estimation" of nonidentifiable parameters. Imposing side conditions amounts to choosing one particular least squares estimate of β. In our earlier discussion of estimation for Example 1.0.2, we presented two sets of parameter estimates. The first estimate, $\hat{\beta}_1$, arbitrarily imposed $\mu = 0$ and $\hat{\beta}_2$ arbitrarily imposed $\alpha_1 = 0$. Side conditions determine a particular least squares estimate by introducing a nonidentifiable, typically a linear nonestimable, constraint on the parameters. With $r \equiv r(X) < p$, one needs $p - r$ individual side conditions to identify the parameters and thus allow "estimation" of the otherwise nonidentifiable parameters. Initially, the model was overparameterized. A linear nonestimable constraint is chosen to remove the ambiguity. Fundamentally, one choice of side conditions is as good as any other. See the discussion near Corollary 3.3.8 for further explication of linear nonestimable constraints. The use of a side condition in one-way ANOVA is also considered in Chapter 4.

When $n < p$, β in never estimable.

Personally, I find it silly to pretend that nonidentifiable functions of the parameters can be estimated. To do so requires strong assumptions that should be made explicit, cf. Christensen (2018). The one good thing about imposing arbitrary side conditions is that they allow computer programs to print out parameter estimates. But different programs use different (equally valid) side conditions, so the printed estimates may differ from program to program. Fortunately, the estimates should agree on all estimable (and, more generally, identifiable) functions of the parameters.

Least squares estimation is not a statistical procedure! Its justification as an optimal estimate is geometric, not statistical. Much of this chapter is devoted to establishing the statistical properties of these geometrically optimal estimates. First, we note that least squares estimates of estimable functions are unbiased.

Proposition 2.2.5 *If $\lambda' = \rho' X$, then* $\mathrm{E}(\rho' MY) = \lambda'\beta$.

Proof $\mathrm{E}(\rho' MY) = \rho' M\mathrm{E}(Y) = \rho' MX\beta = \rho' X\beta = \lambda'\beta$. □

None of our results on least squares estimation have involved the standard assumption that $\mathrm{Cov}(e) = \sigma^2 I$. Least squares provides unique estimates of identifiable functions and unbiased estimates of estimable functions, regardless of the covariance structure. The next three sections establish that least squares estimates have good statistical properties when $\mathrm{Cov}(e) = \sigma^2 I$.

We now consider unbiased estimation of the variance parameter σ^2. First write the *fitted values* (also called the *predicted values*)

$$\hat{Y} \equiv X\hat{\beta} = MY$$

and the *residuals*

$$\hat{e} \equiv Y - X\hat{\beta} = (I - M)Y.$$

The data vector Y can be decomposed as

$$Y = \hat{Y} + \hat{e} = MY + (I - M)Y.$$

The perpendicular projection of Y onto $C(X)$ (i.e., MY) provides an estimate of $X\beta$. Note that $MY = MX\beta + Me = X\beta + Me$ so that MY equals $X\beta$ plus some error where $\mathrm{E}(Me) = M\mathrm{E}(e) = 0$. Similarly, $(I - M)Y = (I - M)X\beta + (I - M)e = (I - M)e$, so $(I - M)Y$ depends only on the error vector e. Since σ^2 is a property of the error vector, it is reasonable to use $(I - M)Y$ to estimate σ^2.

Theorem 2.2.6 *Let $r(X) = r$ and $\mathrm{Cov}(e) = \sigma^2 I$, then $Y'(I - M)Y/(n - r)$ is an unbiased estimate of σ^2.*

Proof From Theorem 1.3.2 and the facts that $\mathrm{E}(Y) = X\beta$ and $\mathrm{Cov}(Y) = \sigma^2 I$, we have
$$\mathrm{E}\big[Y'(I-M)Y\big] = \mathrm{tr}\big[\sigma^2(I-M)\big] + \beta'X'(I-M)X\beta.$$
However, $\mathrm{tr}\big[\sigma^2(I-M)\big] = \sigma^2\,\mathrm{tr}(I-M) = \sigma^2 r(I-M) = \sigma^2\,(n-r)$ and $(I-M)X = 0$, so $\beta'X'(I-M)X\beta = 0$; therefore,
$$\mathrm{E}\big[Y'(I-M)Y\big] = \sigma^2(n-r)$$
and
$$\mathrm{E}\big[Y'(I-M)Y/(n-r)\big] = \sigma^2.$$
□

$Y'(I-M)Y = \hat{e}'\hat{e}$ is called the *sum of squares for error* (*SSE*). It is the squared length of the residual vector $(I-M)Y$. $Y'(I-M)Y/(n-r)$ is called the *mean squared error* (*MSE*). It is the squared length of $(I-M)Y$ divided by the rank of $(I-M)$. In a sense, the *MSE* is the average squared length of $(I-M)Y$, where the average is over the number of dimensions in $C(I-M)$. The rank of $I-M$ is called the *degrees of freedom for error*, denoted dfE.

For Example 1.0.1,
$$SSE = (y_1 - \hat{\beta}_0 - \hat{\beta}_1 1)^2 + (y_2 - \hat{\beta}_0 - \hat{\beta}_1 2)^2 + \cdots + (y_6 - \hat{\beta}_0 - \hat{\beta}_1 6)^2$$
and $MSE = SSE/(6-2)$. For Example 1.0.2,
$$SSE = (y_{11} - \bar{y}_{1\cdot})^2 + (y_{12} - \bar{y}_{1\cdot})^2 + (y_{13} - \bar{y}_{1\cdot})^2$$
$$+ (y_{21} - y_{21})^2 + (y_{31} - \bar{y}_{3\cdot})^2 + (y_{32} - \bar{y}_{3\cdot})^2$$
and $MSE = SSE/(6-3)$.

Finally, one can think about the geometry of least squares estimation in three dimensions. Consider a rectangular table. (Yes, that furniture you have in your kitchen!) Take one corner of the table to be the origin. Take $C(X)$ as the two-dimensional subspace determined by the surface of the table. Y can be any vector originating at the origin, i.e., any point in three-dimensional space. The linear model says that $\mathrm{E}(Y) = X\beta$, which just says that $\mathrm{E}(Y)$ is somewhere on the surface of the table. The least squares estimate $MY = X\hat{\beta}$ is the perpendicular projection of Y onto the table surface. The residual vector $(I-M)Y$ is the vector starting at the origin, perpendicular to the surface of the table, that reaches the same height as Y. Another way to think of the residual vector is to connect the ends of MY and Y with a line segment (that is perpendicular to the surface of the table) but then shift the line segment along the surface (keeping it perpendicular) until the line segment has one end at the origin. The residual vector is the perpendicular projection of Y onto $C(I-M)$, that is, the projection onto the orthogonal complement of the table surface. The orthogonal complement is the one-dimension space in the vertical direction that goes through the origin. Because the orthogonal complement has only one dimension, MSE is just the squared length of the residual vector.

Alternatively, one could take $C(X)$ to be a one-dimensional subspace determined by an edge of the table that includes the origin. The linear model now says that $\mathrm{E}(Y)$ is somewhere on this edge of the table. $MY = X\hat{\beta}$ is found by dropping a perpendicular from Y to the edge of the table. If you connect MY and Y, you essentially get the residual vector $(I - M)Y$, except that the line segment has to be shifted down the edge so that it has one end at the origin. The residual vector is perpendicular to the $C(X)$ edge of the table, but typically would not be perpendicular to the surface of the table. $C(I - M)$ is now the plane that contains everything (through the origin) that is perpendicular to the $C(X)$ edge of the table. In other words, $C(I - M)$ is the two-dimensional space determined by the vertical direction and the *other* edge of the table that goes through the origin. MSE is the squared length of the residual vector divided by 2, because $C(I - M)$ is a two-dimensional space.

2.3 Estimation: Best Linear Unbiased

Another criterion for estimation of $\lambda'\beta$ is to choose the best linear unbiased estimate (*BLUE*) of $\lambda'\beta$. We prove the Gauss–Markov theorem that least squares estimates are best linear unbiased estimates.

Definition 2.3.1 $a'Y$ is a *best linear unbiased estimate* of $\lambda'\beta$ if $a'Y$ is unbiased and if for any other linear unbiased estimate $b'Y$, $\mathrm{Var}(a'Y) \leq \mathrm{Var}(b'Y)$.

Gauss–Markov Theorem 2.3.2 *Consider the standard linear model*

$$Y = X\beta + e, \qquad \mathrm{E}(e) = 0, \qquad \mathrm{Cov}(e) = \sigma^2 I.$$

If $\lambda'\beta$ is estimable, then the least squares estimate of $\lambda'\beta$ is a BLUE of $\lambda'\beta$.

Proof Let M be the perpendicular projection operator onto $C(X)$. Since $\lambda'\beta$ is an estimable function, let $\lambda' = \rho'X$ for some ρ. We need to show that if $a'Y$ is an unbiased estimate of $\lambda'\beta$, then $\mathrm{Var}(a'Y) \geq \mathrm{Var}(\rho'MY)$. Since $a'Y$ is unbiased for $\lambda'\beta$, $\lambda'\beta = \mathrm{E}(a'Y) = a'X\beta$ for any value of β. Therefore $\rho'X = \lambda' = a'X$. Write

$$\begin{aligned}\mathrm{Var}(a'Y) &= \mathrm{Var}(a'Y - \rho'MY + \rho'MY)\\ &= \mathrm{Var}(a'Y - \rho'MY) + \mathrm{Var}(\rho'MY) + 2\mathrm{Cov}\big[(a'Y - \rho'MY), \rho'MY\big].\end{aligned}$$

Since $\mathrm{Var}(a'Y - \rho'MY) \geq 0$, if we show that $\mathrm{Cov}\big[(a'Y - \rho'MY), \rho'MY\big] = 0$, then $\mathrm{Var}(a'Y) \geq \mathrm{Var}(\rho'MY)$ and the theorem holds.

We now show that $\mathrm{Cov}\big[(a'Y - \rho'MY), \rho'MY\big] = 0$.

$$\begin{aligned}\operatorname{Cov}\big[(a'Y - \rho'MY), \rho'MY\big] &= \operatorname{Cov}\big[(a' - \rho'M)Y, \rho'MY\big] \\ &= (a' - \rho'M)\operatorname{Cov}(Y)M\rho \\ &= \sigma^2(a' - \rho'M)M\rho \\ &= \sigma^2(a'M - \rho'M)\rho.\end{aligned}$$

As shown above, $a'X = \rho'X$, and since we can write $M = X(X'X)^{-}X'$, we have $a'M = \rho'M$. It follows that $\sigma^2(a'M - \rho'M)\rho = 0$ as required. □

Corollary 2.3.3 *If $\sigma^2 > 0$, there exists a unique BLUE for any estimable function $\lambda'\beta$.*

Proof Let $\lambda' = \rho'X$, and recall from Section 1 that the vector $\rho'M$ is uniquely determined by λ'. In the proof of Theorem 2.3.2, it was shown that for an arbitrary linear unbiased estimate $a'Y$,

$$\operatorname{Var}(a'Y) = \operatorname{Var}(\rho'MY) + \operatorname{Var}(a'Y - \rho'MY).$$

If $a'Y$ is a BLUE of $\lambda'\beta$, it must be true that $\operatorname{Var}(a'Y - \rho'MY) = 0$. It is easily seen that

$$0 = \operatorname{Var}(a'Y - \rho'MY) = \operatorname{Var}\big[(a' - \rho'M)Y\big] = \sigma^2(a - M\rho)'(a - M\rho).$$

For $\sigma^2 > 0$, this occurs if and only if $a - M\rho = 0$, which is equivalent to the condition $a = M\rho$. □

2.4 Estimation: Maximum Likelihood

Another criterion for choosing estimates of β and σ^2 is maximum likelihood. The likelihood function is derived from the joint density of the observations by considering the parameters as variables and the observations as fixed at their observed values. If we assume $Y \sim N(X\beta, \sigma^2 I)$, then the *maximum likelihood estimates* (*MLE*s) of β and σ^2 are obtained by maximizing

$$(2\pi)^{-n/2}[\det(\sigma^2 I)]^{-1/2}\exp\big[-(Y - X\beta)'(Y - X\beta)/2\sigma^2\big]. \tag{1}$$

Equivalently, the log of the likelihood can be maximized. The log of (1) is

$$\frac{-n}{2}\log(2\pi) - \frac{1}{2}\log[(\sigma^2)^n] - (Y - X\beta)'(Y - X\beta)/2\sigma^2.$$

For every value of σ^2, the log-likelihood is maximized by taking β to minimize $(Y - X\beta)'(Y - X\beta)$, i.e., least squares estimates are MLEs. To estimate σ^2 we can substitute $Y'(I - M)Y = (Y - X\hat{\beta})'(Y - X\hat{\beta})$ for $(Y - X\beta)'(Y - X\beta)$ and differentiate with respect to σ^2 to get $Y'(I - M)Y/n$ as the MLE of σ^2.

The MLE of σ^2 is rarely used in practice. The *MSE* is the standard estimate of σ^2. For almost any purpose except point estimation of σ^2, it is immaterial whether the *MSE* or the MLE is used. They lead to identical confidence intervals and tests for σ^2. They also lead to identical confidence regions and tests for estimable functions of β. It should be emphasized that it is not appropriate to substitute the MLE for the *MSE* and then form confidence intervals and tests as if the *MSE* were being used.

2.5 Estimation: Minimum Variance Unbiased

In Section 3, it was shown that least squares estimates give best estimates among the class of linear unbiased estimates. If the error vector is normally distributed, least squares estimates are best estimates among all unbiased estimates, not just the linear unbiased estimates. In particular, with normal errors, the best estimates happen to be linear estimates. As in Section 3, a best unbiased estimate is taken to be an unbiased estimate with minimum variance.

It is not the purpose of this monograph to develop the theory of minimum variance unbiased estimation. However, we will outline the application of this theory to linear models. See Lehmann (1983, Sections 1.4, 1.5) and Lehmann (1986, Sections 2.6, 2.7, 4.3) for a detailed discussion of the definitions and theorems used here. Our model is

$$Y = X\beta + e, \qquad e \sim N\left(0, \sigma^2 I\right).$$

Definition 2.5.1 A vector-valued sufficient statistic $T(Y)$ is said to be *complete* if $\mathrm{E}[h(T(Y))] = 0$ for all β and σ^2 implies that $\Pr[h(T(Y)) = 0] = 1$ for all β and σ^2.

Theorem 2.5.2 *If $T(Y)$ is a complete sufficient statistic, then $f(T(Y))$ is a minimum variance unbiased estimate (MVUE) of* $\mathrm{E}[f(T(Y))]$.

Proof Suppose $g(Y)$ is an unbiased estimate of $\mathrm{E}[f(T(Y))]$. By the Rao–Blackwell theorem (see Cox and Hinkley 1974),

$$\mathrm{Var}(\mathrm{E}[g(Y)|T(Y)]) \le \mathrm{Var}(g(Y)).$$

Since $\mathrm{E}[g(Y)|T(Y)]$ is unbiased, $\mathrm{E}\{f(T(Y)) - \mathrm{E}[g(Y)|T(Y)]\} = 0$. By completeness of $T(Y)$, $\Pr\{f(T(Y)) = \mathrm{E}[g(Y)|T(Y)]\} = 1$. It follows that $\mathrm{Var}(f(T(Y))) \le \mathrm{Var}(g(Y))$. □

We wish to use the following result from Lehmann (1983, pp. 28, 46):

Theorem 2.5.3 *Let $\theta = (\theta_1, \ldots, \theta_s)'$ and let Y be a random vector with probability density function*

$$f(Y) = c(\theta)\exp\left[\sum_{i=1}^{s} \theta_i T_i(Y)\right] h(Y);$$

then $T(Y) = (T_1(Y), T_2(Y), \ldots, T_s(Y))'$ is a complete sufficient statistic provided that neither θ nor $T(Y)$ satisfy any linear constraints.

Suppose $r(X) = r < p$, then the theorem cannot apply to $X'Y$ because, for $b \perp C(X')$, $Xb = 0$; so $b'X'Y$ is subject to a linear constraint. We need to consider the following reparameterization. Let Z be an $n \times r$ matrix whose columns form a basis for $C(X)$. For some matrix U, we have $X = ZU$. Let $\lambda'\beta$ be an estimable function. Then for some ρ, $\lambda'\beta = \rho'X\beta = \rho'ZU\beta$. Define $\gamma = U\beta$ and consider the linear model

$$Y = Z\gamma + e, \qquad e \sim N\left(0, \sigma^2 I\right).$$

The usual estimate of $\lambda'\beta = \rho'Z\gamma$ is $\rho'MY$ regardless of the parameterization used. We will show that $\rho'MY$ is a minimum variance unbiased estimate of $\lambda'\beta$. The density of Y can be written

$$\begin{aligned} f(Y) &= (2\pi)^{-n/2}\left(\sigma^2\right)^{-n/2}\exp[-(Y - Z\gamma)'(Y - Z\gamma)/2\sigma^2] \\ &= C_1(\sigma^2)\exp\left[-\left(Y'Y - 2\gamma'Z'Y + \gamma'Z'Z\gamma\right)/2\sigma^2\right] \\ &= C_2(\gamma, \sigma^2)\exp\left[(-1/2\sigma^2)Y'Y + (\sigma^{-2}\gamma')(Z'Y)\right]. \end{aligned}$$

This is the form of Theorem 2.5.3. There are no linear constraints on the parameters $(-1/2\sigma^2, \gamma_1/\sigma^2, \ldots, \gamma_r/\sigma^2)$ nor on $(Y'Y, Y'Z)'$, so $(Y'Y, Y'Z)'$ is a complete sufficient statistic. An unbiased estimate of $\lambda'\beta = \rho'X\beta$ is $\rho'MY = \rho'Z(Z'Z)^{-1}Z'Y$. $\rho'MY$ is a function of $Z'Y$, so it is a minimum variance unbiased estimate. Moreover, $Y'(I - M)Y/(n - r)$ is an unbiased estimate of σ^2 and $Y'(I - M)Y = Y'Y - (Y'Z)(Z'Z)^{-1}(Z'Y)$ is a function of the complete sufficient statistic $(Y'Y, Y'Z)'$, so *MSE* is a minimum variance unbiased estimate. We have established the following result:

Theorem 2.5.4 *MSE is a minimum variance unbiased estimate of σ^2 and $\rho'MY$ is a minimum variance unbiased estimate of $\rho'X\beta$ whenever $e \sim N\left(0, \sigma^2 I\right)$.*

2.6 Sampling Distributions of Estimates

If we continue to assume that $Y \sim N(X\beta, \sigma^2 I)$, the distributions of estimates are straightforward. The least squares estimate of $\Lambda'\beta$ where $\Lambda' = P'X$ is $\Lambda'\hat{\beta} = P'MY$. The distribution of $P'MY$ is $N(P'X\beta, \sigma^2 P'MIMP)$ or, equivalently,

$$P'MY \sim N(\Lambda'\beta, \sigma^2 P'MP).$$

Since $M = X(X'X)^- X'$, we can also write

$$\Lambda'\hat{\beta} \sim N(\Lambda'\beta, \sigma^2 \Lambda'(X'X)^- \Lambda).$$

Two special cases are of interest. First, the estimate of $X\beta$ is

$$\hat{Y} \equiv MY \sim N(X\beta, \sigma^2 M).$$

Second, if $(X'X)$ is nonsingular, β is estimable and

$$\hat{\beta} \sim N(\beta, \sigma^2 (X'X)^{-1}).$$

In Section 2 it was shown that the mean square error $Y'(I-M)Y/(n-r)$ is an unbiased estimate of σ^2. We now show that $Y'(I-M)Y/\sigma^2 \sim \chi^2(n-r)$. Clearly $Y/\sigma \sim N(X\beta/\sigma, I)$, so by Theorem 1.3.3

$$Y'(I-M)Y/\sigma^2 \sim \chi^2(r(I-M), \beta'X'(I-M)X\beta/2\sigma^2).$$

We have already shown that $r(I-M) = n-r$ and $\beta'X'(I-M)X\beta/2\sigma^2 = 0$. Moreover, by Theorem 1.3.7, MY and $Y'(I-M)Y$ are independent.

Exercise 2.1 Show that for $\lambda'\beta$ estimable,

$$\frac{\lambda'\hat{\beta} - \lambda'\beta}{\sqrt{MSE\,\lambda'(X'X)^-\lambda}} \sim t(dfE).$$

Find the form of an α level test of $H_0 : \lambda'\beta = 0$ and the form for a $(1-\alpha)100\%$ confidence interval for $\lambda'\beta$. Hint: The test and confidence interval can be found using the methods of Appendix E.

Exercise 2.2 Let $y_{11}, y_{12}, \ldots, y_{1r}$ be $N(\mu_1, \sigma^2)$ and $y_{21}, y_{22}, \ldots, y_{2s}$ be $N(\mu_2, \sigma^2)$ with all y_{ij}s independent. Write this as a linear model. For the rest of the problem use the results of Chapter 2. Find estimates of $\mu_1, \mu_2, \mu_1 - \mu_2$, and σ^2. Using Appendix E and Exercise 2.1, form an $\alpha = 0.01$ test for $H_0 : \mu_1 = \mu_2$. Similarly, form 95% confidence intervals for $\mu_1 - \mu_2$ and μ_1. What is the test for $H_0 : \mu_1 = \mu_2 + \Delta$, where Δ

is some known fixed quantity? How do these results compare with the usual analysis for two independent samples with a common variance?

Exercise 2.3 Let $y_1, y_2, \ldots, y_n$ be independent $N(\mu, \sigma^2)$. Write a linear model for these data. For the rest of the problem use the results of Chapter 2, Appendix E, and Exercise 2.1. Form an $\alpha = 0.01$ test for $H_0 : \mu = \mu_0$, where μ_0 is some known fixed number and form a 95% confidence interval for μ. How do these results compare with the usual analysis for one sample?

Exercise 2.4 Use Corollary 1.3.6a to show that if $\Lambda' = P'X$ then

$$\frac{(\Lambda'\hat{\beta})'[\Lambda'(X'X)^-\Lambda]^-(\Lambda'\hat{\beta})}{\sigma^2} \sim \chi^2[r(\Lambda), (\Lambda'\beta)'[\Lambda'(X'X)^-\Lambda]^-(\Lambda'\beta)]$$

Hints: To show that $\Lambda'\beta \in C(\Lambda'(X'X)^-\Lambda)$ and $r(\Lambda) = r(\Lambda'(X'X)^-\Lambda)$, establish that $C(\Lambda') = C(P'M) = C(P'MP)$. Showing $X'Pb = 0$ iff $MPb = 0$ implies that $C(P'X)^\perp = C(P'M)^\perp$.

2.7 Generalized Least Squares

A slightly more general linear model than the one considered so far is

$$Y = X\beta + e, \qquad \mathrm{E}(e) = 0, \qquad \mathrm{Cov}(e) = \sigma^2 V, \tag{1}$$

where V is some known positive definite matrix. By Corollary B.23, we can write $V = QQ'$ for some nonsingular matrix Q. It follows that $Q^{-1}VQ'^{-1} = I$.

Instead of analyzing model (1), we analyze the equivalent model,

$$\tilde{Y} = \tilde{X}\beta + \tilde{e} \tag{2}$$

where

$$\tilde{Y} \equiv Q^{-1}Y, \quad \tilde{X} \equiv Q^{-1}X, \quad \tilde{e} \equiv Q^{-1}e.$$

For model (2),

$$\mathrm{E}(\tilde{e}) = \mathrm{E}(Q^{-1}e) = 0$$

and

$$\mathrm{Cov}(\tilde{e}) = \mathrm{Cov}(Q^{-1}e) = \sigma^2 Q^{-1}VQ^{-1\prime} = \sigma^2 I.$$

The transformed model (2) satisfies the assumptions made in the previously developed theory. For the transformed model, the least squares estimates minimize

$$\begin{aligned}
\|\tilde{Y} - \tilde{X}\beta\|^2 &\equiv (\tilde{Y} - \tilde{X}\beta)'(\tilde{Y} - \tilde{X}\beta) \\
&= \left(Q^{-1}Y - Q^{-1}X\beta\right)' \left(Q^{-1}Y - Q^{-1}X\beta\right) \\
&= \left[Q^{-1}(Y - X\beta)\right]' \left[Q^{-1}(Y - X\beta)\right] \\
&= (Y - X\beta)' Q^{-1'} Q^{-1} (Y - X\beta) \\
&= (Y - X\beta)' V^{-1} (Y - X\beta) \equiv \|Y - X\beta\|^2_{V^{-1}}.
\end{aligned}$$

As functions of Y (rather than $\tilde{Y}$), the estimates of β that minimize this criterion are called *generalized least squares estimates* because instead of minimizing the squared distance between Y and $X\beta$, a generalized squared distance determined by V^{-1} is minimized. Generalized least squares is a concept in linear model theory and should not be confused with generalized linear models. To differentiate from generalized least squares, the least squares estimation of Section 2 is sometimes called *ordinary least squares (OLS)*.

Theorem 2.7.1

(a) *$\lambda'\beta$ is estimable in model (1) if and only if $\lambda'\beta$ is estimable in model (2).*
(b) *$\hat{\beta}$ is a generalized least squares estimate of β if and only if*

$$X\left(X'V^{-1}X\right)^{-} X'V^{-1}Y = X\hat{\beta}.$$

For any estimable function there exists a unique generalized least squares estimate.

(c) *For an estimable function $\lambda'\beta$, the generalized least squares estimate is the BLUE of $\lambda'\beta$.*
(d) *If $e \sim N(0, \sigma^2 V)$, then for any estimable function $\lambda'\beta$, the generalized least squares estimate is the minimum variance unbiased estimate.*
(e) *If $e \sim N(0, \sigma^2 V)$, then any generalized least squares estimate of β is a maximum likelihood estimate of β.*

Proof

(a) If $\lambda'\beta$ is estimable in model (1), we can write

$$\lambda' = \rho'X = (\rho'Q)Q^{-1}X;$$

so $\lambda'\beta$ is estimable in model (2). If $\lambda'\beta$ is estimable in model (2), then $\lambda' = \rho'Q^{-1}X = (\rho'Q^{-1})X$; so $\lambda'\beta$ is estimable in model (1).

(b) By Theorem 2.2.1, the generalized least squares estimates (i.e., the least squares estimates for model (2)) satisfy the equation

$$Q^{-1}X\left(X'Q^{-1'}Q^{-1}X\right)^{-} X'Q^{-1'}Q^{-1}Y = \tilde{X}(\tilde{X}'\tilde{X})^{-}\tilde{X}'\tilde{Y} = \tilde{X}\hat{\beta} = Q^{-1}X\hat{\beta}.$$

Simplifying and multiplying through on the left by Q gives the equivalent condition

$$X\left(X'V^{-1}X\right)^{-}X'V^{-1}Y = X\hat{\beta}.$$

From Theorem 2.2.3, generalized least squares estimates of estimable functions are unique.

(c) From Theorem 2.3.2 as applied to model (2), the generalized least squares estimate of $\lambda'\beta$ is the BLUE of $\lambda'\beta$ among all unbiased linear combinations of the vector $Q^{-1}Y$. However, any linear combination, in fact any function, of Y can be obtained from $Q^{-1}Y$ because Q^{-1} is invertible. Thus, the generalized least squares estimate is the BLUE.

(d) Applying Theorem 2.5.2 to model (2) establishes that the generalized least squares estimate is the MVUE from among unbiased estimates that are functions of $Q^{-1}Y$. Since Q is nonsingular, any function of Y can be written as a function of $Q^{-1}Y$; so the generalized least squares estimate is the minimum variance unbiased estimate.

(e) The likelihood functions from models (1) and (2) are equal up to a constant of proportionality. From model (2), a generalized least squares estimate $\hat{\beta}$ maximizes the likelihood among all functions of $Q^{-1}Y$, but since Q is nonsingular, $\hat{\beta}$ maximizes the likelihood among all functions of Y. □

Theorem 2.7.1(b) is the generalized least squares equivalent of Theorem 2.2.1. Theorem 2.2.1 relates $X\hat{\beta}$ to the perpendicular projection of Y onto $C(X)$. Theorem 2.7.1(b) also relates $X\hat{\beta}$ to a projection of Y onto $C(X)$, but in Theorem 2.7.1(b) the projection is not the perpendicular projection. If we write

$$A = X\left(X'V^{-1}X\right)^{-}X'V^{-1}, \tag{3}$$

then the condition in Theorem 2.7.1(b) is

$$AY = X\hat{\beta}.$$

We wish to show that A is a projection operator onto $C(X)$. The perpendicular projection operator onto $C(Q^{-1}X)$ is

$$Q^{-1}X\left[(Q^{-1}X)'(Q^{-1}X)\right]^{-}(Q^{-1}X)'.$$

By the definition of a projection operator,

$$Q^{-1}X\left[(Q^{-1}X)'(Q^{-1}X)\right]^{-}(Q^{-1}X)'Q^{-1}X = Q^{-1}X.$$

This can also be written as

$$Q^{-1}AX = Q^{-1}X.$$

Multiplying on the left by Q gives

$$AX = X. \tag{4}$$

From (3) and (4), we immediately have

$$AA = A,$$

so A is a projection matrix. From (3), $C(A) \subset C(X)$ and from (4), $C(X) \subset C(A)$; so $C(A) = C(X)$ and we have proven:

Proposition 2.7.2 *A is a projection operator onto* $C(X)$.

For an estimable function $\lambda'\beta$ with $\lambda' = \rho'X$, the generalized least squares estimate is $\lambda'\hat{\beta} = \rho'AY$. This result is analogous to the ordinary least squares result in Corollary 2.2.3. To obtain tests and confidence intervals for $\lambda'\beta$, we need to know $\text{Cov}(X\hat{\beta})$.

Proposition 2.7.3 $\text{Cov}(X\hat{\beta}) = \sigma^2 X\left(X'V^{-1}X\right)^{-} X'$.

Proof $\text{Cov}(X\hat{\beta}) = \text{Cov}(AY) = \sigma^2 AVA'$. From (3) and (4) it is easily seen (cf. Exercise 2.5) that

$$AVA' = AV = VA'.$$

In particular, $AV = X\left(X'V^{-1}X\right)^{-} X'$. □

Corollary 2.7.4 *If $\lambda'\beta$ is estimable, then the generalized least squares estimate has* $\text{Var}(\lambda'\hat{\beta}) = \sigma^2\lambda'\left(X'V^{-1}X\right)^{-}\lambda$.

Exercise 2.5

(a) Show that $AVA' = AV = VA'$.
(b) Show that $A'V^{-1}A = A'V^{-1} = V^{-1}A$.
(c) Show that A is the same for any choice of $\left(X'V^{-1}X\right)^{-}$.

It is necessary to have an estimate of σ^2. From model (2),

$$\begin{aligned} SSE &= (Q^{-1}Y)'\left[I - Q^{-1}X\left[(Q^{-1}X)'(Q^{-1}X)\right]^{-}(Q^{-1}X)'\right](Q^{-1}Y) \\ &= Y'V^{-1}Y - Y'V^{-1}X\left(X'V^{-1}X\right)^{-}X'V^{-1}Y \\ &= Y'(I-A)'V^{-1}(I-A)Y = \|(I-A)Y\|^2_{V^{-1}}. \end{aligned}$$

Note that $(I - A)Y$ is the vector of residuals $Y - X\hat{\beta}$, so the SSE is a quadratic form in the residuals. Because Q is nonsingular, $r(Q^{-1}X) = r(X)$. It follows from model (2) that an unbiased estimate of σ^2 is obtained from

$$MSE = Y'(I - A)'V^{-1}(I - A)Y / [n - r(X)].$$

With normal errors, this is also the minimum variance unbiased estimate of σ^2.

Suppose that e is normally distributed. From Theorem 1.3.7 applied to model (2), the MSE is independent of $Q^{-1}X\hat{\beta}$. Since $X\hat{\beta}$ is a function of $Q^{-1}X\hat{\beta}$, the MSE is independent of $X\hat{\beta}$. Moreover, $X\hat{\beta}$ is normally distributed and SSE/σ^2 has a chi-squared distribution.

A particular application of these results is that, for an estimable function $\lambda'\beta$,

$$\frac{\lambda'\hat{\beta} - \lambda'\beta}{\sqrt{MSE\,\lambda'\left(X'V^{-1}X\right)^{-}\lambda}} \sim t(n - r(X)).$$

Given this distribution, tests and confidence intervals involving $\lambda'\beta$ can be obtained as in Appendix E.

We now give a result that determines when generalized least squares estimates are (ordinary) least squares estimates. The result will be generalized in Theorem 10.4.5. The generalization changes it to an if and only if statement for arbitrary covariance matrices.

Proposition 2.7.5 *If V is nonsingular and $C(VX) \subset C(X)$, then least squares estimates are BLUEs.*

Proof The proof proceeds by showing that $A \equiv X(X'V^{-1}X)^{-}X'V^{-1}$ is the perpendicular projection operator onto $C(X)$. We already know that A is a projection operator onto $C(X)$, so all we need to establish is that if $w \perp C(X)$, then $Aw = 0$.

V being nonsingular implies that the null spaces of VX and X are identical, so $r(VX) = r(X)$, cf. Exercise 2.11.7. With $C(VX) \subset C(X)$, we must have $C(VX) = C(X)$. $C(VX) = C(X)$ implies that for some matrices B_1 and B_2, $VXB_1 = X$ and $VX = XB_2$. Multiplying through by V^{-1} in both equations gives $XB_1 = V^{-1}X$ and $X = V^{-1}XB_2$, so $C(X) = C(V^{-1}X)$. It follows immediately that $C(X)^{\perp} = C(V^{-1}X)^{\perp}$. Now, $w \perp C(X)$ if and only if $w \perp C(V^{-1}X)$, so

$$Aw = \left[X(X'V^{-1}X)^{-}X'V^{-1}\right]w = X(X'V^{-1}X)^{-}\left[X'V^{-1}w\right] = 0. \qquad \Box$$

Frequently in regression analysis, V is a diagonal matrix, in which case generalized least squares is referred to as *weighted least squares (WLS)*. Considerable simplification results.

The following result will be useful in Section 10. It is essentially the Pythagorean theorem and can be used directly to show Theorem 2.7.1(b), that $X\hat{\beta}$ is a generalized least squares estimate if and only if $X\hat{\beta} = AY$.

Lemma 2.7.6 *Let* $A = X(X'V^{-1}X)^{-}X'V^{-1}$, *then*

$$(Y - X\beta)V^{-1}(Y - X\beta) = (Y - AY)'V^{-1}(Y - AY) + (AY - X\beta)V^{-1}(AY - X\beta)$$
$$= (Y - AY)'V^{-1}(Y - AY) + (\hat{\beta} - \beta)'(X'V^{-1}X)(\hat{\beta} - \beta)$$

where $\hat{\beta} = (X'V^{-1}X)^{-}X'V^{-1}Y$.

Proof Following the proof of Theorem 2.2.2, write $(Y - X\beta) = (Y - AY) + (AY - X\beta)$ and eliminate cross product terms using Exercise 2.5 and

$$(I - A)'V^{-1}(AY - X\beta) = V^{-1}(I - A)(AY - X\beta) = 0.$$ □

Exercise 2.6

(a) Show that A is the projection operator onto $C(X)$ along $C\left(V^{-1}X\right)^{\perp}$.

(b) Show that A is the perpendicular projection operator onto $C(X)$ when the inner product between two vectors x and y is defined as $x'V^{-1}y$. Hint: Recall the discussion after Definition B.50.

(c) Show that $C(VX) \subset C(X)$ if and only if $V = XU_0X' + WU_1W'$ for some U_0 and U_1 nonnegative definite with $X'W = 0$ and $r(W) = n - r(X)$. This is *Rao's simple covariance structure*.

(d) Show that $\text{Cov}[MY, (I - M)Y] = 0$ when Rao's simple covariance structure holds.

2.8 Normal Equations

An alternative method for finding least squares estimates of the parameter β in the model

$$Y = X\beta + e, \qquad \text{E}(e) = 0$$

is to find solutions of what are called the *normal equations*. The normal equations are defined as

$$X'X\beta = X'Y.$$

They are usually arrived at by setting equal to zero the partial derivatives of $(Y - X\beta)'(Y - X\beta)$ with respect to β.

Corollary 2.8.2 shows that solutions of the normal equations are least squares estimates of β. Recall that, by Theorem 2.2.1, least squares estimates are solutions of $X\beta = MY$.

Theorem 2.8.1 $\hat{\beta}$ *is a least squares estimate of* β *if and only if* $(Y - X\hat{\beta}) \perp C(X)$.

Proof Since M is the perpendicular projection operator onto $C(X)$, $(Y - X\hat{\beta}) \perp C(X)$ if and only if $M(Y - X\hat{\beta}) = 0$, i.e, if and only if $MY = X\hat{\beta}$. □

Corollary 2.8.2 *$\hat{\beta}$ is a least squares estimate of β if and only if $X'X\hat{\beta} = X'Y$.*

Proof $X'X\hat{\beta} = X'Y$ if and only if $X'(Y - X\hat{\beta}) = 0$, which occurs if and only if $(Y - X\hat{\beta}) \perp C(X)$. □

For generalized least squares problems, the normal equations are found from model (2.7.2). The normal equations simplify to

$$X'V^{-1}X\beta = X'V^{-1}Y.$$

2.9 Variance-Bias Tradeoff

For a standard linear model the least squares estimates are the best linear unbiased estimates and for multivariate normal data they are the best unbiased estimates. They are best in the sense of having smallest variances. However, it turns out that you can often get better point estimates by incorporating a little bias. A little bit of bias can sometimes eliminate a great deal of variance, making for an overall better estimate.

Suppose the standard linear model

$$Y = X\beta + e, \quad \mathrm{E}(e) = 0, \quad \mathrm{Cov}(e) = \sigma^2 I, \tag{1}$$

is correct and consider fitting a *reduced* model

$$Y = X_0\gamma + e, \quad \text{with } C(X_0) \subset C(X). \tag{2}$$

If $\mathrm{E}(Y) \neq X_0\gamma$, using the reduced linear model to estimate $X\beta$ creates bias. In this section we examine how even incorrect reduced models can improve estimation if the bias they introduce is small relative to the variability in the model.

Example 2.9.1 Consider fitting a linear model with an intercept and three predictors. I am going to fit the full model using ordinary least squares. You, however, think that the regression coefficients for the second and third variable should be the same and that they should be half of the coefficient for the first variable. You incorporate that into your model. If you are correct, your fitted values will be twice as good as mine! But even if you are wrong, if you are close to being correct, your fitted values will still be better than mine. We now explore these claims in some generality.

Under the standard linear model (1), the best fitted values one could ever have are $X\beta$ but we don't know β. For estimated fitted values, say, $F(Y)$, their quality can be measured by looking at

$$\mathrm{E}\left\{[F(Y) - X\beta]'[F(Y) - X\beta]\right\}.$$

For least squares estimates $X\hat{\beta} = MY$, using Theorem 1.3.2 gives

$$\mathrm{E}\left[(MY - X\beta)'(MY - X\beta)\right] = \mathrm{E}\left[(Y - X\beta)'M(Y - X\beta)\right] = \mathrm{tr}[M\sigma^2 I] = \sigma^2 r(X).$$

Now consider the reduced model (2) with M_0 the ppo onto $C(X_0)$. If we estimate the fitted values from the reduced model, i.e. $X_0\hat{\gamma} = M_0 Y$, and the reduced model is true, i.e., $X\beta = X_0\gamma$,

$$\mathrm{E}\left[(M_0Y - X\beta)'(M_0Y - X\beta)\right] = \mathrm{E}\left[(Y - X_0\gamma)'M_0(Y - X_0\gamma)\right]$$
$$= \mathrm{tr}[M_0\sigma^2 I] = \sigma^2 r(X_0).$$

If the reduced model is true, since $r(X_0) \leq r(X)$, we are better off using the reduced model.

In Example 2.9.1, my fitting the full model with $X = [J, X_1, X_2, X_3]$ gives $\mathrm{E}\left[(MY - X\beta)'(MY - X\beta)\right] = 4\sigma^2$. With your reduced model of $X_0 = [J, 2X_1 + X_2 + X_3]$, you get $\mathrm{E}\left[(M_0Y - X\beta)'(M_0Y - X\beta)\right] = 2\sigma^2$, hence the conclusion that your fitted values are twice as good as mine when your model is correct. (One could argue that it would be more appropriate to look at the square roots, so $\sqrt{2}$ times better?)

What about the conclusion that you don't have to be correct to do better than me, you only have to be close to correct? You being correct is $\mathrm{E}(Y) = X_0\gamma$. You will be close to correct if the true $\mathrm{E}(Y)$ is close to $X_0\gamma$, i.e., if $X\beta \doteq X_0\gamma$ for some γ. In particular, you will be close to correct if the true mean $X\beta$ is close to the perpendicular projection of $X\beta$ onto $C(X_0)$, i.e., if $X\beta \doteq M_0X\beta$. In general, because $M_0Y - X\beta = M_0(Y - X\beta) - (I - M_0)X\beta$ where $M_0(Y - X\beta) \perp (I - M_0)X\beta$,

$$\begin{aligned}
&\mathrm{E}\|M_0Y - X\beta\|^2 \\
&= \mathrm{E}\left[(M_0Y - X\beta)'(M_0Y - X\beta)\right] \\
&= \mathrm{E}\left\{[M_0(Y - X\beta) - (I - M_0)X\beta]'[M_0(Y - X\beta) - (I - M_0)X\beta]\right\} \\
&= \mathrm{E}\left\{[M_0(Y - X\beta)]'[M_0(Y - X\beta)]\right\} + \mathrm{E}\left\{[(I - M_0)X\beta]'[(I - M_0)X\beta]\right\} \\
&= \mathrm{E}\left\{(Y - X\beta)'M_0(Y - X\beta)\right\} + \beta'X'(I - M_0)X\beta \\
&= \mathrm{tr}[M_0\sigma^2 I] + \beta'X'(I - M_0)X\beta \\
&= \sigma^2 r(X_0) + \|X\beta - M_0X\beta\|^2.
\end{aligned}$$

We have written the expected squared distance as a variance term that is the product of the observation variance σ^2 and the model size $r(X_0)$ plus a bias term that measures the squared distance of how far the reduced model is from being true. If the reduced model is true, the bias is zero. But even when the reduced model is not true, if a reduced model with $r(X_0)$ substantially smaller than $r(X)$ is close to being true, specifically if

$$\|X\beta - M_0 X\beta\|^2 < \sigma^2[r(X) - r(X_0)],$$

the fitted values of the reduced model will be better estimates than the original least squares estimates. And don't forget that if the full model is not a regression, the estimates of all (estimable) identifiable functions have to be (linear) functions of whatever fitted values we use.

In Example 2.9.1, if the squared distance between the truth, $X\beta$, and the expected value of your reduced model estimate, $M_0 X\beta$, is less than $2\sigma^2$, you will do better than me. Of course we cannot know how close the truth is to the reduced model expected value, but in Subsection 14.1.3 we will see that Mallow's C_p statistic estimates $r(X_0) + \|X\beta - M_0 X\beta\|^2/\sigma^2$, so it gives us an idea about how much better (or worse) a reduced model is doing than the full model. In the context of variable selection, dropping predictor variables with regression coefficients close to zero should result in improved fitted values because the reduced model without those predictors should have $M_0 X\beta \doteq X\beta$.

Most biased estimation methods used in practice are immensely more complicated than this. The variable selection methods discussed in Chapter 14 are perhaps the most widely used methods of creating biased estimates. They use the data to determine an appropriate reduced model, so X_0 is actually a function of Y, say $X_0(Y)$. The computations made here for a fixed X_0 no longer apply. The models considered in Section 9.5 have model matrices that depend on Y in a particular fashion that makes them somewhat more tractable.

The alternative estimates discussed in Chapter 13 and *ALM-III*, Chapter 2 are also biased. If the choice of components in principal component regression is made without reference to Y, then computations similar to those made here are possible. Ridge regression is also relatively tractable. *ALM-III*, Chapter 2 discusses other penalized least squares estimates. Their effects are harder to evaluate.

The Bayesian methods discussed in the next section, when using a proper prior on β, also provide biased estimates. Whether or not they actually improve the estimates, in the sense discussed here, depends on how well the prior reflects reality.

Exercise 2.7 Consider the prediction of further observations Y_{new} from our standard linear model, i.e., new observations that are independent of the original observations. We assume that for some new model matrix X_{new}, $\mathrm{E}(Y_{new}) = X_{new}\beta$ and $\mathrm{Cov}(Y_{new}) = \sigma^2 I$ for an appropriate sized identity matrix. A predictor of Y_{new} is, say, $G(Y)$. If the best predictor minimizes $\mathrm{E}\left\{[Y_{new} - G(Y)]'[Y_{new} - G(Y)]\right\}$, show that the best predictor minimizes $\mathrm{E}\left\{[G(Y) - X_{new}\beta]'[G(Y) - X_{new}\beta]\right\}$ so that the prediction problem reduces to an estimation problem.

2.9.1 Estimable Functions

Suppose we want to estimate some vector $\Lambda'\beta$. In Exercise 2.7, $\Lambda' = X_{new}$. For this vector to be estimable, we need $\Lambda' = P'X$ for some P. Recall that any such P can be replaced by the uniquely determined matrix MP.

The mean squared error of estimation from an arbitrary estimate $\tilde{\beta}$ is

$$\mathrm{E}\left[(\Lambda'\tilde{\beta} - \Lambda'\beta)'(\Lambda'\tilde{\beta} - \Lambda'\beta)\right].$$

Using the full model (1), the least squares estimate is $\Lambda'\hat{\beta} = P'MY$ with

$$\begin{aligned}
\mathrm{E}\left[(\Lambda'\hat{\beta} - \Lambda'\beta)'(\Lambda'\hat{\beta} - \Lambda'\beta)\right] &= \mathrm{E}\|P'M(Y - X\beta)\|^2 \\
&= \mathrm{tr}[M'P(\sigma^2 I)P'M] \\
&= \sigma^2 \mathrm{tr}[P'MP] \\
&= \sigma^2 \mathrm{tr}[\Lambda'(X'X)^{-}\Lambda].
\end{aligned}$$

Fitting the reduced model (2), a possibly biased estimate of $\Lambda'\beta$ is $P'X_0\hat{\gamma} = P'M_0Y$ with

$$\begin{aligned}
\mathrm{E}\|P'(M_0Y - X\beta)\|^2 &= \sigma^2 \mathrm{tr}[M_0PP'M_0] + \beta'X'(I - M_0)PP'(I - M_0)X\beta \\
&= \sigma^2 \mathrm{tr}[P'M_0P] + \|P'(I - M_0)X\beta\|^2.
\end{aligned}$$

The reduced model provides better estimates if

$$\beta'X'(I - M_0)PP'(I - M_0)X\beta < \sigma^2 \mathrm{tr}[P'(M - M_0)P].$$

If $C(MP) \subset C(X_0)$, the reduced model provides an unbiased estimate of $\Lambda'\beta$ and the reduced model gives no worse an estimate than the full model. If $C(MP) \perp C(X_0)$, the reduced model estimate is 0, so the variance term in the mean squared error is 0 but the bias of the expected squared error of estimation is $\|\Lambda'\beta\|^2$. The "estimate" 0 has better expected squared estimation error if $\|\Lambda'\beta\|^2 < \sigma^2 \mathrm{tr}[\Lambda'(X'X)^{-}\Lambda]$. In general some combination of these phenomena occur because an arbitrary MP can be written as $MP = MP_1 + MP_2$ with $C(MP_1) \subset C(X_0)$ and $C(MP_2) \perp C(X_0)$.

Exercise 2.8 Show that if $X_0\hat{\gamma}$ provides improved estimates relative to $X\hat{\beta}$, then for any P, $P'M_0Y$ is at least as good as $P'MY$.

2.10 Bayesian Estimation

Bayesian estimation incorporates the analyst's subjective information about a problem into the analysis. It appears to be the only logically consistent method of analysis, but not the only useful one. Some people object to the loss of objectivity that results from using the analyst's subjective information, but either the data are strong enough for reasonable people to agree on their interpretation or, if not, analysts should be using their subjective (prior) information for making appropriate decisions related to the data.

There is a vast literature on Bayesian statistics. Three fundamental works are de Finetti (1974,1975), Jeffreys (1961), and Savage (1954). Good elementary introductions to the subject are Lindley (1971) and Berry (1996). A few of the well-known books on the subject are Berger (1993), Box and Tiao (1973), DeGroot (1970), Geisser (1993), Gelman et al. (2013), Raiffa and Schlaifer (1961), Robert (2007), and Zellner (1971). My favorite is now Christensen, Johnson, Branscum, and Hanson (2010), which also includes many more references to many more excellent books. Christensen et al. contains a far more extensive discussion of Bayesian linear models than this relatively short section.

Consider the linear model

$$Y = X\beta + e, \quad e \sim N(0, \sigma^2 I),$$

where $r(X) = r$. It will be convenient to consider a full rank reparameterization of this model,

$$Y = Z\gamma + e, \quad e \sim N(0, \sigma^2 I),$$

where $C(X) = C(Z)$. As in Sections 1.2 and 2.4, this determines a density for Y given γ and σ^2, say, $f(Y|\gamma, \sigma^2)$. For a Bayesian analysis, we must have a joint density for γ and σ^2, say $p(\gamma, \sigma^2)$. This distribution reflects the analyst's beliefs, prior to collecting data, about the process of generating the data. We will actually specify this distribution conditionally as $p(\gamma, \sigma^2) = p(\gamma|\sigma^2)p(\sigma^2)$. In practice, it is difficult to specify these distributions for γ given σ^2 and σ^2. Convenient choices that have minimal impact on (most aspects of) the analysis are the (improper) reference priors $p(\gamma|\sigma^2) = 1$ and $p(\sigma^2) = 1/\sigma^2$. These are improper in that neither prior density integrates to 1. Although these priors are convenient, a true Bayesian analysis requires the specification of proper prior distributions.

Specifying prior information is difficult, particularly about such abstract quantities as regression coefficients. A useful tool in specifying prior information is to think in terms of the mean of potential observations. For example, we could specify a vector of predictor variables, say $\tilde{z}_i$, and specify the distribution for the mean of observations having those predictor variables. With covariates $\tilde{z}_i$, the mean of potential observables is $\tilde{z}_i'\gamma$. Typically, we assume that $\tilde{z}_i'\gamma$ has a $N(\tilde{y}_i, \sigma^2/\tilde{w}_i)$ distribution. One way to think about $\tilde{y}_i$ and $\tilde{w}_i$ is that $\tilde{y}_i$ is a prior guess for what one would see with covariates $\tilde{z}_i$, and $\tilde{w}_i$ is the number of observations this guess is worth. To specify a proper prior distribution for γ given σ^2, we specify independent priors at vectors $\tilde{z}_i, i = 1, \ldots, r$, where r is the dimension of γ. As will be seen later, under a mild condition this prior specification leads easily to a proper prior distribution on γ. Although choosing the $\tilde{z}_i$s and specifying the priors may be difficult to do, it is much easier to do than trying to specify an intelligent joint prior directly on γ. If one wishes to specify only partial prior information, one can simply choose fewer than r vectors $\tilde{z}_i$ and the analysis follows as outlined below. In fact, using the reference prior for γ amounts to not choosing any $\tilde{z}_i$s. Again, the analysis follows as outlined below. Bedrick et al. (1996) and Christensen et al. (2010) discuss these techniques in more detail.

I believe that the most reasonable way to specify a proper prior on σ^2 is to think in terms of the variability of potential observables around some fixed mean. Unfortunately, the implications of this idea are not currently as well understood as they are for the related technique of specifying the prior for γ indirectly through priors on means of potential observables. For now, we will simply consider priors for σ^2 that are inverse gamma distributions, i.e., distributions in which $1/\sigma^2$ has a gamma distribution. An inverse gamma distribution has two parameters, a and b. One can think of the prior as being the equivalent of $2a$ (prior) observations with a prior guess for σ^2 of b/a.

It is convenient to write $\tilde{Y} = (\tilde{y}_1, \ldots, \tilde{y}_r)'$ and $\tilde{Z}$ as the $r \times r$ matrix with ith row $\tilde{z}_i'$. In summary, our distributional assumptions are

$$\begin{aligned} Y|\gamma, \sigma^2 &\sim N(Z\gamma, \sigma^2 I), \\ \tilde{Z}\gamma|\sigma^2 &\sim N(\tilde{Y}, \sigma^2 D^{-1}(\tilde{w})), \\ \sigma^2 &\sim InvGa(a, b). \end{aligned}$$

We assume that $\tilde{Z}$ is nonsingular, so that the second of these induces the distribution

$$\gamma|\sigma^2 \sim N(\tilde{Z}^{-1}\tilde{Y}, \sigma^2 \tilde{Z}^{-1} D^{-1}(\tilde{w}) \tilde{Z}^{-1\prime}).$$

Actually, any multivariate normal distribution for γ given σ^2 will lead to essentially the same analysis as given here.

A Bayesian analysis is based on finding the distribution of the parameters given the data, i.e., $p(\gamma, \sigma^2|Y)$. This is accomplished by using *Bayes's theorem*, which states that

$$p(\gamma, \sigma^2|Y) = \frac{f(Y|\gamma, \sigma^2)p(\gamma, \sigma^2)}{f(Y)}.$$

If we know the numerator of the fraction, we can obtain the denominator by

$$f(Y) = \int f(Y|\gamma, \sigma^2)p(\gamma, \sigma^2)\, d\gamma\, d\sigma^2.$$

In fact, because of this relationship, we only need to know the numerator up to a constant multiple, because any multiple will cancel in the fraction.

Later in this section we will show that

$$p(\gamma,\sigma^2|Y) \propto \left(\sigma^2\right)^{-(n+r)/2} p(\sigma^2)$$
$$\times \exp\left\{\frac{-1}{2\sigma^2}\left[(\gamma-\hat{\gamma})'(Z'Z+\tilde{Z}'D(\tilde{w})\tilde{Z})(\gamma-\hat{\gamma})\right]\right\} \qquad (1)$$
$$\times \exp\left\{\frac{-1}{2\sigma^2}\left[(Y-Z\hat{\gamma})'(Y-Z\hat{\gamma})+(\tilde{Y}-\tilde{Z}\hat{\gamma})'D(\tilde{w})(\tilde{Y}-\tilde{Z}\hat{\gamma})\right]\right\},$$

where

$$\hat{\gamma} = \left(Z'Z+\tilde{Z}'D(\tilde{w})\tilde{Z}\right)^{-1}\left[Z'Y+\tilde{Z}'D(\tilde{w})\tilde{Y}\right]. \qquad (2)$$

The joint posterior (post data) density is the righthand side of (1) divided by its integral with respect to γ and σ^2.

The form (1) for the joint distribution given the data is not particularly useful. What we really need are the marginal distributions of σ^2 and functions $\rho'Z\gamma \equiv \rho'X\beta$, and predictive distributions for new observations.

As will be shown, the Bayesian analysis turns out to be quite consistent with the frequentist analysis. For the time being, we use the reference prior $p(\sigma^2) = 1/\sigma^2$. In our model, we have $\tilde{Z}\gamma$ random, but it is convenient to think of $\tilde{Y}$ as being r independent prior observations with mean $\tilde{Z}\gamma$ and weights $\tilde{w}$. Now consider the generalized least squares model

$$\begin{bmatrix} Y \\ \tilde{Y} \end{bmatrix} = \begin{bmatrix} Z \\ \tilde{Z} \end{bmatrix}\gamma + \begin{bmatrix} e \\ \tilde{e} \end{bmatrix}, \quad \begin{bmatrix} e \\ \tilde{e} \end{bmatrix} \sim N\left(\begin{bmatrix} 0_{n\times 1} \\ 0_{r\times 1} \end{bmatrix}, \sigma^2\begin{bmatrix} I_n & 0 \\ 0 & D^{-1}(\tilde{w}) \end{bmatrix}\right). \qquad (3)$$

This generalized least squares model can also be written as, say,

$$Y_* = Z_*\gamma + e_*, \quad e_* \sim N(0,\sigma^2 V_*).$$

The generalized least squares estimate from this model is $\hat{\gamma}$ as given in (2). In the Bayesian analysis, $\hat{\gamma}$ is the expected value of γ given Y. Let *BMSE* denote the mean squared error from the (Bayesian) generalized least squares model with *BdfE* degrees of freedom for error.

In the frequentist generalized least squares analysis, for fixed γ with random $\hat{\gamma}$ and *BMSE*,

$$\frac{\lambda'\hat{\gamma}-\lambda'\gamma}{\sqrt{BMSE\,\lambda'\left(Z'Z+\tilde{Z}'D(\tilde{w})\tilde{Z}\right)^{-1}\lambda}} \sim t(BdfE).$$

In the Bayesian analysis the same distribution holds, but for fixed $\hat{\gamma}$ and *BMSE* with random γ. Frequentist confidence intervals for $\lambda'\gamma$ are identical to Bayesian posterior probability intervals for $\lambda'\gamma$. Note that for estimating a function $\rho' X\beta$, simply write it as $\rho' X\beta = \rho' Z\gamma$ and take $\lambda' = \rho' Z$.

In the frequentist generalized least squares analysis, for fixed σ^2 and random *BMSE*,

$$\frac{(BdfE)BMSE}{\sigma^2} \sim \chi^2(BdfE).$$

In the Bayesian analysis, the same distribution holds, but for fixed *BMSE* and random σ^2. Confidence intervals for σ^2 are identical to Bayesian posterior probability intervals for σ^2.

In the frequentist generalized least squares analysis, a prediction interval for a future independent observation y_0 with predictor vector z_0 and weight 1 is based on the distribution

$$\frac{y_0 - z_0'\hat{\gamma}}{\sqrt{BMSE\left[1 + z_0'\left(Z'Z + \tilde{Z}'D(\tilde{w})\tilde{Z}\right)^{-1} z_0\right]}} \sim t(BdfE), \tag{4}$$

where $\hat{\gamma}$ and *BMSE* are random and y_0 is independent of Y for given γ and σ^2, see Exercise 2.11.1. In the Bayesian analysis, the same distribution holds, but for fixed $\hat{\gamma}$ and *BMSE*. Standard prediction intervals for y_0 are identical to Bayesian prediction intervals.

If we specify an improper prior on γ using fewer than r vectors $\tilde{z}_i$, these relationships between generalized least squares and the Bayesian analysis remain valid. In fact, for the reference prior on γ, i.e., choosing no $\tilde{z}_i$s, the generalized least squares model reduces to the usual model $Y = X\beta + e$, $e \sim N(0, \sigma^2 I)$, and the Bayesian analysis becomes analogous to the usual ordinary least squares analysis.

In the generalized least squares model (3), $BdfE = n$. If we take $\sigma^2 \sim InvGa(a, b)$, the only changes in the Bayesian analysis are that *BMSE* changes to $[(BdfE)\, BMSE + 2b]/(BdfE + 2a)$ and the degrees of freedom for the t and χ^2 distributions change from *BdfE* to $BdfE + 2a$. With reference priors for both γ and σ^2, $BdfE = n - r$, as in ordinary least squares.

Example 2.10.1 Schafer (1987) presented data on 93 individuals at the Harris Bank of Chicago in 1977. The response is beginning salary in hundreds of dollars. There are four covariates: sex, years of education, denoted EDUC, years of experience, denoted EXP, and time at hiring as measured in months after 1-1-69, denoted TIME. This is a regression, so we can take $Z \equiv X$, $\gamma \equiv \beta$, and write $\tilde{X} \equiv \tilde{Z}$. With an intercept in the model, Johnson, Bedrick, and I began by specifying five covariate vectors $\tilde{x}_i' = (1, SEX_i, EDUC_i, EXP_i, TIME_i)$, say, (1, 0, 8, 0, 0), (1, 1, 8, 0, 0), (1, 1, 16, 0, 0), (1, 1, 16, 30, 0), and (1, 1, 8, 0, 36), where a SEX value of 1 indicates a male. For example, the vector (1, 0, 8, 0, 0) corresponds to a male with 8 years of education, no previous experience, and starting work on 1-1-69. Thinking about the mean salary for each set of covariates, we chose $\tilde{y}' = (40, 40, 60, 70, 50)$, which reflects a prior belief that starting salaries are the same for equally qualified men and women and a belief that salary is increasing as a function of EDUC, EXP, and TIME. The weights $\tilde{w}_i$ are all chosen to be 0.4, so that in total the prior carries the same weight as two sampled observations. The induced prior on β given σ^2 has mean vector $(20, 0, 2.50, 0.33, 0.28)'$ and standard deviation vector $(2.74, 2.24, 0.28, 0.07, 0.06)'$. (This analysis can be performed on standard regression software using the weights of prior observations and weights of 1 for actual observations.)

To illustrate partial prior information, we consider the same example, only with the fifth "prior observation" deleted. In this instance, the prior does not reflect any information about the response at TIMEs other than 0. The prior is informative about the mean responses at the first four covariate vectors but is noninformative (the prior is constant) for the mean response at the fifth covariate vector. Moreover, the prior is constant for the mean response with any other choice of fifth vector, provided this vector is linearly independent of the other four. (All such vectors must have a nonzero value for the time component.) In this example, the induced improper distribution on β has the same means and standard deviations for β_1, β_2, β_3, and β_4, but is flat for β_5.

We specify a prior on σ^2 worth $2a = 2$ observations with a prior guess for σ^2 of $b/a = 25$. The prior guess reflects our belief that a typical salary has a standard deviation of 5.

Using our informative prior, the posterior mean of β is

$$\hat{\beta} = (33.68, 6.96, 1.02, 0.18, 0.23)'$$

with $BdfE = 95$ and $BMSE = 2404/95 = 25.3053$. The standard deviations for β are $\sqrt{95/93}(310, 114, 23.43, 6.76, 5.01)'/100$. In the partially informative case discussed above, the posterior mean is

$$\hat{\beta} = (34.04, 7.11, 0.99, 0.17, 0.23),$$

$BdfE = 94$, $BMSE = 2383/94 = 25.3511$, and the standard deviations for β are $\sqrt{94/92}(313, 116, 23.78, 6.80, 5.06)'/100$. Using the standard reference prior, i.e., using ordinary least squares without prior data, $\hat{\beta} = (35.26, 7.22, 0.90, 0.15, 0.23)'$, $BdfE = 88$, $BMSE = 2266/88 = 25.75$, and the standard deviations for β are $\sqrt{88/86}(328, 118, 24.70, 7.05, 5.20)'/100$. The 95% prediction interval for $x_0 = (1, 0, 10, 3.67, 7)'$ with a weight of 1 is (35.97, 56.34) for the informative prior, (35.99, 56.38) with partial prior information, and (36.27, 56.76) for the reference prior.

2.10.1 Distribution Theory

For the time being, we will assume relation (1) for the joint posterior and use it to arrive at marginal distributions. Afterwards, we will establish relation (1).

The distribution theory for the Bayesian analysis involves computations unlike anything else done in this book. It requires knowledge of some basic facts.

The density of a gamma distribution with parameters $a > 0$ and $b > 0$ is

$$g(\tau) = \frac{b^a}{\Gamma(a)} \tau^{a-1} \exp[-b\tau]$$

for $\tau > 0$. A $Gamma(n/2, 1/2)$ distribution is the same as a $\chi^2(n)$ distribution. We will not need the density of an inverse gamma, only the fact that y has a $Gamma(a, b)$ distribution if and only if $1/y$ has an $InvGa(a, b)$ distribution. The improper reference distribution corresponds to $a = 0, b = 0$.

The t distribution is defined in Appendix C. The density of a $t(n)$ distribution is

$$g(w) = \left[1 + \frac{w^2}{n}\right]^{-(n+1)/2} \Gamma\left(\frac{n+1}{2}\right) \Big/ \left[\Gamma\left(\frac{n}{2}\right)\sqrt{n\pi}\right].$$

Eliminating the constants required to make the density integrate to 1 gives

$$g(w) \propto \left[1 + \frac{w^2}{n}\right]^{-(n+1)/2}.$$

Bayesian linear models involve multivariate t distributions, cf. DeGroot (1970, Section 5.6). Let $W \sim N(\mu, V)$, $Q \sim \chi^2(n)$, with W and Q independent. Then if

$$Y \sim (W - \mu)\frac{1}{\sqrt{Q/n}} + \mu,$$

by definition

$$Y \sim t(n, \mu, V).$$

For an r vector Y, a multivariate $t(n)$ distribution with center μ and dispersion matrix V has density

$$g(y) = \left[1 + \frac{1}{n}(y - \mu)'V^{-1}(y - \mu)\right]^{-(n+r)/2} \times \Gamma\left(\frac{n+r}{2}\right) \Big/ \left\{\Gamma\left(\frac{n}{2}\right)(n\pi)^{r/2}[\det(V)]^{1/2}\right\}.$$

This distribution has mean μ for $n > 1$ and covariance matrix $[n/(n-2)]V$ for $n > 2$. To get noninteger degrees of freedom a, just replace Q/n in the definition with bT/a, where $T \sim Gamma(a/2, b/2)$ independent of W.

Note that from the definition of a multivariate t,

$$\frac{\lambda'Y - \lambda'\mu}{\sqrt{\lambda'V\lambda}} \sim \frac{(\lambda'W - \lambda'\mu)/\sqrt{\lambda'V\lambda}}{\sqrt{Q/n}} \sim t(n). \tag{5}$$

Proceeding with the Bayesian analysis, to find the marginal posterior of γ let

$$Q = \left[(Y - Z\hat{\gamma})'(Y - Z\hat{\gamma}) + (\tilde{Y} - \tilde{Z}\hat{\gamma})'D(\tilde{w})(\tilde{Y} - \tilde{Z}\hat{\gamma})\right] + \left[(\gamma - \hat{\gamma})' \ (Z'Z + \tilde{Z}'D(\tilde{w})\tilde{Z})(\gamma - \hat{\gamma})\right].$$

From (1),

$$p(\gamma|Y) \propto \int \left(\sigma^2\right)^{-(n+r)/2} p(\sigma^2) \exp\left\{\frac{-1}{2\sigma^2} Q\right\} d\sigma^2.$$

Transforming σ^2 into $\tau = 1/\sigma^2$ gives $\sigma^2 = 1/\tau$ and $d\sigma^2 = |-\tau^{-2}|d\tau$. Thus

$$p(\gamma|Y) \propto \int (\tau)^{(n+r)/2} p(1/\tau) \exp\{-\tau Q/2\} \tau^{-2} d\tau.$$

Note that if σ^2 has an inverse gamma distribution with parameters a and b, then τ has a gamma distribution with parameters a and b; so $p(1/\tau)\tau^{-2}$ is a gamma density and

$$p(\gamma|Y) \propto \int (\tau)^{(n+r+2a-2)/2} \exp\{-\tau(Q + 2b)/2\} d\tau.$$

The integral is a gamma integral, e.g., the gamma density given earlier integrates to 1, so

$$p(\gamma|Y) \propto \Gamma[(n + r + 2a)/2] \Big/ [(Q + 2b)/2]^{(n+r+2a)/2}$$

or

$$p(\gamma|Y) \propto [Q + 2b]^{-(n+r+2a)/2}.$$

We can rewrite this as

$$p(\gamma|Y) \propto \left[(BdfE)(BMSE) + 2b + (\gamma - \hat{\gamma})'\left(Z'Z + \tilde{Z}'D(\tilde{w})\tilde{Z}\right)(\gamma - \hat{\gamma})\right]^{-(n+r+2a)/2}$$

$$\propto \left[1 + \frac{1}{n + 2a} \frac{(\gamma - \hat{\gamma})'\left(Z'Z + \tilde{Z}'D(\tilde{w})\tilde{Z}\right)(\gamma - \hat{\gamma})}{[(BdfE)(BMSE) + 2b]/(n + 2a)}\right]^{-(n+2a+r)/2},$$

so

$$\gamma|Y \sim t\left(n + 2a, \hat{\gamma}, \frac{(BdfE)(BMSE) + 2b}{n + 2a}\left(Z'Z + \tilde{Z}'D(\tilde{w})\tilde{Z}\right)^{-1}\right).$$

Together with (5), this provides the posterior distribution of $\lambda'\gamma$.

Now consider the marginal (posterior) distribution of σ^2.

$$p(\sigma^2|Y) \propto \left(\sigma^2\right)^{-n/2} p(\sigma^2)$$
$$\times \exp\left\{\frac{-1}{2\sigma^2}\left[(Y - Z\hat{\gamma})'(Y - Z\hat{\gamma}) + (\tilde{Y} - \tilde{Z}\hat{\gamma})'D(\tilde{w})(\tilde{Y} - \tilde{Z}\hat{\gamma})\right]\right\}$$
$$\times \int \left(\sigma^2\right)^{-r/2} \exp\left\{\frac{-1}{2\sigma^2}\left[(\gamma - \hat{\gamma})'(Z'Z + \tilde{Z}'D(\tilde{w})\tilde{Z})(\gamma - \hat{\gamma})\right]\right\} d\gamma.$$

The term being integrated is proportional to a normal density, so the integral is a constant that does not depend on σ^2. Hence,

$$p(\sigma^2|Y) \propto \left(\sigma^2\right)^{-n/2} p(\sigma^2)$$
$$\times \exp\left\{\frac{-1}{2\sigma^2}\left[(Y - Z\hat{\gamma})'(Y - Z\hat{\gamma}) + (\tilde{Y} - \tilde{Z}\hat{\gamma})'D(\tilde{w})(\tilde{Y} - \tilde{Z}\hat{\gamma})\right]\right\}$$

or, using the generalized least squares notation,

$$p(\sigma^2|Y) \propto \left(\sigma^2\right)^{-n/2} p(\sigma^2) \exp\left[\frac{-1}{2\sigma^2}(BdfE)(BMSE)\right].$$

We transform to the *precision*, $\tau \equiv 1/\sigma^2$. The $InvGa(a, b)$ distribution for σ^2 yields

$$p(\tau|Y) \propto (\tau)^{n/2} (\tau)^{a-1} \exp[-b\tau] \exp\left[\frac{-\tau}{2}(BdfE)(BMSE)\right]$$
$$= (\tau)^{[(n+2a)/2]-1} \exp\left[-\frac{2b + (BdfE)(BMSE)}{2}\tau\right];$$

so

$$\tau|Y \sim Gamma\left(\frac{n+2a}{2}, \frac{2b + (BdfE)(BMSE)}{2}\right).$$

It is not difficult to show that

$$[2b + (BdfE)(BMSE)]\tau|Y \sim Gamma\left(\frac{n+2a}{2}, \frac{1}{2}\right),$$

i.e.,

$$\frac{2b + (BdfE)(BMSE)}{\sigma^2}\bigg|Y \sim Gamma\left(\frac{n+2a}{2}, \frac{1}{2}\right).$$

As mentioned earlier, for the reference distribution with $a = 0, b = 0$,

$$\frac{(BdfE)(BMSE)}{\sigma^2}\bigg|Y \sim Gamma\left(\frac{n}{2}, \frac{1}{2}\right) = \chi^2(n).$$

Finally, we establish relation (1).

$$
\begin{aligned}
p(\gamma, \sigma^2|Y) &\propto f(Y|\gamma, \sigma^2)p(\gamma|\sigma^2)p(\sigma^2) \\
&\propto \left\{\left(\sigma^2\right)^{-n/2} \exp\left[-(Y - Z\gamma)'(Y - Z\gamma)/2\sigma^2\right]\right\} \\
&\quad \times \left\{\left(\sigma^2\right)^{-r/2} \exp\left[-\left(\gamma - \tilde{Z}^{-1}\tilde{Y}\right)'\left(\tilde{Z}'D(\tilde{w})\tilde{Z}\right)\left(\gamma - \tilde{Z}^{-1}\tilde{Y}\right)/2\sigma^2\right]\right\} \\
&\quad \times p(\sigma^2)
\end{aligned}
$$

Most of the work involves simplifying the terms in the exponents. We isolate those terms, deleting the multiple $-1/2\sigma^2$.

$$
\begin{aligned}
&(Y - Z\gamma)'(Y - Z\gamma) + \left(\gamma - \tilde{Z}^{-1}\tilde{Y}\right)'\left(\tilde{Z}'D(\tilde{w})\tilde{Z}\right)\left(\gamma - \tilde{Z}^{-1}\tilde{Y}\right) \\
&= (Y - Z\gamma)'(Y - Z\gamma) + (\tilde{Y} - \tilde{Z}\gamma)'D(\tilde{w})(\tilde{Y} - \tilde{Z}\gamma) \\
&= \left(\begin{bmatrix} Y \\ \tilde{Y} \end{bmatrix} - \begin{bmatrix} Z \\ \tilde{Z} \end{bmatrix}\gamma\right)' \begin{bmatrix} I & 0 \\ 0 & D(\tilde{w}) \end{bmatrix} \left(\begin{bmatrix} Y \\ \tilde{Y} \end{bmatrix} - \begin{bmatrix} Z \\ \tilde{Z} \end{bmatrix}\gamma\right).
\end{aligned}
$$

Write

$$
Y_* = \begin{bmatrix} Y \\ \tilde{Y} \end{bmatrix}, \quad Z_* = \begin{bmatrix} Z \\ \tilde{Z} \end{bmatrix}, \quad V_* = \begin{bmatrix} I & 0 \\ 0 & D^{-1}(\tilde{w}) \end{bmatrix}
$$

and apply Lemma 2.7.6 to get

$$
\begin{aligned}
&(Y - Z\gamma)'(Y - Z\gamma) + \left(\gamma - \tilde{Z}^{-1}\tilde{Y}\right)'\left(\tilde{Z}'D(\tilde{w})\tilde{Z}\right)\left(\gamma - \tilde{Z}^{-1}\tilde{Y}\right) \\
&= (Y_* - Z_*\gamma)'V_*^{-1}(Y_* - Z_*\gamma) \\
&= (Y_* - A_*Y_*)'V_*^{-1}(Y_* - A_*Y_*) + (\hat{\gamma} - \gamma)'(Z_*'V_*^{-1}Z_*)(\hat{\gamma} - \gamma),
\end{aligned}
$$

where $A_* = Z_*(Z_*'V_*^{-1}Z_*)^{-1}Z_*'V_*^{-1}$ and

$$
\begin{aligned}
\hat{\gamma} &= (Z_*'V_*^{-1}Z_*)^{-1}Z_*'V_*^{-1}Y_* \\
&= \left(Z'Z + \tilde{Z}'D(\tilde{w})\tilde{Z}\right)^{-1}\left[Z'Y + \tilde{Z}'D(\tilde{w})\tilde{Y}\right].
\end{aligned}
$$

Finally, observe that

$$
\begin{aligned}
&(Y_* - A_*Y_*)'V_*^{-1}(Y_* - A_*Y_*) + (\hat{\gamma} - \gamma)'(Z_*'V_*^{-1}Z_*)(\hat{\gamma} - \gamma) \\
&= \left[(Y - Z\hat{\gamma})'(Y - Z\hat{\gamma}) + (\tilde{Y} - \tilde{Z}\hat{\gamma})'D(\tilde{w})(\tilde{Y} - \tilde{Z}\hat{\gamma})\right] \\
&+ \left[(\gamma - \hat{\gamma})'(Z'Z + \tilde{Z}'D(\tilde{w})\tilde{Z})(\gamma - \hat{\gamma})\right].
\end{aligned}
$$

Substitution gives (1).

Exercise 2.9 Prove relation (4).

2.11 Additional Exercises

Exercise 2.11.1 *Prediction*. Consider a regression model $Y = X\beta + e$, $e \sim N(0, \sigma^2 I)$ and suppose that we want to predict the value of a future observation, say y_0, that will be independent of Y and be distributed $N(x_0'\beta, \sigma^2)$.

(a) Find the distribution of

$$\frac{y_0 - x_0'\hat{\beta}}{\sqrt{MSE\left[1 + x_0'(X'X)^{-1}x_0\right]}}.$$

(b) Find a 95% *prediction interval* for y_0.

Hint: A prediction interval is similar to a confidence interval except that, rather than finding parameter values that are consistent with the data and the model, one finds new observations y_0 that are consistent with the data and the model as determined by an α level test.

(c) Let $\eta \in (0, 0.5]$. The 100ηth percentile of the distribution of y_0 is, say, $\gamma(\eta) = x_0'\beta + z(\eta)\sigma$. (Note that $z(\eta)$ is a negative number.) Find a $(1-\alpha)100\%$ lower confidence bound for $\gamma(\eta)$. In reference to the distribution of y_0, this lower confidence bound is referred to as a lower η *tolerance point* with confidence coefficient $(1-\alpha)100\%$. For example, if $\eta = 0.1$, $\alpha = 0.05$, and y_0 is the octane value of a batch of gasoline manufactured under conditions x_0, then we are 95% confident that no more than 10% of all batches produced under x_0 will have an octane value below the tolerance point.

Hint: Use a noncentral t distribution based on $x_0'\hat{\beta} - \gamma(\eta)$.

Comment: For more detailed discussions of prediction and tolerance (and we all know that tolerance is a great virtue), see Geisser (1993), Aitchison and Dunsmore (1975), and Guttman (1970).

Exercise 2.11.2 Consider the model

$$Y = X\beta + Xb + e, \qquad \mathrm{E}(e) = 0, \qquad \mathrm{Cov}(e) = \sigma^2 I,$$

where b is a known vector and Xb is often called an *offset*. Show that Proposition 2.1.3 is not valid for this model by producing a linear unbiased estimate of $\rho'X\beta$, say $a_0 + a'Y$, for which $a_0 \neq 0$.

Hint: Modify $\rho'MY$.

Exercise 2.11.3 Consider the model $y_i = \beta_1 x_{i1} + \beta_2 x_{i2} + e_i$, e_is i.i.d. $N(0, \sigma^2)$. Use the data given below to answer (a) through (d). Show your work, i.e., do not use a regression or general linear models computer program.

(a) Estimate β_1, β_2, and σ^2.

(b) Give 95% confidence intervals for β_1 and $\beta_1+\beta_2$.
(c) Perform an $\alpha = 0.01$ test for $H_0 : \beta_2 = 3$.
(d) Find an appropriate P value for the test of $H_0 : \beta_1 - \beta_2 = 0$.

obs.	1	2	3	4	5	6	7	8
y	82	79	74	83	80	81	84	81
x_1	10	9	9	11	11	10	10	12
x_2	15	14	13	15	14	14	16	13

Exercise 2.11.4 Consider the model $y_i = \beta_1 x_{i1} + \beta_2 x_{i2} + e_i$, e_is i.i.d. $N(0, \sigma^2)$. There are 15 observations and the sum of the squared observations is $Y'Y = 3.03$. Use the normal equations given below to answer parts (a) through (c).
(a) Estimate β_1, β_2, and σ^2.
(b) Give 98% confidence intervals for β_2 and $\beta_2 - \beta_1$.
(c) Perform an $\alpha = 0.05$ test for $H_0 : \beta_1 = 0.5$.
The normal equations are

$$\begin{bmatrix} 15.00 & 374.50 \\ 374.50 & 9482.75 \end{bmatrix} \begin{bmatrix} \beta_1 \\ \beta_2 \end{bmatrix} = \begin{bmatrix} 6.03 \\ 158.25 \end{bmatrix}.$$

Exercise 2.11.5 Consider the model

$$y_i = \beta_0 + \beta_1 x_{i1} + \beta_2 x_{i2} + \beta_{11} x_{i1}^2 + \beta_{22} x_{i2}^2 + \beta_{12} x_{i1} x_{i2} + e_i,$$

where the predictor variables take on the following values.

i	1	2	3	4	5	6	7
x_{i1}	1	1	−1	−1	0	0	0
x_{i2}	1	−1	1	−1	0	0	0

Show that β_0, β_1, β_2, $\beta_{11} + \beta_{22}$, β_{12} are estimable and find (nonmatrix) algebraic forms for the estimates of these parameters. Find the *MSE* and the standard errors of the estimates.

Exercise 2.11.6 Consider the linear model

$$\begin{bmatrix} Y_1 \\ Y_2 \end{bmatrix} = \begin{bmatrix} J_{N_1} \\ J_{N_2} \end{bmatrix} \mu + e, \quad \mathrm{E}(e) = 0, \quad \mathrm{Cov}\left(\begin{bmatrix} e_1 \\ e_2 \end{bmatrix} \right) = \begin{bmatrix} \sigma_1^2 I_{N_1} & 0 \\ 0 & \sigma_2^2 I_{N_2} \end{bmatrix}.$$

Find the BLUE $\hat{\mu}$ and $\mathrm{Var}(\hat{\mu})$ assuming that the variance parameters are known. How does the result change if σ_1^2 is unknown but σ_2^2/σ_1^2 is known?

Exercise 2.11.7 For V positive definite and $C(VX) \subset C(X)$ use Proposition B.12 and the results in Section B.4 to show that

$$r(VX) = r(VXX') = r(XX') = r(X).$$

Hint: Show $\mathcal{N}(VXX') = \mathcal{N}(XX')$.

Exercise 2.11.8 Use Theorem 2.2.4 and its proof to show that there is a unique least squares estimate in $C(X')$, say $\hat{\beta}_0$, and that all other least squares estimates are obtained by $\hat{\beta} = \hat{\beta}_0 + v$ for some $v \perp C(X')$.

References

Aitchison, J., & Dunsmore, I. R. (1975). *Statistical prediction analysis*. Cambridge: Cambridge University Press.

Arnold, S. F. (1981). *The theory of linear models and multivariate analysis*. New York: Wiley.

Bedrick, E. J., Christensen, R., & Johnson, W. (1996). A new perspective on priors for generalized linear models. *Journal of the American Statistical Association, 91*, 1450–1460.

Berger, J. O. (1993). *Statistical decision theory and Bayesian analysis* (Revised 2nd ed.). New York: Springer.

Berry, D. A. (1996). *Statistics: A Bayesian perspective*. Belmont: Duxbery.

Box, G. E. P., & Tiao, G. C. (1973). *Bayesian inference in statistical analysis*. New York: Wiley.

Christensen, R. (2018). Another look at linear hypothesis testing in dense high-dimensional linear models. http://www.stat.unm.edu/~fletcher/AnotherLook.pdf

Christensen, R., Johnson, W., Branscum, A., & Hanson, T. E. (2010). *Bayesian ideas and data analysis: An introduction for scientists and statisticians*. Boca Raton: Chapman and Hall/CRC Press.

Cox, D. R., & Hinkley, D. V. (1974). *Theoretical statistics*. London: Chapman and Hall.

de Finetti, B. (1974, 1975). *Theory of probability* (Vols. 1, 2). New York: Wiley.

DeGroot, M. H. (1970). *Optimal statistical decisions*. New York: McGraw-Hill.

Eaton, M. L. (1983). *Multivariate statistics: A vector space approach*. New York: Wiley. Reprinted in 2007 by IMS Lecture Notes–Monograph Series.

Geisser, S. (1993). *Predictive inference: An introduction*. New York: Chapman and Hall.

Gelman, A., Carlin, J. B., Stern, H. S., Dunson, D. B., Vehtari, A., & Rubin, D. B. (2013). *Bayesian data analysis* (3rd ed.). Boca Raton: Chapman and Hall/CRC.

Graybill, F. A. (1976). *Theory and application of the linear model*. North Scituate: Duxbury Press.

Guttman, I. (1970). *Statistical tolerance regions*. New York: Hafner Press.

Harville, D. A. (2018). *Linear models and the relevant distributions and matrix algebra*. Boca Raton: CRC Press.

Jeffreys, H. (1961). *Theory of probability* (3rd ed.). London: Oxford University Press.

Lehmann, E. L. (1983). *Theory of point estimation*. New York: Wiley.

Lehmann, E. L. (1986). *Testing statistical hypotheses* (2nd ed.). New York: Wiley.

Lindley, D. V. (1971). *Bayesian statistics: A review*. Philadelphia: SIAM.

Monahan, J. F. (2008). *A primer on linear models*. Boca Raton: Chapman & Hall/CRC Press.

Raiffa, H., & Schlaifer, R. (1961). *Applied statistical decision theory*. Boston: Division of Research, Graduate School of Business Administration, Harvard University.

Rao, C. R. (1973). *Linear statistical inference and its applications* (2nd ed.). New York: Wiley.

Ravishanker, N., & Dey, D. (2002). *A first course in linear model theory*. Boca Raton: Chapman and Hall/CRC Press.

Rencher, A. C., & Schaalje, G. B. (2008). *Linear models in statistics* (2nd ed.). New York: Wiley.

Robert, C. P. (2007). *The Bayesian choice: From decision-theoretic foundations to computational implementation* (2nd ed.). New York: Springer.

Savage, L. J. (1954). *The foundations of statistics*. New York: Wiley.

Schafer, D. W. (1987). Measurement error diagnostics and the sex discrimination problem. *Journal of Business and Economic Statistics, 5*, 529–537.

Scheffé, H. (1959). *The analysis of variance*. New York: Wiley.

Searle, S. R. (1971). *Linear models*. New York: Wiley.

Seber, G. A. F. (1966). *The linear hypothesis: A general theory*. London: Griffin.

Seber, G. A. F. (1977). *Linear regression analysis*. New York: Wiley.

Seber, G. A. F. (2015). *The linear model and hypothesis: A general theory*. New York: Springer.

Wichura, M. J. (2006). *The coordinate-free approach to linear models*. New York: Cambridge University Press.

Zellner, A. (1971). *An introduction to Bayesian inference in econometrics*. New York: Wiley.

Chapter 3
Testing

Abstract This chapter considers two approaches to testing linear models. The approaches are identical in that a test under either approach is a well-defined test under the other. The two methods differ only conceptually. One approach is that of testing models; the other approach involves testing linear parametric functions.

Section 1 discusses the notion that a linear model depends fundamentally on $C(X)$ and that the vector of parameters β is of secondary importance. Section 2 discusses testing different models against each other. Section 3 discusses testing linear functions of the β vector. Section 4 presents a brief discussion of the relative merits of testing models versus testing linear parametric functions. Section 5 examines the problem of testing parametric functions that put a constraint on a given subspace of $C(X)$. In particular, Section 5 establishes that estimates and tests can be obtained by using the projection operator onto the subspace. This result is valuable when using one-way ANOVA methods to analyze balanced multifactor ANOVA models. Section 6 considers the problem of breaking sums of squares into independent components. This is a general discussion that relates to breaking ANOVA treatment sums of squares into sums of squares for orthogonal contrasts and also relates to the issue of writing multifactor ANOVA tables with independent sums of squares. Section 7 discusses the construction of confidence regions for estimable linear parametric functions. Section 8 presents testing procedures for the generalized least squares model of Section 2.7.

3.1 More About Models

For estimation in the standard linear model

$$Y = X\beta + e, \quad \mathrm{E}(e) = 0, \quad \mathrm{Cov}(e) = \sigma^2 I,$$

R. Christensen, *Plane Answers to Complex Questions*, Springer Texts in Statistics,
https://doi.org/10.1007/978-3-030-32097-3_3

we have found that the crucial item needed is M, the perpendicular projection operator onto $C(X)$. For convenience, we will call $C(X)$ the *estimation space* and $C(X)^{\perp}$ the *error space*. $I - M$ is the perpendicular projection operator onto the error space. In a profound sense, any two linear models with the same estimation space are the same model. For example, any two such models will give the same fitted (predicted) values for the observations, hence the same residuals and the same estimate of σ^2.

Suppose we have two linear models for a vector of observations, say $Y = X_1\beta_1 + e_1$ and $Y = X_2\beta_2 + e_2$ with $C(X_1) = C(X_2)$. For these alternative parameterizations, i.e., reparameterizations, M does not depend on which of X_1 or X_2 is used; it depends only on $C(X_1)[= C(X_2)]$. Thus, the *MSE* does not change, and the least squares estimate of $\mathrm{E}(Y)$ is $\hat{Y} = MY = X_1\hat{\beta}_1 = X_2\hat{\beta}_2$.

The expected values of the observations are fundamental to estimation in linear models. Identifiable parameteric functions are functions of the expected values. Attention is often restricted to estimable functions, i.e., functions $\rho' X\beta$ where $X\beta = \mathrm{E}(Y)$. The key idea in estimability is restricting estimation to linear combinations of the rows of $\mathrm{E}(Y)$. $\mathrm{E}(Y)$ depends only on the choice of $C(X)$, whereas the vector β depends on the particular choice of X.

Consider again the two models discussed above. If $\lambda_1'\beta_1$ is estimable, then $\lambda_1'\beta_1 = \rho' X_1\beta_1 = \rho' E(Y)$ for some ρ. This estimable function is the same linear combination of the rows of $\mathrm{E}(Y)$ as $\rho' E(Y) = \rho' X_2\beta_2 = \lambda_2'\beta_2$. These are really the same estimable function, but they are written with different parameters. This estimable function has a unique least squares estimate, $\rho' MY$.

Example 3.1.1 One-Way ANOVA.
Two parameterizations for a one-way ANOVA are commonly used. They are

$$y_{ij} = \mu + \alpha_i + e_{ij}$$

and

$$y_{ij} = \mu_i + e_{ij}.$$

It is easily seen that these models determine the same estimation space. The estimates of σ^2 and $\mathrm{E}(y_{ij})$ are identical in the two models. One convenient aspect of these models is that the relationships between the two sets of parameters are easily identified. In particular,

$$\mu_i = \mathrm{E}(y_{ij}) = \mu + \alpha_i.$$

It follows that the mean of the μ_is equals μ plus the mean of the α_is, i.e., $\bar{\mu}_\cdot = \mu + \bar{\alpha}_\cdot$. It also follows that $\mu_1 - \mu_2 = \alpha_1 - \alpha_2$, etc. The parameters in the two models are different, but they are related. Any estimable function in one model determines a corresponding estimable function in the other model. These functions have the same estimate. Chapter 4 contains a detailed examination of these models.

Example 3.1.2 *Simple Linear Regression.*
The models

$$y_i = \beta_0 + \beta_1 x_i + e_i$$

and

$$y_i = \gamma_0 + \gamma_1 (x_i - \bar{x}_\cdot) + e_i$$

have the same estimation space ($\bar{x}_\cdot$ is the mean of the x_is). Since

$$\beta_0 + \beta_1 x_i = \mathrm{E}(y_i) = \gamma_0 + \gamma_1 (x_i - \bar{x}_\cdot) \tag{1}$$

for all i, it is easily seen from averaging over i that

$$\beta_0 + \beta_1 \bar{x}_\cdot = \gamma_0.$$

Substituting $\beta_0 = \gamma_0 - \beta_1 \bar{x}_\cdot$ into (1) leads to

$$\beta_1 (x_i - \bar{x}_\cdot) = \gamma_1 (x_i - \bar{x}_\cdot)$$

and, if the x_is are not all identical,

$$\beta_1 = \gamma_1.$$

These models are examined in detail in Section 6.1.

When the estimation spaces $C(X_1)$ and $C(X_2)$ are the same, write $X_1 = X_2 T$ to get

$$X_1 \beta_1 = X_2 T \beta_1 = X_2 \beta_2. \tag{2}$$

Estimable functions are equivalent in the two models: $\Lambda_1' \beta_1 = P' X_1 \beta_1 = P' X_2 \beta_2 = \Lambda_2' \beta_2$. It also follows from equation (2) that the parameterizations must satisfy the relation

$$\beta_2 = T \beta_1 + v \tag{3}$$

for some $v \in C(X_2')^\perp$. In general, neither of the parameter vectors β_1 or β_2 is uniquely defined but, to the extent that either parameter vector is defined, equation (3) establishes the relationship between them. A unique parameterization for, say, the X_2 model occurs if and only if $X_2' X_2$ is nonsingular. In such a case, the columns of X_2 form a basis for $C(X_2)$, so the matrix T is uniquely defined. In this case, the vector v must be zero because $C(X_2')^\perp = \{0\}$. An alternative and detailed presentation of equivalent linear models, both the reparameterizations considered here and the equivalences between constrained and unconstrained models considered in subsequent sections, is given by Peixoto (1993).

Basically, the β parameters in

$$Y = X\beta + e, \quad \mathrm{E}(e) = 0, \quad \mathrm{Cov}(e) = \sigma^2 I$$

are either a convenience or a nuisance, depending on what we are trying to do. Having $\mathrm{E}(e) = 0$ gives $\mathrm{E}(Y) = X\beta$, but since β is unknown, this is merely saying that $\mathrm{E}(Y)$ is *some* linear combination of the columns of X. The essence of the model is that

$$\mathrm{E}(Y) \in C(X), \quad \mathrm{Cov}(Y) = \sigma^2 I.$$

As long as we do not change $C(X)$, we can change X itself to suit our convenience.

3.2 Testing Models

In this section, the basic theory for testing a linear model against a reduced model is presented. A generalization of the basic procedure is also presented.

Testing in linear models typically reduces to putting a constraint on the estimation space. We start with a (full) model that we know (assume) to be valid,

$$Y = X\beta + e, \quad e \sim N(0, \sigma^2 I). \tag{1}$$

Our wish is to reduce this model, i.e., we wish to know if some simpler model gives an acceptable fit to the data. Consider whether the model

$$Y = X_0\gamma + e, \quad e \sim N(0, \sigma^2 I), \quad C(X_0) \subset C(X) \tag{2}$$

is acceptable. Clearly, if model (2) is correct, then model (1) is also correct. The question is whether (2) is correct.

The procedure of testing full and reduced models is a commonly used method in statistics.

Example 3.2.0

(a) *One-Way ANOVA.*

The full model is

$$y_{ij} = \mu + \alpha_i + e_{ij}.$$

To test for no treatment effects, i.e., to test that the α_is are extraneous, the reduced model simply eliminates the treatment effects. The reduced model is

$$y_{ij} = \gamma + e_{ij}.$$

Additionally, consider testing $H_0 : \alpha_1 - \alpha_3 = 0$ in Example 1.0.2. The full model is

$$\begin{bmatrix} y_{11} \\ y_{12} \\ y_{13} \\ y_{21} \\ y_{31} \\ y_{32} \end{bmatrix} = \begin{bmatrix} 1 & 1 & 0 & 0 \\ 1 & 1 & 0 & 0 \\ 1 & 1 & 0 & 0 \\ 1 & 0 & 1 & 0 \\ 1 & 0 & 0 & 1 \\ 1 & 0 & 0 & 1 \end{bmatrix} \begin{bmatrix} \mu \\ \alpha_1 \\ \alpha_2 \\ \alpha_3 \end{bmatrix} + \begin{bmatrix} e_{11} \\ e_{12} \\ e_{13} \\ e_{21} \\ e_{31} \\ e_{32} \end{bmatrix}.$$

We can rewrite this as

$$Y = \mu \begin{bmatrix} 1 \\ 1 \\ 1 \\ 1 \\ 1 \\ 1 \end{bmatrix} + \alpha_1 \begin{bmatrix} 1 \\ 1 \\ 1 \\ 0 \\ 0 \\ 0 \end{bmatrix} + \alpha_2 \begin{bmatrix} 0 \\ 0 \\ 0 \\ 1 \\ 0 \\ 0 \end{bmatrix} + \alpha_3 \begin{bmatrix} 0 \\ 0 \\ 0 \\ 0 \\ 1 \\ 1 \end{bmatrix} + e.$$

If we impose the constraint $H_0 : \alpha_1 - \alpha_3 = 0$, i.e., $\alpha_1 = \alpha_3$, we get

$$Y = \mu J + \alpha_1 \begin{bmatrix} 1 \\ 1 \\ 1 \\ 0 \\ 0 \\ 0 \end{bmatrix} + \alpha_2 \begin{bmatrix} 0 \\ 0 \\ 0 \\ 1 \\ 0 \\ 0 \end{bmatrix} + \alpha_1 \begin{bmatrix} 0 \\ 0 \\ 0 \\ 0 \\ 1 \\ 1 \end{bmatrix} + e,$$

or

$$Y = \mu J + \alpha_1 \left(\begin{bmatrix} 1 \\ 1 \\ 1 \\ 0 \\ 0 \\ 0 \end{bmatrix} + \begin{bmatrix} 0 \\ 0 \\ 0 \\ 0 \\ 1 \\ 1 \end{bmatrix} \right) + \alpha_2 \begin{bmatrix} 0 \\ 0 \\ 0 \\ 1 \\ 0 \\ 0 \end{bmatrix} + e,$$

or

$$\begin{bmatrix} y_{11} \\ y_{12} \\ y_{13} \\ y_{21} \\ y_{31} \\ y_{32} \end{bmatrix} = \begin{bmatrix} 1 & 1 & 0 \\ 1 & 1 & 0 \\ 1 & 1 & 0 \\ 1 & 0 & 1 \\ 1 & 1 & 0 \\ 1 & 1 & 0 \end{bmatrix} \begin{bmatrix} \mu \\ \alpha_1 \\ \alpha_2 \end{bmatrix} + \begin{bmatrix} e_{11} \\ e_{12} \\ e_{13} \\ e_{21} \\ e_{31} \\ e_{32} \end{bmatrix}.$$

This is the reduced model determined by $H_0 : \alpha_1 - \alpha_3 = 0$. However, the parameters μ, α_1, and α_2 no longer mean what they did in the full model.

(b) *Multiple Regression.*

Consider the full model

$$y_i = \beta_0 + \beta_1 x_{i1} + \beta_2 x_{i2} + \beta_3 x_{i3} + e_i.$$

For a simultaneous test of whether the variables x_1 and x_3 are adding significantly to the explanatory capability of the regression model, simply eliminate the variables x_1 and x_3 from the model. The reduced model is

$$y_i = \gamma_0 + \gamma_2 x_{i2} + e_i.$$

Now write the original model matrix as $X = [J, X_1, X_2, X_3]$, so

$$Y = [J, X_1, X_2, X_3]\begin{bmatrix}\beta_0\\ \beta_1\\ \beta_2\\ \beta_3\end{bmatrix} + e = \beta_0 J + \beta_1 X_1 + \beta_2 X_2 + \beta_3 X_3 + e.$$

Consider the hypothesis $H_0 : \beta_2 - \beta_3 = 0$ or $H_0 : \beta_2 = \beta_3$. The reduced model is

$$\begin{aligned} Y &= \beta_0 J + \beta_1 X_1 + \beta_2 X_2 + \beta_2 X_3 + e \\ &= \beta_0 J + \beta_1 X_1 + \beta_2 (X_2 + X_3) + e \\ &= [J, X_1, X_2 + X_3]\begin{bmatrix}\beta_0\\ \beta_1\\ \beta_2\end{bmatrix} + e. \end{aligned}$$

However, these β parameters no longer mean what they did in the original model, so it is better to write the model as

$$Y = [J, X_1, X_2 + X_3]\begin{bmatrix}\gamma_0\\ \gamma_1\\ \gamma_2\end{bmatrix} + e.$$

The distribution theory for testing models is given in Theorem 3.2.1. Before stating those results, we discuss the intuitive background of the test based only on the assumptions about the first two moments. Let M and M_0 be the perpendicular projection operators onto $C(X)$ and $C(X_0)$, respectively. Note that with $C(X_0) \subset C(X)$, $M - M_0$ is the perpendicular projection operator onto the orthogonal complement of $C(X_0)$ with respect to $C(X)$, that is, onto $C(M - M_0) = C(X_0)^{\perp}_{C(X)}$, see Theorem B.47.

Under model (1), the estimate of $\mathrm{E}(Y)$ is $X\hat{\beta} = MY$. Under model (2), the estimate is $X_0\hat{\gamma} = M_0 Y$. Recall that the validity of model (2) implies the validity of model (1); so if model (2) is true, MY and $M_0 Y$ are estimates of the same quantity. This suggests that the difference between the two estimates, $X\hat{\beta} - X_0\hat{\gamma} = MY - M_0 Y = (M - M_0)Y$, should be reasonably small. Under model (2), the difference is just error because $\mathrm{E}[(M - M_0)Y] = 0$.

On the other hand, a large difference between the estimates suggests that MY and $M_0 Y$ are estimating different things. By assumption, MY is always an estimate of $\mathrm{E}(Y)$; so $M_0 Y$ must be estimating something different, namely, $M_0\mathrm{E}(Y) \neq \mathrm{E}(Y)$. If $M_0 Y$ is not estimating $\mathrm{E}(Y)$, model (2) cannot be true because model (2) implies that $M_0 Y$ is an estimate of $\mathrm{E}(Y)$.

The decision about whether model (2) is appropriate hinges on deciding whether the vector $X\hat{\beta} - X_0\hat{\gamma} = (M - M_0)Y$ is large. An obvious measure of the size of $(M - M_0)Y$ is its squared length, $[(M - M_0)Y]'[(M - M_0)Y] = Y'(M - M_0)Y$. However, the length of $(M - M_0)Y$ is also related to the relative sizes of $C(M)$ and

$C(M_0)$. It is convenient (not crucial) to adjust for this factor. As a measure of the size of $(M-M_0)Y$, we use the value

$$Y'(M-M_0)Y\big/r(M-M_0).$$

Even though we have an appropriate measure of the size of $(M-M_0)Y$, we still need some idea of how large the measure will be both when model (2) is true and when model (2) is not true. Using only the assumption that model (1) is true, Theorem 1.3.2 implies that

$$\mathrm{E}[Y'(M-M_0)Y\big/r(M-M_0)] = \sigma^2 + \beta'X'(M-M_0)X\beta\big/r(M-M_0).$$

If model (2) is also true, $X\beta = X_0\gamma$ and $(M-M_0)X_0 = 0$; so the expected value of $Y'(M-M_0)Y\big/r(M-M_0)$ is σ^2. If σ^2 were known, our intuitive argument would be complete. If $Y'(M-M_0)Y\big/r(M-M_0)$ is not much larger than σ^2, then we have observed something that is consistent with the validity of model (2). Values that are much larger than σ^2 indicate that model (2) is false because they suggest that $\beta'X'(M-M_0)X\beta\big/r(M-M_0) > 0$, i.e., $X\beta \neq M_0X\beta$.

Typically, we do not know σ^2, so the obvious thing to do is to estimate it. Since model (1) is assumed to be true, the obvious estimate is the $MSE = Y'(I-M)Y/r(I-M)$. Now, values of $Y'(M-M_0)Y\big/r(M-M_0)$ that are much larger than MSE cause us to doubt the validity of model (2). Equivalently, values of the *test statistic*

$$\frac{Y'(M-M_0)Y\big/r(M-M_0)}{MSE}$$

that are considerably larger than 1 cause precisely the same doubts.

We now examine the behavior of this test statistic when model (2) is not correct but model (1) is. $Y'(M-M_0)Y/r(M-M_0)$ and MSE are each estimates of their expected values, so the test statistic obviously provides an estimate of the ratio of their expected values. Recalling $\mathrm{E}[Y'(M-M_0)Y/r(M-M_0)]$ from above and that $\mathrm{E}(MSE) = \sigma^2$, the test statistic gives an estimate of $1+\beta'X'(M-M_0)X\beta\big/r(M-M_0)\sigma^2$. The term $\beta'X'(M-M_0)X\beta$ is crucial to evaluating the behavior of the test statistic when model (1) is valid but model (2) is not, cf. the noncentrality parameter in Theorem 3.2.1, part *i*. Note that $\beta'X'(M-M_0)X\beta = [X\beta - M_0X\beta]'[X\beta - M_0X\beta]$ is the squared length of the difference between $X\beta$ (i.e., E(Y)) and $M_0X\beta$ (the projection of $X\beta$ onto $C(X_0)$). If $X\beta - M_0X\beta$ is large (relative to σ^2), then model (2) is very far from being correct, and the test statistic will tend to be large. On the other hand, if $X\beta - M_0X\beta$ is small (relative to σ^2), then model (2), although not correct, is a good approximation to the correct model. In this case the test statistic will tend to be a little larger than it would be if model (2) were correct, but the effect will be very slight. In other words, if $\beta'X'(M-M_0)X\beta\big/r(M-M_0)\sigma^2$ is small, it is unreasonable to expect any test to work very well.

One can think about the geometry of all this in three dimensions. As in Section 2.2, consider a rectangular table. Take one corner of the table to be the origin. Take $C(X)$ as the two-dimensional subspace determined by the surface of the table and take $C(X_0)$ to be a one-dimensional subspace determined by an edge of the table that includes the origin. Y can be any vector originating at the origin, i.e., any point in three-dimensional space. The full model (1) says that $\mathrm{E}(Y) = X\beta$, which just says that $\mathrm{E}(Y)$ is somewhere on the surface of the table. The reduced model (2) says that $\mathrm{E}(Y)$ is somewhere on the $C(X_0)$ edge of the table. $MY = X\hat{\beta}$ is the perpendicular projection of Y onto the table surface. $M_0Y = X_0\hat{\gamma}$ is the perpendicular projection of Y onto the $C(X_0)$ edge of the table. The residual vector $(I - M)Y$ is the perpendicular projection of Y onto the vertical line through the origin.

If MY is close to the $C(X_0)$ edge of the table, it must be close to M_0Y. This is the behavior we would expect if the reduced model is true, i.e., if $X\beta = X_0\gamma$. The difference between the two estimates, $MY - M_0Y$, is a vector that is on the table, but perpendicular to the $C(X_0)$ edge. In fact, the table edge through the origin that is perpendicular to the $C(X_0)$ edge is the orthogonal complement of $C(X_0)$ with respect to $C(X)$, that is, it is $C(X_0)^{\perp}_{C(X)} = C(M - M_0)$. The difference between the two estimates is $MY - M_0Y = (M - M_0)Y$, which is the perpendicular projection of Y onto $C(M - M_0)$. If $(M - M_0)Y$ is large, it suggests that the reduced model is not true. To decide if $(M - M_0)Y$ is large, we find its average (mean) squared length, where the average is computed relative to the dimension of $C(M - M_0)$, and compare that to the averaged squared length of the residual vector $(I - M)Y$ (i.e., the *MSE*). In our three-dimensional example, the dimensions of both $C(M - M_0)$ and $C(I - M)$ are 1. If $(M - M_0)Y$ is, on average, much larger than $(I - M)Y$, we reject the reduced model.

If the true (but unknown) $X\beta$ happens to be far from the $C(X_0)$ edge of the table, it will be very easy to see that the reduced model is not true. This occurs because MY will be near $X\beta$, which is far from anything in $C(X_0)$ so, in particular, it will be far from M_0Y. Remember that the meaning of "far" depends on σ^2 which is estimated by the *MSE*. On the other hand, if $X\beta$ happens to be near, but not on, the $C(X_0)$ edge of the table, it will be very hard to tell that the reduced model is not true because MY and M_0Y will tend to be close together. On the other hand, if $X\beta$ is near, but not on, the $C(X_0)$ edge of the table, using the incorrect reduced model may not create great problems, cf. Section 2.9.

To this point, the discussion has been based entirely on the assumptions $Y = X\beta + e$, $\mathrm{E}(e) = 0$, $\mathrm{Cov}(e) = \sigma^2 I$. We now quantify the precise behavior of the test statistic for normal errors.

Theorem 3.2.1 *Consider a full model*

$$Y = X\beta + e, \quad e \sim N(0, \sigma^2 I)$$

that holds for some values of β and σ^2 and a reduced model

$$Y = X_0\gamma + e, \quad e \sim N(0, \sigma^2 I), \quad C(X_0) \subset C(X).$$

(i) If the full model is true,

$$\frac{Y'(M - M_0)Y/r(M - M_0)}{Y'(I - M)Y/r(I - M)} \sim F\left(r(M - M_0), r(I - M), \beta' X'(M - M_0)X\beta/2\sigma^2\right).$$

(ii) If the reduced model is true,

$$\frac{Y'(M - M_0)Y/r(M - M_0)}{Y'(I - M)Y/r(I - M)} \sim F(r(M - M_0), r(I - M), 0).$$

When the full model is true, this distribution holds only if the reduced model is true.

Proof (i) Since M and M_0 are the perpendicular projection matrices onto $C(X)$ and $C(X_0)$, $M - M_0$ is the perpendicular projection matrix onto $C(M - M_0)$, cf. Theorem B.47. As in Section 2.6,

$$\frac{Y'(I - M)Y}{\sigma^2} \sim \chi^2(r(I - M))$$

and from Theorem 1.3.3 on the distribution of quadratic forms,

$$\frac{Y'(M - M_0)Y}{\sigma^2} \sim \chi^2\left(r(M - M_0), \beta' X'(M - M_0)X\beta/2\sigma^2\right).$$

Theorem 1.3.7 establishes that $Y'(M - M_0)Y$ and $Y'(I - M)Y$ are independent because

$$\begin{aligned}(M - M_0)(I - M) &= M - M_0 - M + M_0 M \\ &= M - M_0 - M + M_0 = 0.\end{aligned}$$

Finally, part (i) of the theorem holds by Definition C.3.

(ii) It suffices to show that $\beta' X'(M - M_0)X\beta = 0$ if and only if $\mathrm{E}(Y) = X_0\gamma$ for some γ.

$\Leftarrow$ If $\mathrm{E}(Y) = X_0\gamma$, we have $\mathrm{E}(Y) = X\beta$ for some β because $C(X_0) \subset C(X)$. In particular, $\beta' X'(M - M_0)X\beta = \gamma' X_0'(M - M_0)X_0\gamma$, but since $(M - M_0)X_0 = X_0 - X_0 = 0$, we have $\beta' X'(M - M_0)X\beta = 0$.

$\Rightarrow$ If $\beta' X'(M - M_0)X\beta = 0$, then $[(M - M_0)X\beta]'[(M - M_0)X\beta] = 0$. Since for any x, $x'x = 0$ if and only if $x = 0$, we have $(M - M_0)X\beta = 0$ or $X\beta = M_0 X\beta = X_0(X_0'X_0)^- X_0'X\beta$. Taking $\gamma = (X_0'X_0)^- X_0'X\beta$, we have $\mathrm{E}(Y) = X_0\gamma$. □

People typically reject the hypothesis

$$H_0 : \mathrm{E}(Y) = X_0\gamma \quad \text{for some } \gamma,$$

for large observed values of the test statistic. The informal second moment arguments given prior to Theorem 3.2.1 suggest rejecting large values and the existence a positive noncentrality parameter in Theorem 3.2.1(i) would shift the (central) F distribution to the right which also suggests rejecting large values. Both of these arguments depend on the full model being true. Theorem 3.2.1(ii) provides a distribution for the test statistic under the reduced (null) model, so under the conditions of the theorem this test of

$$H_0 : \mathrm{E}(Y) \in C(X_0)$$

rejects H_0 at level α if

$$\frac{Y'(M - M_0)Y/r(M - M_0)}{Y'(I - M)Y/r(I - M)} > F(1 - \alpha, r(M - M_0), r(I - M)) .$$

P values are then *reported* as the probability from an $F(r(M - M_0), r(I - M))$ distribution of being at least as large as the observed value of the test statistic.

This test procedure is "non-Fisherian" in that it assumes more than just the null (reduced) model being true. The decision on when to reject the null model depends on the full model being true. In fact, test statistic values near 0 (reported P values near 1) can be just as interesting as large test statistic values (reported P values near 0). Large reported P values often need to be closer to 1 to be interesting than small reported P values need to be close to 0.

Personally, I don't consider these *reported P* values to be real P values, although they are not without their uses. *Appendix F discusses the significance of small test statistics and some foundational issues related to this test.* For those willing to assume that the full model is true, the F test is the *generalized likelihood ratio test* (see Exercise 3.1) and the *uniformly most powerful invariant (UMPI) test* (see Lehmann, 1986, Chapter 7 and http://www.stat.unm.edu/~fletcher/UMPI.pdf).

In practice it is often easiest to use the following approach to obtain the test statistic: Observe that $(M - M_0) = (I - M_0) - (I - M)$. If we can find the error sums of squares, $Y'(I - M_0)Y$ from the model $Y = X_0\gamma + e$ and $Y'(I - M)Y$ from the model $Y = X\beta + e$, then the difference is $Y'(I - M_0)Y - Y'(I - M)Y = Y'(M - M_0)Y$, which is the numerator sum of squares for the F test. Unless there is some simplifying structure to the model matrix (as in cases we will examine later), it is usually easier to obtain the error sums of squares for the two models than to find $Y'(M - M_0)Y$ directly.

Example 3.2.2 Consider the model matrix given at the end of this example. It is for the unbalanced analysis of variance $y_{ijk} = \mu + \alpha_i + \eta_j + e_{ijk}$, where $i = 1, 2, 3$, $j = 1, 2, 3$, $k = 1, ..., N_{ij}$, $N_{11} = N_{12} = N_{21} = 3$, $N_{13} = N_{22} = N_{23} = N_{31} = N_{32} = N_{33} = 2$. Here we have written $Y = X\beta + e$ with $Y = [y_{111}, y_{112}, \ldots, y_{332}]'$, $\beta = [\mu, \alpha_1, \alpha_2, \alpha_3, \eta_1, \eta_2, \eta_3]'$, and $e = [e_{111}, e_{112}, \ldots, e_{332}]'$. We can now test to see if the model $y_{ijk} = \mu + \alpha_i + e_{ijk}$ is an adequate representation of the data simply by dropping the last three columns from the model matrix. We can also test $y_{ijk} =$

$\mu + e_{ijk}$ by dropping the last six columns of the model matrix. In either case, the test is based on comparing the error sum of squares for the reduced model with that of the full model.

$$X = \begin{bmatrix} 1 & 1 & 0 & 0 & 1 & 0 & 0 \\ 1 & 1 & 0 & 0 & 1 & 0 & 0 \\ 1 & 1 & 0 & 0 & 1 & 0 & 0 \\ 1 & 1 & 0 & 0 & 0 & 1 & 0 \\ 1 & 1 & 0 & 0 & 0 & 1 & 0 \\ 1 & 1 & 0 & 0 & 0 & 1 & 0 \\ 1 & 1 & 0 & 0 & 0 & 0 & 1 \\ 1 & 1 & 0 & 0 & 0 & 0 & 1 \\ 1 & 0 & 1 & 0 & 1 & 0 & 0 \\ 1 & 0 & 1 & 0 & 1 & 0 & 0 \\ 1 & 0 & 1 & 0 & 1 & 0 & 0 \\ 1 & 0 & 1 & 0 & 0 & 1 & 0 \\ 1 & 0 & 1 & 0 & 0 & 1 & 0 \\ 1 & 0 & 1 & 0 & 0 & 0 & 1 \\ 1 & 0 & 1 & 0 & 0 & 0 & 1 \\ 1 & 0 & 0 & 1 & 1 & 0 & 0 \\ 1 & 0 & 0 & 1 & 1 & 0 & 0 \\ 1 & 0 & 0 & 1 & 0 & 1 & 0 \\ 1 & 0 & 0 & 1 & 0 & 1 & 0 \\ 1 & 0 & 0 & 1 & 0 & 0 & 1 \\ 1 & 0 & 0 & 1 & 0 & 0 & 1 \end{bmatrix}.$$

3.2.1 Small Test Statistics

When testing the reduced model, what happens when the full model $Y = X\beta + e$ is not true? In particular, what happens if $\mathrm{E}(Y) \in C(X_0) + C(I - M)$, i.e., $\mathrm{E}(Y) = X_0\gamma + W_0\delta_0$ where $W_0\delta_0 \in C(I - M)$? In that case,

$$\begin{aligned}\mathrm{E}[Y'(M - M_0)Y] &= \sigma^2\mathrm{tr}(M - M_0) + (X_0\gamma + W_0\delta_0)'(M - M_0)(X_0\gamma + W_0\delta_0) \\ &= \sigma^2[r(X) - r(X_0)] + 0\end{aligned}$$

but

$$\begin{aligned}\mathrm{E}[Y'(I - M)Y] &= \sigma^2\mathrm{tr}(I - M) + (X_0\gamma + W_0\delta_0)'(I - M)(X_0\gamma + W_0\delta_0) \\ &= \sigma^2[n - r(X)] + (W_0\delta_0)'(W_0\delta_0),\end{aligned}$$

so the F statistic is an estimate of $1/\{1 + \delta_0' W_0' W_0\delta_0/\sigma^2[n - r(X)]\}$ which gets small (close to 0) when $\| W_0\delta_0 \|^2$ is large. Appendix F goes into more detail about small F statistics.

3.2.2 A Generalized Test Procedure

Before considering tests of parametric functions, we consider a generalization of the test procedure outlined earlier. Assume that the model $Y = X\beta + e$ is correct. We want to test the adequacy of a model $Y = X_0\gamma + Xb + e$, where $C(X_0) \subset C(X)$ and Xb is some known vector. In generalized linear model terminology, Xb is called an *offset*.

Example 3.2.3 *Multiple Regression.*
Consider the model

$$Y = [J, X_1, X_2, X_3] \begin{bmatrix} \beta_0 \\ \beta_1 \\ \beta_2 \\ \beta_3 \end{bmatrix} + e = \beta_0 J + \beta_1 X_1 + \beta_2 X_2 + \beta_3 X_3 + e.$$

To test $H_0 : \beta_2 = \beta_3 + 5, \beta_1 = 0$, write the reduced model as

$$\begin{aligned} Y &= \beta_0 J + (\beta_3 + 5)X_2 + \beta_3 X_3 + e \\ &= \beta_0 J + \beta_3(X_2 + X_3) + 5X_2 + e \\ &= [J, X_2 + X_3] \begin{bmatrix} \beta_0 \\ \beta_3 \end{bmatrix} + 5X_2 + e. \end{aligned} \tag{3}$$

Alternatively, we could write the reduced model as

$$\begin{aligned} Y &= \beta_0 J + \beta_2 X_2 + (\beta_2 - 5)X_3 + e \\ &= \beta_0 J + \beta_2(X_2 + X_3) - 5X_3 + e \\ &= [J, X_2 + X_3] \begin{bmatrix} \beta_0 \\ \beta_2 \end{bmatrix} - 5X_3 + e. \end{aligned} \tag{4}$$

We will see that both reduced models lead to the same test.

The model $Y = X\beta + e$ can be rewritten as $Y - Xb = X\beta - Xb + e$. Since $Xb \in C(X)$, this amounts to a reparameterization, $Y - Xb = X\beta^* + e$, where $\beta^* = \beta - b$. Since Xb is known, $Y - Xb$ is still an observable random variable.

The reduced model $Y = X_0\gamma + Xb + e$ can be rewritten as $Y - Xb = X_0\gamma + e$. The question of testing the adequacy of the reduced model is now a straightforward application of our previous theory. The distribution of the test statistic is

$$\frac{(Y - Xb)'(M - M_0)(Y - Xb)/r(M - M_0)}{(Y - Xb)'(I - M)(Y - Xb)/r(I - M)} \sim F\big(r(M - M_0), r(I - M), \beta^{*\prime}X'(M - M_0)X\beta^*/2\sigma^2\big).$$

The noncentrality parameter is zero if and only if $0 = \beta^{*\prime}X'(M - M_0)X\beta^* = [(M - M_0)(X\beta - Xb)]'[(M - M_0)(X\beta - Xb)]$, which occurs if and only if $(M -$

$M_0)(X\beta - Xb) = 0$ or $X\beta = M_0(X\beta - Xb) + Xb$. The last condition is nothing more or less than that the reduced model is valid with $\gamma = (X_0'X_0)^- X_0(X\beta - Xb)$, a fixed unknown parameter.

Note also that, since $(I - M)X = 0$, in the denominator of the test statistic $(Y - Xb)'(I - M)(Y - Xb) = Y'(I - M)Y$, which is the *SSE* from the original full model. Moreover, the numerator sum of squares is

$$(Y - Xb)'(M - M_0)(Y - Xb) = (Y - Xb)'(I - M_0)(Y - Xb) - Y'(I - M)Y,$$

which can be obtained by subtracting the *SSE* of the original full model from the *SSE* of the reduced model. To see this, write $I - M_0 = (I - M) + (M - M_0)$.

Example 3.2.3 Continued. The numerator sum of squares for testing model (4) is $(Y + 5X_3)'(M - M_0)(Y + 5X_3)$. But $(M - M_0)[X_2 + X_3] = 0$, so, upon observing that $5X_3 = -5X_2 + 5[X_2 + X_3]$,

$$\begin{aligned}
&(Y + 5X_3)'(M - M_0)(Y + 5X_3)\\
&= (Y - 5X_2 + 5[X_2 + X_3])'(M - M_0)(Y - 5X_2 + 5[X_2 + X_3])\\
&= (Y - 5X_2)'(M - M_0)(Y - 5X_2),
\end{aligned}$$

which is the numerator sum of squares for testing model (3). In fact, models (3) and (4) are equivalent because the only thing different about them is that one uses $5X_2$ and the other uses $-5X_3$; but the only difference between these terms is $5[X_2 + X_3] \in C(X_0)$.

The phenomenon illustrated in this example is a special case of a general result. Consider the model $Y = X_0\gamma + Xb + e$ for some unknown γ and known b and suppose $X(b - b_*) \in C(X_0)$ for known b_*. The model $\mathrm{E}(Y) = X_0\gamma + Xb$ holds if and only if $\mathrm{E}(Y) = X_0\gamma + X(b - b_*) + Xb_*$, which holds if and only if $\mathrm{E}(Y) = X_0\gamma_* + Xb_*$ for some unknown γ_*.

Exercise 3.1 (a) Show that the F test developed in the first part of this section is equivalent to the (generalized) likelihood ratio test for the reduced versus full models, cf. Casella and Berger (2002, Subsection 8.2.1). (b) Find an F test for $H_0 : X\beta = X\beta_0$ where β_0 is known. (c) Construct a full versus reduced model test when σ^2 has a known value σ_0^2.

Exercise 3.2 Redo the tests in Exercise 2.2 using the theory of Section 3.2. Write down the models and explain the procedure.

Exercise 3.3 Redo the tests in Exercise 2.3 using the procedures of Section 3.2. Write down the models and explain the procedure.

Hints: (a) Let A be a matrix of zeros, the generalized inverse of A, A^-, can be anything at all because $AA^-A = A$ for any choice of A^-. (b) There is no reason why X_0 cannot be a matrix of zeros.

Exercise 3.4 When testing submodels of a largest model it is considered good practice to use the mean squared error from the largest model in the denominator of the test. Consider three linear models $\mathrm{E}(Y) = X_0\gamma$, $\mathrm{E}(Y) = X_1\delta$, $\mathrm{E}(Y) = X\beta$ with $C(X_0) \subset C(X_1) \subset C(X)$ and ppos M_0, M_1, M, respectively. If $Y \sim N(X_1\delta, \sigma^2 I)$, find the distribution of

$$\frac{Y'(M_1 - M_0)Y/r(M_1 - M_0)}{Y'(I - M)Y/r(I - M)}.$$

3.3 Testing Linear Parametric Functions

In this section, the theory of testing linear parametric functions is presented. A basic test procedure and a generalized test procedure are given. These procedures are analogous to the model testing procedures of Section 2. In the course of this presentation, the important concept of the constraint imposed by an hypothesis is introduced. Finally, a class of hypotheses that is rarely used for linear models but commonly used with log-linear models is given along with results that define the appropriate testing procedure.

Consider a general linear model

$$Y = X\beta + e \tag{1}$$

with X an $n \times p$ matrix. A key aspect of this model is that β is allowed to be any vector in $\mathbf{R}^p$. Additionally, consider an hypothesis concerning a linear function, say $\Lambda'\beta = 0$. The null model to be tested is

$$H_0 : Y = X\beta + e \quad \text{and} \quad \Lambda'\beta = 0.$$

We need to find a reduced model that corresponds to this.

The constraint $\Lambda'\beta = 0$ can be stated in an infinite number of ways. Observe that $\Lambda'\beta = 0$ holds if and only if $\beta \perp C(\Lambda)$; so if Γ is another matrix with $C(\Gamma) = C(\Lambda)$, the constraint can also be written as $\beta \perp C(\Gamma)$ or $\Gamma'\beta = 0$.

To identify the reduced model, pick a matrix U such that

$$C(U) = C(\Lambda)^\perp,$$

then $\Lambda'\beta = 0$ if and only if $\beta \perp C(\Lambda)$ if and only if $\beta \in C(U)$, which occurs if and only if for some vector γ,

$$\beta = U\gamma. \tag{2}$$

Substituting (2) into the linear model gives the reduced model

$$Y = XU\gamma + e$$

or, letting $X_0 \equiv XU$,

$$Y = X_0\gamma + e. \tag{3}$$

Note that $C(X_0) \subset C(X)$. Proposition B.54 establishes that the particular choice of U is irrelevant to defining $C(X_0)$. Moreover, the reduced model does not depend on $\mathrm{Cov}(Y)$ or the exact distribution of e, it only depends on $\mathrm{E}(Y) = X\beta$ and the constraint $\Lambda'\beta = 0$.

If $e \sim N(0, \sigma^2 I)$, the reduced model (3) allows us to test $\Lambda'\beta = 0$ by applying the results of Section 2. If $C(X_0) = C(X)$, the constraint involves only a reparameterization and there is nothing to test. In other words, if $C(X_0) = C(X)$, then $\Lambda'\beta = 0$ involves only arbitrary side conditions that do not affect the model.

Example 3.3.1 Consider the one-way analysis of variance model

$$\begin{bmatrix} y_{11} \\ y_{12} \\ y_{13} \\ y_{21} \\ y_{31} \\ y_{32} \end{bmatrix} = \begin{bmatrix} 1 & 1 & 0 & 0 \\ 1 & 1 & 0 & 0 \\ 1 & 1 & 0 & 0 \\ 1 & 0 & 1 & 0 \\ 1 & 0 & 0 & 1 \\ 1 & 0 & 0 & 1 \end{bmatrix} \beta + e,$$

where $\beta = (\mu, \alpha_1, \alpha_2, \alpha_3)'$. The parameters in this model are not uniquely defined because the rank of X is less than the number of columns.

Let $\lambda_1' = (0, 1, 0, -1)$. The contrast $\lambda_1'\beta$ is estimable, so the hypothesis $\alpha_1 - \alpha_3 = \lambda_1'\beta = 0$ determines an estimable constraint. To obtain $C(\lambda_1)^\perp = C(U)$, one can pick

$$U = \begin{bmatrix} 1 & 0 & 0 \\ 0 & 1 & 0 \\ 0 & 0 & 1 \\ 0 & 1 & 0 \end{bmatrix},$$

which yields

$$XU = \begin{bmatrix} 1 & 1 & 0 \\ 1 & 1 & 0 \\ 1 & 1 & 0 \\ 1 & 0 & 1 \\ 1 & 1 & 0 \\ 1 & 1 & 0 \end{bmatrix}. \tag{4}$$

This is a real restriction on $C(X)$, i.e., $C(XU) \neq C(X)$.

Let $\lambda_2' = (0, 1, 1, 1)$. A nonestimable linear constraint for a one-way analysis of variance is that

$$\alpha_1 + \alpha_2 + \alpha_3 = \lambda_2'\beta = 0.$$

Consider two choices for U with $C(\lambda_2)^\perp = C(U)$, i.e.,

$$U_1 = \begin{bmatrix} 1 & 0 & 0 \\ 0 & 1 & 1 \\ 0 & -1 & 1 \\ 0 & 0 & -2 \end{bmatrix} \quad \text{and} \quad U_2 = \begin{bmatrix} 1 & 0 & 0 \\ 0 & 1 & 1 \\ 0 & 0 & -2 \\ 0 & -1 & 1 \end{bmatrix}.$$

These yield

$$XU_1 = \begin{bmatrix} 1 & 1 & 1 \\ 1 & 1 & 1 \\ 1 & 1 & 1 \\ 1 & -1 & 1 \\ 1 & 0 & -2 \\ 1 & 0 & -2 \end{bmatrix} \quad \text{and} \quad XU_2 = \begin{bmatrix} 1 & 1 & 1 \\ 1 & 1 & 1 \\ 1 & 1 & 1 \\ 1 & 0 & -2 \\ 1 & -1 & 1 \\ 1 & -1 & 1 \end{bmatrix}.$$

Note that $C(X) = C(XU_1) = C(XU_2)$. The models determined by XU_1 and XU_2 are equivalent linear models but have different parameterizations, say $Y = XU_1\xi_1 + e$ and $Y = XU_2\xi_2 + e$, with $\xi_i = (\xi_{i1}, \xi_{i2}, \xi_{i3})'$. Transform ξ_i to the original β parameterization using (2), for example,

$$\begin{bmatrix} \mu \\ \alpha_1 \\ \alpha_2 \\ \alpha_3 \end{bmatrix} = \beta = U_1\xi_1 = \begin{bmatrix} \xi_{11} \\ \xi_{12} + \xi_{13} \\ -\xi_{12} + \xi_{13} \\ -2\xi_{13} \end{bmatrix}.$$

Both ξ_i parameterizations lead to $\alpha_1 + \alpha_2 + \alpha_3 = 0$. Thus both determine the same specific choice for the parameterization of the original one-way analysis of variance model.

Similar results hold for alternative nonidentifiable constraints such as $\alpha_1 = 0$. As will be established later, any nonidentifiable constraint leaves the estimation space unchanged and therefore yields the same estimates of estimable functions.

Now consider the joint constraint $\Lambda_1'\beta = 0$, where

$$\Lambda_1' = \begin{bmatrix} \lambda_1' \\ \lambda_2' \end{bmatrix}.$$

$\lambda_1'\beta$ is a contrast, so it is estimable; therefore $\Lambda_1'\beta$ has estimable aspects. One choice of U with $C(\Lambda_1)^\perp = C(U)$ is

$$U_3 = \begin{bmatrix} 1 & 0 \\ 0 & 1 \\ 0 & -2 \\ 0 & 1 \end{bmatrix}.$$

This gives

$$XU_3 = \begin{bmatrix} 1 & 1 \\ 1 & 1 \\ 1 & 1 \\ 1 & -2 \\ 1 & 1 \\ 1 & 1 \end{bmatrix},$$

and the estimation space is the same as in (4), where only the contrast $\lambda_1'\beta$ was assumed equal to zero.

A constraint equivalent to $\Lambda_1'\beta = 0$ is $\Lambda_2'\beta = 0$, where

$$\Lambda_2' = \begin{bmatrix} \lambda_3' \\ \lambda_2' \end{bmatrix} = \begin{bmatrix} 0 & 2 & 1 & 0 \\ 0 & 1 & 1 & 1 \end{bmatrix}.$$

The constraints are equivalent because $C(\Lambda_1) = C(\Lambda_2)$. (Note that $\lambda_3 - \lambda_2 = \lambda_1$.) Neither $\lambda_3'\beta$ nor $\lambda_2'\beta$ is estimable, so separately they would affect only the parameterization of the model. However, $\Lambda_1'\beta = 0$ involves an estimable constraint, so $\Lambda_2'\beta = 0$ also has an estimable part to the constraint. The concept of the estimable part of a constraint will be examined in detail later.

Estimable Constraints

We have established a perfectly general method for identifying the reduced model determined by an arbitrary linear constraint $\Lambda'\beta = 0$, and thus we have a general method for testing $\Lambda'\beta = 0$ by applying the results of Section 2. Next, we will examine the form of the test statistic when $\Lambda'\beta$ is estimable. Afterwards, we present results showing that there is little reason to consider nonestimable linear constraints.

When $\Lambda'\beta$ is estimable, so that $\Lambda' = P'X$ for some P, rather than finding U we can find the numerator projection operator for testing $\Lambda'\beta = 0$ in terms of P and M. Better yet, we can find the numerator sum of squares in terms of Λ and any least squares estimate $\hat{\beta}$. Recall from Section 2 that the numerator sum of squares is $Y'(M - M_0)Y$, where $M - M_0$ is the perpendicular projection operator onto the orthogonal complement of $C(X_0)$ with respect to $C(X)$. In other words, $M - M_0$ is a perpendicular projection operator with $C(M - M_0) = C(X_0)^{\perp}_{C(X)}$. For testing an estimable parametric hypothesis with $\Lambda' = P'X$, we now show that the perpendicular projection operator onto $C(MP)$ is also the perpendicular projection operator onto the orthogonal complement of $C(X_0)$ with respect to $C(X)$, i.e., that $C(MP) = C(X_0)^{\perp}_{C(X)}$. It follows immediately that the numerator sum of squares in the test is $Y'M_{MP}Y$, where $M_{MP} \equiv MP(P'MP)^{-}P'M$ is the perpendicular projection operator onto $C(MP)$. In particular, from Section 2

$$\frac{Y'M_{MP}Y/r(M_{MP})}{Y'(I - M)Y/r(I - M)} \sim F\big(r(M_{MP}), r(I - M), \beta'X'M_{MP}X\beta/2\sigma^2\big). \quad (5)$$

Proposition 3.3.2 provides a formal proof of the necessary result. However, after the proof, we give an alternative justification based on finding the reduced model associated with the constraint. This reduced model argument differs from the one given at the beginning of the section in that it only applies to estimable constraints.

Proposition 3.3.2 *With U and P defined for* $\Lambda'\beta = 0$,

$$C(M - M_0) = C(X_0)^{\perp}_{C(X)} \equiv C(XU)^{\perp}_{C(X)} = C(MP).$$

Proof From Section 2, we already know that $C(M - M_0) = C(X_0)^{\perp}_{C(X)}$. Since $X_0 \equiv XU$, we need only establish that $C(XU)^{\perp}_{C(X)} = C(MP)$.

If $x \in C(XU)^{\perp}_{C(X)}$, then $0 = x'XU$, so $X'x \perp C(U)$ and $X'x \in C(\Lambda) = C(X'P)$. It follows that

$$x = Mx = [X(X'X)^{-}]X'x \in C([X(X'X)^{-}]X'P) = C(MP).$$

Conversely, if $x \in C(MP)$, then $x = MPb$ for some b and

$$x'XU = b'P'MXU = b'P'XU = b'\Lambda'U = 0,$$

so $x \in C(XU)^{\perp}_{C(X)}$. □

Earlier, we found the reduced model matrix $X_0 = XU$ directly and then, for $\Lambda' = P'X$, we showed that $C(MP) = C(X_0)^{\perp}_{C(X)}$, which led to the numerator sum of squares. An alternative derivation of the test arrives at $C(M - M_0) = C(X_0)^{\perp}_{C(X)} = C(MP)$ more directly for estimable constraints. The reduced model is

$$Y = X\beta + e \quad \text{and} \quad P'X\beta = 0,$$

or

$$Y = X\beta + e \quad \text{and} \quad P'MX\beta = 0,$$

or

$$\mathrm{E}(Y) \in C(X) \quad \text{and} \quad \mathrm{E}(Y) \perp C(MP),$$

or

$$\mathrm{E}(Y) \in C(X) \cap C(MP)^{\perp}.$$

The reduced model matrix X_0 must satisfy $C(X_0) = C(X) \cap C(MP)^{\perp} \equiv C(MP)^{\perp}_{C(X)}$. It follows immediately that $C(X_0)^{\perp}_{C(X)} = C(MP)$. Moreover, it is easily seen that X_0 can be taken as $X_0 = (I - M_{MP})X$.

Theorem 3.3.3 $C[(I - M_{MP})X] = C(X) \cap C(MP)^{\perp}$.

Proof First, assume $x \in C(X)$ and $x \perp C(MP)$. Write $x = Xb$ for some b and note that $M_{MP}x = 0$. It follows that $x = (I - M_{MP})x = (I - M_{MP})Xb$, so $x \in C[(I - M_{MP})X]$.

Conversely, if $x = (I - M_{MP})Xb$ for some b, then clearly $x \in C(X)$ and $x'MP = b'X'(I - M_{MP})MP = 0$ because $(I - M_{MP})MP = 0$ □

Note also that $C(X) \cap C(MP)^{\perp} = C(X) \cap C(P)^{\perp}$.

Example 3.3.4 To illustrate these ideas, consider testing $H_0 : \alpha_1 - \alpha_3 = 0$ in Example 3.3.1. The constraint can be written

$$0 = \alpha_1 - \alpha_3 = \left(\frac{1}{3}, \frac{1}{3}, \frac{1}{3}, 0, \frac{-1}{2}, \frac{-1}{2}\right) \begin{bmatrix} 1 & 1 & 0 & 0 \\ 1 & 1 & 0 & 0 \\ 1 & 1 & 0 & 0 \\ 1 & 0 & 1 & 0 \\ 1 & 0 & 0 & 1 \\ 1 & 0 & 0 & 1 \end{bmatrix} \begin{bmatrix} \mu \\ \alpha_1 \\ \alpha_2 \\ \alpha_3 \end{bmatrix},$$

so $P' = (1/3, 1/3, 1/3, 0, -1/2, -1/2)$. We need $C(X) \cap C(MP)^{\perp}$. Note that vectors in $C(X)$ have the form $(a, a, a, b, c, c)'$ for any a, b, c, so $P = MP$. Vectors in $C(MP)^{\perp}$ are $v = (v_{11}, v_{12}, v_{13}, v_{21}, v_{31}, v_{32})'$ with $P'v = \bar{v}_{1\cdot} - \bar{v}_{3\cdot} = 0$. Vectors in $C(X) \cap C(MP)^{\perp}$ have the first three elements identical, the last two elements identical, and the average of the first three equal to the average of the last two, i.e., they have the form $(a, a, a, b, a, a)'$. A spanning set for this space is given by the columns of

$$X_0 = \begin{bmatrix} 1 & 1 & 0 \\ 1 & 1 & 0 \\ 1 & 1 & 0 \\ 1 & 0 & 1 \\ 1 & 1 & 0 \\ 1 & 1 & 0 \end{bmatrix}.$$

As seen earlier, another choice for P is $(1, 0, 0, 0, -1, 0)'$. Using M from Exercise 1.5.8b, this choice for P also leads to $MP = (1/3, 1/3, 1/3, 0, -1/2, -1/2)'$. To compute $(I - M_{MP})X$, observe that

$$M_{MP} = \frac{1}{(1/3) + (1/2)} \begin{bmatrix} 1/9 & 1/9 & 1/9 & 0 & -1/6 & -1/6 \\ 1/9 & 1/9 & 1/9 & 0 & -1/6 & -1/6 \\ 1/9 & 1/9 & 1/9 & 0 & -1/6 & -1/6 \\ 0 & 0 & 0 & 0 & 0 & 0 \\ -1/6 & -1/6 & -1/6 & 0 & 1/4 & 1/4 \\ -1/6 & -1/6 & -1/6 & 0 & 1/4 & 1/4 \end{bmatrix}$$

$$= \frac{1}{5}\begin{bmatrix} 2/3 & 2/3 & 2/3 & 0 & -1 & -1 \\ 2/3 & 2/3 & 2/3 & 0 & -1 & -1 \\ 2/3 & 2/3 & 2/3 & 0 & -1 & -1 \\ 0 & 0 & 0 & 0 & 0 & 0 \\ -1 & -1 & -1 & 0 & 3/2 & 3/2 \\ -1 & -1 & -1 & 0 & 3/2 & 3/2 \end{bmatrix}.$$

Then

$$(I - M_{MP})X = X - M_{MP}X$$

$$= X - \frac{1}{5}\begin{bmatrix} 0 & 2 & 0 & -2 \\ 0 & 2 & 0 & -2 \\ 0 & 2 & 0 & -2 \\ 0 & 0 & 0 & 0 \\ 0 & -3 & 0 & 3 \\ 0 & -3 & 0 & 3 \end{bmatrix} = \begin{bmatrix} 1 & 3/5 & 0 & 2/5 \\ 1 & 3/5 & 0 & 2/5 \\ 1 & 3/5 & 0 & 2/5 \\ 1 & 0 & 1 & 0 \\ 1 & 3/5 & 0 & 2/5 \\ 1 & 3/5 & 0 & 2/5 \end{bmatrix},$$

which has the same column space as X_0 given earlier.

We have reduced the problem of finding X_0 to that of finding $C(X) \cap C(MP)^\perp$, which is just the orthogonal complement of $C(MP)$ with respect to $C(X)$. By Corollary B.48, $C(X) \cap C(MP)^\perp = C(M - M_{MP})$, so $M - M_{MP}$ is another valid choice for X_0. For Example 3.3.1 with $H_0 : \alpha_1 - \alpha_3 = 0$, M was given in Exercise 1.5.8b and M_{MP} was given earlier, so

$$M - M_{MP} = \begin{bmatrix} 1/5 & 1/5 & 1/5 & 0 & 1/5 & 1/5 \\ 1/5 & 1/5 & 1/5 & 0 & 1/5 & 1/5 \\ 1/5 & 1/5 & 1/5 & 0 & 1/5 & 1/5 \\ 0 & 0 & 0 & 1 & 0 & 0 \\ 1/5 & 1/5 & 1/5 & 0 & 1/5 & 1/5 \\ 1/5 & 1/5 & 1/5 & 0 & 1/5 & 1/5 \end{bmatrix}.$$

This matrix has the same column space as the other choices of X_0 that have been given.

For $\Lambda'\beta$ estimable, we now rewrite the test statistic in (5) in terms of Λ and $\hat\beta$. First, we wish to show that $r(\Lambda) = r(M_{MP})$. It suffices to show that $r(\Lambda) = r(MP)$. Writing $\Lambda = X'P$, we see that for any vector b, $X'Pb = 0$ if and only if $Pb \perp C(X)$, which occurs if and only if $MPb = 0$. It follows that $C(P'X)^\perp = C(P'M)^\perp$ so that $C(P'X) = C(P'M)$, $r(P'X) = r(P'M)$, and $r(\Lambda) = r(X'P) = r(MP)$.

Now rewrite the quadratic form $Y'M_{MP}Y$. Recall that since $X\hat\beta = MY$, we have $\Lambda'\hat\beta = P'X\hat\beta = P'MY$. Substitution gives

$$\begin{aligned} Y'M_{MP}Y &= Y'MP(P'MP)^- P'MY \\ &= \hat\beta'\Lambda(P'X(X'X)^- X'P)^- \Lambda'\hat\beta \\ &= \hat\beta'\Lambda[\Lambda'(X'X)^-\Lambda]^- \Lambda'\hat\beta. \end{aligned}$$

The test statistic in (5) becomes

$$\frac{\hat{\beta}'\Lambda[\Lambda'(X'X)^{-}\Lambda]^{-}\Lambda'\hat{\beta}/r(\Lambda)}{MSE}.$$

A similar argument shows that the noncentrality parameter in (5) can be written as $\beta'\Lambda[\Lambda'(X'X)^{-}\Lambda]^{-}\Lambda'\beta/2\sigma^2$. The test statistic consists of three main parts: *MSE*, $\Lambda'\hat{\beta}$, and the generalized inverse of $\Lambda'(X'X)^{-}\Lambda$. Note that $\sigma^2\Lambda'(X'X)^{-}\Lambda = \mathrm{Cov}(\Lambda'\hat{\beta})$. *These facts give an alternative method of deriving tests. One can simply find the estimate $\Lambda'\hat{\beta}$, the covariance matrix of the estimate, and the MSE.*

For a single degree of freedom hypothesis $H_0 : \lambda'\beta = 0$, the numerator takes the especially nice form

$$\hat{\beta}'\lambda[\lambda'(X'X)^{-}\lambda]^{-1}\lambda'\hat{\beta} = (\lambda'\hat{\beta})^2/[\lambda'(X'X)^{-}\lambda];$$

so the F test becomes: reject $H_0 : \lambda'\beta = 0$ if

$$\frac{(\lambda'\hat{\beta})^2}{MSE[\lambda'(X'X)^{-}\lambda]} > F(1-\alpha, 1, dfE),$$

which is just the square of the t test that could be derived from the sampling distributions of the least squares estimate and the *MSE*, cf. Exercise 2.1. The *sum of squares for testing* $H_0 : \lambda'\beta = 0$ is defined as

$$SS(\lambda'\beta) \equiv \frac{(\lambda'\hat{\beta})^2}{\lambda'(X'X)^{-}\lambda} = \frac{(\rho'MY)^2}{\rho'M\rho} = Y'[M\rho(\rho'M\rho)^{-1}\rho'M]Y \equiv Y'M_{M\rho}Y,$$

so that the test statistic is $SS(\lambda'\beta)/MSE$.

Definition 3.3.5 The condition $\mathrm{E}(Y) \perp C(MP)$ is called the *constraint* on the model caused (imposed) by $\Lambda'\beta = 0$, where $\Lambda' = P'X$. As a shorthand, we will call $C(MP)$ the constraint caused by $\Lambda'\beta = 0$. If $C(MP) \subset \mathcal{M} \subset C(X)$, we say that $C(MP)$ is the constraint on $\mathcal{M}$ caused by $\Lambda'\beta = 0$. If $\Lambda'\beta = 0$ puts a constraint on $\mathcal{M}$, we say that $\Lambda'\beta = 0$ is an hypothesis in $\mathcal{M}$.

Exercise 3.5
(a) Show that $\beta'X'M_{MP}X\beta = 0$ if and only if $\Lambda'\beta = 0$.
(b) Show that a necessary and sufficient condition for $\rho_1'X\beta = 0$ and $\rho_2'X\beta = 0$ to determine orthogonal constraints on the model is that $\rho_1'M\rho_2 = 0$.

Exercise 3.6 In testing a reduced model $Y = X_0\gamma + e$ against a full model $Y = X\beta + e$, what linear parametric function of the parameters is being tested?

Theoretical Complements

If, rather than testing the constraint $\Lambda'\beta = 0$, our desire is to estimate β subject to the constraint, simply estimate γ in model (3) and use $\hat{\beta} = U\hat{\gamma}$. In this *constrained estimation*, the estimates automatically satisfy $\Lambda'\hat{\beta} = 0$. Moreover, estimable functions $\Gamma'\beta = Q'X\beta$ are equivalent to $\Gamma'U\gamma = Q'XU\gamma$, and optimal estimates of γ are transformed into optimal estimates of β.

We now examine the implications of testing $\Lambda'\beta = 0$ when $\Lambda'\beta$ is not estimable. Recall that we began this section by finding the reduced model associated with such a constraint, so we already have a general method for performing such tests.

The first key result is that in defining a linear constraint there is no reason to use anything but estimable functions, because only estimable functions induce a real constraint on $C(X)$. Theorem 3.3.6 identifies the *estimable part* of $\Lambda'\beta$, say $\Lambda_0'\beta$, and implies that $\Lambda_0'\beta = 0$ gives the same reduced model as $\Lambda'\beta = 0$. Λ_0 is a matrix chosen so that $C(\Lambda) \cap C(X') = C(\Lambda_0)$. With such a choice, $\Lambda'\beta = 0$ implies that $\Lambda_0'\beta = 0$ but $\Lambda_0'\beta$ is estimable because $C(\Lambda_0) \subset C(X')$, so $\Lambda_0' = P_0'X$ for some P_0.

Theorem 3.3.6 *If $C(\Lambda) \cap C(X') = C(\Lambda_0)$ and $C(U_0) = C(\Lambda_0)^\perp$, then $C(XU) = C(XU_0)$. Thus $\Lambda'\beta = 0$ and $\Lambda_0'\beta = 0$ induce the same reduced model.*

Proof $C(\Lambda_0) \subset C(\Lambda)$, so $C(U) \equiv C(\Lambda)^\perp \subset C(\Lambda_0)^\perp = C(U_0)$ and $C(XU) \subset C(XU_0)$.

To complete the proof, we show that there cannot be any vectors in $C(XU_0)$ that are not in $C(XU)$. In particular, we show that there are no nontrivial vectors in $C(XU_0)$ that are orthogonal to $C(XU)$, i.e., if $v \in C(XU)^\perp_{C(XU_0)}$ then $v = 0$. If $v \in C(XU)^\perp_{C(XU_0)}$, then $v'XU = 0$, so $X'v \perp C(U)$ and $X'v \in C(\Lambda)$. But also note that $X'v \in C(X')$, so $X'v \in C(\Lambda) \cap C(X') = C(\Lambda_0)$. This implies that $X'v \perp C(U_0)$, so $v \perp C(XU_0)$. We have shown that the vector v which, by assumption, is in $C(XU_0)$, is also orthogonal to $C(XU_0)$. The only such vector is the 0 vector. □

Nontrivial estimable constraints always induce a real constraint on the column space.

Proposition 3.3.7 *If $\Lambda'\beta$ is estimable and $\Lambda \neq 0$, then $\Lambda'\beta = 0$ implies that $C(XU) \neq C(X)$.*

Proof With $\Lambda' = P'X$, the definition of U gives $0 = \Lambda'U = P'XU = P'MXU$, so $C(XU) \perp C(MP)$. Both are subspaces of $C(X)$; therefore if $C(MP) \neq \{0\}$, we have $C(X) \neq C(XU)$. To see that $C(MP) \neq \{0\}$, note $P'MX = \Lambda' \neq 0$, so $C(MP)$ is not orthogonal to $C(X)$ and $C(MP) \neq \{0\}$. □

Note that Proposition 3.3.7 also implies that whenever the estimable part Λ_0 is different from 0, there is always a real constraint on the column space.

Corollary 3.3.8 establishes that $\Lambda'\beta$ has no estimable part if and only if the constraint does not affect the model. If the constraint does not affect the model, it merely defines a reparameterization, in other words, it merely specifies arbitrary side conditions. The corollary follows from the observation made about Λ_0 after Proposition 3.3.7 and taking $\Lambda_0 = 0$ in Theorem 3.3.6.

Corollary 3.3.8 $C(\Lambda) \cap C(X') = \{0\}$ *if and only if* $C(XU) = C(X)$.

In particular, if $\Lambda'\beta$ is not estimable, we can obtain the numerator sum of squares for testing $\Lambda'\beta = 0$ either by finding $X_0 = XU$ directly and using it to get $M - M_0$, or by finding Λ_0, writing $\Lambda_0' = P_0'X$, and using M_{MP_0}. But as noted earlier, there is no reason to have $\Lambda'\beta$ not estimable.

One final point worth noting. Each component of $\Lambda'\beta$ may be nonestimable but the joint constraint may contain an estimable component. For example, in one-way ANOVA, $y_{ij} = \mu + \alpha_i + e_{ij}$, by itself the constraint $\alpha_1 = 0$ is nonestimable. Similarly, $\alpha_2 = 0$ is a nonestimable constraint. But together they imply that the estimable contrast $\alpha_1 - \alpha_2$ equals 0.

3.3.1 A Generalized Test Procedure

We now consider hypotheses of the form $\Lambda'\beta = d$ where $d \in C(\Lambda')$ so that the equation $\Lambda'\beta = d$ is solvable. (If the rows of Λ' are linearly independent, the equation is always solvable.) Let b be such a solution. Note that

$$\Lambda'\beta = \Lambda'b = d$$

if and only if

$$\Lambda'(\beta - b) = 0$$

if and only if

$$(\beta - b) \perp C(\Lambda).$$

Again picking a matrix U such that

$$C(U) = C(\Lambda)^{\perp},$$

$\Lambda'(\beta - b) = 0$ if and only if

$$(\beta - b) \in C(U),$$

which occurs if and only if for some vector γ

$$(\beta - b) = U\gamma.$$

Multiplying both sides by X gives

$$X\beta - Xb = XU\gamma$$

or

$$X\beta = XU\gamma + Xb.$$

We can now substitute this into the linear model to get the reduced model

$$Y = XU\gamma + Xb + e,$$

or, letting $X_0 \equiv XU$,

$$Y = X_0\gamma + Xb + e. \tag{6}$$

Recall that b is a vector we can find, so Xb is a known (offset) vector. Exercise 3.9.8 establishes that the particular choice of b is irrelevant to the test. To estimate β subject to the linear constraint $\Lambda'\beta = d$, take $\hat{\beta} = U\hat{\gamma} + b$.

The test for reduced models such as (6) was developed in Section 2. For nonestimable linear hypotheses, use that theory directly. If $\Lambda' = P'X$, then $C(X_0)^{\perp}_{C(X)} = C(MP)$ and the test statistic is easily seen to be

$$\frac{(Y - Xb)'M_{MP}(Y - Xb)/r(M_{MP})}{(Y - Xb)'(I - M)(Y - Xb)/r(I - M)}.$$

Note that $\Lambda'\beta = d$ imposes the constraint $\mathrm{E}(Y - Xb) \perp C(MP)$, so once again we could refer to $C(MP)$ as the constraint imposed by the hypothesis.

We did not specify the solution b to $\Lambda'\beta = d$ that should be used. For $\Lambda'\beta$ estimable, it is easy (unlike the general case) to see that the test does not depend on the choice of b. As mentioned in the previous section, $(Y - Xb)'(I - M)(Y - Xb) = Y'(I - M)Y$, so the denominator of the test is just the MSE and does not depend on b. The numerator term $(Y - Xb)'M_{MP}(Y - Xb)$ equals $(\Lambda'\hat{\beta} - d)'[\Lambda'(X'X)^-\Lambda]^-(\Lambda'\hat{\beta} - d)$. The test statistic can be written as

$$\frac{(\Lambda'\hat{\beta} - d)'[\Lambda'(X'X)^-\Lambda]^-(\Lambda'\hat{\beta} - d)/r(\Lambda)}{MSE}. \tag{7}$$

For $\Lambda'\beta$ estimable, the linear model $Y = X_0\gamma + Xb + e$ implies that $\Lambda'\beta = d$, but for nonestimable linear constraints, there are infinitely many constraints that result in the same reduced model. (If you think of nonestimable linear constraints as including arbitrary side conditions, that is not surprising.) In particular, if $\Lambda'\beta = d$, the same reduced model results if we take $\Lambda'\beta = d_0$ where $d_0 = d + \Lambda'v$ and $v \perp C(X')$. Note that, in this construction, if $\Lambda'\beta$ is estimable, $d_0 = d$ for any v.

We now present an application of the test statistic (7). The results are given without justification, but they should seem similar to results from a statistical methods course.

Example 3.3.9 Consider the balanced two-way ANOVA without interaction model

$$y_{ijk} = \mu + \alpha_i + \eta_j + e_{ijk},$$

$i = 1, \ldots, a$, $j = 1, \ldots, b$, $k = 1, \ldots, N$. (The analysis for this model is presented in Section 7.1.) We examine the test of the null hypothesis

$$H_0 : \sum_{i=1}^{a} \lambda_i \alpha_i = 4 \text{ and } \sum_{j=1}^{b} \gamma_j \eta_j = 7,$$

where $\sum_{i=1}^{a} \lambda_i = 0 = \sum_{j=1}^{b} \gamma_j$. The hypothesis is simultaneously specifying the values of a contrast in the α_is and a contrast in the η_js.

In terms of the model $Y = X\beta + e$, we have

$$\beta = [\mu, \alpha_1, \ldots, \alpha_a, \eta_1, \ldots, \eta_b]'$$

$$\Lambda' = \begin{bmatrix} 0 & \lambda_1 & \cdots & \lambda_a & 0 & \cdots & 0 \\ 0 & 0 & \cdots & 0 & \gamma_1 & \cdots & \gamma_b \end{bmatrix}$$

$$d = \begin{bmatrix} 4 \\ 7 \end{bmatrix}$$

$$\Lambda'\hat{\beta} = \begin{bmatrix} \sum_{i=1}^{a} \lambda_i \bar{y}_{i\cdot\cdot} \\ \sum_{j=1}^{b} \gamma_j \bar{y}_{\cdot j\cdot} \end{bmatrix}$$

$$\mathrm{Cov}(\Lambda'\hat{\beta})/\sigma^2 = \Lambda'(X'X)^{-}\Lambda = \begin{bmatrix} \sum_{i=1}^{a} \lambda_i^2/bN & 0 \\ 0 & \sum_{j=1}^{b} \gamma_j^2/aN \end{bmatrix}.$$

The diagonal elements of the covariance matrix are just the variances of the estimated contrasts. The off-diagonal elements are zero because this is a balanced two-way ANOVA, hence the estimates of the α contrast and the η contrast are independent. We will see in Chapter 7 that these contrasts define orthogonal constraints in the sense of Definition 3.3.5, so they are often referred to as being orthogonal parameters.

There are two linearly independent contrasts being tested, so $r(\Lambda) = 2$. The test statistic is

$$\frac{1}{2MSE} \begin{bmatrix} \sum_{i=1}^{a} \lambda_i \bar{y}_{i\cdot\cdot} - 4 & \sum_{j=1}^{b} \gamma_j \bar{y}_{\cdot j\cdot} - 7 \end{bmatrix}$$
$$\times \begin{bmatrix} \frac{bN}{\sum_{i=1}^{a} \lambda_i^2} & 0 \\ 0 & \frac{aN}{\sum_{j=1}^{b} \gamma_j^2} \end{bmatrix} \begin{bmatrix} \sum_{i=1}^{a} \lambda_i \bar{y}_{i\cdot\cdot} - 4 \\ \sum_{j=1}^{b} \gamma_j \bar{y}_{\cdot j\cdot} - 7 \end{bmatrix}$$

or

$$\frac{1}{2MSE}\left[\frac{\left(\sum_{i=1}^{a}\lambda_i\bar{y}_{i\cdot\cdot}-4\right)^2}{\sum_{i=1}^{a}\lambda_i^2/bN}+\frac{\left(\sum_{j=1}^{b}\gamma_j\bar{y}_{\cdot j\cdot}-7\right)^2}{\sum_{j=1}^{b}\gamma_j^2/aN}\right].$$

Note that the term $\left(\sum_{i=1}^{a}\lambda_i\bar{y}_{i\cdot\cdot}-4\right)^2/\left(\sum_{i=1}^{a}\lambda_i^2/bN\right)$ is, except for subtracting the 4, the sum of squares for testing $\sum_{i=1}^{a}\lambda_i\alpha_i=0$. We are subtracting the 4 because we are testing $\sum_{i=1}^{a}\lambda_i\alpha_i=4$. Similarly, we have a term that is very similar to the sum of squares for testing $\sum_{j=1}^{b}\gamma_j\eta_j=0$. The test statistic takes the average of these sums of squares and divides by the *MSE*. The test is then defined by reference to an $F(2, dfE, 0)$ distribution.

3.3.2 Testing an Unusual Class of Hypotheses

Occasionally, a valid linear hypothesis $\Lambda'\beta=d$ is considered where d is not completely known but involves other parameters. (This is the linear structure involved in creating a logistic regression model from a log-linear model.) For $\Lambda'\beta=d$ to give a valid linear hypothesis, $\Lambda'\beta=d$ must put a restriction on $C(X)$, so that when the hypothesis is true, E(Y) lies in some subspace of $C(X)$.

Let X_1 be such that $C(X_1)\subset C(X)$ and consider an hypothesis

$$P'X\beta=P'X_1\delta$$

for some parameter vector δ. We seek an appropriate reduced model for such an hypothesis.

Note that the hypothesis occurs if and only if

$$P'M(X\beta-X_1\delta)=0,$$

which occurs if and only if

$$(X\beta-X_1\delta)\perp C(MP),$$

which occurs if and only if

$$(X\beta-X_1\delta)\in C(MP)^{\perp}_{C(X)}.$$

As discussed earlier in this section, we choose X_0 so that $C(MP)^{\perp}_{C(X)}=C(X_0)$. The choice of X_0 does not depend on X_1. Using X_0, the hypothesis occurs if and only if for some γ

$$(X\beta-X_1\delta)=X_0\gamma.$$

Rewriting these terms, we see that

$$X\beta=X_0\gamma+X_1\delta$$

which is the mean structure for the reduced model. In other words, assuming the null hypothesis is equivalent to assuming a reduced model

$$Y = X_0\gamma + X_1\delta + e.$$

To illustrate, consider a linear model for pairs of observations (y_{1j}, y_{2j}), $j = 1, \ldots, N$. Write $Y = (y_{11}, \ldots, y_{1N}, y_{21}, \ldots, y_{2N})'$. Initially, we will impose no structure on the means so that $\mathrm{E}(y_{ij}) = \mu_{ij}$. We are going to consider an hypothesis for the differences between the pairs,

$$\mu_{1j} - \mu_{2j} = z_j'\delta$$

for some known predictor vector z_j. Of course we could just fit a linear model to the differences $y_{1j} - y_{2j}$, but we want to think about comparing such a model to models that are not based on the differences.

The conditions just specified correspond to a linear model $Y = X\beta + e$ in which $X = I_{2N}$ and $\beta = (\mu_{11}, \ldots, \mu_{2N})'$. Write

$$P = \begin{bmatrix} I_N \\ -I_N \end{bmatrix} \quad \text{and} \quad X_1 = \begin{bmatrix} Z \\ 0 \end{bmatrix}$$

where $Z' = [z_1, \ldots, z_N]$. Then the hypothesis for the differences can be specified as

$$P'X\beta = P'X_1\delta = Z\delta.$$

Finally, it is not difficult to see that a valid choice of X_0 is

$$X_0 = \begin{bmatrix} I_N \\ I_N \end{bmatrix}.$$

It follows that, under the reduced model

$$X\beta \equiv I\beta = \begin{bmatrix} I & Z \\ I & 0 \end{bmatrix} \begin{bmatrix} \gamma \\ \delta \end{bmatrix}$$

or that the reduced model is

$$Y = \begin{bmatrix} I & Z \\ I & 0 \end{bmatrix} \begin{bmatrix} \gamma \\ \delta \end{bmatrix} + e.$$

This relationship is of particular importance in the analysis of frequency data. The model $y_{ij} = \mu_{ij} + e_{ij}$ is analogous to a saturated log-linear model. The hypothesis $\mu_{1j} - \mu_{2j} = \alpha_0 + \alpha_1 t_j \equiv z_j'\delta$ is analogous to the hypothesis that a simple linear logit model in a predictor t holds. We have found the vector space such that restricting the (saturated) log-linear model to that space gives the logit model, see Christensen (1997) for more details.

3.4 Discussion

The reason that we considered testing models first and then discussed testing parametric functions by showing them to be changes in models is because, in *general,* only model testing is ever performed. This is not to say that parametric functions are not tested as such, but that parametric functions are only tested in special cases. In particular, parametric functions can easily be tested in balanced ANOVA problems and one-way ANOVAs. Multifactor ANOVA designs with unequal numbers of observations in the treatment cells, as illustrated in Example 3.2.2, are best analyzed by considering alternative models. In fact, Christensen (2015) was written (largely) to illustrate that balanced ANOVA methods can be applied to unbalanced ANOVA by recasting the procedures as identifying relevant reduced models.

Even in regression models, where all the parameters are estimable, it is often more enlightening to think in terms of model selection. Of course, in regression there is a special relationship between the parameters and the model matrix. For the model $y_i = \beta_0 + \beta_1 x_{i1} + \beta_2 x_{i2} + \cdots + \beta_{p-1} x_{i\,p-1} + e$, the model matrix can be written as $X = [J, X_1, \ldots, X_{p-1}]$, where $X_j = [x_{1j}, x_{2j}, \ldots, x_{nj}]'$. The test of $H_0 : \beta_j = 0$ is obtained by just leaving X_j out of the model matrix.

Another advantage of the method of testing models is that it is often easy in simple but nontrivial cases to see immediately what new model is generated by a null hypothesis. This was illustrated in Examples 3.2.0 and 3.2.3.

Example 3.4.1 *One-Way ANOVA.*
Consider the model $y_{ij} = \mu + \alpha_i + e_{ij}$, $i = 1, 2, 3$, $j = 1, \ldots, N_i$, $N_1 = N_3 = 3$, $N_2 = 2$. In matrix terms this is

$$Y = \begin{bmatrix} 1&1&0&0 \\ 1&1&0&0 \\ 1&1&0&0 \\ 1&0&1&0 \\ 1&0&1&0 \\ 1&0&0&1 \\ 1&0&0&1 \\ 1&0&0&1 \end{bmatrix} \begin{bmatrix} \mu \\ \alpha_1 \\ \alpha_2 \\ \alpha_3 \end{bmatrix} + e.$$

Let the null hypothesis be $\alpha_1 = \mu + 2\alpha_2$. Writing $X = [J, X_1, X_2, X_3]$ and

$$\mathrm{E}(Y) = X\beta = \mu J + \alpha_1 X_1 + \alpha_2 X_2 + \alpha_3 X_3,$$

the reduced model is easily found by substituting $\mu + 2\alpha_2$ for α_1 which leads to

$$\mathrm{E}(Y) = \mu(J + X_1) + \alpha_2(2X_1 + X_2) + \alpha_3 X_3.$$

This gives the reduced model

$$Y = \begin{bmatrix} 2\,2\,0 \\ 2\,2\,0 \\ 2\,2\,0 \\ 1\,1\,0 \\ 1\,1\,0 \\ 1\,0\,1 \\ 1\,0\,1 \\ 1\,0\,1 \end{bmatrix} \begin{bmatrix} \gamma_0 \\ \gamma_1 \\ \gamma_2 \end{bmatrix} + e.$$

For Examples 3.2.0, 3.2.3, and 3.4.1, it would be considerable work to go through the procedure developed in Section 3 to test the hypotheses. In fairness, it should be added that for these special cases, there is no need to go through the general procedures of Section 3 to get the tests (assuming that you get the necessary computer output for the regression problem).

3.5 Testing Single Degrees of Freedom in a Given Subspace

Consider a two-way ANOVA model $y_{ijk} = \mu + \alpha_i + \eta_j + e_{ijk}$. Suppose we want to look at contrasts in the α_is and η_js. For analyzing a balanced two-way ANOVA it would be very convenient if estimates and tests for contrasts in the α_is, say, could be based on the projection operator associated with dropping the α_is out of the one-way ANOVA model $y_{ijk} = \mu + \alpha_i + e_{ijk}$ rather than the projection operator for the two-way ANOVA model. One convenience is that the projection operator for the one-way model turns out to be much simpler than the projection operator for the two-way model. A second convenience is that orthogonality of the projection operators for dropping the α_is and η_js in the balanced two-way model leads to independence between estimates of contrasts in the α_is and η_js. We would also like to establish that orthogonal contrasts (contrasts that define orthogonal constraints) in the α_is, say, depend only on the projection operator for dropping the α_is in the one-way model.

With these ultimate goals in mind, we now examine, in general, estimates, tests, and orthogonality relationships between single degree of freedom hypotheses that put a constraint on a particular subspace.

Consider a perpendicular projection operator M_* used in the numerator of a test statistic. In the situation of testing a model $Y = X\beta + e$ against a reduced model $Y = X_0\gamma + e$ with $C(X_0) \subset C(X)$, if M and M_0 are the perpendicular projection operators onto $C(X)$ and $C(X_0)$, respectively, then $M_* = M - M_0$. For testing the estimable parametric hypothesis $\Lambda'\beta = 0$, if $\Lambda' = P'X$, then $M_* = M_{MP}$.

We want to examine the problem of testing a single degree of freedom hypothesis in $C(M_*)$. Let $\lambda' = \rho'X$. Then, by Definition 3.3.2, $\lambda'\beta = 0$ puts a constraint on $C(M_*)$ if and only if $M\rho \in C(M_*)$. If $M\rho \in C(M_*)$, then $M\rho = M_*M\rho = M_*\rho$ because $MM_* = M_*$. It follows that the estimate of $\lambda'\beta$ is $\rho'M_*Y$ because $\rho'M_*Y = \rho'MY$. From Section 3, the test statistic for $H_0 : \lambda'\beta = 0$ is

$$\frac{Y'M_*\rho(\rho'M_*\rho)^{-1}\rho'M_*Y}{MSE} = \frac{(\rho'M_*Y)^2/\rho'M_*\rho}{MSE},$$

where $MSE = Y'(I-M)Y/r(I-M)$ and $r(M_*\rho(\rho'M_*\rho)^{-1}\rho'M_*) = 1$.

Let $\lambda_1' = \rho_1'X$ and $\lambda_2' = \rho_2'X$, and let the hypotheses $\lambda_1'\beta = 0$ and $\lambda_2'\beta = 0$ define orthogonal constraints on the model. The constraints are, respectively, $\mathrm{E}(Y) \perp M\rho_1$ and $\mathrm{E}(Y) \perp M\rho_2$. These constraints are said to be orthogonal if the vectors $M\rho_1$ and $M\rho_2$ are orthogonal. This occurs if and only if $\rho_1'M\rho_2 = 0$. If $\lambda_1'\beta = 0$ and $\lambda_2'\beta = 0$ both put constraints on $C(M_*)$, then orthogonality is equivalent to $0 = \rho_1'M\rho_2 = \rho_1'M_*\rho_2$.

We have now shown that for any estimable functions that put constraints on $C(M_*)$, estimates, tests, and finding orthogonal constraints in $C(M_*)$ require only the projection operator M_* and the MSE.

Exercise 3.7a Show that $\rho'MY = \rho'[M\rho(\rho'M\rho)^-\rho'M]Y$ so that to estimate $\rho'X\beta$, one only needs the perpendicular projection of Y onto $C(M\rho)$.

3.6 Breaking a Sum of Squares into Independent Components

We now present a general theory that includes, as special cases, the breaking down of the treatment sum of squares in a one-way ANOVA into sums of squares for orthogonal contrasts and the breaking of the sum of squares for the model into independent sums of squares as in an ANOVA table. This is an important device in statistical analyses.

Frequently, a reduced model matrix X_0 is a submatrix of X. This is true for the initial hypotheses considered in both cases of Example 3.2.0 and for Example 3.2.2. If we can write $X = [X_0, X_1]$, it is convenient to write $SSR(X_1|X_0) \equiv Y'(M - M_0)Y$. $SSR(X_1|X_0)$ is called the sum of squares for regressing X_1 after X_0. We will also write $SSR(X) \equiv Y'MY$, the sum of squares for regressing on X. Similarly, $SSR(X_0) \equiv Y'M_0Y$. The $SSR(\cdot)$ notation is one way of identifying sums of squares for tests. Other notations exist, and one alternative will soon be introduced. Note that $SSR(X) = SSR(X_0) + SSR(X_1|X_0)$, which constitutes a breakdown of $SSR(X)$ into two parts. If $e \sim N(0, \sigma^2 I)$, these two parts are independent.

We begin with a general theory and conclude with a discussion of breaking down the sums of squares in a two-way ANOVA model $y_{ijk} = \mu + \alpha_i + \eta_j + e_{ijk}$. The projection operators used in the numerator sums of squares for dropping the α_is and η_js are orthogonal if and only if the numerator sum of squares for dropping the η_js out of the two-way model is the same as the numerator sum of squares for dropping the η_js out of the one-way ANOVA model $y_{ijk} = \mu + \eta_j + e_{ijk}$.

3.6.1 General Theory

We now present a general theory that is based on finding an orthonormal basis for a subspace of the estimation space. (This subspace could be the entire estimation space.) We discuss two methods of doing this. The first is a direct method involving identifying the subspace and choosing an orthonormal basis. The second method determines an orthonormal basis indirectly by examining single degree of freedom hypotheses and the constraints imposed by those hypotheses.

Our general model is $Y = X\beta + e$ with M the perpendicular projection operator onto $C(X)$. Let M_* be any perpendicular projection operator with $C(M_*) \subset C(X)$. Then M_* defines a test statistic

$$\frac{Y'M_*Y/r(M_*)}{Y'(I-M)Y/r(I-M)}$$

for testing the reduced model, say, $Y = (M - M_*)\gamma + e$. If $r(M_*) = r$, then we will show that we can break the sum of squares based on M_* (i.e., $Y'M_*Y$) into as many as r independent sums of squares whose sum will equal $Y'M_*Y$. By using M_* in the numerator of the test, we are testing whether the subspace $C(M_*)$ is adding anything to the predictive (estimative) ability of the model. What we have done is break $C(X)$ into two orthogonal parts, $C(M_*)$ and $C(M - M_*)$. In this case, $C(M - M_*)$ is the estimation space under H_0 and we can call $C(M_*)$ the *test space*. $C(M_*)$ is a space that will contain only error if H_0 is true but which is part of the estimation space under the full model. Note that the error space under H_0 is $C(I - (M - M_*))$, but $I - (M - M_*) = (I - M) + M_*$ so that $C(I - M)$ is part of the error space under both models.

We now break $C(M_*)$ into r orthogonal subspaces. Take an orthonormal basis for $C(M_*)$, say $R_1, R_2, \ldots, R_r$. Note that, using Gram–Schmidt, R_1 can be any normalized vector in $C(M_*)$. It is the statistician's choice. R_2 can then be any normalized vector in $C(M_*)$ orthogonal to R_1, etc. Let $R = [R_1, R_2, \ldots, R_r]$, then as in Theorem B.35,

$$M_* = RR' = [R_1, \ldots, R_r] \begin{bmatrix} R_1' \\ \vdots \\ R_r' \end{bmatrix} = \sum_{i=1}^{r} R_i R_i'.$$

Let $M_i \equiv R_i R_i'$, then M_i is a perpendicular projection operator in its own right and $M_i M_j = 0$ for $i \neq j$ because of the orthogonality of the R_is.

The goal of this section is to break up the sum of squares into independent components. By Theorem 1.3.7, the sums of squares $Y'M_iY$ and $Y'M_jY$ are independent for any $i \neq j$ because $M_iM_j = 0$. Also, $Y'M_*Y = \sum_{i=1}^{r} Y'M_iY$ simply because $M_* = \sum_{i=1}^{r} M_i$. Moreover, since $r(M_i) = 1$,

$$\frac{Y'M_iY}{Y'(I-M)Y/r(I-M)} \sim F(1, r(I-M), \beta'X'M_iX\beta/2\sigma^2).$$

In a one-way ANOVA, $Y'M_*Y$ corresponds to the treatment sum of squares while the $Y'M_iY$s correspond to the sums of squares for a set of orthogonal contrasts, cf. Example 3.6.2 below.

We now consider the correspondence between the hypothesis tested using $Y'M_*Y$ and those tested using the $Y'M_iY$s. Because M_* and the M_is are nonnegative definite,

$$0 = \beta'X'M_*X\beta = \sum_{i=1}^{r} \beta'X'M_iX\beta$$

if and only if $\beta'X'M_iX\beta = 0$ for all i if and only if $R_i'X\beta = 0$ for all i. Thus, the null hypothesis that corresponds to the test based on M_* is true if and only if the null hypotheses $R_i'X\beta = 0$ corresponding to all the M_is are true. Equivalently, if the null hypothesis corresponding to M_* is not true, we have

$$0 < \beta'X'M_*X\beta = \sum_{i=1}^{r} \beta'X'M_iX\beta.$$

Again, since M_* and the M_is are nonnegative definite, this occurs if and only if at least one of the terms $\beta'X'M_iX\beta$ is greater than zero. Thus the null hypothesis corresponding to M_* is not true if and only if at least one of the hypotheses corresponding to the M_is is not true. Thinking in terms of a one-way ANOVA, these results correspond to stating that (1) the hypothesis of no treatment effects is true if and only if all the contrasts in a set of orthogonal contrasts are zero or, equivalently, (2) the hypothesis of no treatment effects is not true if and only if at least one contrast in a set of orthogonal contrasts is not zero.

We have broken $Y'M_*Y$ into r independent parts. It is easy to see how to break it into less than r parts. Suppose $r = 7$. We can break $Y'M_*Y$ into three parts by looking at projections onto only three subspaces. For example, $Y'M_*Y = Y'(M_1 + M_3 + M_6)Y + Y'(M_2 + M_7)Y + Y'(M_4 + M_5)Y$, where we have used three projection operators $M_1 + M_3 + M_6$, $M_2 + M_7$, and $M_4 + M_5$. Note that these three projection operators are orthogonal, so the sums of squares are independent. By properly choosing R, an ANOVA table can be developed using this idea.

Example 3.6.1 *One-Way ANOVA.*

In this example we examine breaking up the treatment sum of squares in a one-way ANOVA. Consider the model $y_{ij} = \mu + \alpha_i + e_{ij}$, $i = 1, 2, 3$, $j = 1, 2, 3$. In matrix terms this is

$$Y = \begin{bmatrix} 1 & 1 & 0 & 0 \\ 1 & 1 & 0 & 0 \\ 1 & 1 & 0 & 0 \\ 1 & 0 & 1 & 0 \\ 1 & 0 & 1 & 0 \\ 1 & 0 & 1 & 0 \\ 1 & 0 & 0 & 1 \\ 1 & 0 & 0 & 1 \\ 1 & 0 & 0 & 1 \end{bmatrix} \begin{bmatrix} \mu \\ \alpha_1 \\ \alpha_2 \\ \alpha_3 \end{bmatrix} + e. \tag{1}$$

Denote the model matrix $X = [J, X_1, X_2, X_3]$. To test $H_0 : \alpha_1 = \alpha_2 = \alpha_3$, the reduced model is clearly

$$Y = J\mu + e.$$

The projection operator for the test is $M_* = M - [1/n]JJ'$. The test space is $C(M_*) = C(M - [1/n]JJ')$, i.e., the test space is the set of all vectors in $C(X)$ that are orthogonal to a column of ones. The test space can be obtained by using Gram–Schmidt to remove the effect of J from the last three columns of the model matrix, that is, $C(M_*)$ is spanned by the columns of

$$\begin{bmatrix} 2 & -1 & -1 \\ 2 & -1 & -1 \\ 2 & -1 & -1 \\ -1 & 2 & -1 \\ -1 & 2 & -1 \\ -1 & 2 & -1 \\ -1 & -1 & 2 \\ -1 & -1 & 2 \\ -1 & -1 & 2 \end{bmatrix},$$

which is a rank 2 matrix. The statistician is free to choose R_1 within $C(M_*)$. R_1 could be a normalized version of

$$\begin{bmatrix} 2 \\ 2 \\ 2 \\ -1 \\ -1 \\ -1 \\ -1 \\ -1 \\ -1 \end{bmatrix} + \begin{bmatrix} -1 \\ -1 \\ -1 \\ 2 \\ 2 \\ 2 \\ -1 \\ -1 \\ -1 \end{bmatrix} = \begin{bmatrix} 1 \\ 1 \\ 1 \\ 1 \\ 1 \\ 1 \\ -2 \\ -2 \\ -2 \end{bmatrix},$$

which was chosen as $X_1 + X_2$ with the effect of J removed. R_2 must be the only normalized vector left in $C(M_*)$ that is orthogonal to R_1. R_2 is a normalized version of $[1, 1, 1, -1, -1, -1, 0, 0, 0]'$. The sum of squares for testing $H_0 : \alpha_1 = \alpha_2 = \alpha_3$ is $Y'R_1R_1'Y + Y'R_2R_2'Y$.

Using the specified form of R_1, $Y'M_1Y$ is the numerator sum of squares for testing

$$0 = R_1'X\beta \propto (0, 3, 3, -6) \begin{bmatrix} \mu \\ \alpha_1 \\ \alpha_2 \\ \alpha_3 \end{bmatrix} = 6\left(\frac{\alpha_1 + \alpha_2}{2} - \alpha_3\right).$$

Similarly,

$$R_2'X\beta \propto \alpha_1 - \alpha_2.$$

R_1 was chosen so that $Y'R_2R_2'Y$ would be the sum of squares for testing $H_0 : \alpha_1 = \alpha_2$.

The discussion thus far has concerned itself with directly choosing an orthonormal basis for $C(M_*)$. An equivalent approach to the problem of finding an orthogonal breakdown is in terms of single degree of freedom hypotheses $\lambda'\beta = 0$.

If we choose any r single degree of freedom hypotheses $\lambda_1'\beta = \cdots = \lambda_r'\beta = 0$ with $\rho_k'X = \lambda_k'$, $M\rho_k \in C(M_*)$, and $\rho_k'M\rho_h = 0$ for all $k \neq h$, then the vectors $M\rho_k/\sqrt{\rho_k'M\rho_k}$ form an orthonormal basis for $C(M_*)$. The projection operators are $M_k \equiv M\rho_k(\rho_k'M\rho_k)^{-1}\rho_k'M$. The sums of squares for these hypotheses,

$$Y'M_kY = Y'M\rho_k(\rho_k'M\rho_k)^{-1}\rho_k'MY = \frac{(\rho_k'MY)^2}{\rho_k'M\rho_k} = \frac{(\rho_k M_* Y)^2}{\rho_k'M_*\rho_k},$$

form an orthogonal breakdown of $Y'M_*Y$.

As shown in Section 3, the sum of squares for testing $\lambda_k'\beta = 0$ can be found from λ_k, $\hat{\beta}$, and $(X'X)^-$ as $(\lambda_k'\hat{\beta})/\lambda_k'(X'X)^-\lambda_k$. In many ANOVA problems, the condition $\rho_k'M\rho_h = 0$ can be checked by considering an appropriate function of λ_k and λ_h. It follows that, in many problems, an orthogonal breakdown can be obtained without actually finding the vectors $\rho_1, \ldots, \rho_r$.

Example 3.6.2 *One-Way ANOVA.*

Consider the model $y_{ij} = \mu + \alpha_i + e_{ij}, i = 1, \ldots, t, j = 1, \ldots, N_i$. Let $Y'M_*Y$ correspond to the sum of squares for treatments (i.e., the sum of squares for testing $\alpha_1 = \cdots = \alpha_t$). The hypotheses $\lambda_k'\beta = 0$ correspond to contrasts $c_{k1}\alpha_1 + \cdots + c_{kt}\alpha_t = 0$, where $c_{k1} + \cdots + c_{kt} = 0$. In Chapter 4, it will be shown that contrasts are estimable functions and that any contrast imposes a constraint on the space for testing equality of treatments. In other words, Chapter 4 shows that the $\lambda_k'\beta$s can be contrasts and that if they are contrasts, then $M\rho_k \in C(M_*)$. In Chapter 4 it will also be shown that the condition for orthogonality, $\rho_k'M\rho_h = 0$, reduces to the condition $c_{k1}c_{h1}/N_1 + \cdots + c_{kt}c_{ht}/N_t = 0$. If the contrasts are orthogonal, then the sums of squares for the contrasts add up to the sums of squares for treatments, and the sums of squares for the contrasts are independent.

3.6.2 Two-Way ANOVA

We discuss the technique of breaking up sums of squares as it applies to the two-way ANOVA model of Example 3.2.2. The results really apply to any two-way ANOVA with unequal numbers. The sum of squares for the full model is $Y'MY$ (by definition). We can break this up into three parts, one for fitting the η_js after having fit the α_is and μ, one for fitting the α_is after fitting μ, and one for fitting μ. In Example 3.2.2, the model is $y_{ijk} = \mu + \alpha_i + \eta_j + e_{ijk}$, $i = 1, 2, 3$, $j = 1, 2, 3$. The seven columns of X correspond to the elements of $\beta = [\mu, \alpha_1, \alpha_2, \alpha_3, \eta_1, \eta_2, \eta_3]'$ and were given earlier. Let $J = (1, \ldots, 1)'$ be the first column of X. Let X_0 be a matrix consisting of the first four columns of X, those corresponding to μ, α_1, α_2, and α_3. Take M and M_0 corresponding to X and X_0. It is easy to see that $(1/21)JJ'$ is the perpendicular projection operator onto $C(J)$.

Since $J \in C(X_0) \subset C(X)$, we can write, with $n = 21$,

$$Y'MY = Y'\frac{1}{n}JJ'Y + Y'\left(M_0 - \frac{1}{n}JJ'\right)Y + Y'(M - M_0)Y,$$

where $(1/n)JJ'$, $M_0 - (1/n)JJ'$, and $M - M_0$ are all perpendicular projection matrices. Since X_0 is obtained from X by dropping the columns corresponding to the η_js, $Y'(M - M_0)Y$ is the sum of squares used to test the full model against the reduced model with the η_js left out. Recalling our technique of looking at the differences in error sums of squares, we write $Y'(M - M_0)Y \equiv R(\eta|\alpha, \mu)$. $R(\eta|\alpha, \mu)$ is the reduction in (error) sum of squares due to fitting the η_js after fitting μ and the α_is, or, more simply, the sum of squares due to fitting the η_js after the α_is and μ. Similarly, if we wanted to test the model $y_{ijk} = \mu + e_{ijk}$ against $y_{ijk} = \mu + \alpha_i + e_{ijk}$, we would use $Y'(M_0 - [1/n]JJ')Y \equiv R(\alpha|\mu)$, i.e. the sum of squares for fitting the α_is after μ. Finally, to test $y_{ijk} = \mu + e_{ijk}$ against $y_{ijk} = \mu + \alpha_i + \eta_j + e_{ijk}$, we would use $Y'(M - [1/n]JJ')Y \equiv R(\alpha, \eta|\mu)$. Note that $R(\alpha, \eta|\mu) = R(\eta|\alpha, \mu) + R(\alpha|\mu)$.

The notations $SSR(\cdot)$ and $R(\cdot)$ are different notations for essentially the same idea. The $SSR(\cdot)$ notation emphasizes variables and is often used in regression problems. The $R(\cdot)$ notation emphasizes parameters and is frequently used in analysis of variance problems.

Alternatively, we could have chosen to develop the results in this discussion by comparing the model $y_{ijk} = \mu + \alpha_i + \eta_j + e_{ijk}$ to the model $y_{ijk} = \mu + \eta_j + e_{ijk}$. Then we would have taken X_0 as columns 1, 5, 6, and 7 of the X matrix instead of columns 1, 2, 3, and 4. This would have led to terms such as $R(\eta|\mu)$, $R(\alpha|\eta, \mu)$, and $R(\alpha, \eta|\mu)$. In general, these two analyses *will not* be the same. Typically, $R(\eta|\alpha, \mu) \neq R(\eta|\mu)$ and $R(\alpha|\mu) \neq R(\alpha|\eta, \mu)$. There do exist cases (e.g., balanced two-way ANOVA models) where the order of the analysis has no effect. Specifically, if the columns of the X matrix associated with α and those associated with η are orthogonal after somehow fitting μ, then $R(\eta|\alpha, \mu) = R(\eta|\mu)$ and $R(\alpha|\mu) = R(\alpha|\eta, \mu)$. In Section 7.4 we establish that these relationships hold if and only if the data display *proportional numbers*.

As mentioned, the preceding discussion applies to all two-way ANOVA models. We now state precisely the sense in which the columns for α and η need to be orthogonal. Let X_0 be the columns of X associated with μ and the α_is, and let X_1 be the columns of X associated with μ and the η_js. Let M_0 and M_1 be the projection operators onto $C(X_0)$ and $C(X_1)$, respectively. We will show that $R(\eta|\alpha,\mu) = R(\eta|\mu)$ for all Y if and only if $C(M_1 - [1/n]JJ') \perp C(M_0 - [1/n]JJ')$, i.e., $(M_1 - [1/n]JJ')(M_0 - [1/n]JJ') = 0$.

Since $R(\eta|\mu) = Y'(M_1 - [1/n]JJ')Y$ and $R(\eta|\alpha,\mu) = Y'(M - M_0)Y$, it suffices to show the next proposition.

Proposition 3.6.3 *In two-way ANOVA, $(M_1 - [1/n]JJ') = (M - M_0)$ if and only if $(M_1 - [1/n]JJ')(M_0 - [1/n]JJ') = 0$.*

Proof $\Rightarrow$ If $(M_1 - [1/n]JJ') = (M - M_0)$, then

$$(M_1 - [1/n]JJ')(M_0 - [1/n]JJ') = (M - M_0)(M_0 - [1/n]JJ') = 0$$

because $J \in C(M_0) \subset C(M)$.

$\Leftarrow$ To simplify notation, let

$$M_\alpha \equiv (M_0 - [1/n]JJ') \quad \text{and} \quad M_\eta \equiv (M_1 - [1/n]JJ').$$

We know that $M = [1/n]JJ' + M_\alpha + (M - M_0)$. If we could show that $M = [1/n]JJ' + M_\alpha + M_\eta$, we would be done.

$[1/n]JJ' + M_\alpha + M_\eta$ is symmetric and is easily seen to be idempotent since $0 = M_\eta M_\alpha = M_\alpha M_\eta$. It suffices to show that $C\big[(1/n)JJ' + M_\alpha + M_\eta\big] = C(X)$. Clearly, $C\big[(1/n)JJ' + M_\alpha + M_\eta\big] \subset C(X)$.

Suppose now that $v \in C(X)$. Since $C(M_0) = C(X_0)$ and $C(M_1) = C(X_1)$, if we let $Z = [M_0, M_1]$, then $C(Z) = C(X)$ and $v = Zb = M_0 b_0 + M_1 b_1$. Since $J \in C(X_0)$ and $J \in C(X_1)$, it is easily seen that $M_\alpha M_1 = M_\alpha M_\eta = 0$ and $M_\eta M_0 = 0$. Observe that

$$\begin{aligned}\left[\frac{1}{n}JJ' + M_\alpha + M_\eta\right] v &= \big[M_0 + M_\eta\big] M_0 b_0 + [M_1 + M_\alpha] M_1 b_1 \\ &= M_0 b_0 + M_1 b_1 = v,\end{aligned}$$

so $C(X) \subset C\big[(1/n)JJ' + M_\alpha + M_\eta\big]$. □

The condition $(M_1 - [1/n]JJ')(M_0 - [1/n]JJ') = 0$ is equivalent to what follows. Using the Gram–Schmidt orthogonalization algorithm, make all the columns corresponding to the αs and ηs orthogonal to J. Now, if the transformed α columns are orthogonal to the transformed η columns, then $R(\eta|\alpha,\mu) = R(\eta|\mu)$. In other words, check the condition $X_0'(I - [1/n]JJ')X_1 = 0$. In particular, this occurs in a balanced two-way ANOVA model, see Section 7.1, and is the crux of the proportional numbers argument in Section 7.4.

From the symmetry of the problem, it follows that $R(\alpha|\eta,\mu) = R(\alpha|\mu)$ whenever $R(\eta|\alpha,\mu) = R(\eta|\mu)$.

3.7 Confidence Regions

Consider the problem of finding a confidence region for the estimable parametric vector $\Lambda'\beta$. A $(1-\alpha)100\%$ confidence region for $\Lambda'\beta$ consists of all the vectors d that would not be rejected by an α level test of $\Lambda'\beta = d$. That is to say, a $(1-\alpha)100\%$ confidence region for $\Lambda'\beta$ consists of all the vectors d that are consistent with the data and the full model as determined by an α level test of $\Lambda'\beta = d$. Based on the distribution theory of Section 2 and the algebraic simplifications of Section 3, the $(1-\alpha)100\%$ confidence region consists of all the vectors d that satisfy the inequality

$$\frac{[\Lambda'\hat{\beta} - d]'[\Lambda'(X'X)^-\Lambda]^-[\Lambda'\hat{\beta} - d]/r(\Lambda)}{MSE} \le F(1-\alpha, r(\Lambda), r(I-M)). \quad (1)$$

These vectors form an ellipsoid that is degenerate when $r(\Lambda)$ is less than its number of columns.

Alternative forms for the confidence region are

$$[\Lambda'\hat{\beta} - d]'[\Lambda'(X'X)^-\Lambda]^-[\Lambda'\hat{\beta} - d] \le r(\Lambda)\ MSE\ F(1-\alpha, r(\Lambda), r(I-M))$$

and

$$[P'MY - d]'(P'MP)^-[P'MY - d] \le r(MP)\ MSE\ F(1-\alpha, r(MP), r(I-M)).$$

For regression problems we can get a considerable simplification. If we take $P' = (X'X)^{-1}X'$, then we have $\Lambda' = P'X = I_p$ and $\Lambda'\beta = \beta = d$. Using these in (1) and renaming the placeholder variable d as β gives

$$\begin{aligned}[\Lambda'\hat{\beta} - d]'[\Lambda'(X'X)^-\Lambda]^-[\Lambda'\hat{\beta} - d] &= (\hat{\beta} - \beta)'[(X'X)^{-1}]^{-1}(\hat{\beta} - \beta) \\ &= (\hat{\beta} - \beta)'(X'X)(\hat{\beta} - \beta)\end{aligned}$$

with $r(\Lambda) = r(I_p) = r(X)$. The confidence region is thus the set of all βs satisfying

$$(\hat{\beta} - \beta)'(X'X)(\hat{\beta} - \beta) \le p\ MSE\ F(1-\alpha, p, n-p).$$

Exercise 3.7b *Fieller's method.*
For a one-dimensional estimable function $\lambda'\beta$, use the quadratic formula to show that the confidence region agrees with the usual $t(dfE)$ confidence interval discussed in Section 2.6.

3.8 Tests for Generalized Least Squares Models

We now consider the problem of deriving tests for the model of Section 2.7. For testing, we take the generalized least squares model as

$$Y = X\beta + e, \quad e \sim N\left(0, \sigma^2 V\right), \tag{1}$$

where V is a known positive definite matrix. As in Section 2.7, we can write $V = QQ'$ for Q nonsingular. The model

$$\tilde{Y} = \tilde{X}\beta + \tilde{e}, \quad \tilde{e} \sim N\left(0, \sigma^2 I\right) \tag{2}$$

where

$$\tilde{Y} \equiv Q^{-1}Y, \quad \tilde{X} \equiv Q^{-1}X, \quad \tilde{e} \equiv Q^{-1}e.$$

is analyzed instead of model (1).

First consider the problem of testing model (1) against a reduced model, say

$$Y = X_0\beta_0 + e, \quad e \sim N\left(0, \sigma^2 V\right), \quad C(X_0) \subset C(X). \tag{3}$$

The reduced model can be transformed to

$$\tilde{Y} = \tilde{X}_0\beta_0 + \tilde{e}, \quad \tilde{e} \sim N\left(0, \sigma^2 I\right) \tag{4}$$

where

$$\tilde{X}_0 \equiv Q^{-1}X.$$

The test of model (3) against model (1) is performed by testing model (4) against model (2). To test model (4) against model (2), we need to know that model (4) is a reduced model relative to model (2). In other words, we need to show that $C(Q^{-1}X_0) \subset C(Q^{-1}X)$. From model (3), $C(X_0) \subset C(X)$, so there exists a matrix U such that $X_0 = XU$. It follows immediately that $Q^{-1}X_0 = Q^{-1}XU$; hence $C(\tilde{X}_0) = C(Q^{-1}X_0) \subset C(Q^{-1}X) = C(\tilde{X})$.

Recall from Section 2.7 that $A = X(X'V^{-1}X)^{-}X'V^{-1}$ and that for model (1)

$$MSE = Y'(I - A)'V^{-1}(I - A)Y \Big/ [n - r(X)].$$

Define $A_0 = X_0(X_0'V^{-1}X_0)^{-}X_0'V^{-1}$. The test comes from the following distributional result.

Theorem 3.8.1

(i) $$\frac{Y'(A - A_0)'V^{-1}(A - A_0)Y/[r(X) - r(X_0)]}{MSE} \sim F(r(X) - r(X_0), n - r(X), \pi),$$

where $\pi = \beta' X'(A - A_0)' V^{-1} (A - A_0) X\beta / 2\sigma^2$.
(ii) $\beta' X'(A - A_0)' V^{-1} (A - A_0) X\beta = 0$ *if and only if* $\mathrm{E}(Y) \in C(X_0)$.

Proof (i) Theorem 3.2.1 applied to models (2) and (4) gives the appropriate test statistic. It remains to show that part (i) involves the same test statistic. Exercise 3.8 is to show that $Y'(A - A_0)' V^{-1} (A - A_0) Y / [r(X) - r(X_0)]$ is the appropriate numerator mean square.

(ii) From part (i) and Theorem 3.2.1 applied to models (2) and (4),

$$\beta' X'(A - A_0)' V^{-1} (A - A_0) X\beta = 0$$

if and only if $\mathrm{E}(Q^{-1}Y) \in C(Q^{-1}X_0)$. $\mathrm{E}(Q^{-1}Y) \in C(Q^{-1}X_0)$ if and only if $\mathrm{E}(Y) \in C(X_0)$. □

Exercise 3.8 Show that $Y'(A - A_0)' V^{-1} (A - A_0) Y / [r(X) - r(X_0)]$ is the appropriate numerator mean square for testing model (4) against model (2).

The intuition behind the test based on Theorem 3.8.1 is essentially the same as that behind the usual test (which was discussed in Section 2). The usual test is based on the difference $MY - M_0Y = (M - M_0)Y$. MY is the estimate of $\mathrm{E}(Y)$ from the full model, and M_0Y is the estimate of $\mathrm{E}(Y)$ from the reduced model. The difference between these estimates indicates how well the reduced model fits. If the difference is large, the reduced model fits poorly; if the difference is small, the reduced model fits relatively well. To determine whether the difference vector is large or small, the squared length of the vector, as measured in Euclidean distance, is used. The squared length of $(M - M_0)Y$ reduces to the usual form $Y'(M - M_0)Y$. The basis of the test is to quantify how large the difference vector must be before there is some assurance that the difference between the vectors is due to more than just the variability of the data.

For generalized least squares models, the estimate of $\mathrm{E}(Y)$ from the full model is AY and the estimate of $\mathrm{E}(Y)$ from the reduced model is A_0Y. The difference between these vectors, $AY - A_0Y = (A - A_0)Y$, indicates how well the reduced model fits. The test is based on the squared length of the vector $(A - A_0)Y$, but the length of the vector is no longer measured in terms of Euclidean distance. The inverse of the covariance matrix is used to define a distance measure appropriate to generalized least squares models. Specifically, the squared distance between two vectors u and v is defined to be $(u - v)' V^{-1} (u - v)$. Note that with this distance measure, the generalized least squares estimate AY is the vector in $C(X)$ that is closest to Y, i.e., AY is the perpendicular projection onto $C(X)$ (cf. Section 2.7).

It should be noted that if $V = I$, then $A = M$, $A_0 = M_0$, and the test is exactly as in Section 2. Also as in Section 2, the key term in the numerator of the test statistic, $Y'(A - A_0)' V^{-1} (A - A_0) Y$, can be obtained as the difference between the *SSE* for the reduced model and the *SSE* for the full model.

Now consider testing parametric functions. If $\Lambda'\beta$ is an estimable parametric vector, then $\Lambda' = P'X$, $\Lambda'\hat{\beta} = P'AY$, and the test of the hypothesis $\Lambda'\beta = 0$ can be obtained from the following result:

Theorem 3.8.2

$$\text{(i)}\quad \frac{\hat{\beta}'\Lambda\left[\Lambda'(X'V^{-1}X)^{-}\Lambda\right]^{-}\Lambda'\hat{\beta}/r(\Lambda)}{MSE} \sim F(r(\Lambda), n - r(X), \pi),$$

where $\pi = \beta'\Lambda\left[\Lambda'(X'V^{-1}X)^{-}\Lambda\right]^{-}\Lambda'\beta/2\sigma^2$.

(ii) $\beta'\Lambda\left[\Lambda'(X'V^{-1}X)^{-}\Lambda\right]^{-}\Lambda'\beta = 0$ *if and only if* $\Lambda'\beta = 0$.

Proof $\Lambda'\beta$ is estimable in model (1) if and only if $\Lambda'\beta$ is estimable in model (2). $\Lambda'\hat{\beta}$ is the least squares estimate of $\Lambda'\beta$ from model (2), and $\sigma^2\Lambda'(X'V^{-1}X)^{-}\Lambda$ is the covariance matrix of $\Lambda'\hat{\beta}$. The result follows immediately from Section 3 applied to model (2). □

Note that $\Lambda'\beta = 0$ defines the same reduced model as in Section 3 but the test of the reduced model changes. Just as in Section 3 for ordinary least squares models, Theorem 3.8.2 provides a method of finding tests for generalized least squares models. To test $\Lambda'\beta = 0$, one need only find $\Lambda'\hat{\beta}$, $\text{Cov}(\Lambda'\hat{\beta})$, and *MSE*. If these can be found, the test follows immediately.

We have assumed that V is a known matrix. Since the results depend on V, they would seem to be of little use if V were not known. Nevertheless, the validity of the results does not depend on V being known. In Chapter 11, we will consider cases where V is not known, but where V and X are related in such a way that the results of this section can be used. In Chapter 11, we will need the distribution of the numerators of the test statistics.

Theorem 3.8.3

(i) $Y'(A - A_0)'V^{-1}(A - A_0)Y/\sigma^2 \sim \chi^2(r(X) - r(X_0), \pi)$,
where $\pi \equiv \beta'X'(A - A_0)'V^{-1}(A - A_0)X\beta/2\sigma^2$, *and* $\text{E}(Y) \in C(X_0)$ *if and only if* $\pi = 0$.

(ii) For $\Lambda'\beta$ *estimable,* $\hat{\beta}'\Lambda\left[\Lambda'(X'V^{-1}X)^{-}\Lambda\right]^{-}\Lambda'\hat{\beta}/\sigma^2 \sim \chi^2(r(\Lambda), \pi)$,
where $\pi \equiv \beta'\Lambda\left[\Lambda'(X'V^{-1}X)^{-}\Lambda\right]^{-}\Lambda'\beta/2\sigma^2$, *and* $\Lambda'\beta = 0$ *if and only if* $\pi = 0$.

Proof The results follow from Sections 3.2 and 3.3 applied to model (2). □

Exercise 3.9 Show that $Y'(A - A_0)'V^{-1}(A - A_0)Y$ equals the difference in the *SSE*s for models (3) and (1).

3.8.1 Conditions for Simpler Procedures

Just as Proposition 2.7.5 establishes that least squares estimates can be BLUEs even when $\text{Cov}(Y) \equiv \sigma^2 V \neq \sigma^2 I$, there exist conditions where $\text{Cov}(Y) \equiv \sigma^2 V \neq \sigma^2 I$ but under which the F statistic of Section 3.2 still has an $F(r(M - M_0), r(I - M))$ distribution under the null model with multivariate normal errors. In particular, when testing $Y = X\beta + e$ versus the reduced model $Y = X_0\gamma + e$ where $\text{Cov}(Y) \equiv V = \sigma^2[I + X_0B' + BX_0']$, the standard central F distribution continues to hold under the null hypothesis. In fact, the matrix B can even contain unknown parameters without affecting the validity of this result.

To obtain the result on F tests, one need only check that the usual numerator and denominator have the same independent χ^2 distributions under the null hypothesis as established in the proof of Theorem 3.2.1. This can be demonstrated by applying Theorems 1.3.6 and 1.3.8. In particular, $Y'(M - M_0)Y/\sigma^2 \sim \chi^2(r(M - M_0))$ because, with this V,

$$\left[\frac{1}{\sigma^2}(M - M_0)\right] V \left[\frac{1}{\sigma^2}(M - M_0)\right] = \frac{1}{\sigma^2}(M - M_0).$$

Similarly, $Y'(I - M)Y/\sigma^2 \sim \chi^2(r(I - M))$ because

$$\left[\frac{1}{\sigma^2}(I - M)\right] V \left[\frac{1}{\sigma^2}(I - M)\right] = \frac{1}{\sigma^2}(I - M).$$

Finally, independence follows from the fact that

$$\left[\frac{1}{\sigma^2}(M - M_0)\right] V \left[\frac{1}{\sigma^2}(I - M)\right] = 0.$$

Moreover, the arguments given in Huynh and Feldt (1970) should generalize to establish that the F distribution holds only if V has the form indicated.

Of course, it is not clear whether $\sigma^2[I + X_0B' + BX_0']$ is positive definite, as good covariance matrices should be. However, $X_0B' + BX_0'$ is symmetric, so it has real eigenvalues, and if the negative of its smallest eigenvalue is less than 1, $I + X_0B' + BX_0'$ will be positive definite.

A special case of this covariance structure has $X_0B' + BX_0' = X_0B_0X_0'$ for some B_0. In this special case, it is sufficient (but not necessary) to have B_0 nonnegative definite. In fact, if a positive definite V has Rao's simple covariance structure from Exercise 2.6 relative to the reduced model, it also has the $I + X_0B_0X_0'$ structure. In this special case, not only do standard F tests apply, but least squares estimates are BLUEs in both models because Proposition 2.7.5 applies. Yet, in general with $V = I + X_0B' + BX_0'$, it is possible to use the standard F tests even though least squares does not give BLUEs.

To illustrate the ideas, consider a balanced two-way ANOVA without interaction or replication, $y_{ij} = \mu + \alpha_i + \eta_j + e_{ij}$, $i = 1, \dots, a$, $j = 1, \dots, b$. *In this context, we think about the α_is as block effects, so there are a blocks and b treatments.* We explore situations in which observations within each block are correlated, but the usual F test for treatment effects continues to apply. Write the linear model in matrix form as

$$Y = X\beta + e = [J_{ab}, X_\alpha, X_\eta] \begin{bmatrix} \mu \\ \alpha \\ \eta \end{bmatrix} + e.$$

Here $\alpha = (\alpha_1, \dots, \alpha_a)'$ and $\eta = (\eta_1, \dots, \eta_b)'$. The test of no treatment effects uses the reduced model

$$Y = X_0\gamma + e = [J_{ab}, X_\alpha] \begin{bmatrix} \mu \\ \alpha \end{bmatrix} + e.$$

In the first illustration given below, $V = I + X_\alpha B_{0*} X'_\alpha$ for some B_{0*}. In the second illustration, $V = I + X_\alpha B'_* + B_* X'_\alpha$. In both cases it suffices to write V using X_α rather than X_0. This follows because $C(X_0) = C(X_\alpha)$, so we can always write $X_0 B' = X_\alpha B'_*$.

One covariance structure that is commonly used involves *compound symmetry*, that is, independence between blocks, homoscedasticity, and constant correlation within blocks. In other words,

$$\mathrm{Cov}(y_{ij}, y_{i'j'}) = \begin{cases} \sigma_*^2 & \text{if } i = i', j = j' \\ \sigma_*^2 \rho & \text{if } i = i', j \neq j' \\ 0 & \text{if } i \neq i'. \end{cases}$$

One way to write this covariance matrix is as

$$\sigma^2 V = \sigma_*^2(1-\rho)I + \sigma_*^2 \rho X_\alpha X'_\alpha.$$

In the context, σ^2 from the general theory is $\sigma_*^2(1-\rho)$ from the example and $B_{0*} \equiv [\rho/(1-\rho)]I_a$.

A more general covariance structure is

$$\mathrm{Cov}(y_{ij}, y_{i'j'}) = \begin{cases} \sigma^2(1+2\delta_j) & \text{if } i = i', j = j' \\ \sigma^2(\delta_j + \delta_{j'}) & \text{if, } i = i', j \neq j' \\ 0 & \text{if } i \neq i'. \end{cases}$$

We want to find B so that this covariance structure can be written as $\sigma^2[I + X_0 B' + BX'_0]$. It suffices to show that for some B_* we can write $V = I + X_\alpha B'_* + B_* X'_\alpha$. In the balanced two-way ANOVA without interaction or replication, when $Y = [y_{11}, y_{12}, \dots, y_{ab}]'$, X_α can be written using Kronecker products as

$$X_\alpha = [I_a \otimes J_b] = \begin{bmatrix} J_b & 0 & \cdots & 0 \\ 0 & J_b & & \vdots \\ \vdots & & \ddots & \\ 0 & \cdots & & J_b \end{bmatrix}.$$

Now define $\delta = (\delta_1, \ldots, \delta_b)'$ and take

$$B_* = [I_a \otimes \delta] = \begin{bmatrix} \delta & 0 & \cdots & 0 \\ 0 & \delta & & \vdots \\ \vdots & & \ddots & \\ 0 & \cdots & & \delta \end{bmatrix}.$$

With these choices, it is not difficult to see that the covariance structure specified earlier has

$$V = I_{ab} + [I_a \otimes J_b][I_a \otimes \delta]' + [I_a \otimes \delta][I_a \otimes J_b]'.$$

This second illustration is similar to a discussion in Huynh and Feldt (1970). The split plot models of Chapter 11 involve covariance matrices with compound symmetry, so they are *similar* in form to the first illustration that involved $\sigma^2 I + X_0 B_0 X_0'$. The results here establish that the F tests in the subplot analyses of Chapter 11 could still be obtained when using the more general covariance structures considered here.

3.9 Additional Exercises

Exercise 3.9.1 Consider the model $y_i = \beta_0 + \beta_1 x_{i1} + \beta_2 x_{i2} + e_i$, e_is i.i.d. $N(0, \sigma^2)$. Use the data given below to answer (a) and (b).

Obs.	1	2	3	4	5	6
y	−2	7	2	5	8	−1
x_1	4	−1	2	0	−2	3
x_2	2	−3	0	−2	−4	1

(a) Find $SSR(X_1, X_2|J) = R(\beta_1, \beta_2|\beta_0)$.
(b) Are β_0, β_1, and β_2 estimable?

Exercise 3.9.2 For a standard linear model, find the form of the generalized likelihood ratio test of $H_0 : \sigma^2 = \sigma_0^2$ versus $H_A : \sigma^2 \neq \sigma_0^2$ in terms of rejecting H_0

when some function of SSE/σ_0^2 is small. Show that the test makes sense in that it rejects for both large and small values of SSE/σ_0^2.

Exercise 3.9.3 Consider a set of seemingly unrelated regression equations

$$Y_i = X_i\beta_i + e_i, \quad e_i \sim N(0, \sigma^2 I),$$

$i = 1, \ldots, r$, where X_i is an $n_i \times p$ matrix and the e_is are independent. Find the test for $H_0 : \beta_1 = \cdots = \beta_r$.

Exercise 3.9.4

(a) Using the notation of Section 3.3 and in particular of model (3.3.6), show that the least squares estimate of β subject to the constraint $\Lambda'\beta = d$ is $\hat{\beta} = U\hat{\gamma} + b$.

(b) What happens to the test of $\Lambda'\beta = d$ if $\Lambda'\beta$ has no estimable part?

Exercise 3.9.5 Consider the model

$$Y = X\beta + e, \quad \mathrm{E}(e) = 0, \quad \mathrm{Cov}(e) = \sigma^2 I, \tag{1}$$

with the additional restriction

$$\Lambda'\beta = d,$$

where $d = \Lambda' b$ for some (known) vector b and $\Lambda' = P'X$. Model (1) with the additional restriction is equivalent to the model

$$(Y - Xb) = (M - M_{MP})\gamma + e. \tag{2}$$

If the parameterization of model (1) is particularly appropriate, then we might be interested in estimating $X\beta$ subject to the restriction $\Lambda'\beta = d$. To do this, write

$$X\beta = \mathrm{E}(Y) = (M - M_{MP})\gamma + Xb,$$

and define the BLUE of $\lambda'\beta = \rho'X\beta$ in the restricted version of (1) to be $\rho'(M - M_{MP})\hat{\gamma} + \rho'Xb$, where $\rho'(M - M_{MP})\hat{\gamma}$ is the BLUE of $\rho'(M - M_{MP})\gamma$ in model (2). Let $\hat{\beta}_1$ be the least squares estimate of β in the unrestricted version of model (1). Show that the BLUE of $\lambda'\beta$ in the restricted version of model (1) is

$$\lambda'\hat{\beta}_1 - \left[\mathrm{Cov}(\lambda'\hat{\beta}_1, \Lambda'\hat{\beta}_1)\right]\left[\mathrm{Cov}(\Lambda'\hat{\beta}_1)\right]^{-}(\Lambda'\hat{\beta}_1 - d), \tag{3}$$

where the covariance matrices are computed as in the unrestricted version of model (1).

Hint: This exercise is actually nothing more than simplifying the terms in (3) to show that it equals $\rho'(M - M_{MP})\hat{\gamma} + \rho'Xb$.

Note: The result in (3) is closely related to best linear prediction, cf. Section 6.3.

Exercise 3.9.6 Discuss how the testing procedures from this chapter would change if you actually knew the variance σ^2.

Exercise 3.9.7 Consider a linear model $Y = X\beta + e$, $\mathrm{E}(e) = 0$ with the (not necessarily estimable) linear constraint $\Lambda'\beta = d$. Consider two solutions to the constraint, b_1 and b_2, so that $\Lambda' b_k = d$, $k = 1, 2$. Define appropriate least squares fitted values $\hat{Y}_k$ from the model $Y = X_0\gamma + Xb_k + e$ where $X_0 = XU$ with $C(U) = C(\Lambda)^{\perp}$. Show that $\hat{Y}_1 = \hat{Y}_2$. Hint: After finding $\hat{Y}_k$, show that $(I - M_0)X(b_1 - b_2) = 0$.

Exercise 3.9.8 Under the conditions of Exercise 3.9.7, show that $(Y - Xb_1)'(M - M_0)(Y - Xb_1) = (Y - Xb_2)'(M - M_0)(Y - Xb_2)$.

References

Casella, G., & Berger, R. L. (2002). *Statistical inference* (2nd ed.). Pacific Grove: Duxbury Press.
Christensen, R. (1997). *Log-linear models and logistic regression* (2nd ed.). New York: Springer.
Christensen, R. (2015). *Analysis of variance, design, and regression: Linear modeling for unbalanced data* (2nd ed.). Boca Raton: Chapman and Hall/CRC Press.
Huynh, H., & Feldt, L. S. (1970). Conditions under which mean square ratios in repeated measurements designs have exact F-distributions. *Journal of the American Statistical Association, 65*, 1582–1589.
Peixoto, J. L. (1993). Four equivalent definitions of reparameterizations and restrictions in linear models. *Communications in Statistics, A, 22*, 283–299.

Chapter 4
One-Way ANOVA

Abstract This chapter considers the analysis of the one-way ANOVA models originally exploited by R.A. Fisher.

In this and the following chapters, we apply the general theory of linear models to various special cases. This chapter considers the analysis of one-way ANOVA models. A one-way ANOVA model can be written

$$y_{ij} = \mu + \alpha_i + e_{ij}, \qquad i = 1, \ldots, t, \quad j = 1, \ldots, N_i, \tag{4.1}$$

where $\mathrm{E}(e_{ij}) = 0$, $\mathrm{Var}(e_{ij}) = \sigma^2$, and $\mathrm{Cov}(e_{ij}, e_{i'j'}) = 0$ when $(i, j) \neq (i', j')$. For finding tests and confidence intervals, the e_{ij}s are assumed to have a multivariate normal distribution. Here α_i is an effect for y_{ij} belonging to the ith group of observations. Group effects are often called *treatment effects* because one-way ANOVA models are used to analyze completely randomized experimental designs.

Section 1 is devoted primarily to deriving the ANOVA table for a one-way ANOVA. The ANOVA table in this case is a device for presenting the sums of squares necessary for testing the reduced model

$$y_{ij} = \mu + e_{ij}, \qquad i = 1, \ldots, t, \quad j = 1, \ldots, N_i, \tag{4.2}$$

against model (1). This test is equivalent to testing the hypothesis $H_0 : \alpha_1 = \cdots = \alpha_t$.

The main tool needed for deriving the analysis of model (1) is the perpendicular projection operator. The first part of Section 1 is devoted to finding M. Since the ys in model (1) are identified with two subscripts, it will be necessary to develop notation that allows the rows of a vector to be denoted by two subscripts. Once M is found, some comments are made about estimation and the role of side conditions in estimation. Finally, the perpendicular projection operator for testing $H_0 : \alpha_1 = \cdots = \alpha_t$ is found and the ANOVA table is presented. Section 2 is an examination of contrasts. First, contrasts are defined and discussed. Estimation and testing procedures are presented. Orthogonal contrasts are defined and applications of Sections 3.5 and 3.6 are

R. Christensen, *Plane Answers to Complex Questions*, Springer Texts in Statistics,
https://doi.org/10.1007/978-3-030-32097-3_4

given. Fortunately, many balanced multifactor analysis of variance problems can be analyzed by repeatedly using the analysis for a one-way analysis of variance. For that reason, the results of this chapter are particularly important.

4.1 Analysis of Variance

In linear model theory, the main tools we need are perpendicular projection matrices. Our first project in this section is finding the perpendicular projection matrix for a one-way ANOVA model. We will then discuss estimation, side conditions, and the ANOVA table.

Usually, the one-way ANOVA model is written

$$y_{ij} = \mu + \alpha_i + e_{ij}, \quad i = 1, \ldots, t, \quad j = 1, \ldots, N_i.$$

Let $n = \sum_{i=1}^{t} N_i$. Although the notation N_i is standard, we will sometimes use $N(i)$ instead. Thus, $N(i) \equiv N_i$. We proceed to find the perpendicular projection matrix $M = X(X'X)^{-}X'$.

Example 4.1.1 In any particular example, the matrix manipulations necessary for finding M are simple. Suppose $t = 3$, $N_1 = 5$, $N_2 = 3$, $N_3 = 3$. In matrix notation the model can be written

$$\begin{bmatrix} y_{11} \\ y_{12} \\ y_{13} \\ y_{14} \\ y_{15} \\ y_{21} \\ y_{22} \\ y_{23} \\ y_{31} \\ y_{32} \\ y_{33} \end{bmatrix} = \begin{bmatrix} 1 & 1 & 0 & 0 \\ 1 & 1 & 0 & 0 \\ 1 & 1 & 0 & 0 \\ 1 & 1 & 0 & 0 \\ 1 & 1 & 0 & 0 \\ 1 & 0 & 1 & 0 \\ 1 & 0 & 1 & 0 \\ 1 & 0 & 1 & 0 \\ 1 & 0 & 0 & 1 \\ 1 & 0 & 0 & 1 \\ 1 & 0 & 0 & 1 \end{bmatrix} \begin{bmatrix} \mu \\ \alpha_1 \\ \alpha_2 \\ \alpha_3 \end{bmatrix} + e.$$

To find the perpendicular projection matrix M, first find

$$X'X = \begin{bmatrix} 11 & 5 & 3 & 3 \\ 5 & 5 & 0 & 0 \\ 3 & 0 & 3 & 0 \\ 3 & 0 & 0 & 3 \end{bmatrix}.$$

By checking that $(X'X)(X'X)^-(X'X) = X'X$, it is easy to verify that

$$(X'X)^- = \begin{bmatrix} 0 & 0 & 0 & 0 \\ 0 & 1/5 & 0 & 0 \\ 0 & 0 & 1/3 & 0 \\ 0 & 0 & 0 & 1/3 \end{bmatrix}.$$

Then

$$\begin{aligned} M &= X(X'X)^- X' \\ &= X \begin{bmatrix} 0 & 0 & 0 & 0 & 0 & 0 & 0 & 0 & 0 & 0 & 0 \\ 1/5 & 1/5 & 1/5 & 1/5 & 1/5 & 0 & 0 & 0 & 0 & 0 & 0 \\ 0 & 0 & 0 & 0 & 0 & 1/3 & 1/3 & 1/3 & 0 & 0 & 0 \\ 0 & 0 & 0 & 0 & 0 & 0 & 0 & 0 & 1/3 & 1/3 & 1/3 \end{bmatrix} \\ &= \begin{bmatrix} 1/5 & 1/5 & 1/5 & 1/5 & 1/5 & 0 & 0 & 0 & 0 & 0 & 0 \\ 1/5 & 1/5 & 1/5 & 1/5 & 1/5 & 0 & 0 & 0 & 0 & 0 & 0 \\ 1/5 & 1/5 & 1/5 & 1/5 & 1/5 & 0 & 0 & 0 & 0 & 0 & 0 \\ 1/5 & 1/5 & 1/5 & 1/5 & 1/5 & 0 & 0 & 0 & 0 & 0 & 0 \\ 1/5 & 1/5 & 1/5 & 1/5 & 1/5 & 0 & 0 & 0 & 0 & 0 & 0 \\ 0 & 0 & 0 & 0 & 0 & 1/3 & 1/3 & 1/3 & 0 & 0 & 0 \\ 0 & 0 & 0 & 0 & 0 & 1/3 & 1/3 & 1/3 & 0 & 0 & 0 \\ 0 & 0 & 0 & 0 & 0 & 1/3 & 1/3 & 1/3 & 0 & 0 & 0 \\ 0 & 0 & 0 & 0 & 0 & 0 & 0 & 0 & 1/3 & 1/3 & 1/3 \\ 0 & 0 & 0 & 0 & 0 & 0 & 0 & 0 & 1/3 & 1/3 & 1/3 \\ 0 & 0 & 0 & 0 & 0 & 0 & 0 & 0 & 1/3 & 1/3 & 1/3 \end{bmatrix}. \end{aligned}$$

Thus, in this example, M is Blk diag$[N_i^{-1} J_{N(i)}^{N(i)}]$, where J_r^c is a matrix of 1s with r rows and c columns. In fact, we will see below that this is the general form for M in a one-way ANOVA when the observation vector Y has subscripts changing fastest on the right.

A somewhat easier way of finding M is as follows. Let Z be the model matrix for the alternative one-way analysis of variance model

$$y_{ij} = \mu_i + e_{ij},$$

$i = 1, \ldots, t$, $j = 1, \ldots, N_i$. (See Example 3.1.1.) Z is then just a matrix consisting of the last t columns of X, i.e., $X = [J, Z]$. Clearly $C(X) = C(Z)$, $Z'Z = \text{Diag}(N_1, N_2, \ldots, N_t)$, and $(Z'Z)^{-1} = \text{Diag}(N_1^{-1}, N_2^{-1}, \ldots, N_t^{-1})$. It is easy to see that $Z(Z'Z)^- Z' = \text{Blk diag}[N_i^{-1} J_{N(i)}^{N(i)}]$.

In particular, write the observations in the ith group as

$$Y_i \equiv \begin{bmatrix} y_{i1} \\ \vdots \\ y_{iN(i)} \end{bmatrix}.$$

Now we can write the one-way ANOVA linear model as

$$\begin{bmatrix} Y_1 \\ Y_2 \\ \vdots \\ Y_t \end{bmatrix} = \begin{bmatrix} J_{N(1)} & J_{N(1)} & 0 & \cdots & 0 \\ J_{N(2)} & 0 & J_{N(2)} & & 0 \\ \vdots & \vdots & & \ddots & \\ J_{N(t)} & 0 & 0 & & J_{N(t)} \end{bmatrix} \begin{bmatrix} \mu \\ \alpha_1 \\ \alpha_2 \\ \vdots \\ \alpha_t \end{bmatrix} + e$$

or

$$\begin{bmatrix} Y_1 \\ Y_2 \\ \vdots \\ Y_t \end{bmatrix} = \begin{bmatrix} J_{N(1)} & 0 & \cdots & 0 \\ 0 & J_{N(2)} & & 0 \\ \vdots & & \ddots & \\ 0 & 0 & & J_{N(t)} \end{bmatrix} \begin{bmatrix} \mu_1 \\ \mu_2 \\ \vdots \\ \mu_t \end{bmatrix} + e.$$

It is not difficult to see that

$$M = \begin{bmatrix} \frac{1}{N(1)} J_{N(1)}^{N(1)} & 0 & \cdots & 0 \\ 0 & \frac{1}{N(2)} J_{N(2)}^{N(2)} & & 0 \\ \vdots & & \ddots & \\ 0 & 0 & & \frac{1}{N(t)} J_{N(t)}^{N(t)} \end{bmatrix}.$$

It follows that $X\hat{\beta} = MY$ reduces to, depending on the parameterization,

$$\begin{bmatrix} (\hat{\mu} + \hat{\alpha}_1) J_{N(1)} \\ (\hat{\mu} + \hat{\alpha}_2) J_{N(2)} \\ \vdots \\ (\hat{\mu} + \hat{\alpha}_t) J_{N(t)} \end{bmatrix} = \begin{bmatrix} J_{N(1)} & J_{N(1)} & 0 & \cdots & 0 \\ J_{N(2)} & 0 & J_{N(2)} & & 0 \\ \vdots & \vdots & & \ddots & \\ J_{N(t)} & 0 & 0 & & J_{N(t)} \end{bmatrix} \begin{bmatrix} \hat{\mu} \\ \alpha_1 \\ \vdots \\ \alpha_t \end{bmatrix}$$

$$= \begin{bmatrix} \frac{1}{N(1)} J_{N(1)}^{N(1)} & 0 & \cdots & 0 \\ 0 & \frac{1}{N(2)} J_{N(2)} N(2) & & 0 \\ \vdots & & \ddots & \\ 0 & 0 & & \frac{1}{N(t)} J_{N(t)}^{N(t)} \end{bmatrix} \begin{bmatrix} Y_1 \\ Y_2 \\ \vdots \\ Y_t \end{bmatrix} = \begin{bmatrix} \bar{y}_{1\cdot} J_{N(1)} \\ \bar{y}_{2\cdot} J_{N(2)} \\ \vdots \\ \bar{y}_{t\cdot} J_{N(t)} \end{bmatrix}$$

or

$$\begin{bmatrix} \hat{\mu}_1 J_{N(1)} \\ \hat{\mu}_2 J_{N(2)} \\ \vdots \\ \hat{\mu}_t J_{N(t)} \end{bmatrix} = \begin{bmatrix} J_{N(1)} & 0 & \cdots & 0 \\ 0 & J_{N(2)} & & 0 \\ \vdots & & \ddots & \\ 0 & 0 & & J_{N(t)} \end{bmatrix} \begin{bmatrix} \hat{\mu}_1 \\ \hat{\mu}_2 \\ \vdots \\ \hat{\mu}_t \end{bmatrix}$$

$$= \begin{bmatrix} \frac{1}{N(1)} J_{N(1)}^{N(1)} & 0 & \cdots & 0 \\ 0 & \frac{1}{N(2)} J_{N(2)} N(2) & & 0 \\ \vdots & & \ddots & \\ 0 & 0 & & \frac{1}{N(t)} J_{N(t)}^{N(t)} \end{bmatrix} \begin{bmatrix} Y_1 \\ Y_2 \\ \vdots \\ Y_t \end{bmatrix} = \begin{bmatrix} \bar{y}_{1\cdot} J_{N(1)} \\ \bar{y}_{2\cdot} J_{N(2)} \\ \vdots \\ \bar{y}_{t\cdot} J_{N(t)} \end{bmatrix}.$$

We now present a rigorous derivation of these results that does not require the Y vector to be written in any particular order. (This is important for our later examination of multifactor ANOVA.) The ideas involved in Example 4.1.1 are perfectly general. A similar computation can be performed for any values of t and the N_is. The difficulty in a rigorous general presentation lies entirely in being able to write down the model in matrix form. The elements of Y are the y_{ij}s. The y_{ij}s have two subscripts, so a pair of subscripts must be used to specify each row of the vector Y. The elements of the model matrices X and Z are determined entirely by knowing the order in which the y_{ij}s have been listed in the Y vector. For example, the row of Z corresponding to y_{ij} would have a 1 in the ith column and 0s everywhere else. Clearly, it will also be convenient to use a pair of subscripts to specify the rows of the model matrices.

If

$$Y' = (y_{11}, y_{12}, \ldots, y_{1N(1)}, y_{21}, \ldots, y_{tN(t)}),$$

then y_{21} is the $N_1 + 1$ row of Y and y_{ij} is the $(N_1 + \cdots + N_{i-1} + j)$th row of Y. In conformance with Y, write any other vector S as $S = [s_{ij}]$, where it is understood that for $j = 1, \ldots, N_i$, s_{ij} denotes the $(N_1 + \cdots + N_{i-1} + j)$th row of S. But the $S = [s_{ij}]$ notation for a vector does not require a particular ordering for the entries of Y. The vector $Y = [y_{ij}]$ can have any ordering and other vectors S use the same ordering. The discussion of tensors in Appendix B may help the reader feel more comfortable with this use of subscripts.

To specify the model matrices X and Z we must identify the columns of X and Z. Write $X = [J, X_1, X_2, \ldots, X_t]$ and $Z = [X_1, X_2, \ldots, X_t]$. Note that the kth column of Z can be written

$$X_k = [t_{ij}], \quad \text{where } t_{ij} = \delta_{ik} \tag{1}$$

with δ_{ik} equal to 0 if $i \neq k$ and 1 if $i = k$. This means that if the observation in the ij row belongs to the kth group, the ij row of X_k is 1. If not, the ij row of X_k is zero.

Our goal is to find $M = Z(Z'Z)^{-1}Z'$. To do this we need to find $(Z'Z)$ and $(Z'Z)^{-1}$. Noting that $(X_k)'(X_q)$ is a real number, we can write the elements of $Z'Z$ as

$$(Z'Z) = \left[(X_k)'(X_q)\right]_{t \times t}.$$

Now, from (1)

$$(X_k)'(X_k) = \sum_{ij} \delta_{ik}\delta_{ik} = \sum_{i=1}^{t}\sum_{j=1}^{N_i} \delta_{ik} = \sum_{i=1}^{t} N_i \delta_{ik} = N_k$$

and for $k \neq q$

$$(X_k)'(X_q) = \sum_{i=1}^{t}\sum_{j=1}^{N_i} \delta_{ik}\delta_{iq} = \sum_{i=1}^{t} N_i \delta_{ik}\delta_{iq} = 0.$$

It follows that

$$(Z'Z) = \mathrm{Diag}(N_i)$$

and clearly

$$(Z'Z)^{-1} = \text{Diag}(N_i^{-1}).$$

We can now find $Z(Z'Z)^{-1}$.

$$\begin{aligned} Z(Z'Z)^{-1} &= [X_1, X_2, \ldots, X_t]\text{Diag}(N_i^{-1}) \\ &= \left[N_1^{-1}X_1, N_2^{-1}X_2, \ldots, N_t^{-1}X_t\right]. \end{aligned}$$

Finally, we are in a position to find $M = Z(Z'Z)^{-1}Z'$. We denote the columns of an $n \times n$ matrix using the convention introduced above for denoting rows, i.e., by using two subscripts. Then the matrix M can be written

$$M = [m_{ij,i'j'}].$$

We now find the entries of this matrix. Note that $m_{ij,i'j'}$ is the ij row of $Z(Z'Z)^{-1}$ times the $i'j'$ column of Z' (i.e., the $i'j'$ row of Z). The ij row of $Z(Z'Z)^{-1}$ is $(N_1^{-1}\delta_{i1}, \ldots, N_t^{-1}\delta_{it})$. The $i'j'$ row of Z is $(\delta_{i'1}, \ldots, \delta_{i't})$. The product is

$$\begin{aligned} m_{ij,i'j'} &= \sum_{k=1}^{t} N_k^{-1}\delta_{ik}\delta_{i'k} \\ &= N_i^{-1}\delta_{ii'}. \end{aligned}$$

If Y is listed in the usual order with the second subscript changing fastest, these values of $m_{ij,i'j'}$ determine a block diagonal matrix

$$M = \text{Blk diag}(N_i^{-1}J_{N(i)}^{N(i)})$$

just as in Example 4.1.1, but these values apply to any ordering of the entries in Y.

To reiterate, the notation and methods developed above are somewhat unusual, but they are necessary for giving a rigorous treatment of ANOVA models. The device of indicating the rows of vectors with multiple subscripts will be used extensively in later discussions of multifactor ANOVA. The arguments given above apply to any order of specifying the entries in a vector $S = [s_{ij}]$; they do not depend on having $S = (s_{11}, s_{12}, \ldots, s_{tN(t)})'$. If we specified some other ordering, we would still get the perpendicular projection matrix M; however, M might no longer be block diagonal.

Exercise 4.1 To develop some facility with this notation, let

$$T_r = [t_{ij}], \qquad \text{where } t_{ij} = \delta_{ir} - \frac{N_r}{n}$$

for $r = 1, \ldots, t$. Find $T_r'T_r$, $T_r'T_s$ for $s \neq r$, and $J'T_r$.

A very important application of this notation is in characterizing the vector MY. As discussed in Section 3.1, the vector MY is the base from which all estimates of parametric functions are found. A second important application involves the projection operator

$$M_\alpha = M - \frac{1}{n}J_n^n .$$

M_α is useful in testing hypotheses and is especially important in the analysis of multifactor ANOVAs. It is therefore necessary to have a characterization of M_α.

Exercise 4.2 Show that

$$MY = [t_{ij}], \qquad \text{where } t_{ij} = \bar{y}_{i\cdot}$$

and

$$M_\alpha Y = [u_{ij}], \qquad \text{where } u_{ij} = \bar{y}_{i\cdot} - \bar{y}_{\cdot\cdot} .$$

Hint: Write $M_\alpha Y = MY - (\frac{1}{n}J_n^n)Y$.

These characterizations MY and $M_\alpha Y$ tell us how to find Mv and $M_\alpha v$ for any vector v. In fact, they completely characterize the perpendicular projection operators M and M_α.

Example 4.1.1 Continued. In this example,

$$MY = (\bar{y}_{1\cdot}, \bar{y}_{1\cdot}, \bar{y}_{1\cdot}, \bar{y}_{1\cdot}, \bar{y}_{1\cdot}, \bar{y}_{2\cdot}, \bar{y}_{2\cdot}, \bar{y}_{2\cdot}, \bar{y}_{3\cdot}, \bar{y}_{3\cdot}, \bar{y}_{3\cdot})'$$

and

$$M_\alpha Y = (\bar{y}_{1\cdot} - \bar{y}_{\cdot\cdot}, \bar{y}_{1\cdot} - \bar{y}_{\cdot\cdot}, \bar{y}_{1\cdot} - \bar{y}_{\cdot\cdot}, \bar{y}_{1\cdot} - \bar{y}_{\cdot\cdot}, \bar{y}_{1\cdot} - \bar{y}_{\cdot\cdot},$$
$$\bar{y}_{2\cdot} - \bar{y}_{\cdot\cdot}, \bar{y}_{2\cdot} - \bar{y}_{\cdot\cdot}, \bar{y}_{2\cdot} - \bar{y}_{\cdot\cdot}, \bar{y}_{3\cdot} - \bar{y}_{\cdot\cdot}, \bar{y}_{3\cdot} - \bar{y}_{\cdot\cdot}, \bar{y}_{3\cdot} - \bar{y}_{\cdot\cdot})'.$$

We can now obtain a variety of estimates. Recall that estimable functions are linear combinations of the rows of $X\beta$, e.g., $\rho' X\beta$. Since

$$X\beta = \big(\mu+\alpha_1, \mu+\alpha_1, \mu+\alpha_1, \mu+\alpha_1, \mu+\alpha_1,$$
$$\mu+\alpha_2, \mu+\alpha_2, \mu+\alpha_2, \mu+\alpha_3, \mu+\alpha_3, \mu+\alpha_3\big)',$$

if ρ' is taken to be $\rho' = (1,0,0,0,0,0,0,0,0,0,0)'$, then it is easily seen that $\mu+\alpha_1 = \rho' X\beta$ is estimable. The estimate of $\mu+\alpha_1$ is $\rho' MY = \bar{y}_{1\cdot}$. Similarly, the estimates of $\mu+\alpha_2$ and $\mu+\alpha_3$ are $\bar{y}_{2\cdot}$ and $\bar{y}_{3\cdot}$, respectively. The contrast $\alpha_1 - \alpha_2$ can be obtained as $\rho' X\beta$ using $\rho' = (1,0,0,0,0,-1,0,0,0,0,0)'$. The estimate of $\alpha_1 - \alpha_2$ is $\rho' MY = \bar{y}_{1\cdot} - \bar{y}_{2\cdot}$. Note that for this contrast $\rho' MY = \rho' M_\alpha Y$.

Estimation is as easy in a general one-way ANOVA as it is in Example 4.1.1. We have found M and MY, and it is an easy matter to see that, for instance, $\mu + \alpha_i$ is estimable and the estimate of $\mu + \alpha_i$ is

$$\{\hat{\mu} + \hat{\alpha}_i\} = \bar{y}_{i\cdot}\,.$$

The notation $\{\hat{\mu} + \hat{\alpha}_i\}$ will be used throughout this chapter to denote the estimate of $\mu + \alpha_i$.

For computational purposes, it is often convenient to present one particular set of least squares estimates. In one-way ANOVA, the traditional *side condition* on the parameters is $\sum_{i=1}^{t} N_i \alpha_i = 0$. With this condition, one obtains

$$\mu = \frac{1}{n} \sum_{i=1}^{t} N_i(\mu + \alpha_i)$$

and an estimate

$$\hat{\mu} = \frac{1}{n} \sum_{i=1}^{t} N_i \{\hat{\mu} + \hat{\alpha}_i\} = \frac{1}{n} \sum_{i=1}^{t} \frac{N_i}{N_i} \sum_{j=1}^{N_i} y_{ij} = \bar{y}_{\cdot\cdot},$$

which is the mean of all the observations. Similarly,

$$\hat{\alpha}_i = \{\hat{\mu} + \hat{\alpha}_i\} - \hat{\mu} = \bar{y}_{i\cdot} - \bar{y}_{\cdot\cdot}\,.$$

The traditional side condition is no longer the most often used. Exercise 4.9 involves finding parameter estimates that satisfy a different commonly used side condition. Fortunately, all side conditions lead to the same estimates of identifiable functions, so one choice of a side condition is as good as any other. The best choice of a side condition is the most convenient choice. However, different side conditions do lead to different "estimates" of nonidentifiable parameters. Do not be lulled into believing that an arbitrary side condition allows you to say anything meaningful about nonidentifiable parameters. That is just silly!

We now derive the analysis of variance table for the one-way ANOVA. The analysis of variance table is a device for displaying an orthogonal breakdown of the *total sum of squares* of the data (*SSTot*), i.e., $Y'Y$. More often, the total sum of squares corrected for fitting the *grand mean* is broken down. The *sum of squares for fitting the grand mean* (*SSGM*) is just the sum of squares accounted for by the model

$$Y = J\mu + e \tag{2}$$

This is also known as the *correction factor* C so

$$Y'[1/n]J_n^n Y = SSGM = C = SSR(J) = R(\mu).$$

The *total sum of squares corrected for the grand mean* ($SSTot - C$) is the error sum of squares in model (2), i.e., $Y'\left(I - [1/n]J_n^n\right)Y = Y'Y - C$. Included in an ANOVA table is information to identify the sums of squares (Source), the degrees of freedom for the sums of squares (df), the sums of squares (*SS*), and the mean squares (*MS*). The mean squares are just the sums of squares divided by their degrees of freedom. Sometimes the expected values of the mean squares are included. From the expected mean squares, the hypotheses tested by the various sums of squares can be identified. Recall that, when divided by σ^2, the sums of squares have χ^2 distributions and that there is a very close relationship between the expected mean square, the expected sum of squares, the noncentrality parameter of the χ^2 distribution, and the noncentrality parameter of an F distribution with the mean square in the numerator. In particular, if the expected mean square is $\sigma^2 + \pi/df$, then the noncentrality parameter is $\pi/2\sigma^2$. Assuming the full model is true, the null hypothesis being tested is that the noncentrality parameter of the F distribution is zero.

The usual orthogonal breakdown for a one-way ANOVA is to isolate the effect of the grand mean (μ), and then the effect of fitting the groups ($\alpha_i s$) after fitting the mean. The *sum of squares for groups* (*SSGrps*) is just what is left after removing the sum of squares for μ from the sum of squares for the model. In other words, the sum of squares for groups is the sum of squares for testing the reduced model (4.0.2) against model (4.0.1). As we have seen earlier, the projection operator for fitting the grand mean is based on the first column of X, i.e., J. The projection operator is $(1/n)J_n^n = (1/n)JJ'$. The projection operator for the groups sum of squares is then

$$M_\alpha = M - \frac{1}{n}J_n^n.$$

The sum of squares for fitting groups after μ, $Y'M_\alpha Y$, is the difference between the sum of squares for fitting the full model, $Y'MY$, and the sum of squares for fitting the model with just the mean, $Y'\left([1/n]J_n^n\right)Y$.

Table 4.1 gives an ANOVA table and indicates some common notation for the entries.

Exercise 4.3 Verify that the estimate of $\mu + \alpha_i$ is $\bar{y}_{i\cdot}$ and that the algebraic formulas for the sums of squares in the ANOVA table are correct.

Hint: To find, for example, $Y\left(M - [1/n]J_n^n\right)Y = Y'M_\alpha Y$, use Exercise 4.2 to get $M_\alpha Y$ and recall that $Y'M_\alpha Y = [M_\alpha Y]'[M_\alpha Y]$.

Exercise 4.4 Verify that the formulas for expected mean squares in the ANOVA table are correct. Hint: Use Theorem 1.3.2 and Exercise 4.3.

The techniques suggested for Exercises 4.3 and 4.4 are very useful. The reader should make a point of remembering them.

Table 4.1 One-way analysis of variance table

Matrix notation

Source	df	SS
Grand mean	1	$Y'\left(\frac{1}{n}J_n^n\right)Y$
Groups	$t-1$	$Y'\left(M-\frac{1}{n}J_n^n\right)Y$
Error	$n-t$	$Y'(I-M)Y$
Total	n	$Y'Y$

Source	SS	E(MS)
Grand mean	$SSGM$	$\sigma^2+\beta'X'\left(\frac{1}{n}J_n^n\right)X\beta$
Groups	$SSGrps$	$\sigma^2+\beta'X'\left(M-\frac{1}{n}J_n^n\right)X\beta/(t-1)$
Error	SSE	σ^2
Total	$SSTot$	

Algebraic notation

Source	df	SS
Grand mean	$dfGM$	$n^{-1}y_{\cdot\cdot}^2=n\bar{y}_{\cdot\cdot}^2$
Groups	$dfGrps$	$\sum_{i=1}^t N_i\left(\bar{y}_{i\cdot}-\bar{y}_{\cdot\cdot}\right)^2$
Error	dfE	$\sum_{i=1}^t\sum_{j=1}^{N_i}\left(y_{ij}-\bar{y}_{i\cdot}\right)^2$
Total	$dfTot$	$\sum_{i=1}^t\sum_{j=1}^{N_i}y_{ij}^2$

Source	MS	E(MS)*
Grand mean	$SSGM$	$\sigma^2+n(\mu+\bar{\alpha}_\cdot)^2$
Groups	$SSGrps/(t-1)$	$\sigma^2+\sum_{i=1}^t N_i\left(\alpha_i-\bar{\alpha}_\cdot\right)^2/(t-1)$
Error	$SSE/(n-t)$	σ^2
Total		

$^*\bar{\alpha}_\cdot=\sum_{i=1}^t N_i\alpha_i/n$

4.2 Estimating and Testing Contrasts

In this section, contrasts are defined and characterizations of contrasts are given. Estimates of contrasts are found. The numerator sum of squares necessary for doing an F test of a contrast and the form of the F test are given. The form of a confidence interval for a contrast is presented and the idea of orthogonal contrasts is discussed. Finally, the results of Section 3.5 are reviewed by deriving them anew for the one-way ANOVA model.

A contrast in the one-way ANOVA (4.0.1) is a function $\sum_{i=1}^t\lambda_i\alpha_i$, with $\sum_{i=1}^t\lambda_i=0$. In other words, the vector λ' in $\lambda'\beta$ is $(0,\lambda_1,\lambda_2,\ldots,\lambda_t)$ and $\lambda'J_{t+1}=0$. To establish that $\lambda'\beta$ is estimable, we need to find ρ such that $\rho'X=\lambda'$. Write

$$\begin{aligned}\rho' &= (\lambda_1/N_1, \ldots, \lambda_1/N_1, \lambda_2/N_2, \ldots, \lambda_2/N_2, \lambda_3/N_3, \ldots, \lambda_t/N_t) \\ &= [(\lambda_1/N_1)J'_{N(1)}, (\lambda_2/N_2)J'_{N(2)}, \ldots, (\lambda_t/N_t)J'_{N(t)}]\end{aligned}$$

where ρ' is a $1 \times n$ vector and the string of λ_i/N_is is N_i long. In the alternate notation that uses two subscripts to denote a row of a vector, we have

$$\rho = [t_{ij}], \qquad \text{where } t_{ij} = \lambda_i/N_i. \tag{1}$$

Recall from Section 2.1 that, while other choices of ρ may exist with $\rho'X = \lambda'$, the vector $M\rho$ is unique. As shown in Exercise 4.10, for ρ as in (1), $\rho \in C(X)$; so $\rho = M\rho$. Thus, for any contrast $\lambda'\beta$, the vector $M\rho$ has the structure

$$M\rho = [t_{ij}], \qquad \text{where } t_{ij} = \lambda_i/N_i. \tag{2}$$

We now show that the contrasts are precisely the estimable functions that do not involve μ. Note that since J is the column of X associated with μ, $\rho'X\beta$ does not involve μ if and only if $\rho'J = 0$.

Proposition 4.2.1 *$\rho'X\beta$ is a contrast if and only if $\rho'J = 0$.*

Proof Clearly, a contrast does not involve μ, so $\rho'J = 0$. Conversely, if $\rho'J = 0$, then $\rho'X\beta = \rho'[J, Z]\beta$ does not involve μ; so we need only show that $0 = \rho'XJ_{t+1}$. This follows because $XJ_{t+1} = 2J_n$, and we know that $\rho'J_n = 0$. □

We now show that the contrasts are the estimable functions that impose constraints on $C(M_\alpha)$. Recall that the constraint imposed on $C(X)$ by $\rho'X\beta = 0$ is that $\mathrm{E}(Y) \in C(X)$ and $\mathrm{E}(Y) \perp M\rho$, i.e., $\mathrm{E}(Y)$ is constrained to be orthogonal to $M\rho$. By definition, $\rho'X\beta$ puts a constraint on $C(M_\alpha)$ if $M\rho \in C(M_\alpha)$.

Proposition 4.2.2 *$\rho'X\beta$ is a contrast if and only if $M\rho \in C(M_\alpha)$.*

Proof Using Proposition 4.2.1 and $J \in C(X)$, we see that $\rho'X\beta$ is a contrast if and only if $0 = \rho'J = \rho'MJ$, i.e., $J \perp M\rho$. However, $C(M_\alpha)$ is everything in $C(X)$ that is orthogonal to J; thus $J \perp M\rho$ if and only if $M\rho \in C(M_\alpha)$. □

Finally, we can characterize $C(M_\alpha)$.

Proposition 4.2.3 $C(M_\alpha) = \left\{\rho \,\middle|\, \rho = [t_{ij}],\ t_{ij} = \lambda_i/N_i,\ \sum_{i=1}^t \lambda_i = 0\right\}.$

Proof Any vector ρ with the structure of (1) and $\sum_i \lambda_i = 0$ has $\rho'J = 0$ and by Proposition 4.2.1 determines a contrast $\rho'X\beta$. By Proposition 4.2.2, $M\rho \in C(M_\alpha)$. However, vectors that satisfy (1) also satisfy $M\rho = \rho$, so $\rho \in C(M_\alpha)$. Conversely, if $\rho \in C(M_\alpha)$, then $\rho'J = 0$; so $\rho'X\beta$ determines a contrast. It follows that $M\rho$ must be of the form (2), where $\lambda_1 + \cdots + \lambda_t = 0$. However, since $\rho \in C(M_\alpha)$, $M\rho = \rho$; so ρ must be of the form (1) with $\lambda_1 + \cdots + \lambda_t = 0$.

Exercise 4.5 Show that $\alpha_1 = \alpha_2 = \cdots = \alpha_t$ if and only if all contrasts are zero.

We now consider estimation and testing for contrasts. The least squares estimate of a contrast $\sum_{i=1}^t \lambda_i \alpha_i$ is easily obtained. Let $\hat{\mu}, \hat{\alpha}_1, \ldots, \hat{\alpha}_t$ be any choice of least squares estimates for the nonidentifiable parameters $\mu, \alpha_1, \ldots, \alpha_t$. Since $\sum_{i=1}^t \lambda_i = 0$, we can write

$$\sum_{i=1}^t \lambda_i \hat{\alpha}_i = \sum_{i=1}^t \lambda_i \{\hat{\mu} + \hat{\alpha}_i\} = \sum_{i=1}^t \lambda_i \bar{y}_{i\cdot},$$

because $\mu + \alpha_i$ is estimable and its unique least squares estimate is $\bar{y}_{i\cdot}$. This result can also be seen by examining $\rho' Y = \rho' MY$ for the ρ given earlier in (1). To test the hypothesis that $\lambda' \beta = 0$, we have seen that the numerator of the F test statistic is $(\rho' MY)^2 / \rho' M \rho$. However, $\rho' MY = \rho' X \hat{\beta} = \lambda' \hat{\beta} = \sum_{i=1}^t \lambda_i \bar{y}_{i\cdot}$. We also need to find $\rho' M \rho$. The easiest way is to observe that, since $M\rho$ has the structure of (2),

$$\begin{aligned} \rho' M \rho &= [M\rho]'[M\rho] \\ &= \sum_{i=1}^t \sum_{j=1}^{N_i} \lambda_i^2 / N_i^2 \\ &= \sum_{i=1}^t \lambda_i^2 / N_i. \end{aligned}$$

The numerator sum of squares for testing the contrast is

$$SS(\lambda' \beta) \equiv \left(\sum_{i=1}^t \lambda_i \bar{y}_{i\cdot} \right)^2 \Bigg/ \left(\sum_{i=1}^t \lambda_i^2 / N_i \right).$$

The α level test for $H_0 : \sum_{i=1}^t \lambda_i \alpha_i = 0$ is to reject H_0 if

$$\frac{\left(\sum_{i=1}^t \lambda_i \bar{y}_{i\cdot} \right)^2 \Big/ \left(\sum_{i=1}^t \lambda_i^2 / N_i \right)}{MSE} > F(1 - \alpha, 1, dfE).$$

Equivalently, $\sum_{i=1}^t \lambda_i \bar{y}_{i\cdot}$ has a normal distribution, $\mathrm{E}\left(\sum_{i=1}^t \lambda_i \bar{y}_{i\cdot} \right) = \sum_{i=1}^t \lambda_i (\mu + \alpha_i) = \sum_{i=1}^t \lambda_i \alpha_i$, and $\mathrm{Var}\left(\sum_{i=1}^t \lambda_i \bar{y}_{i\cdot} \right) = \sum_{i=1}^t \lambda_i^2 \mathrm{Var}(\bar{y}_{i\cdot}) = \sigma^2 \sum_{i=1}^t \lambda_i^2 / N_i$, so we have a t test available. The α level test is to reject H_0 if

$$\frac{\left| \sum_{i=1}^t \lambda_i \bar{y}_{i\cdot} \right|}{\sqrt{MSE \sum_{i=1}^t \lambda_i^2 / N_i}} > t\left(1 - \frac{\alpha}{2}, dfE \right).$$

Note that since $\sum_{i=1}^t \lambda_i \bar{y}_{i\cdot} = \rho' MY$ is a function of MY and MSE is a function of $Y'(I - M)Y$, we have the necessary independence for the t test. In fact, all tests and confidence intervals follow as in Exercise 2.1.

In order to break up the sums of squares for groups into $t-1$ orthogonal single degree of freedom sums of squares, we need to find $t-1$ contrasts $\lambda_1'\beta, \ldots, \lambda_{t-1}'\beta$ with the property that $\rho_r' M \rho_s = 0$ for $r \neq s$, where $\rho_r' X = \lambda_r'$ (see Section 3.6). Let $\lambda_r' = (0, \lambda_{r1}, \ldots, \lambda_{rt})$ and recall that $M\rho_r$ has the structure of (2). The condition required is

$$\begin{aligned} 0 &= \rho_r' M \rho_s \\ &= [M\rho_r]'[M\rho_s] \\ &= \sum_{i=1}^{t}\sum_{j=1}^{N_i} (\lambda_{ri}/N_i)(\lambda_{si}/N_i) \\ &= \sum_{i=1}^{t} \lambda_{ri}\lambda_{si}/N_i . \end{aligned}$$

With any set of contrasts $\sum_{i=1}^{t} \lambda_{ri}\alpha_i$, $r = 1, \ldots, t-1$, for which $0 = \sum_{i=1}^{t} \lambda_{ri}\lambda_{si}/N_i$ for all $r \neq s$, we have a set of $t-1$ orthogonal constraints on the test space so that the sums of squares for the contrasts add up to the *SSGrps*. Contrasts that determine orthogonal constraints are referred to as *orthogonal contrasts*.

In later analyses, we will need to use the fact that the analysis developed here depends only on the projection matrix onto the space for testing $\alpha_1 = \alpha_2 = \cdots = \alpha_t$. That projection matrix is $M_\alpha = M - (1/n)J_n^n$. Note that $M = (1/n)J_n^n + M_\alpha$. For any contrast $\lambda'\beta$ with $\rho' X = \lambda'$, we know that $\rho' J_n = 0$. It follows that $\rho' M = \rho'(1/n)J_n^n + \rho' M_\alpha = \rho' M_\alpha$. There are two main uses for this fact. First,

$$\sum_{i=1}^{t} \lambda_i \hat{\alpha}_i = \sum_{i=1}^{t} \lambda_i \bar{y}_{i\cdot} = \rho' M Y = \rho' M_\alpha Y,$$

$$\sum_{i=1}^{t} \lambda_i^2/N_i = \lambda'(X'X)^- \lambda = \rho' M \rho = \rho' M_\alpha \rho,$$

so estimation, and therefore tests, depend only on the projection M_α. Second, the condition for contrasts to give an orthogonal breakdown of *SSGrps* is

$$0 = \sum_{i=1}^{t} \lambda_{ri}\lambda_{si}/N_i = \rho_r' M \rho_s = \rho_r' M_\alpha \rho_s,$$

which depends only on M_α. This is just a specific example of the theory of Section 3.5.

Exercise 4.6 Using the theories of Sections 3.3 and 2.6, respectively, find the F test and the t test for the hypothesis $H_0 : \sum_{i=1}^{t} \lambda_i \alpha_i = d$ in terms of the MSE, the $\bar{y}_{i\cdot}$s, and the λ_is.

Exercise 4.7 Suppose $N_1 = N_2 = \cdots = N_t \equiv N$. Rewrite the ANOVA table incorporating any simplifications due to this assumption.

Exercise 4.8 If $N_1 = N_2 = \cdots = N_t \equiv N$, show that two contrasts $\lambda_1' \beta$ and $\lambda_2' \beta$ are orthogonal if and only if $\lambda_1' \lambda_2 = 0$.

Exercise 4.9 Find the least squares estimates of μ, α_1, and α_t using the side condition $\alpha_1 = 0$. (The other commonly used side condition is $\alpha_t = 0$.)

Exercise 4.10 Using ρ as defined by (1) and X as defined in Section 1, especially (4.1.1), show that

(a) $\rho' X = \lambda'$, where $\lambda' = (0, \lambda_1, \ldots, \lambda_t)$.
(b) $\rho \in C(X)$.
(c) For any scalar k, the test of $k\lambda'\beta = 0$ is the same as the test for $\lambda'\beta = 0$.

4.3 Additional Exercises

Exercise 4.3.1 An experiment was conducted to see which of four brands of blue jeans were most resistant to wearing out as a result of students kneeling before their linear models instructor begging for additional test points. In a class of 32 students, 8 students were randomly assigned to each brand of jeans. Before being informed of their test score, each student was required to fall to his/her knees and crawl 3 m to the instructor's desk. This was done after each of 5 mid-quarter and 3 final exams. (The jeans were distributed along with each of the 8 test forms and were collected again 36 h after grades were posted.) A fabric wear score was determined for each pair of jeans. The scores are listed below.

Brand 1:	3.41	1.83	2.69	2.04	2.83	2.46	1.84	2.34
Brand 2:	3.58	3.83	2.64	3.00	3.19	3.57	3.04	3.09
Brand 3:	3.32	2.62	3.92	3.88	2.50	3.30	2.28	3.57
Brand 4:	3.22	2.61	2.07	2.58	2.80	2.98	2.30	1.66

(a) Give an ANOVA table for these data, and perform and interpret the F test for the differences between brands.

(b) Brands 2 and 3 were relatively inexpensive, while Brands 1 and 4 were very costly. Based on these facts, determine an appropriate set of orthogonal contrasts to consider in this problem. Find the sums of squares for the contrasts.

(c) What conclusions can be drawn from these data? Perform any additional computations that may be necessary

Exercise 4.3.2 After the final exam of spring quarter, 30 of the subjects of the previous experiment decided to test the sturdiness of 3 brands of sport coats and 2 brands of shirts. In this study, sturdiness was measured as the length of time before tearing when the instructor was hung by his collar out of his second-story office window. Each brand was randomly assigned to 6 students, but the instructor was occasionally dropped before his collar tore, resulting in some missing data. The data are listed below.

Coat 1:	2.34	2.46	2.83	2.04	2.69	
Coat 2:	2.64	3.00	3.19	3.83		
Coat 3:	2.61	2.07	2.80	2.58	2.98	2.30
Shirt 1:	1.32	1.62	1.92	0.88	1.50	1.30
Shirt 2:	0.41	0.83	0.53	0.32	1.62	

(a) Give an ANOVA table for these data, and perform and interpret the F test for the differences between brands.

(b) Test whether, on average, these brands of coats are sturdier than these brands of shirts.

(c) Give three contrasts that are mutually orthogonal and orthogonal to the contrast used in (b). Compute the sums of squares for all four contrasts.

(d) Give a 95% confidence interval for the difference in sturdiness between shirt Brands 1 and 2. Is one brand significantly sturdier than the other?

Chapter 5
Multiple Comparison Techniques

Abstract This chapter introduces the multiple testing methods most commonly used with linear models.

In analyzing a linear model we can examine as many single degree of freedom hypotheses as we want. If we test all of these hypotheses at, say, the 0.05 level, then the (weak) *experimentwise error rate* (the probability of rejecting at least one of these hypotheses when all are true) will be greater than 0.05. Multiple comparison techniques are methods of performing the tests so that if all the hypotheses are true, then the probability of rejecting any of the hypotheses is no greater than some specified value, i.e., the experimentwise error rate is controlled.

A multiple comparison method can be said to be more powerful than a competing method if both methods control the experimentwise error rate at the same level, but the method in question rejects hypotheses more often than its competitor. Being more powerful, in this sense, is a mixed blessing. If one admits the idea that a null hypothesis really can be true (an idea that I am often loath to admit), then the purpose of a multiple comparison procedure is to identify which hypotheses are true and which are false. The more powerful of two multiple comparison procedures will be more likely to correctly identify hypotheses that are false as being false. It will also be more likely to incorrectly identify hypotheses that are true as being false.

A related issue is that of examining the data before deciding on the hypotheses. If the data have been examined, an hypothesis may be chosen to test because it looks as if it is likely to be significant. The nominal significance level of such a test is invalid. In fact, when doing multiple tests by any standard method, nearly all the nominal significance levels are invalid. For some methods, however, selecting the hypotheses after examining the data make the error levels intolerably bad.

The sections of this chapter contain discussions of individual multiple comparison methods. Section 1 provides a more formal background to the ideas just introduced. The specific methods discussed are Scheffé's method, the Least Significant Difference (LSD) method, the Bonferroni method, Tukey's Honest Significant Difference

R. Christensen, *Plane Answers to Complex Questions*, Springer Texts in Statistics,
https://doi.org/10.1007/978-3-030-32097-3_5

(HSD) method, and multiple range tests. The section on multiple range tests examines both the Newman–Keuls method and Duncan's method. The final section of the chapter compares the various methods. For a more complete discussion of multiple comparison methods, see Miller (1981), Hochberg and Tamhane (1987), Hsu (1996), or Bretz, Hothorn, and Westfall (2011). Christensen (1996) discusses the methods of this chapter at a more applied level, discusses Dunnett's method for comparing treatments with a control, and discusses Ott's analysis of means method.

5.1 Basic Ideas

The Neyman–Pearson theory of hypothesis testing is based on the idea of controlling the probability of Type I error, i.e., the probability of rejecting a null hypothesis when it is true. If you have multiple hypotheses to test, you have multiple chances for Type I error, so one might be interested in controlling the probability of rejecting any of the null hypotheses when they are true. Multiple testing methods seek to do this. A multiple testing method declares each hypothesis as accepted or rejected but it seeks to place some control on the overall rate of making Type I errors.

Probability of Type I error is an hypothesis testing (Neyman–Pearson) concept, not a significance testing (Fisherian) concept, so we feel free to discuss accepting null hypotheses. In this book the term *multiple comparisons* is used synonymously with multiple testing, although the former term is more properly restricted to analysis of variance problems.

Multiple testing is a complicated problem. Suppose we are interested in a collection of single degree of freedom null hypotheses $H_{0j} : \lambda'_j\beta = 0$, $j = 1, \ldots, s$. Obviously, we would like to know which hypotheses are true and which are false. There are 2^s possibilities as to which are true and false. For each of those possibilities, we can draw either the correct conclusion or an incorrect conclusion. Thus there are 4^s different conditions a set of tests for these hypotheses can take. Ideally, for each j we would like to know the probability of rejecting when H_{0j} is true and of accepting when H_{0j} is false. But we would also like to know what the collective probability of rejecting any of the true null hypotheses is, as well as knowing the collective probability of accepting any of the false hypotheses. Alas, that would require us to know which ones are true and which ones are not true. If we knew that, we would not be testing.

Rather than trying to deal with all of the possibilities, we focus attention on one specific aspect of this problem. Assuming that all of $H_{0j} : \lambda'_j\beta = 0$, $j = 1, \ldots, s$ are true, we look at methods for which the probability of rejecting any of them will be less than or equal to a specified number α. This is often referred to as controlling the *(weak) experimentwise error rate* at α.

Define $\Lambda = [\lambda_1, \ldots, \lambda_s]$. Clearly, $\Lambda'\beta = 0$ if and only if $\lambda'_j\beta = 0$, $j = 1, \ldots, s$. Moreover, both of these conditions are equivalent to $\lambda'\beta = 0$ for every $\lambda \in C(\Lambda)$. Equivalently, there exists a $\lambda \in C(\Lambda)$ such that $\lambda'\beta \neq 0$ if and only if $\Lambda'\beta \neq 0$ which, since the λ_js are a spanning set for $C(\Lambda)$, occurs if and only if there exists a j such that $\lambda_j'\beta \neq 0$. Indeed, we do not even need to define $\Lambda = [\lambda_1, \dots, \lambda_s]$ as long as we pick a Λ with

$$C(\Lambda) = \text{span}\{\lambda_1, \dots, \lambda_s\}.$$

Henceforth, *any reference within this chapter to $\lambda'\beta$ presumes that $\lambda \in C(\Lambda)$.*

If you are testing $\lambda'\beta = 0$ for any $\lambda \in C(\Lambda)$, you need to be sure that your procedure behaves in a reasonable way when λ is chosen as some projection of $\hat{\beta}$ into $C(\Lambda)$. This is far less of a problem when you are only looking at a prespecified collection of hypotheses $\lambda_j'\beta = 0$, $j = 1, \dots, s$. More on this later.

All of the multiple testing methods of which I am aware fit into the following algorithm:

1. Test $H_0 : \Lambda'\beta = 0$ at level α.
2. If this overall test is not significant, quit and go home. None of the $\lambda'\beta$s are declared significant.
3. If this overall test is significant, use some *reasonable* method to decide which $\lambda'\beta$s to declare significant.

Whether the method used in step 3 is "reasonable" or not, this algorithm controls the experimentwise error rate for testing $\lambda_j'\beta = 0$, $j = 1, \dots, s$ or even for testing $\lambda'\beta = 0$ for any $\lambda \in C(\Lambda)$. Based on step 1, if the null hypotheses are all true, the probability of rejecting any of them is no more than α. The trick is to find a method for declaring significance in step 3 that behaves in ways that you like. You also need to choose a test for step 1. Typically, the procedure in step 3 is related to the choice of test in step 1, but there is no compelling reason that it needs to be.

The difficulty in step 3 arises when some js have $\lambda_j'\beta = 0$ but some have $\lambda_j'\beta \neq 0$. The latter may get you into step 3 where you would like to get the decision correct for each $\lambda_j'\beta$. More generally, the λ_js with $\lambda_j'\beta = 0$ define some subspace of $C(\Lambda)$ where $\lambda'\beta = 0$ but any $\lambda \in C(\Lambda)$ outside of that subspace has $\lambda'\beta \neq 0$. Ideally we would like to make the correct decision on every $\lambda'\beta$. Choosing a multiple testing method is largely a choice about how hard you want to make it to reject an hypothesis $\lambda'\beta = 0$. If you make it hard to reject, you will get most of the $\lambda'\beta = 0$s correct but can make many mistakes when $\lambda'\beta \neq 0$. If you make it easy to reject an hypothesis $\lambda'\beta = 0$, you will get more of the $\lambda'\beta \neq 0$s correct but make more mistakes when $\lambda'\beta = 0$.

To illustrate the algorithm, we assume that $\Lambda'\beta$ is estimable. This implies that each $\lambda_j'\beta$ is estimable and that $\lambda'\beta$ is estimable whenever $\lambda \in C(\Lambda)$. Recall that the sum of squares for testing an estimable $\lambda'\beta = 0$ is

$$SS(\lambda'\beta) \equiv \frac{(\lambda'\hat{\beta})^2}{\lambda'(X'X)^-\lambda} = \frac{(\rho'MY)^2}{\rho'M\rho} = Y'[M\rho(\rho'M\rho)^{-1}\rho'M]Y \equiv Y'M_{M\rho}Y,$$

where $\lambda' = \rho'X$.

The *Least Significant Difference (LSD)* method often (I think incorrectly) associated with Fisher, amounts to

1. Check the truth of

$$\frac{(\Lambda'\hat{\beta})'[\Lambda'(X'X)^-\Lambda]^-(\Lambda'\hat{\beta})/r(\Lambda)}{MSE} > F[1-\alpha, r(\Lambda), n-r(X)].$$

2. If the inequality is not true, quit and go home. None of the $\lambda'\beta$s are declared significant.
3. If the inequality is true, declare $\lambda'\beta$ significantly different from 0 if and only if

$$\frac{SS(\lambda'\beta)}{MSE} > F[1-\alpha, 1, n-r(X)].$$

Note that step 3 is just the usual α level F test for $\lambda'\beta = 0$. Certainly, this step 3 qualifies as a reasonable method but it makes it easy to reject. Collective experience indicates that the method works reasonably well if you restrict yourself in step 3 to testing only a prespecified finite collection $\lambda'_j\beta = 0$, $j = 1, \ldots, s$. Experience indicates that this method performs very badly, in the sense that it rejects too many hypotheses $\lambda'\beta = 0$ that are true, if you use it to test arbitrarily chosen $\lambda'\beta$.

Scheffé's method amounts to

1. Check the truth of

$$\frac{(\Lambda'\hat{\beta})'[\Lambda'(X'X)^-\Lambda]^-(\Lambda'\hat{\beta})/r(\Lambda)}{MSE} > F[1-\alpha, r(\Lambda), n-r(X)]. \quad (1)$$

2. If the inequality is not true, quit and go home. None of the $\lambda'\beta$s are declared significant.
3. If the inequality is true, declare $\lambda'\beta$ significantly different from 0 if and only if

$$\frac{SS(\lambda'\beta)/r(\Lambda)}{MSE} > F[1-\alpha, r(\Lambda), n-r(X)].$$

The basis of Scheffé's method is replacing $(\Lambda'\hat{\beta})'[\Lambda'(X'X)^-\Lambda]^-(\Lambda'\hat{\beta})'$ in (1) with $SS(\lambda'\beta)$. We will show later that

$$(\Lambda'\hat{\beta})'[\Lambda'(X'X)^-\Lambda]^-(\Lambda'\hat{\beta})' \geq SS(\lambda'\beta), \quad (2)$$

so it is impossible to reject $\lambda'\beta = 0$ unless you have previously rejected $\Lambda'\beta = 0$. In fact, you can skip steps 1 and 2 of the algorithm because they take care of themselves. We will also show that there exists a $\lambda \in C(\Lambda)$ for which equality occurs in (2). This ensures that if you get past step 1, in step 3 there always exists a $\lambda'\beta = 0$ that would be rejected. Unfortunately, there is no assurance that any of the specific $\lambda'_j\beta = 0$ hypotheses will be rejected, even though we are reasonably confident that they cannot all be true.

Scheffé's procedure makes it very difficult to reject a $\lambda'\beta = 0$. Collective experience indicates that this method performs well if you use it to test arbitrary $\lambda'\beta$

(including ones suggested by the data) but is too conservative if your interest is only in testing a prespecified finite collection $\lambda_j'\beta = 0$, $j = 1, \ldots, s$. Too conservative means not finding enough of the $\lambda_j'\beta$s that are different from 0.

The Bonferroni method fits the algorithm somewhat awkwardly. Pick numbers $\alpha_j > 0$ such that $\sum_{j=1}^{s} \alpha_j = \alpha$. Typically, one picks $\alpha_j = \alpha/s$. The α level test of $\Lambda'\beta = 0$ is rejected if any of the following statements are true:

$$\frac{SS(\lambda_j'\beta)}{MSE} > F[1 - \alpha_j, 1, n - r(X)],$$

$j = 1, \ldots, s$. As will be shown later, typically this is not an exact α level test but has a null hypothesis probability of rejecting $\Lambda'\beta = 0$ that is less than or equal to α. The multiple testing method is

1. Check if any of the following statements are true:

$$\frac{SS(\lambda_j'\beta)}{MSE} > F[1 - \alpha_j, 1, n - r(X)], \tag{3}$$

   $j = 1, \ldots, s$
2. If none of the inequalities is true, quit and go home. None of the $\lambda_j'\beta$s are declared significant.
3. For each j, if inequality (3) is true, declare $\lambda_j'\beta$ significantly different from 0,

I have never seen this used to test arbitrary $\lambda'\beta = 0$ for $\lambda \in C(\Lambda)$ although if you take $\alpha_j \equiv \alpha/s$ it is obvious how to do so. Typically, Bonferroni makes it harder to reject $\lambda'\beta = 0$ than LSD but easier to reject than Scheffé.

As discussed in Christensen (1996, Chapter 6) [but *not* in the second edition, Christensen (2015)], for a balanced one-way ANOVA with $\Lambda'\beta = 0$ denoting equality of group means, there are several other tests available for $\Lambda'\beta = 0$. The best known of these is the test based on the Studentized range statistic which leads naturally to Tukey's *Honest Significant Difference (HSD)* method but also to the Newman-Keuls multiple range method. The difference between these two is that they use different reasonable methods at step 3. Dunnett's many-one *t* statistic method is based on a test designed to have good power for detecting differences between the mean of a control group and any of the other group means. Otts' analysis of means method, which is closely related to Shewhart's control charts, is based on a version of the Studentized maximum modulus statistic that is designed to have good power for detecting differences between the mean of a group and the average of the other group means. The test is appropriate when most groups should have the same mean, but occasionally a weird group pops up. Typically, these tests are used only for a finite number of hypotheses that are naturally associated with the form of the test statistic, specifically, $\mu_j - \mu_k$, $j \neq k$ for Tukey, $\mu_1 - \mu_j$, $j \neq 1$ for Dunnet, and $\mu_j - \bar{\mu}. \propto \mu_j - \text{mean}_{k \neq j}\mu_k$. In fact, step 3 is typically only defined for the naturally associated hypotheses. Scheffé (1959) discusses how to extend Tukey's method to testing arbitrary contrasts $\lambda'\beta$, and finds that Tukey's method works worse than his for arbitrary contrasts.

To be honest, all of these multiple testing methods except LSD are defined in such a say that you never have to think about steps 1 or 2. They are defined so that it is impossible to reject in step 3 unless you have already rejected in step one. The value of the algorithm is that it emphasizes how much freedom one has to choose a reasonable procedure in step 3.

A more general multiple testing problem evaluates $H_{0j} : \lambda_j'\beta = d_j, j = 1, \ldots, s$, for known values d_j. For this problem to make sense we need to know that there exists, and that we can actually find, a vector b such that $d_j = \lambda_j'b$, $j = 1, \ldots, s$. Collectively, the testing problem is $\Lambda'\beta = d$ where $d = (d_1, \ldots, d_s)'$ and there exists b such that $\Lambda'b = d$. The three procedures that we outlined continue to hold with the following substitutions:

$$(\Lambda'\hat{\beta})'[\Lambda'(X'X)^-\Lambda]^-(\Lambda'\hat{\beta})' \quad \rightarrow \quad (\Lambda'\hat{\beta} - d)'[\Lambda'(X'X)^-\Lambda]^-(\Lambda'\hat{\beta} - d)$$

and

$$SS(\lambda'\beta) \quad \rightarrow \quad \frac{[\lambda'(\hat{\beta} - b)]^2}{\lambda'(X'X)^-\lambda} = \frac{[\rho'M(Y - Xb)]^2}{\rho'M\rho}$$

so that

$$SS(\lambda_j'\beta) = \frac{(\lambda_j'\hat{\beta} - d_j)^2}{\lambda'(X'X)^-\lambda}.$$

We can define multiple confidence intervals for the $\lambda_j'\beta$s. The interval for $\lambda_j'\beta$ is the set of all values d_j for which $\lambda_j'\beta = d_j$ is not rejected. Unfortunately, the intersection of such intervals can contain nonsensical values. If $\lambda_1 = 2\lambda_2$, and if the interval for $\lambda_1'\beta$ is (a, b), then the interval for $\lambda_2'\beta$ is $(2a, 2b)$. Collectively these define a rectangle of reasonable values for the pair $(\lambda_1'\beta, \lambda_2'\beta)$ defined by the endpoints $(a, 2a)$, $(a, 2b)$, $(b, 2b)$, and $(b, 2a)$. For our choice of λ_1 and λ_2, every point in that rectangle in nonsensical except for those on the line segment from $(a, 2a)$ to $(b, 2b)$.

5.2 Scheffé's Method

Scheffé's method of multiple comparisons is an omnibus technique that allows one to test any and all single degree of freedom hypotheses that put constraints on a given subspace. It provides the assurance that the experimentwise error rate will not exceed a given level α. Typically, this subspace will be for fitting a set of parameters after fitting the mean, and in ANOVA problems is some sort of treatment space.

It is, of course, easy to find silly methods of doing multiple comparisons. One could, for example, always accept the null hypothesis. However, if the subspace is of value in fitting the model, Scheffé's method assures us that there is at least one hypothesis in the subspace that will be rejected. That is, if the F test is significant for testing that the subspace adds to the model, then there exists a linear hypothesis, putting a constraint on the subspace, that will be deemed significant by Scheffé's method.

We look at single degree of freedom estimable hypotheses $\lambda'\beta = 0$ that put constraints on a given subspace. Call that subspace the test space, say, $\mathscr{M}_T \subset C(X)$ with ppo M_T. A test space can result from testing a reduced model $Y = X_0\gamma + e$ with $C(X_0) \subset C(X)$ and $\mathscr{M}_T = C(X_0)^{\perp}_{C(X)}$ so that $M_T = M - M_0$. Another convenient way of defining the test space is by defining an estimable hypothesis $\Lambda'\beta = 0$ with $\Lambda' = P'X$. In that case $\mathscr{M}_T = C(MP)$ so that $M_T = M_{MP}$. In either case we restrict ourselves to looking at $\lambda'\beta$ with $\lambda' = \rho'X$ and $M\rho \in \mathscr{M}_T$. In the former case, λ must have $M\rho \perp C(X_0)$ and in the later case $\lambda \in C(\Lambda)$.

Scheffé's method is that an hypothesis $H_0 : \lambda'\beta = 0$ is rejected if

$$\frac{SS(\lambda'\beta)/s}{MSE} > F(1-\alpha, s, dfE),$$

where $SS(\lambda'\beta)$ is the sum of squares for the usual test of the hypothesis, s is the dimension of the subspace $\mathscr{M}_T$, and $\lambda'\beta = 0$ is assumed to put a constraint on the subspace.

In terms of a one-way analysis of variance where the subspace is the space for testing equality of the group means, Scheffé's method applies to testing whether all contrasts equal zero. With t groups, a contrast is deemed significantly different from zero at the α level if the sum of squares for the contrast divided by $t-1$ and the MSE is greater than $F(1-\alpha, t-1, dfE)$.

Theorem 5.1.1 given below leads immediately to the key properties of Scheffé's method. Recall that if $\rho'X\beta = 0$ puts a constraint on a subspace, then $M\rho$ is an element of that subspace. Theorem 5.1.1 shows that the F test for the subspace rejects if and only if the Scheffé test rejects the single degree of freedom hypothesis $\rho'X\beta = 0$ for some ρ with $M\rho$ in the subspace. The proof is accomplished by finding a vector in the subspace having the property that the sum of squares for testing the corresponding one degree of freedom hypothesis equals the sum of squares for the entire space. Of course, the particular vector that has this property depends on Y. To emphasize this dependence on Y, the vector is denoted m_Y. In the proof of the theorem, m_Y is seen to be just the projection of Y onto the subspace (hence the use of the letter m in the notation).

It follows that for a one-way ANOVA there is always a contrast for which the contrast sum of squares equals the sum of squares for groups. The exact nature of this contrast depends on Y and often the contrast is completely uninterpretable. Nevertheless, the existence of such a contrast establishes that Scheffé's method rejects for some contrast if and only if the test for equality of group means is rejected.

Theorem 5.2.1 *Consider the linear model $Y = X\beta + e$ and let M_T be the perpendicular projection operator onto some subspace of $C(X)$. Let $r(M_T) = s$. Then*

$$\frac{Y'M_TY/s}{MSE} > F(1-\alpha, s, dfE)$$

if and only if there exists a vector m_Y such that $Mm_Y \in C(M_T)$ and

$$\frac{SS(m_Y'X\beta)/s}{MSE} > F(1-\alpha, s, dfE).$$

Proof $\Rightarrow$ We want to find a vector m_Y so that if the F test for the subspace is rejected, then Mm_Y is in $C(M_T)$, and the hypothesis $m_Y'X\beta = 0$ is rejected by Scheffé's method. If we find m_Y within $C(M_T)$ and $SS(m_Y'X\beta) = Y'M_TY$, we are done. Let $m_Y = M_TY$.

As in Section 3.5, $SS(m_Y'X\beta) = Y'M_Tm_Y[m_Y'M_Tm_Y]^{-1}m_Y'M_TY$. Since $M_Tm_Y = M_TM_TY = M_TY = m_Y$, we have $SS(m_Y'X\beta) = Y'm_Y[m_Y'm_Y]^{-1}m_Y'Y = (Y'M_TY)^2/Y'M_TY = Y'M_TY$, and we are finished.

$\Leftarrow$ We prove the contrapositive, i.e., if $Y'M_TY/s\ MSE \le F(1-\alpha, s, dfE)$, then for any ρ such that $M\rho \in C(M_T)$, we have $SS(\rho'X\beta)/s\ MSE \le F(1-\alpha, s, dfE)$. To see this, observe that

$$SS(\rho'X\beta) = Y'[M\rho(\rho'M\rho)^{-1}\rho'M]Y.$$

Since $[M\rho(\rho'M\rho)^{-1}\rho'M]$ is the perpendicular projection matrix onto a subspace of $C(M_T)$,

$$Y'[M\rho(\rho'M\rho)^{-1}\rho'M]Y \le Y'M_TY$$

and we are done. □

For an application of this Theorem, see Exercise 5.8.5

$Y'M_TY$ is the sum of squares for testing the reduced model $Y = (M - M_T)\gamma + e$. If this null model is true,

$$\Pr\left[\frac{Y'M_TY}{s\ MSE} > F(1-\alpha, s, dfE)\right] = \alpha.$$

The theorem therefore implies that the experimentwise error rate for testing all hypotheses $\rho'X\beta = 0$ with $M\rho \in C(M_T)$ is exactly α. More technically, we wish to test the hypotheses

$$H_0 : \lambda'\beta = 0 \quad \text{for } \lambda \in \{\lambda | \lambda' = \rho'X \text{ with } M\rho \in C(M_T)\}.$$

The theorem implies that

$$\Pr\left[\frac{SS(\lambda'\beta)/s}{MSE} > F(1-\alpha, s, dfE) \text{ for some } \lambda, \lambda' = \rho'X, M\rho \in C(M_T)\right] = \alpha,$$

so the experimentwise error rate is α. The theorem also implies that if the omnibus F test rejects, there exists some single degree of freedom test that will be rejected. Note that which single degree of freedom tests are rejected depends on what the data are, as should be expected.

Scheffé's method can also be used for testing a subset of the set of all hypotheses putting a constraint on $C(M_T)$. For testing a subset, the experimentwise error rate will be no greater than α and typically much below α. The primary problem with using Scheffé's method is that, for testing a finite number of hypotheses, the experimentwise error rate is so much below the nominal rate of α that the procedure has very little power. (On the other hand, you can be extra confident, when rejecting with Scheffé's method, that you are not making a Type I error.)

Suppose that we want to test

$$H_0 : \lambda_k'\beta = 0, \quad k = 1, \ldots, r.$$

The constraints imposed by these hypotheses are $M\rho_k$, $k = 1, \ldots, r$, where $\lambda_k' = \rho_k'X$. If $C(M_T)$ is chosen so that $C(M\rho_1, \ldots, M\rho_r) \subset C(M_T)$, then by the previous paragraph, if H_0 is true,

$$\Pr\left[\frac{SS(\lambda_k'\beta)/s}{MSE} > F(1-\alpha, s, dfE) \text{ for some } k,\ k = 1, \ldots, r\right] \le \alpha.$$

For testing a finite number of hypotheses, it is possible to reject the overall F test but not reject for any of the specific hypotheses.

We now show that the most efficient procedure is to choose $C(M\rho_1, \ldots, M\rho_r) = C(M_T)$. In particular, given that a subspace contains the necessary constraints, the smaller the rank of the subspace, the more powerful is Scheffé's procedure. Consider two subspaces, one of rank s and another of rank t, where $s > t$. Both procedures guarantee that the experimentwise error rate is no greater than α. The more powerful procedure is the one that rejects more often. Based on the rank s subspace, Scheffé's method rejects if

$$SS(\lambda'\beta)/MSE > sF(1-\alpha, s, dfE).$$

For the rank t subspace, the method rejects if

$$SS(\lambda'\beta)/MSE > tF(1-\alpha, t, dfE).$$

With $s > t$, by Theorem C.4,

$$sF(1-\alpha, s, dfE) \ge tF(1-\alpha, t, dfE).$$

One gets more rejections with the rank t space, hence it gives a more powerful procedure.

Example 5.2.2 *One-Way ANOVA.*
Consider the model

$$y_{ij} = \mu + \alpha_i + e_{ij}, \quad e_{ij}\text{s i.i.d. } N(0, \sigma^2),$$

$i = 1, 2, 3, 4$, $j = 1, \ldots, N$. To test the three contrast hypotheses

$$\lambda_1'\beta = \alpha_1 + \alpha_2 - \alpha_3 - \alpha_4 = 0,$$
$$\lambda_2'\beta = \alpha_1 - \alpha_2 + \alpha_3 - \alpha_4 = 0,$$
$$\lambda_3'\beta = \alpha_1 + 0 + 0 - \alpha_4 = 0,$$

we can observe that the contrasts put constraints on the space for testing $H_0 : \alpha_1 = \alpha_2 = \alpha_3 = \alpha_4$ which has rank 3. We can apply Scheffé's method: reject $H_0 : \lambda_k'\beta = 0$ if

$$\frac{SS(\lambda_k'\beta)/3}{MSE} > F(1-\alpha, 3, 4(N-1)).$$

A more efficient method is to notice that $\lambda_1'\beta + \lambda_2'\beta = 2\lambda_3'\beta$. This is true for any β, so $\lambda_1' + \lambda_2' = 2\lambda_3'$ and, using (4.2.2), $M\rho_1 + M\rho_2 = 2M\rho_3$. Since λ_1 and λ_2 are linearly independent, $M\rho_1$ and $M\rho_2$ are also; thus $C(M\rho_1, M\rho_2, M\rho_3)$ is a rank 2 space and Scheffé's method can be applied as: reject $H_0 : \lambda_k'\beta = 0$ if

$$\frac{SS(\lambda_k'\beta)/2}{MSE} > F(1-\alpha, 2, 4(N-1)).$$

One virtue of Scheffé's method is that since it is really a test of all the hypotheses in a subspace, you can look at the data to help you pick an hypothesis and the test remains valid.

Scheffé's method can also be used to find simultaneous confidence intervals. To show this we need some additional structure for the problem. Let $X = [X_0, X_1]$ and let $\beta' = [\beta_0', \beta_1']$, so that

$$Y = X_0\beta_0 + X_1\beta_1 + e.$$

Let M and M_0 be the perpendicular projection operators onto $C(X)$ and $C(X_0)$, respectively, and let $M_T = M - M_0$. We seek to find simultaneous confidence intervals for all estimable functions $\rho' X_1\beta_1$. Note that $\rho' X_1\beta_1$ is estimable if and only if $\rho' X_0 = 0$, which occurs if and only if $0 = M_0\rho = M\rho - M_T\rho$, i.e., $M\rho = M_T\rho$. It follows that if $\rho' X_1\beta_1$ is an estimable function, then $\rho' X_1\beta_1 = \rho' M X_1\beta_1 = \rho' M_T X_1\beta_1$. Conversely, for any vector ρ, $\rho' M_T(X_0\beta_0 + X_1\beta_1) = \rho' M_T X_1\beta_1$ is an estimable function. Proceeding as in Section 3.7, and observing that $M_T X\beta = M_T X_0\beta_0 + M_T X_1\beta_1 = M_T X_1\beta_1$, we have

$$\frac{(Y - X_1\beta_1)' M_T (Y - X_1\beta_1)/r(M_T)}{MSE} \sim F(r(M_T), dfE, 0),$$

so that

$$\Pr\left[\frac{(Y - X_1\beta_1)' M_T (Y - X_1\beta_1)/r(M_T)}{MSE} \leq F(1-\alpha, r(M_T), dfE)\right] = 1 - \alpha$$

or, equivalently,

$$
\begin{aligned}
1-\alpha &= \Pr\left[\frac{(Y-X_1\beta_1)'M_T\rho(\rho'M_T\rho)^{-1}\rho'M_T(Y-X_1\beta_1)/r(M_T)}{MSE}\right. \\
&\qquad\qquad \left.\le F(1-\alpha, r(M_T), dfE) \text{ for all } \rho\right] \\
&= \Pr\big[\,|\rho'M_TY-\rho'M_TX_1\beta_1| \\
&\qquad\qquad \le \sqrt{(\rho'M_T\rho)(MSE)r(M_T)F(1-\alpha, r(M_T), dfE)} \text{ for all } \rho\big].
\end{aligned}
$$

This leads to obvious confidence intervals for all functions $\rho'M_TX_1\beta_1$ and thus to confidence intervals for arbitrary estimable functions $\rho'X_1\beta_1$.

Example 5.2.3 *One-Way ANOVA.*
Consider the model $y_{ij} = \mu + \alpha_i + e_{ij}$, e_{ij}s independent $N(0, \sigma^2)$, $i = 1, \ldots, t$, $j = 1, \ldots, N_i$, and the space for testing $\alpha_1 = \alpha_2 = \cdots = \alpha_t$. The linear functions that put constraints on that space are the contrasts. Scheffé's method indicates that $H_0 : \sum_{i=1}^t \lambda_i\alpha_i = 0$ should be rejected if

$$
\frac{(\sum \lambda_i \bar{y}_{i\cdot})^2/(\sum \lambda_i^2/N_i)}{(t-1)MSE} > F(1-\alpha, t-1, dfE).
$$

To find confidence intervals for contrasts, write $X = [J, X_1]$ and $\beta' = [\mu, \beta_1']$, where $\beta_1' = [\alpha_1, \ldots, \alpha_t]$. We can get simultaneous confidence intervals for estimable functions $\rho'X_1\beta_1$. As discussed in Chapter 4, the estimable functions $\rho'X_1\beta_1$ are precisely the contrasts. The simultaneous $(1-\alpha)100\%$ confidence intervals have limits

$$
\sum \lambda_i \bar{y}_{i\cdot} \pm \sqrt{(t-1)F(1-\alpha, t-1, dfE)MSE\left(\sum \lambda_i^2/N_i\right)}.
$$

5.3 Least Significant Difference Method

The *Least Significant Difference (LSD)* method is a general technique for testing a fixed number of hypotheses $\lambda_k'\beta = 0$, $k = 1, \ldots, r$, chosen without looking at the data. The constraints imposed by these hypotheses generate some subspace. (Commonly, one identifies the subspace first and picks hypotheses that will generate it.) The technique involves three steps. First, do an α level F test for whether the subspace adds to the model. If this omnibus F test is not significant, we can conclude that the data are consistent with $\lambda_k'\beta = 0$, $k = 1, \ldots, r$. If the F test is significant, we want to identify which hypotheses are not true. To do this, test each hypothesis $\lambda_k'\beta = 0$ with a t test (or an equivalent F test) at the α level.

The experimentwise error rate is controlled by using the F test for the subspace. When all of the hypotheses are true, the probability of identifying any of them as false is no more than α, because α is the probability of rejecting the omnibus F test. Although the omnibus F test is precisely a test of $\lambda_k'\beta = 0$, $k = 1, \ldots, r$, even if the F test is rejected, the LSD method may not reject any of the specific hypotheses being considered. For this reason, the experimentwise error rate is less than α.

The LSD method is more powerful than Scheffé's method. If the hypotheses generate a space of rank s, then Scheffé's method rejects if $SS(\lambda_k'\beta)/MSE > sF(1-\alpha, s, dfE)$. The LSD rejects if $SS(\lambda_k'\beta)/MSE > F(1-\alpha, 1, dfE)$. By Theorem C.4, $sF(1-\alpha, s, dfE) > F(1-\alpha, 1, dfE)$, so the LSD method will reject more often than Scheffé's method. Generally, the LSD method is more powerful than other methods for detecting when $\lambda_k'\beta \neq 0$; but if $\lambda_k'\beta = 0$, it is more likely than other methods to incorrectly identify the hypothesis as being different from zero.

Note that it is not appropriate to use an F test for a space that is larger than the space generated by the r hypotheses. Such an F test can be significant for reasons completely unrelated to the hypotheses, thus invalidating the experimentwise error rate.

Exercise 5.1 Consider the ANOVA model

$$y_{ij} = \mu + \alpha_i + e_{ij},$$

$i = 1, \ldots, t$, $j = 1, \ldots, N$, with the e_{ij}s independent $N(0, \sigma^2)$. Suppose it is desired to test the hypotheses $\alpha_i = \alpha_{i'}$ for all $i \neq i'$. Show that, if the F test for group effects is significant, there is one number, called the LSD, so that the least significant difference rejects $\alpha_i = \alpha_{i'}$ precisely when

$$|\bar{y}_{i\cdot} - \bar{y}_{i'\cdot}| > LSD.$$

Exercise 5.2 In the model of Exercise 5.1, let $t = 4$. Suppose we want to use the LSD method to test contrasts labeled A, B, and C defined by

Name	λ_1	λ_2	λ_3	λ_4
A	1	1	−1	−1
B	0	0	1	−1
C	1/3	1/3	1/3	−1

Describe the procedure. Give test statistics for each test that is to be performed. Hint: The contrasts define a two-dimensional space.

5.4 Bonferroni Method

Suppose we have chosen, before looking at the data, a set of r hypotheses to test, say, $\lambda_k'\beta = 0, k = 1, \ldots, r$. The *Bonferroni method (BSD)* consists of rejecting H_0 : $\lambda_k'\beta = 0$ if

$$\frac{SS(\lambda_k'\beta)}{MSE} > F\left(1 - \frac{\alpha}{r}, 1, dfE\right).$$

The Bonferroni method simply reduces the significance level of each individual test so that the sum of the significance levels is no greater than α. (In fact, the reduced significance levels do not have to be α/r as long as the sum of the individual significance levels is α.)

This method rests on a Bonferroni inequality. For sets $A_1, \ldots, A_r$, $\Pr(\bigcup_{k=1}^r A_k) \leq \sum_{k=1}^r \Pr(A_k)$. (This inequality is nothing more than the statement that a probability measure is finitely subadditive.) If all the hypotheses $\lambda_k'\beta = 0, k = 1, \ldots, r$ are true, then the experimentwise error rate is

$$\begin{aligned}
&\Pr\left(SS(\lambda_k'\beta) > MSEF\left(1 - \frac{\alpha}{r}, 1, dfE\right) \text{ for some } k\right) \\
&= \Pr\left(\bigcup_{k=1}^r \left[SS(\lambda_k'\beta) > MSEF\left(1 - \frac{\alpha}{r}, 1, dfE\right)\right]\right) \\
&\leq \sum_{k=1}^r \Pr\left(SS(\lambda_k'\beta) > MSEF\left(1 - \frac{\alpha}{r}, 1, dfE\right)\right) \\
&= \sum_{k=1}^r \frac{\alpha}{r} = \alpha.
\end{aligned}$$

If the hypotheses to be tested are chosen after looking at the data, the individual significance levels of α/r are invalid, so the experimentwise error rate has not been controlled.

Given that the subspace F test is rejected, the LSD method is more powerful than the Bonferroni method because $F\left(1 - \frac{\alpha}{r}, 1, dfE\right) > F(1 - \alpha, 1, dfE)$. The Bonferroni method is designed to handle a finite number of hypotheses, so it is not surprising that it is usually a more powerful method than Scheffé's method for testing the r hypotheses if r is not too large.

5.5 Tukey's Method

Tukey's method, also known as the *Honest Significant Difference (HSD)* method, is designed to compare all pairs of means for a set of independent normally distributed random variables with a common variance. (Meaning no disrespect to a great statistician, John Tukey, I've never been able to think of this as anything other than Honest John's Significant Difference.) Let $y_i \sim N(\mu_i, \sigma^2), i = 1, \ldots, t$, let the y_is be independent, and let S^2 be an estimate of σ^2 with S^2 independent of the y_is and

$$\frac{\nu S^2}{\sigma^2} \sim \chi^2(\nu).$$

Tukey's method depends on knowing the distribution of the *Studentized range* when $\mu_1 = \mu_2 = \cdots = \mu_t$, i.e., we need to know that

$$Q \equiv \frac{\max_i y_i - \min_i y_i}{S} \sim Q(t, \nu)$$

and we need to be able to find percentage points of the $Q(t, \nu)$ distribution. These are programmed into a lot of statistical software and are tabled in many books on statistical methods, e.g., Christensen (1996, 2015), Snedecor and Cochran (1980), and Kutner, Nachtsheim, Neter, and Li (2005).

If the observed value of Q is too large, the null hypothesis $H_0 : \mu_1 = \cdots = \mu_t$ should be rejected. That is because any differences in the μ_is will tend to make the range large relative to the distribution of the range when all the μ_is are equal. Since the hypothesis $H_0 : \mu_1 = \cdots = \mu_t$ is equivalent to the hypothesis $H_0 : \mu_i = \mu_j$ for all i and j, we can use the Studentized range test to test all pairs of means. Reject the hypothesis that $H_0 : \mu_i = \mu_j$ if

$$\frac{|y_i - y_j|}{S} > Q(1 - \alpha, t, \nu),$$

where $Q(1 - \alpha, t, \nu)$ is the $(1 - \alpha)100$ percentage point of the $Q(t, \nu)$ distribution. If $H_0 : \mu_i = \mu_j$ for all i and j is true, then at least one of these tests will reject H_0 if and only if

$$\frac{\max_i y_i - \min_i y_i}{S} > Q(1 - \alpha, t, \nu),$$

which happens with probability α. Thus the experimentwise error rate is exactly α.

Example 5.5.1 *Balanced Two-Way ANOVA.*
Consider the model

$$y_{ijk} = \mu + \alpha_i + \beta_j + e_{ijk}, \quad e_{ijk}\text{s i.i.d. } N(0, \sigma^2),$$

$i = 1, \ldots, a$, $j = 1, \ldots, b$, $k = 1, \ldots, N$. Suppose we want to test the hypotheses $H_0 : \beta_j = \beta_{j'}$ for all $j \neq j'$. Consider the $\bar{y}_{\cdot j \cdot}$ values. Here

$$\bar{y}_{\cdot j \cdot} \sim N\big(\mu + \bar{\alpha}_{\cdot} + \beta_j, \sigma^2/aN\big)$$

and the $\bar{y}_{\cdot j \cdot}$s are independent because $\bar{y}_{\cdot j \cdot}$ depends only on $\bar{e}_{\cdot j \cdot}$ for its randomness; and since $\bar{e}_{\cdot j \cdot}$ and $\bar{e}_{\cdot j' \cdot}$ are based on disjoint sets of the e_{ijk}s, they must be independent. We will see in Section 7.1 that the $\bar{y}_{\cdot j \cdot}$s are least squares estimates of the $\mu + \bar{\alpha}_{\cdot} + \beta_j$s, so the $\bar{y}_{\cdot j \cdot}$s must be independent of the MSE. It follows quickly that if $H_0 : \beta_1 = \cdots = \beta_b$ is true, then

$$\frac{\max_j \bar{y}_{\cdot j \cdot} - \min_j \bar{y}_{\cdot j \cdot}}{\sqrt{MSE/aN}} \sim Q(b, dfE);$$

and we reject $H_0 : \beta_j = \beta_{j'}$ if

$$|\bar{y}_{\cdot j \cdot} - \bar{y}_{\cdot j' \cdot}| > Q(1 - \alpha, b, dfE)\sqrt{MSE/aN}.$$

Note that the Studentized range provides a competitor to the usual analysis of variance F test for the hypothesis $H_0 : \beta_1 = \cdots = \beta_b$, cf. http://www.stat.unm.edu/~fletcher/UMPI.pdf. Also, Tukey's method is only applicable when all the means being used are based on the same number of observations.

5.6 Multiple Range Tests: Newman–Keuls and Duncan

The *Newman–Keuls multiple range method* is a competitor to the Tukey method. It looks at all pairs of means. In fact, it amounts to a sequential version of Tukey's method. Using the notation of the previous section, order the y_is from smallest to largest, say

$$y_{(1)} \leq y_{(2)} \leq \cdots \leq y_{(t)},$$

and define $\mu_{(i)} \equiv \mu_j$ when $y_{(i)} = y_j$. Note that the $\mu_{(i)}$s need not be ordered in any particular way. However, the Newman–Keuls method acts as if the $\mu_{(i)}$s are also ordered. With this notation, we can write the Studentized range as

$$Q = \frac{y_{(t)} - y_{(1)}}{S}.$$

The Newman–Keuls method rejects $H_0 : \mu_{(t)} = \mu_{(1)}$ if $y_{(t)} - y_{(1)} > SQ(1 - \alpha, t, \nu)$. If this hypothesis is not rejected, stop. All means are considered equal. If this hypothesis is rejected, we continue.

The next step tests two hypotheses. $H_0 : \mu_{(t-1)} = \mu_{(1)}$ is rejected if $y_{(t-1)} - y_{(1)} > SQ(1 - \alpha, t - 1, \nu)$. $H_0 : \mu_{(t)} = \mu_{(2)}$ is rejected if $y_{(t)} - y_{(2)} > SQ(1 - \alpha, t - 1, \nu)$. If $\mu_{(t-1)} = \mu_{(1)}$ is not rejected, then $\mu_{(1)}, \mu_{(2)}, \ldots, \mu_{(t-1)}$ are assumed to be equal, and no more tests concerning only those means are performed. Similar conclusions hold if $\mu_{(t)} = \mu_{(2)}$ is not rejected. If either hypothesis is rejected, the next round of hypotheses is considered.

The next round of hypotheses includes three hypotheses: $H_0 : \mu_{(t)} = \mu_{(3)}$, $H_0 : \mu_{(t-1)} = \mu_{(2)}$, and $H_0 : \mu_{(t-2)} = \mu_{(1)}$. The hypothesis $H_0 : \mu_{(t-3+i)} = \mu_{(i)}$ is rejected if $y_{(t-3+i)} - y_{(i)} > SQ(1 - \alpha, t - 2, v)$ for $i = 1, 2, 3$ but only if the hypothesis had not been accepted as part of a previous round.

The procedure continues until, at the last round, the hypotheses $H_0 : \mu_{(i)} = \mu_{(i-1)}$, $i = 2, \ldots, t$, are considered. An hypothesis is rejected if $y_{(i)} - y_{(i-1)} > SQ(1 - \alpha, 2, v)$.

Remember that if, say, $H_0 : \mu_{(t-1)} = \mu_{(1)}$ is not rejected, we will never test $H_0 : \mu_{(t-1)} = \mu_{(2)}$ or $H_0 : \mu_{(t-2)} = \mu_{(1)}$ in the next round or any other hypothesis in any other round that involves only $\mu_{(1)}, \mu_{(2)}, \ldots, \mu_{(t-1)}$.

The experimentwise error rate is exactly α because if $H_0 : \mu_1 = \cdots = \mu_t$ is true, the Newman–Keuls procedure will conclude that there is a difference in the means if and only if the Tukey method concludes that there is a difference. Because $Q(1 - \alpha, 2, v) < Q(1 - \alpha, 3, v) < \cdots < Q(1 - \alpha, t - 1, v) < Q(1 - \alpha, t, v)$, the Newman–Keuls method will reject the hypothesis that a pair of means is equal more often than Tukey's method. The Newman–Keuls method is thus more powerful. On the other hand, for pairs of μs that are equal, the Newman–Keuls method will make more mistakes than the Tukey method.

Example 5.6.1 Let $\alpha = 0.01$, $v = 10$, $S = 1$, $t = 5$, and $y_1 = 6.5$, $y_2 = 1.2$, $y_3 = 6.9$, $y_4 = 9.8$, $y_5 = 3.4$. We need the numbers $Q(0.99, 5, 10) = 6.14$, $Q(0.99, 4, 10) = 5.77$, $Q(0.99, 3, 10) = 5.27$, $Q(0.99, 2, 10) = 4.48$. Ordering the y_is gives

i	2	5	1	3	4
y_i	1.2	3.4	6.5	6.9	9.8

To test $H_0 : \mu_4 = \mu_2$, consider $9.8 - 1.2 = 8.6$, which is larger than $SQ(0.99, 5, 10) = 6.14$. There is a difference. Next test $H_0 : \mu_2 = \mu_3$. Since $6.9 - 1.2 = 5.7$ is less than 5.77, we conclude that $\mu_2 = \mu_5 = \mu_1 = \mu_3$. We do no more tests concerning only those means. Now test $H_0 : \mu_5 = \mu_4$. Since $9.8 - 3.4 = 6.4 > 5.77$, we reject the hypothesis.

We have concluded that $\mu_2 = \mu_5 = \mu_1 = \mu_3$, so the next allowable test is $H_0 : \mu_1 = \mu_4$. Since $9.8 - 6.5 = 3.4 < 5.27$, we conclude that $\mu_1 = \mu_3 = \mu_4$.

Drawing lines under the values that give no evidence of a difference in means, we can summarize our results as follows:

i	2	5	1	3	4
y_i	1.2	3.4	6.5	6.9	9.8

Note that if we had concluded that $\mu_2 \neq \mu_3$, we could test $H_0 : \mu_2 = \mu_1$. The test would be $6.5 - 1.2 = 5.3 > 5.27$, so we would have rejected the hypothesis. However, since we concluded that $\mu_2 = \mu_5 = \mu_1 = \mu_3$, we never get to do the test of $\mu_2 = \mu_1$.

Duncan has a multiple range test that is similar to Newman–Keuls but where the α levels for the various rounds of tests keep decreasing. In fact, *Duncan's method* is exactly the same as Newman–Keuls except that the α levels used when taking values from the table of the Studentized range are different. Duncan suggests using a $1-(1-\alpha)^{p-1}$ level test when comparing a set of p means. If there is a total of t means to be compared, Duncan's method only controls the experimentwise error rate at $1-(1-\alpha)^{t-1}$. For $\alpha = 0.05$ and $t = 6$, Duncan's method can only be said to have an experimentwise error rate of 0.23. As Duncan suggests, his method should only be performed when a corresponding omnibus F test has been found significant. This two stage procedure may be a reasonable compromise between the powers of the LSD and Newman–Keuls methods.

5.7 Summary

The emphasis in this chapter has been on controlling the experimentwise error rate. We have made some mention of power and the fact that increased power can be a mixed blessing. The really difficult problem for multiple comparison procedures is not in controlling the experimentwise error rate, but in carefully addressing the issues of power and the sizes of individual tests.

The discussion of Duncan's multiple range test highlights an interesting fact about multiple comparison methods. Any method of rejecting hypotheses, if preceded by an appropriate omnibus F test, is a valid multiple comparison procedure, valid in the sense that the experimentwise error rate is controlled. For example, if you do an F test first and stop if the F test does not reject, you can then (1) reject all individual hypotheses if the analysis is being performed on your mother's birth date, (2) reject no individual hypotheses on other dates. As stupid as this is, the experimentwise error rate is controlled. Intelligent choice of a multiple comparison method also involves consideration of the error rates (probabilities of type I errors) for the individual hypotheses. The main question is: If not all of the hypotheses are true, how many of the various kinds of mistakes do the different methods make?

In a one-way ANOVA with 12 groups and no mean differences, doing unadjusted $\alpha = 0.05$ level tests you could easily find $3 \doteq \binom{12}{2} \times 0.05$ phantom differences that look significant. None of the methods discussed here are likely to do that. If you had 13 groups with exactly 1 mean different from the other 12 (and substantially so), LSD is likely to find the 12 real differences but may also find 3 phantom differences. None of the other discussed methods (except Duncan) is likely find the phantom differences.

A reasonable goal might be to have the experimentwise error rate and the error rates for the individual hypotheses all no greater than α. The Scheffé, LSD, Bonferroni, Tukey, and Newman–Keuls methods all seem to aim at this goal. The Duncan method does not seem to accept this goal.

Suppose we want α level tests of the hypotheses

$$H_0 : \lambda_k' \beta = 0, \qquad k \in \Omega.$$

A reasonable procedure is to reject an hypothesis if

$$SS(\lambda_k' \beta)/MSE > C$$

for some value C. For example, the LSD method takes $C = F(1-\alpha, 1, dfE)$. If Ω consists of all the hypotheses in a t-dimensional space, Scheffé's method takes $C = tF(1-\alpha, t, dfE)$. If Ω is a finite set, say $\Omega = \{1, \ldots, r\}$, then the Bonferroni method takes $C = F(1-\alpha/r, 1, dfE)$.

To control the level of the individual test $H_0 : \lambda_k' \beta = 0$, one needs to pick C as the appropriate percentile of the distribution of $SS(\lambda_k' \beta)/MSE$. At one extreme, if one ignores everything else that is going on and if $\lambda_k' \beta$ was chosen without reference to the data, the appropriate distribution for $SS(\lambda_k' \beta)/MSE$ is $F(1, dfE)$. At the other extreme, if one picks $\lambda_k' \beta$ so that $SS(\lambda_k' \beta)$ is maximized in a t-dimensional space, then the appropriate distribution for $SS(\lambda_k' \beta)/MSE$ is t times an $F(t, dfE)$; it is clear that the probability of rejecting any hypothesis other than that associated with maximizing $SS(\lambda_k' \beta)$ must be less than α. Thus, in the extremes, we are led to the LSD and Scheffé methods. What one really needs is the distribution of $SS(\lambda_k' \beta)/MSE$ given $\lambda_k' \beta = 0$, and all the information contained in knowing $\lambda_j' \beta$ for $j \in \Omega - \{k\}$ and that $SS(\lambda_j' \beta)$ for $j \in \Omega - \{k\}$ will also be observed. Since the desired distribution will depend on the $\lambda_j' \beta$s, and they will never be known, there is no hope of achieving this goal.

The quality of the LSD method depends on how many hypotheses are to be tested. If only one hypothesis is to be tested, LSD is the method of choice. If all of the hypotheses in a subspace are to be tested, LSD is clearly a bad choice for testing the hypothesis that maximizes $SS(\lambda_k' \beta)$ and also a bad choice for testing other hypotheses that look likely to be significant. For testing a reasonably small number of hypotheses that were chosen without looking at the data, the LSD method seems to keep the levels of the individual tests near the nominal level of α. (The fact that the individual hypotheses are tested only if the omnibus F test is rejected helps keep the error rates near their nominal levels.) However, as the number of hypotheses to be tested increases, the error rate of the individual tests can increase greatly. The LSD method is not very responsive to the problem of controlling the error level of each individual test, but it is very powerful in detecting hypotheses that are not zero.

Scheffé's method puts an upper bound of α on the probability of type I error for each test, but for an individual hypothesis chosen without examining the data, the true probability of type I error is typically far below α. Scheffé's method controls the type I error but at a great cost in the power of each test.

The Bonferroni method uses the same distributions for $SS(\lambda_k' \beta)/MSE$, $k \in \Omega$, as the LSD method uses. The difference is in the different ways of controlling the experimentwise error rate. Bonferroni reduces the size of each individual test, while LSD uses an overall F test. The Bonferroni method, since it reduces the size of each test, does a better job of controlling the error rate for each individual hypothesis than does the LSD method. This is done at the cost of reducing the power relative to LSD. For a reasonable number of hypotheses, the Bonferroni method tends to be more

Table 5.1 Summary of Multiple Comparison Methods

Method	Hypotheses	Control	Comments
Scheffé	Any and all hypotheses constraining a particular subspace	*F* test for subspace	Lowest error rate and power of any method. Good for data snooping. HSD better for pairs of means
LSD	Any and all hypotheses constraining a particular subspace	*F* test for subspace	Highest error rate and power of any method. Best suited for a finite number of hypotheses
Bonferroni	Any finite set of hypotheses	Bonferroni inequality	Most flexible method. Often similar to HSD for pairs of means
Tukey's HSD	All differences between pairs of means	Studentized range test	Lowest error rate and power for pairs of means
Newman–Keuls	All differences between pairs of means	Studentized range test	Error rate and power intermediate between HSD and Duncan
Duncan	All differences between pairs of means	Studentized range test or *F* test	Error rate and power intermediate between Newman–Keuls and LSD

powerful than Scheffé's method and tends to have error levels nearer the nominal than Scheffé's method.

A similar evaluation can be made of the methods for distinguishing between pairs of means. The methods that are most powerful have the highest error rates for individual hypotheses. From greatest to least power, the methods seem to rank as LSD, Duncan, Newman–Keuls, Tukey. Scheffé's method should rank after Tukey's. The relative position of Bonferroni's method is unclear.

When deciding on a multiple comparison method, one needs to decide on the importance of correctly identifying nonzero hypotheses (high power) relative to the importance of incorrectly identifying zero hypotheses as being nonzero (controlling the type I error). With high power, one will misidentify some zero hypotheses as being nonzero. When controlling the type I error, one is more likely not to identify some nonzero hypotheses as being nonzero.

Table 5.1 contains a summary of the methods considered in this chapter. It lists the hypotheses for which each method is appropriate, the method by which the experimentwise error rate is controlled, and comments on the relative powers and probabilities of type I error (error rates) for testing individual hypotheses.

Exercise 5.3 Show that for testing all hypotheses in a six-dimensional space with 30 degrees of freedom for error, if the subspace F test is omitted and the nominal LSD level is $\alpha = 0.005$, then the true error rate must be less than 0.25.

Hint: Try to find a Scheffé rejection region that is comparable to the LSD rejection region.

5.7.1 *Fisher Versus Neyman–Pearson*

I have *tried* to maintain a Fisherian view towards statistical inference in this edition of the book. However, I think multiple comparison procedures are fundamentally a tool of Neyman–Pearson testing. Fisherian testing is about measuring the evidence against the null model, while Neyman–Pearson testing is about controlling error rates. Controlling the experimentwise error rate seems anti-Fisherian to me.

Fisher is often credited with (blamed for) the LSD method. However, Fisher (1935, Chapter 24) did not worry about the experimentwise error rate when making multiple comparisons using his least significant difference method in analysis of variance. He did, however, worry about drawing inappropriate conclusions by using an invalid null distribution for tests determined by examining the data. In particular, Fisher proposed a Bonferroni correction when comparing the largest and smallest sample means.

If you are going to look at all pairs of means in a balanced ANOVA, then the appropriate distribution for comparing the largest and smallest sample means is the Studentized range. It gives the approriate P value. An appropriate P value for other comparisons is difficult, but P values based on the Studentized range should be conservative (larger than the true P value). Similar arguments can be made for other procedures.

5.8 Additional Exercises

Exercise 5.8.1 Compare all pairs of means for the blue jeans exercise of Chapter 4. Use the following methods:

(a) Scheffé's method, $\alpha = 0.01$,
(b) LSD method, $\alpha = 0.01$,
(c) Bonferroni method, $\alpha = 0.012$,
(d) Tukey's HSD method, $\alpha = 0.01$,
(e) Newman–Keuls method, $\alpha = 0.01$.

Exercise 5.8.2 Test whether the four orthogonal contrasts you chose for the blue jeans exercise of Chapter 4 equal zero. Use all of the appropriate multiple comparison methods discussed in this chapter to control the experimentwise error rate at $\alpha = 0.05$ (or thereabouts).

Exercise 5.8.3 Compare all pairs of means in the coat–shirt exercise of Chapter 4. Use all of the appropriate multiple comparison methods discussed in this chapter to control the experimentwise error rate at $\alpha = 0.05$ (or thereabouts).

Exercise 5.8.4 Suppose that in a balanced one-way ANOVA the group means $\bar{y}_{1\cdot}, \ldots, \bar{y}_{t\cdot}$ are not independent but have some nondiagonal covariance matrix V. How can Tukey's HSD method be modified to accommodate this situation?

Exercise 5.8.5 For an unbalanced one-way ANOVA, give the contrast coefficients for the contrast whose sum of squares equals the sum of squares for groups. Show the equality of the sums of squares. Hint: Recall that in Exercise 4.2 we found the form of $M_\alpha Y$ and that equation (4.2.2) allows one to read off the coefficients of any contrast determined by a vector in $C(M_\alpha)$.

References

Bretz, F., Hothorn, T., & Westfall, P. (2011). *Multiple comparisons using R*. Boca Raton: Chapman and Hall/CRC.

Christensen, R. (1996). *Analysis of variance, design, and regression: Applied statistical methods*. London: Chapman and Hall.

Christensen, R. (2015). *Analysis of variance, design, and regression: Linear modeling for unbal anced data* (2nd ed.). Boca Raton: Chapman and Hall/CRC Press.

Fisher, R. A. (1935). *The design of experiments*, (9th ed., 1971). New York: Hafner Press.

Hochberg, Y., & Tamhane, A. (1987). *Multiple comparison procedures*. New York: Wiley.

Hsu, J. C. (1996). *Multiple comparisons: Theory and methods*. London: Chapman and Hall.

Kutner, M. H., Nachtsheim, C. J., Neter, J., & Li, W. (2005). *Applied linear statistical models* (5th ed.). New York: McGraw-Hill Irwin.

Miller, R. G, Jr. (1981). *Simultaneous statistical inference* (2nd ed.). New York: Springer.

Scheffé, H. (1959). *The analysis of variance*. New York: Wiley.

Snedecor, G. W., & Cochran, W. G. (1980). *Statistical methods* (7th ed.). Ames: Iowa State University Press.

Chapter 6
Regression Analysis

Abstract Francis Galton, a half-cousin of Charles Darwin, is often credited as the founder of regression analysis, a tool he used for studying heredity and the social sciences. R.A. Fisher seems to be responsible for our current focus of treating the X matrix as fixed and known. In this chapter we give a mathematical, rather than subject matter, definition of regression and discuss many of its standard features. We also introduce prediction theory which is based on having X random. We explore prediction theory's close connection to regression and use it as the basis for defining many standard features of traditional regression.

A regression model is any general linear model $Y = X\beta + e$ in which $X'X$ is nonsingular. $X'X$ is nonsingular if and only if the $n \times p$ matrix X has $r(X) = p \leq n$. In regression models, the parameter vector β is estimable. Let $P' = (X'X)^{-1}X'$, then $\beta = P'X\beta$.

The simple linear regression model considered in Section 1 is similar to the one-way ANOVA model considered in Chapter 4 in that the theory of estimation and testing simplifies to the point that results can be presented in simple algebraic formulae. For the general regression model of Section 2 there is little simplification. Section 2 also contains brief introductions to nonparametric regression and generalized additive models as well as an analysis of a partitioned model. Nonparametric regression and generalized additive models are also discussed in *ALM-III*, Chapter 1 and Christensen (2015, Chapter 8, Section 9.10). The partitioned model is important for two reasons. First, the partitioned model appears in many discussions of regression. Second, the results for the partitioned model are used to establish the correspondence between the standard regression theory presented in Section 2 and an alternative approach to regression, based on *best prediction* and *best linear prediction*, presented in Section 3. The approach given in Section 3 assumes that the rows of the model matrix are a random sample from a population of possible row vectors. Thus, in Section 3, X is a random matrix, whereas in Section 2, X is fixed. Section 3 presents the alternative approach which establishes that best linear prediction also yields the least squares estimates. Sections 4 and 5 discuss some special correlation coefficients related to best predictors and best linear predictors. It is established that

R. Christensen, *Plane Answers to Complex Questions*, Springer Texts in Statistics,
https://doi.org/10.1007/978-3-030-32097-3_6

the natural estimates of these correlation coefficients can be obtained from standard regression results. Section 6 introduces *best linear unbiased prediction (BLUP)*. (In previous editions this material was in Section 12.2.) Section 7 examines testing for *lack of fit*. Finally, Section 8 establishes the basis of the relationship between polynomial regression and polynomial contrasts in one-way ANOVA.

There is additional material, spread throughout the book, that relates to the material in this chapter. Section 2 examines a partitioned model. Partitioned models are treated in general in Sections 9.1 and 9.2. Chapter 9 also contains an exercise that establishes the basis of the sweep operator used in regression computations. The results of Section 7 are extended in Section 7.3 to relate polynomial regression with polynomial contrasts in a two-way ANOVA. Finally, Chapters 12–14 are concerned with topics that are traditionally considered part of regression analysis.

There are a number of fine books available on regression analysis. Those that I refer to most often are Cook and Weisberg (1999), Daniel and Wood (1980), Draper and Smith (1998), and Weisberg (2014).

6.1 Simple Linear Regression

The model for simple linear regression is $y_i = \beta_0 + \beta_1 x_i + e_i$ or $Y = X\beta + e$, where $\beta' = [\beta_0, \beta_1]$ and

$$X' = \begin{bmatrix} 1 & 1 & \cdots & 1 \\ x_1 & x_2 & \cdots & x_n \end{bmatrix}.$$

Often it is easier to work with the alternative model $y_i = \gamma_0 + \gamma_1(x_i - \bar{x}_\cdot) + e_i$ or $Y = X_*\gamma + e$, where $\gamma' = [\gamma_0, \gamma_1]$ and

$$X'_* = \begin{bmatrix} 1 & 1 & \cdots & 1 \\ x_1 - \bar{x}_\cdot & x_2 - \bar{x}_\cdot & \cdots & x_n - \bar{x}_\cdot \end{bmatrix}.$$

Note that $C(X) = C(X_*)$. In fact, letting

$$U = \begin{bmatrix} 1 & -\bar{x}_\cdot \\ 0 & 1 \end{bmatrix},$$

we have $X_* = XU$ and $X_*U^{-1} = X$. Moreover, $\mathrm{E}(Y) = X\beta = X_*\gamma = XU\gamma$. Letting $P' = (X'X)^{-1}X'$ leads to

$$\beta = P'X\beta = P'XU\gamma = U\gamma.$$

In particular,

$$\begin{bmatrix} \beta_0 \\ \beta_1 \end{bmatrix} = \begin{bmatrix} \gamma_0 - \gamma_1\bar{x}_\cdot \\ \gamma_1 \end{bmatrix}.$$

(See also Example 3.1.2.)

To find the least squares estimates and the projection matrix, observe that

$$X_*'X_* = \begin{bmatrix} n & 0 \\ 0 & \sum_{i=1}^n (x_i - \bar{x}_\cdot)^2 \end{bmatrix},$$

$$(X_*'X_*)^{-1} = \begin{bmatrix} 1/n & 0 \\ 0 & 1/\sum_{i=1}^n (x_i - \bar{x}_\cdot)^2 \end{bmatrix},$$

and, since the inverse of $X_*'X_*$ exists, the estimate of γ is

$$\hat{\gamma} = (X_*'X_*)^{-1}X_*'Y = \begin{bmatrix} \bar{y}_\cdot \\ \sum_{i=1}^n (x_i - \bar{x}_\cdot)y_i / \sum_{i=1}^n (x_i - \bar{x}_\cdot)^2 \end{bmatrix}.$$

Moreover,

$$\hat{\beta} = U\hat{\gamma},$$

so the least squares estimate of β is

$$\hat{\beta} = \begin{bmatrix} \bar{y}_\cdot - \hat{\gamma}_1\bar{x}_\cdot \\ \hat{\gamma}_1 \end{bmatrix}.$$

The projection matrix $M = X_*(X_*'X_*)^{-1}X_*'$ is

$$\begin{bmatrix} \frac{1}{n} + \frac{(x_1-\bar{x}_\cdot)^2}{\sum_{i=1}^n (x_i-\bar{x}_\cdot)^2} & \cdots & \frac{1}{n} + \frac{(x_1-\bar{x}_\cdot)(x_n-\bar{x}_\cdot)}{\sum_{i=1}^n (x_i-\bar{x}_\cdot)^2} \\ \vdots & & \vdots \\ \frac{1}{n} + \frac{(x_1-\bar{x}_\cdot)(x_n-\bar{x}_\cdot)}{\sum_{i=1}^n (x_i-\bar{x}_\cdot)^2} & \cdots & \frac{1}{n} + \frac{(x_n-\bar{x}_\cdot)^2}{\sum_{i=1}^n (x_i-\bar{x}_\cdot)^2} \end{bmatrix}.$$

The covariance matrix of $\hat{\gamma}$ is $\sigma^2(X_*'X_*)^{-1}$; for $\hat{\beta}$ it is $\sigma^2(X'X)^{-1}$. The usual tests and confidence intervals follow immediately upon assuming that $e \sim N(0, \sigma^2 I)$.

A natural generalization of the simple linear regression model $y_i = \beta_0 + \beta_1 x_i + e_i$ is to expand it into a polynomial, say

$$y_i = \beta_0 + \beta_1 x_i + \beta_2 x_i^2 + \cdots + \beta_{p-1} x_i^{p-1} + e_i.$$

Although polynomial regression has some special features that will be discussed later, at a fundamental level it is simply a linear model that involves an intercept and $p-1$ predictor variables. It is thus a special case of multiple regression, the model treated in the next section.

Exercise 6.1 For simple linear regression, find the *MSE*, $\text{Var}(\hat{\beta}_0)$, $\text{Var}(\hat{\beta}_1)$, and $\text{Cov}(\hat{\beta}_0, \hat{\beta}_1)$.

Exercise 6.2 Use Scheffé's method of multiple comparisons to derive the *Working–Hotelling simultaneous confidence band* for a simple linear regression line $\mathrm{E}(y) = \beta_0 + \beta_1 x$.

6.2 Multiple Regression

Multiple regression is any regression problem with $p \geq 2$ that is not simple linear regression. If we take as our model $Y = X\beta + e$, we have

$$\begin{aligned} \hat{\beta} &= (X'X)^{-1}X'Y, \\ \mathrm{Cov}\left(\hat{\beta}\right) &= \sigma^2(X'X)^{-1}X'IX(X'X)^{-1} = \sigma^2(X'X)^{-1}, \\ SSR(X) &= Y'MY = \hat{\beta}'(X'X)\hat{\beta}, \\ SSE &= Y'(I-M)Y, \\ dfE &= r(I-M) = n-p. \end{aligned}$$

Since β is estimable, any linear function $\lambda'\beta$ is estimable. If $Y \sim N(X\beta, \sigma^2 I)$, tests and confidence intervals based on

$$\frac{\lambda'\hat{\beta} - \lambda'\beta}{\sqrt{MSE\,\lambda'(X'X)^{-1}\lambda}} \sim t(dfE)$$

are available.

Suppose we write the regression model as

$$Y = [X_1, \ldots, X_p] \begin{bmatrix} \beta_1 \\ \vdots \\ \beta_p \end{bmatrix} + e.$$

If we let $\lambda_j' = (0, \ldots, 0, 1, 0, \ldots, 0)$, with the 1 in the jth place, we have

$$\frac{\hat{\beta}_j - \beta_j}{\sqrt{MSE\,\lambda_j'(X'X)^{-1}\lambda_j}} \sim t(dfE),$$

where $\lambda_j'(X'X)^{-1}\lambda_j$ is the jth diagonal element of $(X'X)^{-1}$. This yields a test of the hypothesis $H_0 : \beta_j = 0$. It is important to remember that this t test is equivalent to the F test for testing the reduced model

$$Y = [X_1, \ldots, X_{j-1}, X_{j+1}, \ldots, X_p] \begin{bmatrix} \beta_1 \\ \vdots \\ \beta_{j-1} \\ \beta_{j+1} \\ \vdots \\ \beta_p \end{bmatrix} + e$$

against the full regression model. The t and F tests for $\beta_j = 0$ depend on all of the other variables in the regression model. Add or delete any other variable in the model and the tests change.

$SSR(X)$ can be broken down into single degree of freedom components:

$$\begin{aligned} SSR(X) &= SSR(X_1) + SSR(X_2|X_1) + SSR(X_3|X_1, X_2) + \cdots + SSR(X_p|X_1, \ldots, X_{p-1}) \\ &= R(\beta_1) + R(\beta_2|\beta_1) + R(\beta_3|\beta_1, \beta_2) + \cdots + R(\beta_p|\beta_1, \ldots, \beta_{p-1}). \end{aligned}$$

Of course, any permutation of the subscripts $1, \ldots, p$ gives another breakdown. The interpretation of these terms is somewhat unusual. For instance, $SSR(X_3|X_1, X_2)$ *is not* the sum of squares for testing any very interesting hypothesis about the full regression model. $SSR(X_3|X_1, X_2)$ is the sum of squares needed for testing the model

$$Y = [X_1, X_2] \begin{bmatrix} \beta_1 \\ \beta_2 \end{bmatrix} + e$$

against the larger model

$$Y = [X_1, X_2, X_3] \begin{bmatrix} \beta_1 \\ \beta_2 \\ \beta_3 \end{bmatrix} + e.$$

This breakdown is useful in that, for instance,

$$SSR(X_{p-1}, X_p|X_1, \ldots, X_{p-2}) = SSR(X_p|X_1, \ldots, X_{p-1}) + SSR(X_{p-1}|X_1, \ldots, X_{p-2}).$$

$SSR(X_{p-1}, X_p|X_1, \ldots, X_{p-2})$ is the sum of squares needed to test $H_0 : \beta_p = \beta_{p-1} = 0$.

The *usual multiple regression model* is assumed to have a column of 1s in the model matrix. In that case, the model can be written

$$y_i = \beta_0 + \beta_1 x_{i1} + \beta_2 x_{i2} + \cdots + \beta_{p-1} x_{ip-1} + e_i.$$

An analysis of variance table is often written for testing this model against the model

$$y_i = \beta_0 + e_i,$$

for which the model matrix consists only of a column of 1s. The table is given below.

ANOVA

Source	df	SS
β_0	1	$Y'\left(\frac{1}{n}J_n^n\right)Y$
Regression	$p-1$	$Y'\left(M-\frac{1}{n}J_n^n\right)Y$
Error	$n-p$	$Y'(I-M)Y$
Total	n	$Y'Y$

The *SSReg* from the table can be rewritten as $\hat{\beta}'(X'X)\hat{\beta} - C$, where C is the correction factor, i.e., $C \equiv Y'([1/n]J_n^n)Y = n(\bar{y}_\cdot)^2$.

Example 6.2.1 Consider the data given in Table 6.1 on the heating requirements for a factory. There are 25 observations on a dependent variable y (the number of pounds of steam used per month) and 2 independent variables, x_1 (the average atmospheric temperature for the month in $^\circ F$) and x_2 (the number of operating days in the month). The predictor variables are from Draper and Smith (1998).

Table 6.1 Steam data

Obs. no.	x_1	x_2	y	Obs. no.	x_1	x_2	y
1	35.3	20	17.8270	14	39.1	19	19.0198
2	29.7	20	17.0443	15	46.8	23	20.6128
3	30.8	23	15.6764	16	48.5	20	20.7972
4	58.8	20	26.0350	17	59.3	22	28.1459
5	61.4	21	28.3908	18	70.0	22	33.2510
6	71.3	22	31.1388	19	70.0	11	30.4711
7	74.4	11	32.9019	20	74.5	23	36.1130
8	76.7	23	37.7660	21	72.1	20	35.3671
9	70.7	21	31.9286	22	58.1	21	25.7301
10	57.5	20	24.8575	23	44.6	20	19.9729
11	46.4	20	21.0482	24	33.4	20	16.6504
12	28.9	21	15.3141	25	28.6	22	16.5597
13	28.1	21	15.2673				

The parameter estimates, standard errors, and t statistics for testing whether each parameter equals zero are

Parameter	Estimate	SE	t
β_0	−1.263	2.423	−0.052
β_1	0.42499	0.01758	24.18
β_2	0.1790	0.1006	1.78

As will be seen from the ANOVA table below, the t statistics have 22 degrees of freedom. There is a substantial effect for variable x_1. The P value for β_2 is approximately 0.10. The estimated covariance matrix for the parameter estimates is $MSE\ (X'X)^{-1}$. MSE is given in the ANOVA table below. The matrix $(X'X)^{-1}$ is

	β_0	β_1	β_2
β_0	2.77875	−0.01124	−0.10610
β_1	−0.01124	0.00015	0.00018
β_2	−0.10610	0.00018	0.00479

The analysis of variance table is

ANOVA

Source	df	SS	MS	F
β_0	1	15270.78	15270.78	
Regression	2	1259.32	629.66	298
Error	22	46.50	2.11	
Total	25	16576.60		

The F statistic is huge. There is a very significant effect due to fitting the regression variables (after fitting a mean value to the data). One breakdown of the sum of squares for regression is

$$SSR(X_1|J) = 1252.62$$
$$SSR(X_2|X_1, J) = 6.70.$$

6.2.1 Partitioned Model

We now partition the usual model matrix and parameter vector in order to get a multiple regression analogue of the alternative model for simple linear regression. This alternative model is often discussed in regression analysis and is necessary for establishing, in the next section, the relationship between multiple regression and best linear prediction. However, the results in this subsection do not actually require the linear model to be a regression, only that the model have an intercept term.

We rewrite the usual model $Y = X\beta + e$ as

$$Y = [J, Z] \begin{bmatrix} \beta_0 \\ \beta_* \end{bmatrix} + e,$$

where $\beta_* = [\beta_1, \ldots, \beta_{p-1}]'$ and

$$Z = \begin{bmatrix} x_{11} & \cdots & x_{1p-1} \\ \vdots & \ddots & \vdots \\ x_{n1} & \cdots & x_{np-1} \end{bmatrix}$$

An alternative way to write the model that is often used is

$$Y = \left[J_n, \left(I - \frac{1}{n}J_n^n\right) Z\right] \begin{bmatrix}\gamma_0 \\ \gamma_*\end{bmatrix} + e.$$

The models are equivalent because $C[J, Z] = C[J, (I - [1/n]J_n^n)Z]$. The second model is correcting all of the variables in Z for their means, i.e., the second model is

$$y_i = \gamma_0 + \gamma_1(x_{i1} - \bar{x}_{\cdot 1}) + \gamma_2(x_{i2} - \bar{x}_{\cdot 2}) + \cdots + \gamma_{p-1}(x_{ip-1} - \bar{x}_{\cdot p-1}) + e_i.$$

This is analogous to the alternative model considered for simple linear regression. We now find formulae for sums of squares and estimation of parameters based on the mean adjusted variables $(I - [1/n]J_n^n)Z$.

The parameters in the two models are related by $\gamma_* = \beta_*$ and $\beta_0 = \gamma_0 - (1/n)J_1^n Z\gamma_*$, cf. Exercise 6.9.10. Since $I - [1/n]J_n^n$ is the ppo onto $C(J_n)^\perp$, it is easily seen that $C[(I - [1/n]J_n^n)Z]$ is the orthogonal complement of $C(J_n)$ with respect to $C(X)$; therefore

$$\begin{aligned} SSReg = SS(Z|J_n) &= Y'(M - [1/n]J_n^n)Y \\ &= Y'(I - [1/n]J_n^n)Z\{Z'(I - [1/n]J_n^n)Z\}^- Z'(I - [1/n]J_n^n)Y, \end{aligned}$$

cf. Exercise B.0a.

In order to parallel the theory of best linear prediction, we use the normal equations to obtain least squares estimates of γ_0 and γ_*. Since

$$\left[J_n, \left(I - \frac{1}{n}J_n^n\right) Z\right]' \left[J_n, \left(I - \frac{1}{n}J_n^n\right) Z\right] = \begin{bmatrix} n & 0 \\ 0 & Z'\left(I - \frac{1}{n}J_n^n\right) Z\end{bmatrix}$$

and

$$\left[J_n, \left(I - \frac{1}{n}J_n^n\right) Z\right]' Y = \begin{bmatrix} \sum_{i=1}^n y_i \\ Z'\left(I - \frac{1}{n}J_n^n\right) Y\end{bmatrix},$$

the least squares estimates of γ_0 and γ_* are solutions to

$$\begin{bmatrix} n & 0 \\ 0 & Z'\left(I - \frac{1}{n}J_n^n\right) Z\end{bmatrix} \begin{bmatrix}\gamma_0 \\ \gamma_*\end{bmatrix} = \begin{bmatrix} \sum_{i=1}^n y_i \\ Z'\left(I - \frac{1}{n}J_n^n\right) Y\end{bmatrix}.$$

Equivalently $\hat{\gamma}_0 = \bar{y}_\cdot$ and $\hat{\gamma}_*$ is a solution to

$$Z'\left(I - \frac{1}{n}J_n^n\right) Z\gamma_* = Z'\left(I - \frac{1}{n}J_n^n\right) Y.$$

Since $\gamma_* = \beta_*$, we have that $\hat{\gamma}_* = \hat{\beta}_*$. Since $\beta_0 = \gamma_0 - (1/n)J_1^n Z\gamma_* = \gamma_0 - \beta_1\bar{x}_{\cdot 1} - \cdots - \beta_{p-1}\bar{x}_{\cdot p-1}$, we have $\hat{\beta}_0 = \hat{\gamma}_0 - \hat{\beta}_1\bar{x}_{\cdot 1} - \cdots - \hat{\beta}_{p-1}\bar{x}_{\cdot p-1}$.

Finally, from the formula for *SSReg* developed earlier, the normal equations, and the fact that $\hat{\gamma}_* = \hat{\beta}_*$, we get

$$\begin{aligned} SSReg &= Y'\left(I - \frac{1}{n}J_n^n\right)Z\left[Z'\left(I - \frac{1}{n}J_n^n\right)Z\right]^- Z'\left(I - \frac{1}{n}J_n^n\right)Y \\ &= \hat{\beta}_*' Z'\left(I - \frac{1}{n}J_n^n\right)Z\left[Z'\left(I - \frac{1}{n}J_n^n\right)Z\right]^- Z'\left(I - \frac{1}{n}J_n^n\right)Z\hat{\beta}_* \\ &= \hat{\beta}_*' Z'\left(I - \frac{1}{n}J_n^n\right)Z\hat{\beta}_*, \end{aligned}$$

where the last equality follows from the definition of a generalized inverse.

6.2.2 Nonparametric Regression and Generalized Additive Models

In general, regression models assume $y_i = m(x_i) + e_i$, $\mathrm{E}(e_i) = 0$, where x_i is a p vector of known predictor variables and $m(\cdot)$ is some function. If $m(x_i) = x_i'\beta$, the model is linear. If $m(x_i) = h(x_i'\beta)$ for a known function h, the model is generalized linear as discussed in Section 1.4. If $m(x_i) = h(x_i; \beta)$, for a known h depending on the predictor variables x_i and some parameters β, the model is nonlinear regression, see Christensen (2015, Chapter 23).

In *nonparametric regression*, m is not assumed to fall into any such parametric family. However, by making weak assumptions about $m(x)$, one can often write it as $m(x) = \sum_{j=1}^{\infty} \beta_j \phi_j(x)$ for some class of known functions $\{\phi_j\}$. When x is a scalar, examples of such classes include polynomials, cosines, and wavelets. In practice, one fits a *linear model*

$$y_i = \sum_{j=1}^{p} \beta_j \phi_j(x_i) + e_i, \tag{1}$$

where p is large enough to capture the interesting behavior of $m(\cdot)$, see *ALM-III*, Chapter 1. Note that in the linear model, each of the terms $\phi_j(x_i)$ is a predictor variable, i.e., the jth column of the model matrix is $[\phi_j(x_1), \cdots, \phi_j(x_n)]'$.

This "*basis function*" approach to nonparametric regression involves fitting one large, complicated linear model. Other approaches involve fitting many simple linear models. For example, the *lowess fit (locally weighted scatterplot smoother)* begins estimation of $m(x)$ by fitting a weighted linear regression to some fraction of the x_is that are nearest x with weights proportional to the distance from x. The linear model is used only to estimate one point, $m(x)$. One performs a separate fit for a large number of points x_g, and to estimate all of $m(x)$ just connect the dots in the graph of $[x_g, \hat{m}(x_g)]$. Kernel estimation is similar but the weights are determined by a kernel function and the estimate is just a weighted average, not the result of fitting a weighted line or plane.

Unfortunately, these methods quickly run into a "curse of dimensionality." With the basis function approach and only one predictor variable, it might take, for example, $p = 8$ functions to get an adequate approximation to $m(\cdot)$. No problem! With 2 predictor variables, we could expect to need approximately $p = 8^2 = 64$ predictor variables in the linear model. Given a few hundred observations, this is doable. However, with 5 predictor variables, we could expect to need about $p = 8^5 = 32{,}768$ functions.

One way to get around this curse of dimensionality is to fit *generalized additive models*. For example, with 3 predictor variables, $x = (x_1, x_2, x_3)'$, we might expect to need $p = 8^3 = 512$ terms to approximate $m(\cdot)$. To simplify the problem, we might assume that $m(\cdot)$ follows a generalized additive model such as

$$m(x) = f_1(x_1) + f_{23}(x_2, x_3) \tag{2}$$

or

$$m(x) = f_{12}(x_1, x_2) + f_{23}(x_2, x_3). \tag{3}$$

If we need 8 terms to approximate $f_1(\cdot)$ and 64 terms to approximate each of the $f_{jk}(\cdot, \cdot)$s, the corresponding linear model for (2) involves fitting only 72 predictor variables, and for model (3) only $128 - 8 = 120$ rather than $8^3 = 512$. (Fitting f_{12} and f_{23} typically duplicates fitting of an f_2 term.)

With the same 8 term approximations and 5 predictor variables, a generalized additive model that includes all of the possible $f_{jk}(\cdot, \cdot)$s involves only 526 terms, rather than the 32,768 required by a full implementation of a nonparametric regression.

I should repeat that my use of 8 terms per dimension is merely an illustration. One might need 5 terms, or 10 terms, or 15 terms. And in two dimensions, one might need more or less than 8^2 terms. But these computations illustrate the magnitude of the problem. It should also be noted that with alternative (nonlinear) methods of fitting generalized additive models it may be necessary to fit lower order terms, i.e., instead of model (3), fit

$$m(x) = f_1(x_1) + f_2(x_2) + f_3(x_3) + f_{12}(x_1, x_2) + f_{23}(x_2, x_3).$$

Another problem with the basis function approach is that it can lead to very strange results if the number of functions p being fitted is too large. It is well known, see for example Christensen (2015, Section 8.2), that fitting high order polynomials can be a bad thing to do. For x a scalar predictor, by a high order polynomial we mean a polynomial in which the number of polynomial terms p is close to the number of distinct x_i values in the data. If the x_i values are all distinct, a high order polynomial will fit every observed data point almost perfectly. The price of this almost perfect fit to the data is that *between the x_i values* the polynomial can do very weird and inappropriate things. Thus, we have a model that will work poorly for future predictions. This behavior is not unique to polynomials. It occurs with all the standard classes of functions. (It seems that the problems can be ameliorated by fitting smaller models on subsets of the data, e.g. wavelets or splines, even though these retain the form in (1).) Similar problems occur when the x_i values are not all distinct.

ALM-III, Chapter 1 contains much more information on nonparametric regression.

6.3 General Prediction Theory

General prediction theory provides another approach to regression analysis. It ties in closely with linear model theory but also with generalized linear models, nonlinear regression, and nonparametric regression.

One of the great things about writing a technical book is that if you do enough good mathematics, people will put up with you spouting off once in a while. In this section, we take up the subject of prediction and its application to linear model theory. To me, prediction is what science is all about, and I cannot resist the urge to spout off. If you just want to get to work, skip down to the subsection headed "General Prediction."

6.3.1 Discussion

There is a fundamental philosophical conflict in statistics between Bayesians (who incorporate subjective beliefs into their analyses) and non-Bayesians. On the philosophical issues, the Bayesians seem to me to have much the better of the argument. (Yes, hard as it may be to believe, the author of this book is a philosophical Bayesian.) However, in the practice of 20th Century statistics, the non-Bayesians carried the day. Perhaps the most difficult philosophical aspect of Bayesian statistics for non-Bayesians to accept is the incorporation of subjective beliefs. Many scientists believe that objectivity is of paramount importance, and classical statistics maintains the semblance of objectivity. (In fact, classical statistics is rife with subjective inputs. Choosing an experimental design, choosing independent variables to consider in regression, and any form of data snooping such as choosing contrasts after looking at the data are examples.)

As Smith (1986) has pointed out, this concept of objectivity is very elusive. Objectivity is almost indistinguishable from the idea of consensus. If all the "clear thinking" people agree on something, then the consensus is (for the time being) "objective" reality.

Fortunately, the essence of science is not objectivity; it is repeatability. The object of scientific statistical inference is not the examination of parameters (too often created by and for statisticians so that they have something to draw inferences about). The object of scientific statistical inference is the (correct) prediction of future observable events. (I bet you were wondering what all this had to do with prediction.) Parameters are at best a convenience, and parameters are at their best when they are closely related to prediction (e.g., probabilities of survival). Geisser (1971, 1993) gives excellent discussions of the predictive approach.

In this book the emphasis has been placed on models rather than on parameters. Now you know why. Models can be used for prediction. They are an endproduct. Parameters are an integral part of most models, but they are a tool and not an end in themselves. Christensen (1995) gives a short discussion on the relation among models, prediction, and testing. Having now convinced a large portion of the statistical community of the unreliability of my ideas, I shall return to the issue at hand.

6.3.2 General Prediction

Suppose we have random variables $y, x_1, x_2, \ldots, x_{p-1}$. Regression can be viewed as the problem of predicting y from the values of $x_1, \ldots, x_{p-1}$. We will examine this problem and consider its relationship to the linear model theory approach to regression. Let x be the vector $x = (x_1, \ldots, x_{p-1})'$. A reasonable criterion for choosing a predictor of y is to pick a predictor $f(x)$ that minimizes the mean squared error, $\mathrm{E}[y - f(x)]^2$. (The *MSE* defined in linear model theory is a function of the observations that estimates the theoretical mean squared error defined here.) Note that, unlike standard linear model theory, the expected value is taken over the joint distribution of y and x.

The use of an expected squared error criterion presupposes the existence of first and second moments for y and $f(x)$. Let $\mathrm{E}(y) \equiv \mu_y$, $\mathrm{Var}(y) \equiv \sigma_y^2 \equiv \sigma_{yy}$ and let $\mathrm{E}[f(x)] \equiv \mu_f$, $\mathrm{Var}[f(x)] \equiv \sigma_{ff}$, and $\mathrm{Cov}[y, f(x)] \equiv \sigma_{yf}$, with similar notations for other functions of x, e.g., $m(x)$ has $\sigma_{ym} \equiv \mathrm{Cov}[y, m(x)]$.

The remainder of this section consists of three subsections. The next examines best predictors, i.e., those functions $f(x)$ that do the best job of predicting y. Without knowing (or being able to estimate) the joint distribution of x and y, we cannot find the best predictor, so in Subsection 4 we examine the best predictors among functions $f(x)$ that are restricted to be linear functions of x. These best linear predictors depend only on the means and covariances of the random variables and are thus relatively easy to estimate. Subsection 4 also examines the relationship between best linear predictors and least squares estimation. Both the best predictor and the best linear predictor can be viewed as orthogonal projections in an appropriate vector space, a subject that is commented on earlier but is amplified in Subsection 5.

6.3.3 Best Prediction

We now establish that the best predictor is the conditional expectation of y given x. See Appendix D for definitions and results about conditional expectations.

Theorem 6.3.1 *Let $m(x) \equiv \mathrm{E}(y|x)$. Then for any other predictor $f(x)$, $\mathrm{E}[y - m(x)]^2 \leq \mathrm{E}[y - f(x)]^2$; thus $m(x)$ is the best predictor of y.*

Proof

$$\begin{aligned}\mathrm{E}[y - f(x)]^2 &= \mathrm{E}[y - m(x) + m(x) - f(x)]^2 \\ &= \mathrm{E}[y - m(x)]^2 + \mathrm{E}[m(x) - f(x)]^2 + 2\mathrm{E}\{[y - m(x)][m(x) - f(x)]\}.\end{aligned}$$

Since both $\mathrm{E}[y - m(x)]^2$ and $\mathrm{E}[m(x) - f(x)]^2$ are nonnegative, it is enough to show that $\mathrm{E}\{[y - m(x)][m(x) - f(x)]\} = 0$. Consider this expectation conditional on x.

$$\begin{aligned}\mathrm{E}\{[y - m(x)][m(x) - f(x)]\} &= \mathrm{E}\big(\mathrm{E}\{[y - m(x)][m(x) - f(x)]|x\}\big) \\ &= \mathrm{E}\big([m(x) - f(x)]\mathrm{E}\{[y - m(x)]|x\}\big) \\ &= \mathrm{E}\big([m(x) - f(x)]0\big) = 0\end{aligned}$$

where $\mathrm{E}\{[y - m(x)]|x\} = 0$ because $\mathrm{E}(y|x) = m(x)$. □

The goal of most predictive procedures is to find, or rather estimate, the function $\mathrm{E}(y|x) \equiv m(x)$. Suppose we have a random sample (x_i', y_i), $i = 1, \ldots, n$. In linear regression, usually $m(x_i) = \alpha + x_i'\beta$ with α and β unknown. A generalized linear model assumes a distribution for y given x and that $\mathrm{E}(y_i|x_i) = m(x_i) = h(\alpha + x_i'\beta)$ for known h and unknown α and β. Here h is just the inverse of the link function. The standard nonlinear regression model is a more general version of these. It uses $m(x_i) = h(x_i; \alpha, \beta)$, where h is some known function but α and β are unknown. The conditional mean structure of all three parametric models is that of the nonlinear regression model: $m(x) = h(x; \alpha, \beta)$, h known. We then become interested in estimating α and β. Evaluating whether we have the correct "known" form for h is a question of whether lack of fit exists, see Section 7. Nonparametric regression is unwilling to assume a functional form $h(x; \alpha, \beta)$. The standard nonparametric regression model is $y_i = m(x_i) + e_i$ where, conditional on the x_is, the e_is are independent with mean 0 and variance σ^2. In nonparametric regression, m is completely unknown, but often it is assumed to be continuous over a closed bounded interval.

All of these versions of regression involve estimating whatever parts of $m(x)$ are not assumed to be known. On the other hand, best prediction *theory* treats $m(x)$ as a known function, so for models involving α and β it treats them as known.

6.3.3.1 Residuals

In Theorem 6.3.3, we present a result that does two things. First, it provides a justification for the residual plots used in Chapter 12 to identify lack of fit. Second, as discussed in Subsection 5, it establishes that $\mathrm{E}(y|x)$ can be viewed as the perpendicular projection of y into the space of random variables, say $f(x)$, that are functions of x alone, have mean $\mathrm{E}[f(x)] = E[y]$, and a finite variance, see also deLaubenfels

(2006). Before doing this, we need to establish some covariance and correlation properties.

Proposition 6.3.2 $\text{Cov}[y, f(x)] = \text{Cov}[m(x), f(x)]$. *In particular,* $\text{Cov}[y, m(x)] = \text{Var}[m(x)] = \sigma_{mm}$ *and* $\text{Corr}^2[y, m(x)] = \sigma_{mm}/\sigma_{yy}$.

Proof Recall that, from the definition of conditional expectation, $\text{E}[m(x)] = \mu_y$.

$$\begin{aligned}
\text{Cov}[y, f(x)] &= \text{E}_{yx}\{[y - \mu_y]f(x)\} \\
&= \text{E}_{yx}\{[y - m(x) + m(x) - \mu_y]f(x)\} \\
&= \text{E}_{yx}\{[y - m(x)]f(x)\} + \text{E}_{yx}\{[m(x) - \mu_y]f(x)\} \\
&= \text{E}_x\left(\text{E}_{y|x}\{[y - m(x)]f(x)\}\right) + \text{E}_x\left(\text{E}_{y|x}\{[m(x) - \mu_y]f(x)\}\right) \\
&= \text{E}_x(0) + \text{E}_x\{[m(x) - \mu_y]f(x)\} \\
&= \text{Cov}[m(x), f(x)].
\end{aligned}$$

□

Now consider an arbitrary predictor $\tilde{y}(x)$ and its residual $y - \tilde{y}(x)$. We show that $\tilde{y}(x)$ is the best predictor if and only if it is unbiased and its residuals are uncorrelated with any function of x.

Theorem 6.3.3 *Let* $\tilde{y}(x)$ *be any predictor, then* $\text{E}[\tilde{y}(x)] = \mu_y$ *and* $\text{Cov}[f(x), y - \tilde{y}(x)] = 0$ *for any function* f *if and only if* $\text{E}(y|x) = \tilde{y}(x)$ *almost surely.*

Proof $\Leftarrow$ If $\text{E}(y|x) \equiv m(x) = \tilde{y}(x)$, $\text{E}[\tilde{y}(x)] = \mu_y$ and an immediate consequence of Proposition 2 is that $\text{Cov}[f(x), y - m(x)] = 0$ for any function f.

$\Rightarrow$ Now suppose that $\text{E}[\tilde{y}(x)] = \mu_y$ and $\text{Cov}[f(x), y - \tilde{y}(x)] = 0$ for any function f. To show that $\tilde{y}(x) = m(x)$ a.s., it is enough to note that $\text{E}[\tilde{y}(x) - m(x)] = \mu_y - \mu_y = 0$ and to show that $\text{Var}[\tilde{y}(x) - m(x)] = 0$. Thinking of $f(x) = \tilde{y}(x) - m(x)$ in the covariance conditions, observe that by Proposition 2 and the assumption of the theorem,

$$\begin{aligned}
\text{Var}[\tilde{y}(x) - m(x)] &= \text{Cov}[\tilde{y}(x) - m(x), \tilde{y}(x) - m(x)] \\
&= \text{Cov}\{[y - m(x)] - [y - \tilde{y}(x)], [\tilde{y}(x) - m(x)]\} \\
&= \text{Cov}\{[y - m(x)], [\tilde{y}(x) - m(x)]\} - \text{Cov}\{[y - \tilde{y}(x)], [\tilde{y}(x) - m(x)]\} \\
&= 0 - 0.
\end{aligned}$$

□

In fitting standard linear models, or any other regression procedure, we typically obtain fitted values $\hat{y}_i$ corresponding to the observed data y_i, from which we can obtain residuals $y_i - \hat{y}_i$. According to Theorem 6.3.3, if the fitted values are coming from the best predictor, plotting the residuals against any function of the predictor vector x_i should display zero correlation. Thus, if we plot the residuals against some function $f(x_i)$ of the predictors and observe a correlation, we obviously do not have the best predictor, so we should try fitting some other model. In particular, regardless of how nonlinear the original regression model might have been, adding a linear term to the model using $f(x_i)$ as the predictor should improve the fit of the model. Unfortunately, there is no reason to think that adding such a linear term will get you to the best predictor. Finally, it should be noted that a common form of lack of fit detected in residual plots is a parabolic shape, which does not necessarily suggest a nonzero correlation with the predictor used in the plot. However, when the residual plot is a parabola, a plot of the residuals versus the (suitably centered) squared predictor will display a nonzero correlation.

Finally, if we plot the residuals against some predictor variable z that is not part of x, the fact that we would make such a plot suggests that we really want the best predictor $m(x, z)$ rather than $m(x)$, although if we originally left z out, we probably suspect that $m(x) = m(x, z)$. A linear relationship between the residuals and z indicates that the estimated regression function $\hat{m}(x)$ is not an adequate estimate of the best predictor $m(x, z)$.

6.3.3.2 Other Loss Functions

Our focus has been, and will continue to be, on squared error prediction loss. However, there are a number of other loss functions (somewhat less) commonly used for prediction. These include weighted squared error $L[y, \tilde{y}(x)] = w(y)[y - \tilde{y}(x)]^2$ wherein $w(y) > 0$, absolute error $L[y, \tilde{y}(x)] = |y - \tilde{y}(x)|$, and, when y is binary, *Hamming* loss $L[y, \tilde{y}(x)] = I[y \neq \tilde{y}(x)]$ wherein I(statement) is 1 if the statement is true and 0 if the statement is false. http://www.stat.unm.edu/~fletcher/DecisionTheory.pdf links to a document "Decision Theory and Prediction" that proves the following statements.

- For data (x', y), $y \in \mathbf{R}$, and $L[y, \tilde{y}(x)] = w(y)[y - \tilde{y}(x)]^2$, the best predictor is $\hat{y} = \mathrm{E}[w(y)y|x]/\mathrm{E}[w(y)|x]$.
- For data (x', y), $y \in \mathbf{R}$, and $L[y, \tilde{y}(x)] = |y - \tilde{y}(x)|$, the best predictor is $\hat{y} = \mathrm{Median}(y|x)$.
- For data (x', y), $y \in \{0, 1\}$ and $L[y, \tilde{y}(x)] = I[y \neq \tilde{y}(x)]$, the best predictor is

$$\hat{y}(x) \equiv \begin{cases} 0 & \text{if } \Pr(y = 0|x) > 0.5 \\ 1 & \text{if } \Pr(y = 0|x) < 0.5. \end{cases}$$

The proofs of these results are identical to proofs that establish the optimal Bayesian estimators for the first two loss functions and the optimal Bayesian hypothesis test in the last case.

Binary logistic regression involves $y \in \{0, 1\}$ and seeks to estimate $\Pr(y = 1|x)$. Since this determines the entire (estimated) conditional distribution of y given x, it leads immediately to estimated best predictors under any of the predictive loss functions.

6.3.4 Best Linear Prediction

The ideas of best linear prediction and best linear unbiased prediction (see Section 6) are very important. As will be seen here (and in *ALM-III*), best linear prediction theory has important applications in standard linear models. The theory has traditionally been taught as part of multivariate analysis (cf. Anderson 2003). It is important for general stochastic processes (cf. Doob 1953), mixed linear models (cf. McCulloch et al. 2008; Hodges 2013; or *ALM-III*, Chapter 5), time series (cf. Shumway and Stoffer 2011; Brockwell and Davis 1991; or *ALM-III*, Chapters 6, 7), spatial data (cf. Cressie 1993; Cressie and Wikle 2011; Ripley 1981; or *ALM-III*, Chapter 8), principal component analysis (cf. Johnson and Wichern 2007 or *ALM-III*, Chapter 14), and it is the basis for linear Bayesian methods (cf. Hartigan 1969).

In order to use the results on best prediction, one needs to know $\mathrm{E}(y|x)$, which generally requires knowledge of the joint distribution of $(y, x_1, x_2, \ldots, x_{p-1})'$. If the joint distribution is not available but the means, variances, and covariances are known, we can find the best linear predictor of y. We seek a linear predictor $\alpha + x'\beta$ that minimizes $\mathrm{E}[y - \alpha - x'\beta]^2$ for all scalars α and $(p-1) \times 1$ vectors β.

In addition to our earlier assumptions that first and second moments for y exist, we now also assume the existence of $\mathrm{E}(x) \equiv \mu_x$, $\mathrm{Cov}(x) \equiv V_{xx}$, and $\mathrm{Cov}(x, y) \equiv V_{xy} = V'_{yx} \equiv \mathrm{Cov}(y, x)'$.

Let β_* be a solution to $V_{xx}\beta = V_{xy}$, then we will show that the function

$$\hat{E}(y|x) \equiv \mu_y + (x - \mu_x)'\beta_*$$

is a best linear predictor of y based on x. $\hat{E}(y|x)$ is also called *the linear expectation of y given x*. (Actually, it is the linear expectation of y given x and a random variable that is constant with probability 1.) Note that when V_{xx} is singular, there are an infinite number of vectors β_* that solve $V_{xx}\beta = V_{xy}$, but by using Lemma 1.3.5 we can show that all such solutions give the same best linear predictor with probability 1. The idea is that since $(x - \mu_x) \in C(V_{xx})$ with probability 1, for some random b, $(x - \mu_x) = V_{xx}b$ with probability 1, so

$$(x - \mu_x)'\beta_* = b'V_{xx}\beta_* = b'V_{xy},$$

which does not depend on the choice of β_* with probability 1.

Theorem 6.3.4 *$\hat{E}(y|x)$ is a best linear predictor of y.*

Proof Define η so that $\alpha = \eta - \mu_x'\beta$. An arbitrary linear predictor is $f(x) \equiv \alpha + x'\beta = \eta + (x - \mu_x)'\beta$.

$$\begin{aligned}
\mathrm{E}[y - f(x)]^2 &= \mathrm{E}[y - \hat{E}(y|x) + \hat{E}(y|x) - f(x)]^2 \\
&= \mathrm{E}[y - \hat{E}(y|x)]^2 + \mathrm{E}[\hat{E}(y|x) - f(x)]^2 \\
&\quad + 2\mathrm{E}\{[y - \hat{E}(y|x)][\hat{E}(y|x) - f(x)]\}.
\end{aligned}$$

If we show that $\mathrm{E}\{[y-\hat{E}(y|x)][\hat{E}(y|x)-f(x)]\}=0$, the result follows almost immediately. In that case,

$$\mathrm{E}[y-f(x)]^2=\mathrm{E}[y-\hat{E}(y|x)]^2+\mathrm{E}[\hat{E}(y|x)-f(x)]^2.$$

To find $f(x)$ that minimizes the left-hand side, observe that both terms on the right are nonnegative, the first term does not depend on $f(x)$, and the second term is minimized by taking $f(x)=\hat{E}(y|x)$.

We now show that $\mathrm{E}\{[y-\hat{E}(y|x)][\hat{E}(y|x)-f(x)]\}=0$.

$$\begin{aligned}
&\mathrm{E}\{[y-\hat{E}(y|x)][\hat{E}(y|x)-f(x)]\}\\
&=\mathrm{E}\big(\{y-[\mu_y+(x-\mu_x)'\beta_*]\}\{[\mu_y+(x-\mu_x)'\beta_*]-[\eta+(x-\mu_x)'\beta]\}\big)\\
&=\mathrm{E}\big(\{(y-\mu_y)-(x-\mu_x)'\beta_*\}\{(\mu_y-\eta)+(x-\mu_x)'(\beta_*-\beta)\}\big)\\
&=\mathrm{E}\big[(y-\mu_y)(\mu_y-\eta)-(x-\mu_x)'\beta_*(\mu_y-\eta)\\
&\quad+(y-\mu_y)(x-\mu_x)'(\beta_*-\beta)-(x-\mu_x)'\beta_*(x-\mu_x)'(\beta_*-\beta)\big]\\
&=(\mu_y-\eta)\mathrm{E}[(y-\mu_y)]-\mathrm{E}[(x-\mu_x)']\beta_*(\mu_y-\eta)\\
&\quad+\mathrm{E}[(y-\mu_y)(x-\mu_x)'](\beta_*-\beta)-\mathrm{E}[\beta_*'(x-\mu_x)(x-\mu_x)'(\beta_*-\beta)]\\
&=0-0+V_{yx}(\beta_*-\beta)-\beta_*'\mathrm{E}[(x-\mu_x)(x-\mu_x)'](\beta_*-\beta)\\
&=V_{yx}(\beta_*-\beta)-\beta_*'V_{xx}(\beta_*-\beta).
\end{aligned}$$

However, by definition, $\beta_*'V_{xx}=V_{yx}$; so

$$V_{yx}(\beta_*-\beta)-\beta_*'V_{xx}(\beta_*-\beta)=V_{yx}(\beta_*-\beta)-V_{yx}(\beta_*-\beta)=0. \qquad \square$$

It is of interest to note that if $(y,x')'$ has a multivariate normal distribution, then the best linear predictor is the best predictor. Morrison (2004) contains a discussion of conditional expectations for multivariate normals.

The following proposition will be used in Section 6 to develop the theory of best linear unbiased prediction.

Proposition 6.3.5 $\mathrm{E}[y-\alpha-x'\beta]^2=\mathrm{E}[y-\hat{E}(y|x)]^2+\mathrm{E}[\hat{E}(y|x)-\alpha-x'\beta]^2.$

Proof The result is part of the proof of Theorem 6.3.4. $\square$

We show that best linear predictors are essentially unique. In other words, we show that to minimize $\mathrm{E}[\hat{E}(y|x)-\eta-(x-\mu_x)'\beta]^2$, you need $\mu_y=\eta$ and β must be a solution to $V_{xx}\beta=V_{xy}$. It is not difficult to show that

$$\mathrm{E}[\hat{E}(y|x)-\eta-(x-\mu_x)'\beta]^2=(\mu_y-\eta)^2+\mathrm{E}[(x-\mu_x)'\beta_*-(x-\mu_x)'\beta]^2.$$

Clearly, minimization requires $\mu_y=\eta$ and $\mathrm{E}[(x-\mu_x)'\beta_*-(x-\mu_x)'\beta]^2=0$. The best linear predictors will be essentially unique if we can show that $\mathrm{E}[(x-\mu_x)'\beta_*-(x-\mu_x)'\beta]^2=0$ implies that β must be a solution to $V_{xx}\beta=V_{xy}$. Observe that

$$
\begin{aligned}
\mathrm{E}[(x-\mu_x)'\beta_* - (x-\mu_x)'\beta]^2 &= \mathrm{E}[(x-\mu_x)'(\beta_*-\beta)]^2 \\
&= \mathrm{Cov}[(x-\mu_x)'(\beta_*-\beta)] \\
&= (\beta_*-\beta)'V_{xx}(\beta_*-\beta).
\end{aligned}
$$

Write $V_{xx} = QQ'$ with $C(V_{xx}) = C(Q)$. Then $(\beta_* - \beta)'V_{xx}(\beta_* - \beta) = 0$ if and only if $(\beta_* - \beta)'QQ'(\beta_* - \beta) = 0$ if and only if $Q'(\beta_* - \beta) = 0$ if and only if $(\beta_* - \beta) \perp C(Q) = C(V_{xx})$ if and only if $V_{xx}(\beta_* - \beta) = 0$ if and only if $V_{xx}\beta = V_{xx}\beta_* = V_{xy}$. So β must be a solution.

The variance of the prediction error $y - \hat{E}(y|x)$ is given in Section 5. (Actually, the covariance matrix for a bivariate prediction is given.)

6.3.4.1 Relation to Least Squares Estimation

Next, we examine the correspondence between this theory and linear model regression theory. Suppose we have n observations on the vector $(y, x')' = (y, x_1, x_2, \ldots, x_{p-1})'$. We can write these as $(y_i, x_i')' = (y_i, x_{i1}, x_{i2}, \ldots, x_{i,p-1})'$, $i = 1, \ldots, n$. In matrix notation write $Y = (y_1, y_2, \ldots, y_n)'$ and $Z = [x_{ij}]$, $i = 1, \ldots, n$, $j = 1, \ldots, p-1$. (Z plays the same role as Z did in Section 2 on multiple regression.) The usual unbiased estimates for V_{xx} and V_{xy} can be written as

$$
S_{xx} = \frac{1}{n-1} Z'\left(I - \frac{1}{n}J_n^n\right)Z = \frac{1}{n-1}\sum_{i=1}^{n}(x_i - \bar{x}_\cdot)(x_i - \bar{x}_\cdot)'
$$

$$
S_{xy} = \frac{1}{n-1} Z'\left(I - \frac{1}{n}J_n^n\right)Y = \frac{1}{n-1}\sum_{i=1}^{n}(x_i - \bar{x}_\cdot)(y_i - \bar{y}_\cdot).
$$

The usual unbiased estimates of μ_y and μ_x are

$$
\bar{y}_\cdot = \frac{1}{n}J_1^n Y = \frac{1}{n}\sum_{i=1}^{n} y_i
$$

$$
\bar{x}_\cdot' = \frac{1}{n}J_1^n Z = \left(\frac{1}{n}\sum_{i=1}^{n} x_{i1}, \ldots, \frac{1}{n}\sum_{i=1}^{n} x_{ip-1}\right)'.
$$

The obvious estimate for the best linear predictor of y is

$$\hat{y} = \bar{y}_{\cdot} + (x - \bar{x}_{\cdot})'\hat{\beta}_*,$$

where $\hat{\beta}_*$ is a solution to $S_{xx}\beta_* = S_{xy}$, i.e., it solves $Z'\left(I - \frac{1}{n}J_n^n\right)Z\beta_* = Z'(I - \frac{1}{n}J_n^n)\,Y$. From the results of Section 2, $\bar{y}_{\cdot} = \hat{\gamma}_0$ and any solution of $Z'\left(I - \frac{1}{n}J_n^n\right)$ $Z\beta_* = Z'\left(I - \frac{1}{n}J_n^n\right)Y$ is a least squares estimate of $\beta_* = \gamma_*$. Thus, the natural estimates of the parameters in the best linear predictor are the least squares estimates from the mean corrected regression model considered in the previous section.

6.3.4.2 Residuals

Finally, we include a result for best linear predictors that is analogous to Theorem 6.3.3 for best predictors. First, if we have residuals from the best linear predictor, they will be uncorrelated with any linear combination of the predictor variables. Second, $\hat{E}(y|x)$ can be viewed as the perpendicular projection of y into the space of random variables $f(x)$ that are linear functions of x and have $\mathrm{E}[f(x)] = E[y]$, cf. Subsection 5.

Theorem 6.3.6 *Suppose $\tilde{y}(x)$ is any linear predictor, then* $\mathrm{E}[\tilde{y}(x)] = \mu_y$ *and* $\mathrm{Cov}[f(x), y - \tilde{y}(x)] = 0$ *for any linear function f if and only if $\hat{E}(y|x) = \tilde{y}(x)$ almost surely.*

Proof Let $f(x) \equiv \eta + (x - \mu_x)'\beta$ for some η and β.
$\Leftarrow$ For $\hat{E}(y|x) = \tilde{y}(x)$, $\mathrm{E}[\tilde{y}(x)] = \mu_y$ and

$$\begin{aligned}\mathrm{Cov}[\eta + (x - \mu_x)'\beta, (y - \mu_y) - (x - \mu_x)'\beta_*] &= \beta' V_{xy} - \beta' V_{xx}\beta_* \\ &= \beta' V_{xy} - \beta' V_{xy} = 0.\end{aligned}$$

$\Rightarrow$ From $\mathrm{E}[\tilde{y}(x)] = \mu_y$, we can write $\tilde{y} = \mu_y + (x - \mu_x)'\delta$ for some δ. If

$$\begin{aligned}0 &= \mathrm{Cov}[\eta + (x - \mu_x)'\beta, (y - \mu_y) - (x - \mu_x)'\delta] \\ &= \beta' V_{xy} - \beta' V_{xx}\delta\end{aligned}$$

for any β, then $V_{xy} = V_{xx}\delta$, so $\tilde{y}(x) = \hat{E}(y|x)$ with probability 1. □

As mentioned earlier, the theory of best linear predictors also comes up in developing the theory of best linear unbiased predictors (BLUPs), which is an important subject in models with dependent y_is, such as random effects models and spatial data models. In random effects models, part of the β vector in $Y = X\beta + e$ is assumed to be random and unobservable. BLUPs are discussed in Section 6. Random effects models and spatial models are mentioned in Section 6 and discussed more fully in *ALM-III*.

The result of the next exercise will be used in Section 5 on partial correlations.

Exercise 6.3 For predicting $y = (y_1, \ldots, y_q)'$ from $x = (x_1, \ldots, x_{p-1})'$ we say that a predictor $f(x)$ is best if the scalar $\mathrm{E}\{[y - f(x)]'[y - f(x)]\}$ is minimized. Show that with simple modifications, Theorems 6.3.1 and 6.3.4 hold for the extended problem, as does Proposition 6.3.5.

6.3.5 *Inner Products and Orthogonal Projections in General Spaces*

In most of this book, we define orthogonality and length using the Euclidean inner product in $\mathbf{R}^n$. For two vectors x and y, the Euclidean inner product is $x'y$, so $x \perp y$ if $x'y = 0$ and the length of x is $\|x\| = \sqrt{x'x}$. In Section B.3 we discussed, in detail, perpendicular projection operators relative to this inner product. We established that the projection of a vector Y into $C(X)$ is $\hat{Y} \equiv MY$. It also follows from Theorems 2.2.1 and 2.8.1 that $\hat{Y}$ is the unique vector in $C(X)$ that satisfies $(Y - \hat{Y}) \perp C(X)$. This last property is sometimes used to define what it means for $\hat{Y}$ to be the perpendicular projection of Y into $C(X)$. We use this concept to extend the application of perpendicular projections to more general vector spaces.

More generally in $\mathbf{R}^n$, we can use any positive definite matrix W to define an inner product between x and y as $x'Wy$. As before, x and y are orthogonal if their inner product $x'Wy$ is zero and the length of x is the square root of its inner product with itself, now $\|x\|_W \equiv \sqrt{x'Wx}$. As argued in Section B.3, any idempotent matrix is always a projection operator, but which one is the perpendicular projection operator depends on the inner product. As can be seen from Proposition 2.7.2 and Exercise 2.5, the matrix $A \equiv X(X'WX)^{-}X'W$ is an oblique projection operator onto $C(X)$ for the Euclidean inner product, but it is the perpendicular projection operator onto $C(X)$ with the inner product defined using the matrix W. It is not too difficult to see that AY is the unique vector in $C(X)$ that satisfies $(Y - AY) \perp_W C(X)$, i.e., $(Y - AY)'WX = 0$.

These ideas can be applied in very general spaces. In particular, they can be applied to the concepts of prediction introduced in this section. For example, we can define the inner product between two random variables y and x with mean 0 and finite variances as the $\mathrm{Cov}(x, y)$. In this case, $\mathrm{Var}(x)$ plays the role of the squared length of the random variable and the standard deviation is the length. Two random variables are orthogonal if they are uncorrelated.

Now consider a vector of random variables $x = (x_1, \ldots, x_{p-1})'$ and the space of all functions $f(x)$ into $\mathbf{R}^1$ that have mean 0 and variances. We showed in Theorem 6.3.3 that $m(x) \equiv \mathrm{E}(y|x)$ is the unique function of x having mean μ_y for which $\mathrm{Cov}[y - m(x), f(x)] = 0$ for any $f(x)$. Thus, as alluded to above, $m(x) - \mu_y$ satisfies a property often used to *define* the perpendicular projection of $y - \mu_y$ into the space of functions of x that have mean 0 and variances. Alternatively, we can think of $m(x)$ as the perpendicular projection of y into the space of functions of x that have mean μ_y and variances.

We can also consider a reduced space of random variables, the linear functions of x, i.e., $f(x) = \alpha + x'\beta$. In Theorem 6.3.6 we show that $\text{Cov}[y - \hat{E}(y|x), \alpha + x'\beta] = 0$ for any linear function of x, so once again, the best linear predictor is the perpendicular projection of y into the linear functions of x with mean μ_y.

We now generalize the definitions of an inner product space, orthogonality, and orthogonal projection.

Definition A.15 (Alternate) A vector space $\mathscr{X}$ is an inner product space if for any $x, y \in \mathscr{X}$, there exists a symmetric bilinear function $\langle x, y \rangle$ into $\mathbf{R}$ with $\langle x, x \rangle > 0$ for any $x \neq 0$ and $\langle x, x \rangle = 0$ when $x = 0$. A bilinear function has the properties that for any scalars a_1, a_2 and any vectors x_1, x_2, y, $\langle a_1x_1 + a_2x_2, y \rangle = a_1\langle x_1, y \rangle + a_2\langle x_2, y \rangle$ and $\langle y, a_1x_1 + a_2x_2 \rangle = a_1\langle y, x_1 \rangle + a_2\langle y, x_2 \rangle$. A symmetric function has $\langle x_1, x_2 \rangle = \langle x_2, x_1 \rangle$. The vectors x and y are orthogonal if $\langle x, y \rangle = 0$ and the squared length of x is $\langle x, x \rangle$. The perpendicular projection of y into a subspace $\mathscr{X}_0$ of $\mathscr{X}$ is defined as the unique vector $y_0 \in \mathscr{X}_0$ with the property that $\langle y - y_0, x \rangle = 0$ for any $x \in \mathscr{X}_0$.

Note that the set of mean zero, finite variance, real-valued functions of random x and y form a vector space under Definition A.1 and an inner product space using the covariance of any two such functions as the inner product. Both the set of mean 0, finite variance functions of x and the set of mean 0 linear functions of x are subspaces, so y can be projected into either subspace.

We now relate Definition A.15 (Alternate) to the concept of a perpendicular projection operator.

Exercise 6.4 Consider an inner product space $\mathscr{X}$ and a subspace $\mathscr{X}_0$. Any vector $y \in \mathscr{X}$ can be written uniquely as $y = y_0 + y_1$ with $y_0 \in \mathscr{X}_0$ and $y_1 \perp \mathscr{X}_0$. Let $M(x)$ be a linear operator on $\mathscr{X}$ in the sense that for any $x \in \mathscr{X}$, $M(x) \in \mathscr{X}$ and for any scalars a_1, a_2 and any vectors x_1, x_2, $M(a_1x_1 + a_2x_2) = a_1M(x_1) + a_2M(x_2)$. $M(x)$ is defined to be a perpendicular projection operator onto $\mathscr{X}_0$ if for any $x_0 \in \mathscr{X}_0$, $M(x_0) = x_0$, and for any $x_1 \perp \mathscr{X}_0$, $M(x_1) = 0$. Using Definition A.15 (Alternate), show that for any vector y, $M(y)$ is the perpendicular projection of y into $\mathscr{X}_0$.

6.4 Multiple Correlation

The coefficient of determination, denoted R^2, is a commonly used measure of the predictive ability of a model. Computationally, it is most often defined as

$$R^2 = \frac{SSReg}{SSTot - C},$$

so it is the proportion of the total variability explained by the independent variables. The greater the proportion of the total variability in the data that is explained by the model, the better the ability to predict with that model. The use and possible abuse of R^2 as a tool for comparing models is discussed in Chapter 14, however it should be noted here that since R^2 is a measure of the predictive ability of a model, R^2 does not give direct information about whether a model fits the data properly. Demonstrably bad models can have very high R^2s and perfect models can have low R^2s.

We now define the multiple correlation, characterize it in terms of the best linear predictor, and show that R^2 is the natural estimate of it. Subsection 6.4.1 generalizes the idea of R^2 from best linear prediction to best prediction.

Recall that the correlation between two random variables, say x_1 and x_2, is

$$\mathrm{Corr}(x_1, x_2) = \mathrm{Cov}(x_1, x_2) \Big/ \sqrt{\mathrm{Var}(x_1)\mathrm{Var}(x_2)}.$$

The multiple correlation of y and $(x_1, x_2, \ldots, x_{p-1})' = x$ is the maximum of $\mathrm{Corr}(y, \alpha + x'\beta)$ over all α and β. Note that

$$\mathrm{Cov}(y, \alpha + x'\beta) = V_{yx}\beta = \beta_*' V_{xx}\beta,$$

where β_* is defined as in Subsection 6.3.4 and

$$\mathrm{Var}(\alpha + x'\beta) = \beta' V_{xx}\beta.$$

In particular, $\mathrm{Cov}(y, \alpha + x'\beta_*) = \beta_*' V_{xx}\beta_* = \mathrm{Var}(\alpha + x'\beta_*)$. The Cauchy–Schwarz inequality (see Exercise B.0) says that

$$\left(\sum_{i=1}^{t} r_i s_i\right)^2 \leq \sum_{i=1}^{t} r_i^2 \sum_{i=1}^{t} s_i^2.$$

Since $V_{xx} = RR'$ for some matrix R, the Cauchy–Schwarz inequality gives

$$\left(\beta_*' V_{xx}\beta\right)^2 \leq \left(\beta' V_{xx}\beta\right)\left(\beta_*' V_{xx}\beta_*\right).$$

Considering the squared correlation gives

$$\begin{aligned}
\mathrm{Corr}^2(y, \alpha + x'\beta) &= \left(\beta_*' V_{xx}\beta\right)^2 \Big/ \left(\beta' V_{xx}\beta\right)\sigma_y^2 \\
&\leq \beta_*' V_{xx}\beta_* \big/ \sigma_y^2 \\
&= \left(\beta_*' V_{xx}\beta_*\right)^2 \Big/ \left(\beta_*' V_{xx}\beta_*\right)\sigma_y^2 \\
&= \mathrm{Corr}^2(y, \alpha + x'\beta_*)
\end{aligned}$$

and the squared multiple correlation between y and x equals

$$\mathrm{Corr}^2(y, \alpha + x'\beta_*) = \beta_*' V_{xx}\beta_* / \sigma_y^2 \equiv \mathscr{R}^2.$$

If we have observations $(y_i, x_{i1}, x_{i2}, \ldots, x_{i,p-1})', i = 1, \ldots, n$, the usual estimate of σ_y^2 can be written as

$$s_y^2 = \frac{1}{n-1} Y' \left(I - \frac{1}{n} J_n^n \right) Y = \frac{1}{n-1} \sum_{i=1}^{n} (y_i - \bar{y})^2.$$

Using equivalences derived earlier, the natural estimate of the squared multiple correlation coefficient between y and x is

$$\frac{\hat{\beta}_*' S_{xx} \hat{\beta}_*}{s_y^2} = \frac{\hat{\beta}_*' Z' \left(I - \frac{1}{n} J_n^n \right) Z \hat{\beta}_*}{Y' \left(I - \frac{1}{n} J_n^n \right) Y} = \frac{SSReg}{SSTot - C}.$$

It is worth noticing that $SSTot - C = SSReg + SSE$ and that

$$\begin{aligned} SSReg/SSE &= SSReg/[(SSE + SSReg) - SSReg] \\ &= SSReg/[SSTot - C - SSReg] \\ &= \frac{SSReg}{[SSTot - C][1 - R^2]} \\ &= R^2/[1 - R^2]. \end{aligned}$$

For normally distributed data, the α level F test for $H_0 : \beta_1 = \beta_2 = \cdots = \beta_{p-1} = 0$ is to reject H_0 if

$$\frac{n-p}{p-1} \frac{R^2}{1 - R^2} > F(1 - \alpha, p - 1, n - p).$$

Example 6.4.1 From the information given in Example 6.2.1, the coefficient of determination can be found. From the ANOVA table in Example 6.2.1, we get

$$SSReg = 1259.32$$

$$SSTot - C = SSTotal - SS(\beta_0) = 16576.60 - 15270.78 = 1305.82$$

and

$$R^2 = 1259.32/1305.82 = 0.964.$$

This is a very high value for R^2 and indicates that the model has very substantial predictive ability. However, before you conclude that the model actually fits the data well, see Example 6.6.3.

Exercise 6.5

1. Show that for a linear model with an intercept, R^2 is simply the square of the sample correlation between the data y_i and the predicted values $\hat{y}_i$.
2. It is a well known fact that if $Q \sim F(r_1, r_2)$ then

$$\frac{r_1 Q/r_2}{1 + r_1 Q/r_2} \sim \text{Beta}(r_1/2, r_2/2).$$

Show that under the intercept only model R^2 has a beta distribution and find its expected value and variance.

3. If p is increasing with n, such that $p/n \to f$, with $0 \le f < 1$, what happens to R^2 as $n, p \to \infty$, assuming that the intercept only model is true?

This discussion has focused on R^2, which estimates the squared multiple correlation coefficient, a measure of the predictive ability of the best linear predictor. We now consider an analogous measure for the best predictor.

6.4.1 Squared Predictive Correlation

As in Subsection 6.3.2, consider an arbitrary predictor $\tilde{y}(x)$. This is a function of x alone and not a function of y.

The *squared predictive correlation* of $\tilde{y}(x)$ is $\text{Corr}^2[y, \tilde{y}(x)]$. The highest squared predictive correlation is obtained by using the best predictor. Note that in the special case where the best predictor $m(x)$ is also the best linear predictor, the highest squared predictive correlation equals the squared multiple correlation coefficient.

Theorem 6.4.2 $\text{Corr}^2[y, \tilde{y}(x)] \le \text{Corr}^2[y, m(x)] \equiv \mathsf{R}^2$.

Proof By Cauchy–Schwarz, $(\sigma_{m\tilde{y}})^2 \le \sigma_{mm}\sigma_{\tilde{y}\tilde{y}}$, so $(\sigma_{m\tilde{y}})^2/\sigma_{\tilde{y}\tilde{y}} \le \sigma_{mm}$. Using Proposition 6.3.2,

$$\text{Corr}^2[y, \tilde{y}(x)] = \frac{(\sigma_{y\tilde{y}})^2}{\sigma_{yy}\sigma_{\tilde{y}\tilde{y}}} = \frac{(\sigma_{m\tilde{y}})^2}{\sigma_{yy}\sigma_{\tilde{y}\tilde{y}}} \le \frac{\sigma_{mm}}{\sigma_{yy}} = \text{Corr}^2[y, m(x)]. \qquad \square$$

Theorem 6.4.2 is also established in Rao (1973, Section 4g.1). From Theorem 6.4.2, the best regression function $m(x)$ has the highest squared predictive correlation. When we have perfect prediction, the highest squared predictive correlation is 1. In other words, if the conditional variance of y given x is 0, then $y = m(x)$ a.s., and the highest squared predictive correlation is the correlation of $m(x)$ with itself, which is 1. On the other hand, if there is no regression relationship, i.e., if $m(x) = \mu_y$ a.s., then $\sigma_{mm} = 0$, and the highest squared predictive correlation is 0.

We would now like to show that as the squared predictive correlation increases, we get increasingly better prediction. First we need to deal with the fact that high squared predictive correlations can be achieved by bad predictors. Just because $\tilde{y}(x)$ is highly correlated with y does not mean that $\tilde{y}(x)$ is actually close to y. Recall that $\tilde{y}(x)$ is simply a random *variable* that is being used to predict y. As such, $\tilde{y}(x)$ is a linear predictor of y, that is, $\tilde{y}(x) = 0 + 1\tilde{y}(x)$. We can apply Theorem 6.3.4 to this random variable to obtain a linear predictor that is at least as good as $\tilde{y}(x)$, namely

$$\hat{y}(x) = \mu_y + \frac{\sigma_{y\tilde{y}}}{\sigma_{\tilde{y}\tilde{y}}}[\tilde{y}(x) - \mu_{\tilde{y}}].$$

We refer to such predictors as *linearized predictors*. Note that $\mathrm{E}[\hat{y}(x)] \equiv \mu_{\hat{y}} = \mu_y$,

$$\sigma_{\hat{y}\hat{y}} \equiv \mathrm{Var}[\hat{y}(x)] = \left(\frac{\sigma_{y\tilde{y}}}{\sigma_{\tilde{y}\tilde{y}}}\right)^2 \sigma_{\tilde{y}\tilde{y}} = \frac{(\sigma_{y\tilde{y}})^2}{\sigma_{\tilde{y}\tilde{y}}},$$

and

$$\sigma_{y\hat{y}} \equiv \mathrm{Cov}[y, \hat{y}(x)] = \frac{\sigma_{y\tilde{y}}}{\sigma_{\tilde{y}\tilde{y}}}\sigma_{y\tilde{y}} = \frac{(\sigma_{y\tilde{y}})^2}{\sigma_{\tilde{y}\tilde{y}}}.$$

In particular, $\sigma_{\hat{y}\hat{y}} = \sigma_{y\hat{y}}$, so the squared predictive correlation of $\hat{y}(x)$ is

$$\mathrm{Corr}^2[y, \hat{y}(x)] = \frac{\sigma_{\hat{y}\hat{y}}}{\sigma_{yy}}.$$

In addition, the direct measure of the goodness of prediction for $\hat{y}(x)$ is

$$\mathrm{E}[y - \hat{y}(x)]^2 = \sigma_{yy} - 2\sigma_{y\hat{y}} + \sigma_{\hat{y}\hat{y}} = \sigma_{yy} - \sigma_{\hat{y}\hat{y}}.$$

This leads directly to the next result.

Theorem 6.4.2 *For two linearized predictors $\hat{y}_1(x)$ and $\hat{y}_2(x)$, the squared predictive correlation of $\hat{y}_2(x)$ is higher if and only if $\hat{y}_2(x)$ is a better predictor.*

Proof $\sigma_{\hat{y}_1\hat{y}_1}/\sigma_{yy} < \sigma_{\hat{y}_2\hat{y}_2}/\sigma_{yy}$ if and only if $\sigma_{\hat{y}_1\hat{y}_1} < \sigma_{\hat{y}_2\hat{y}_2}$ if and only if $\sigma_{yy} - \sigma_{\hat{y}_2\hat{y}_2} < \sigma_{yy} - \sigma_{\hat{y}_1\hat{y}_1}$. □

It should be noted that linearizing $m(x)$ simply returns $m(x)$.

For any predictor $\tilde{y}(x)$, no matter how one arrives at it, to estimate the squared predictive correlation from i.i.d. data (y_i, x_i'), simply compute the squared sample correlation between y_i and its predictor of $\tilde{y}(x_i)$. This should also work fine for y_is that are conditionally uncorrelated and homoscedastic given the x_is.

6.5 Partial Correlation Coefficients

Many regression programs have options available to the user that depend on the values of the sample partial correlation coefficients. The partial correlation is defined in terms of two random variables of interest, say y_1 and y_2, and several auxiliary variables, say $x_1, \ldots, x_{p-1}$. The partial correlation coefficient of y_1 and y_2 given $x_1, \ldots, x_{p-1}$, written $\rho_{y \cdot x}$, is a measure of the linear relationship between y_1 and y_2 after taking the effects of $x_1, \ldots, x_{p-1}$ out of both variables.

Writing $y = (y_1, y_2)'$ and $x = (x_1, \ldots, x_{p-1})'$, Exercise 6.3 indicates that the best linear predictor of y given x is

$$\hat{E}(y|x) = \mu_y + \beta_*'(x - \mu_x),$$

where β_* is a solution of $V_{xx}\beta_* = V_{xy}$ and V_{xy} is now a $(p-1) \times 2$ matrix. We take the effects of x out of y by looking at the prediction error

$$y - \hat{E}(y|x) = (y - \mu_y) - \beta_*'(x - \mu_x),$$

which is a 2×1 random vector. The partial correlation is simply the correlation between the two components of this vector and is readily obtained from the covariance matrix. We now find the prediction error covariance matrix. Let $\mathrm{Cov}(y) \equiv V_{yy}$.

$$\begin{aligned}
\mathrm{Cov}[(y - \mu_y) - \beta_*'(x - \mu_x)] &= \mathrm{Cov}(y - \mu_y) + \beta_*'\mathrm{Cov}(x - \mu_x)\beta_* \\
&\quad - \mathrm{Cov}(y - \mu_y, x - \mu_x)\beta_* \\
&\quad - \beta_*'\mathrm{Cov}(x - \mu_x, y - \mu_y) \\
&= V_{yy} + \beta_*' V_{xx}\beta_* - V_{yx}\beta_* - \beta_*' V_{xy} \\
&= V_{yy} + \beta_*' V_{xx}\beta_* - \beta_*' V_{xx}\beta_* - \beta_*' V_{xx}\beta_* \\
&= V_{yy} - \beta_*' V_{xx}\beta_*.
\end{aligned}$$

Since $V_{xx}\beta_* = V_{xy}$ and, for any generalized inverse, $V_{xx}V_{xx}^- V_{xx} = V_{xx}$,

$$\begin{aligned}
\mathrm{Cov}[(y - \mu_y) - \beta_*'(x - \mu_x)] &= V_{yy} - \beta_*' V_{xx} V_{xx}^- V_{xx}\beta_* \\
&= V_{yy} - V_{yx}V_{xx}^- V_{xy}. \qquad (1)
\end{aligned}$$

If we have a sample of the ys and xs, say $y_{i1}, y_{i2}, x_{i1}, x_{i2}, \ldots, x_{i,p-1}, i = 1, \ldots, n$, we can estimate the covariance matrix in the usual way. The relationship with the linear regression model of Section 2 is as follows: Let

$$Y = \begin{bmatrix} y_{11} & y_{12} \\ \vdots & \vdots \\ y_{n1} & y_{n2} \end{bmatrix} = [Y_1, Y_2]$$

and

$$Z = \begin{bmatrix} x_{11} & \cdots & x_{1p-1} \\ \vdots & \ddots & \vdots \\ x_{n1} & \cdots & x_{np-1} \end{bmatrix}.$$

The usual estimate of $V_{yy} - V_{yx}V_{xx}^{-}V_{xy}$ is $(n-1)^{-1}$ times

$$Y'\left(I - \frac{1}{n}J_n^n\right)Y - Y'\left(I - \frac{1}{n}J_n^n\right)Z\left[Z'\left(I - \frac{1}{n}J_n^n\right)Z\right]^{-} Z'\left(I - \frac{1}{n}J_n^n\right)Y.$$

From Section 2, we know that this is the same as

$$Y'\left(I - \frac{1}{n}J_n^n\right)Y - Y'\left(M - \frac{1}{n}J_n^n\right)Y = Y'(I - M)Y,$$

where M is the perpendicular projection operator onto $C([J, Z])$. Remembering that $Y'(I - M)Y = [(I - M)Y]'[(I - M)Y]$, we can see that the estimate of $\rho_{y \cdot x}$, written $r_{y \cdot x}$ and called the *sample partial correlation coefficient*, is just the sample correlation coefficient between the residuals of fitting $Y_1 = [J, Z]\beta_1 + e_1$ and the residuals of fitting $Y_2 = [J, Z]\beta_2 + e_2$, i.e.,

$$r_{y \cdot x} = \frac{Y_1'(I - M)Y_2}{\sqrt{Y_1'(I - M)Y_1 Y_2'(I - M)Y_2}}.$$

The square of the sample partial correlation coefficient, often called the coefficient of partial determination, has a nice relationship to another linear model. Consider fitting $Y_1 = [J, Z, Y_2]\gamma + e$. Because $C[(I - M)Y_2]$ is the orthogonal complement of $C([J, Z])$ with respect to $C([J, Z, Y_2])$, the sum of squares for testing whether Y_2 adds to the model is

$$SSR(Y_2|J, Z) = Y_1'(I - M)Y_2[Y_2'(I - M)Y_2]^{-1}Y_2'(I - M)Y_1.$$

Since $Y_2'(I - M)Y_2$ is a scalar, it is easily seen that

$$r_{y \cdot x}^2 = \frac{SSR(Y_2|J, Z)}{SSE(J, Z)},$$

where $SSE(J, Z) = Y_1'(I - M)Y_1$, the sum of squares for error when fitting the model $Y_1 = [J, Z]\beta + e$.

Finally, for normally distributed data we can do a t test of the null hypothesis $H_0 : \rho_{y \cdot x} = 0$. If H_0 is true, then

$$\sqrt{n - p - 1}\, r_{y \cdot x} \Big/ \sqrt{1 - r_{y \cdot x}^2} \sim t(n - p - 1).$$

See Exercise 6.7 for a proof of this result.

Example 6.5.1 Using the data of Example 6.2.1, the coefficient of partial determination (squared sample partial correlation coefficient) between y and x_2 given x_1 can be found:

$$SSR(X_2|J, X_1) = 6.70,$$

$$SSE(J, X_1) = SSE(J, X_1, X_2) + SSR(X_2|J, X_1) = 46.50 + 6.70 = 53.20,$$

$$r_{y\cdot x}^2 = 6.70/53.20 = 0.1259.$$

The absolute value of the t statistic for testing whether $\rho_{y2\cdot 1} = 0$ can also be found. In this application we are correcting Y and X_2 for only one variable X_1, so $p-1 = 1$ and $p = 2$. The formula for the absolute t statistic becomes

$$\sqrt{25-2-1}\sqrt{0.1259}\Big/\sqrt{1-0.1259} = 1.78.$$

Note that this is precisely the t statistic reported for β_2 in Example 6.2.1.

Exercise 6.6 Assume that V_{xx} is nonsingular. Show that $\rho_{y\cdot x} = 0$ if and only if the best linear predictor of y_1 based on x and y_2 equals the best linear predictor of y_1 based on x alone.

Exercise 6.7 If $(y_{i1}, y_{i2}, x_{i1}, x_{i2}, \ldots, x_{i,p-1})'$, $i = 1, \ldots, n$ are independent $N(\mu, V)$, find the distribution of $\sqrt{n-p-1}\, r_{y\cdot x}\Big/\sqrt{1-r_{y\cdot x}^2}$ when $\rho_{y\cdot x} = 0$. Hint: Use the linear model $\mathrm{E}(Y_1|X, Y_2) \in C(J, X, Y_2)$, i.e., $Y_1 = [J, X, Y_2]\gamma + e$, to find a distribution conditional on X, Y_2. Note that the distribution does not depend on the values of X and Y_2, so it must also be the unconditional distribution. Note also that from Exercise 6.5 and the equality between the conditional expectation and the best linear predictor for multivariate normals that we have $\rho_{y\cdot x} = 0$ if and only if the regression coefficient of Y_2 is zero.

Finally, the usual concept of partial correlation, which looks at the correlation between the components of $y - \hat{E}(y|x)$, i.e., the residuals based on best linear prediction, can be generalized to a concept based on examining the correlations between the components of $y - \mathrm{E}(y|x)$, the residuals from the best predictor.

6.6 Best Linear Unbiased Prediction

In this section we consider the general theory of best linear unbiased prediction. In a series of examples, this theory will be used to examine prediction in standard linear model theory, Kriging of spatial data, and prediction of random effects in mixed models. The last two subjects are treated in more depth in *ALM-III*.

Consider a set of random variables $y_i, i = 0, 1, \ldots, n$. We want to use $y_1, \ldots, y_n$ to predict y_0. Let $Y = (y_1, \ldots, y_n)'$. In Section 6.3 it was shown that the best predictor (BP) of y_0 is $\mathrm{E}(y_0|Y)$. Typically, the joint distribution of y_0 and Y is not available, so $\mathrm{E}(y_0|Y)$ cannot be found. However, if the means and covariances of the y_is are available, then we can find the best linear predictor (BLP) of y_0. Let $\mathrm{Cov}(Y) \equiv V$, $\mathrm{Cov}(Y, y_0) \equiv V_{y0}$, $\mathrm{E}(y_i) \equiv \mu_i, i = 0, 1, \ldots, n$, and $\mu \equiv (\mu_1, \ldots, \mu_n)'$. Again from Subsection 6.3.4, the BLP of y_0 is

$$\hat{E}(y_0|Y) \equiv \mu_0 + \delta_*'(Y - \mu), \tag{1}$$

where δ_* satisfies $V\delta_* = V_{y0}$.

We now want to weaken the assumption that μ and μ_0 are known. Since the prediction is based on Y and there are as many unknown parameters in μ as there are observations in Y, we need to impose some structure on the mean vector μ before we can generalize the theory. This will be done by specifying a linear model for Y. Since y_0 is being predicted and has not been observed, it is necessary either to know μ_0 or to know that μ_0 is related to μ in a specified manner. In the theory below, it will be assumed that μ_0 is related to the linear model for Y.

Suppose that a vector of known concomitant variables $x_i' = (x_{i1}, \ldots, x_{ip})$ is associated with each random observation y_i, $i = 0, \ldots, n$. We impose structure on the μ_is by assuming that $\mu_i = x_i'\beta$ for some vector of unknown parameters β and all $i = 0, 1, \ldots, n$.

We can now reset our notation in terms of linear model theory. Let

$$X = \begin{bmatrix} x_1' \\ \vdots \\ x_n' \end{bmatrix}.$$

The observed vector Y satisfies the linear model

$$Y = X\beta + e, \quad \mathrm{E}(e) = 0, \quad \mathrm{Cov}(e) = V. \tag{2}$$

With this additional structure, the BLP of y_0 given in (1) becomes

$$\hat{E}(y_0|Y) = x_0'\beta + \delta_*'(Y - X\beta), \tag{3}$$

where again $V\delta_* = V_{y0}$.

The standard assumption that μ and μ_0 are known now amounts to the assumption that β is known. It is this assumption that we renounce. By weakening the assumptions, we consequently weaken the predictor. If β is known, we can find the best linear predictor. When β is unknown, we find the best linear unbiased predictor.

Before proceeding, a technical detail must be mentioned. To satisfy estimability conditions, we need to assume that $x_0' = \rho' X$ for some vector ρ. Frequently, model (2) will be a regression model, so choosing $\rho' = x_0'(X'X)^{-1}X'$ will suffice. In applications to mixed models, $x_0' = 0$, so $\rho = 0$ will suffice.

Definition 6.6.1 A *predictor* $f(Y)$ of y_0 is said to be *unbiased* if

$$\mathrm{E}[f(Y)] = \mathrm{E}(y_0).$$

Although $\mathrm{E}[\mathrm{E}(y_0|Y)] = \mathrm{E}(y_0)$ and $\mathrm{E}[\hat{E}(y_0|Y)] = \mathrm{E}(y_0)$, in this context neither of $\mathrm{E}(y_0|Y)$ and $\hat{E}(y_0|Y)$ are unbiased predictors because they are not predictors. We do not know what they are, so we cannot use them to predict.

Definition 6.6.2 $a_0 + a'Y$ is a *best linear unbiased predictor* of y_0 if $a_0 + a'Y$ is unbiased and if, for any other unbiased predictor $b_0 + b'Y$,

$$\mathrm{E}\big[y_0 - a_0 - a'Y\big]^2 \leq \mathrm{E}\big[y_0 - b_0 - b'Y\big]^2.$$

Theorem 6.6.3 *The best linear unbiased predictor of y_0 is $x_0'\hat{\beta} + \delta_*'(Y - X\hat{\beta})$, where $V\delta_* = V_{y0}$ and $X\hat{\beta}$ is a BLUE of $X\beta$.*

Proof The technique of the proof is to change the prediction problem into an estimation problem and then to use the theory of best linear unbiased estimation. Consider an arbitrary linear unbiased predictor of y_0, say $b_0 + b'Y$. By Proposition 6.3.5,

$$\mathrm{E}[y_0 - b_0 - b'Y]^2 = \mathrm{E}[y_0 - \hat{E}(y_0|Y)]^2 + \mathrm{E}[\hat{E}(y_0|Y) - b_0 - b'Y]^2;$$

so it is enough to find $b_0 + b'Y$ that minimizes $\mathrm{E}[\hat{E}(y_0|Y) - b_0 - b'Y]^2$.

From the definition of $\hat{E}(y_0|Y)$ and unbiasedness of $b_0 + b'Y$, we have

$$0 = \mathrm{E}[\hat{E}(y_0|Y) - b_0 - b'Y].$$

Substituting from equation (3) gives

$$0 = \mathrm{E}[x_0'\beta + \delta_*'(Y - X\beta) - b_0 - b'Y].$$

This relationship holds if and only if $b_0 + (b - \delta_*)'Y$ is a linear unbiased estimate of $x_0'\beta - \delta_*'X\beta$. By Proposition 2.1.9,

$$b_0 = 0;$$

so the term we are trying to minimize is

$$\begin{aligned} \mathrm{E}[\hat{E}(y_0|Y) - b_0 - b'Y]^2 &= \mathrm{E}[(b-\delta_*)'Y - (x_0'\beta - \delta_*'X\beta)]^2 \\ &= \mathrm{Var}\big[(b-\delta_*)'Y\big]. \end{aligned}$$

Because $(b-\delta_*)'Y$ is a linear unbiased estimate, to minimize the variance choose b so that $(b-\delta_*)'Y = x_0'\hat{\beta} - \delta_*'X\hat{\beta}$ is a BLUE of $x_0'\beta - \delta_*'X\beta$. It follows that the best linear unbiased predictor of y_0 is

$$b'Y = x_0'\hat{\beta} + \delta_*'(Y - X\hat{\beta}). \qquad \Box$$

Of course if we did not know the covariance matrices, the BLUP would not be a predictor either. It should not be overlooked that both $\hat{\beta}$ and δ_* depend crucially on V. When all inverses exist, $\hat{\beta} = (X'V^{-1}X)^{-1}X'V^{-1}Y$ and $\delta_* = V^{-1}V_{y0}$. However, this proof remains valid when either inverse does not exist.

It is also of value to know the prediction variance of the BLUP. Let $\mathrm{Var}(y_0) \equiv \sigma_0^2$.

$$\begin{aligned} &\mathrm{E}\Big[y_0 - x_0'\hat{\beta} - \delta_*'(Y - X\hat{\beta})\Big]^2 \\ &= \mathrm{E}\Big[y_0 - \hat{E}(y_0|Y)\Big]^2 + \mathrm{E}\Big[\hat{E}(y_0|Y) - x_0'\hat{\beta} - \delta_*'(Y - X\hat{\beta})\Big]^2 \\ &= \mathrm{E}\Big[y_0 - \hat{E}(y_0|Y)\Big]^2 + \mathrm{Var}\Big[x_0'\hat{\beta} - \delta_*'X\hat{\beta}\Big] \\ &= \sigma_0^2 - V_{y0}'V^-V_{y0} + (x_0' - \delta_*'X)(X'V^-X)^-(x_0 - X'\delta_*), \end{aligned}$$

or, writing the BLUP as $b'Y$, the prediction variance becomes

$$\mathrm{E}\left[y_0 - b'Y\right]^2 = \sigma_0^2 - 2b'V_{y0} + b'Vb.$$

If for some reason $\mathrm{E}(y_0)$ is known, we can still find a BLUP. If $\mathrm{E}(y_0) = 0$, the theory holds merely by taking $x_0 = 0$. If $\mathrm{E}(y_0) = \mu_0$, the more precise way to proceed is to go back into the proof replacing $x_0'\beta$ with μ_0 and observe that the proof continues to hold if we choose $b_0 = \mu_0$. A simpler but less precise way to proceed is to instead predict $y_* \equiv y_0 - \mu_0$ which has mean zero and notice that

$$\hat{y}_0 - \mu_0 \equiv \hat{y}_* = \delta_*'(Y - X\hat{\beta})$$

so that

$$\hat{y}_0 = \mu_0 + \delta_*'(Y - X\hat{\beta}).$$

In either case, the prediction variance is

$$\mathrm{E}\Big[y_0 - \mu_0 - \delta_*'(Y - X\hat{\beta})\Big]^2 = \sigma_0^2 - V_{y0}'V^-V_{y0} + \delta_*'X(X'V^-X)^-X'\delta_*.$$

Example 6.6.4 Prediction in Standard Linear Models.
In the standard linear model the y_is have zero covariance and identical variances. Thus, model (2) is satisfied with $V = \sigma^2 I$. A new observation y_0 would typically have zero covariance with the previous data, so $V_{y0} = 0$. It follows that $\delta_* = 0$ and the BLUP of y_0 is $x_0'\hat{\beta}$. This is just the BLUE of $\mathrm{E}(y_0) = x_0'\beta$.

Example 6.6.5 Spatial Data and Kriging.
If y_i is an observation taken at some point in space, then x_i' typically contains the coordinates of the point. In dealing with spatial data, finding the BLUP is often called *Kriging*. The real challenge with spatial data is getting some idea of the covariance matrices V and V_{y0}. See *ALM-III*, Chapter 8 for a more detailed discussion with additional references.

Exercise 6.8 Prove three facts about Kriging.

(a) If $b'Y$ is the BLUP of y_0 and if $x_{i1} = 1, i = 0, 1, \ldots, n$, then $b'J = 1$.

(b) If $(y_0, x_0') = (y_i, x_i')$ for some $i \geq 1$, then the BLUP of y_0 is just y_i.

(c) If V is nonsingular and $b'Y$ is the BLUP of y_0, then there exists a vector γ such that the following equation is satisfied:

$$\begin{bmatrix} V & X \\ X' & 0 \end{bmatrix} \begin{bmatrix} b \\ \gamma \end{bmatrix} = \begin{bmatrix} V_{y0} \\ x_0 \end{bmatrix}.$$

Typically, this equation will have a unique solution.

Hint: Recall that $x_0' = \rho' X$ and that $X(X'V^{-1}X)^{-}X'V^{-1}$ is a projection operator onto $C(X)$.

Example 6.6.6 Mixed Model Prediction.
A mixed model (is similar to the analysis of covariance models of Chapter 9 but) has

$$Y = X\beta + Z\gamma + e, \tag{4}$$

where X and Z are known matrices, β is an unobservable vector of fixed effects, but where γ is an unobservable vector of random effects with $\mathrm{E}(\gamma) = 0$, $\mathrm{Cov}(\gamma) \equiv D$, and $\mathrm{Cov}(\gamma, e) = 0$. Let $\mathrm{Cov}(e) \equiv R$.

We wish to find the BLUP of $\lambda'\gamma$ based on the data Y. The vector λ can be any known vector. Note that $\mathrm{E}(\lambda'\gamma) = 0$; so we can let $y_0 = \lambda'\gamma$ and $x_0' = 0$. Also,

$$V \equiv \mathrm{Cov}(Y) = \mathrm{Cov}(Z\gamma + e) = ZDZ' + R$$

and

$$\begin{aligned} V_{y0} &= \text{Cov}(Y, \lambda'\gamma) \\ &= \text{Cov}(Z\gamma + e, \lambda'\gamma) \\ &= \text{Cov}(Z\gamma, \lambda'\gamma) \\ &= ZD\lambda. \end{aligned}$$

The BLUP of $\lambda'\gamma$ is

$$\delta_*'(Y - X\hat{\beta}),$$

where $X\hat{\beta}$ is the BLUE of $X\beta$ and δ_* satisfies $(ZDZ' + R)\delta_* = ZD\lambda$. The matrices X, Z, and λ are all known. As in Kriging, the practical challenge is to get some idea of the covariance matrices. In this case, we need D and R. Estimates of D and R are one byproduct of variance component estimation. *ALM-III*, Chapter 5 contains much more information on mixed models.

Exercise 6.9 If $\lambda_1'\beta$ is estimable, find the best linear unbiased predictor of $\lambda_1'\beta + \lambda_2'\gamma$. For this problem, $b_0 + b'Y$ is unbiased if $\text{E}(b_0 + b'Y) = \text{E}(\lambda_1'\beta + \lambda_2'\gamma)$. The best predictor minimizes $\text{E}[b_0 + b'Y - \lambda_1'\beta - \lambda_2'\gamma]^2$.

Exercise 6.10 Assuming the results of Exercise 6.3, show that the BLUP of the random vector $\Lambda'\gamma$ is $Q'(Y - X\hat{\beta})$, where $VQ = ZD\Lambda$.

6.7 Testing Lack of Fit

Suppose we have a linear model $Y = X\beta + e$ and we suspect that the model is an inadequate explanation of the data. The obvious thing to do to correct the problem is to add more predictor variables to the model, i.e, fit a model $Y = Z\gamma + e$, where $C(X) \subset C(Z)$. Two questions present themselves: (1) how does one choose Z, and (2) is there really a lack of fit? Given a choice for Z, the second of these questions can be addressed by testing $Y = X\beta + e$ against $Y = Z\gamma + e$. This is referred to as a *test for lack of fit*. Since Z is chosen so that $Y = Z\gamma + e$ will actually fit the data (or at least fit the data better than $Y = X\beta + e$), the error sum of squares for the model $Y = Z\gamma + e$, say $SSE(Z)$, can be called the *sum of squares for pure error, SSPE*. The difference $SSE(X) - SSE(Z)$ is used for testing lack of fit, so $SSE(X) - SSE(Z)$ is called the *sum of squares for lack of fit, SSLF*.

In general, there are few theoretical guidelines for choosing Z. A common situation is where there are measured variables not currently included in the model and it is necessary to select additional variables to include in the model. Variable selection techniques are discussed in Chapter 14. In this section, we discuss the problem of testing lack of fit when there are no other variables available for inclusion in the model. With no other variables available, the model matrix X must be used as the basis for choosing Z. We will present four approaches. The first is the traditional method based on having a model matrix X in which some of the rows are identical. A second approach is based on identifying clusters of rows in X that are nearly identical. A third approach examines different subsets of the data. Finally, we briefly mention a nonparametric regression approach to testing lack of fit. Christensen (2015) covers similar material at a more applied level.

One final note. This section is in the chapter on regression because testing lack of fit has traditionally been considered as a topic in regression analysis. Nowhere in this section do we assume that $X'X$ is nonsingular. *The entire discussion holds for general linear models.*

6.7.1 The Traditional Test

To discuss the traditional approach that originated with Fisher (1922), we require notation for identifying which rows of the model matrix are identical. A model with replications can be written

$$y_{ij} = x_i'\beta + e_{ij},$$

where β is the vector of parameters, $x_i' = (x_{i1}, \ldots, x_{ip})$, $i = 1, \cdots, c$, and $j = 1, \cdots, N_i$. We presume that $x_i' \neq x_k'$ for $i \neq k$. Using the notation of Chapter 4 in which a pair of subscripts is used to denote a row of a vector or a row of the model matrix, we have $Y = [y_{ij}]$ and

$$X = [w_{ij}'], \qquad \text{where } w_{ij}' = x_i'.$$

The idea of pure error, when there are rows of X that are replicated, is that if several observations have the same mean value, the variability about that mean value is in some sense pure error. The problem is to estimate the mean value. If we estimate the mean value in the ith group with $x_i'\hat{\beta}$, then estimate the variance for the group by looking at the deviations about the estimated mean value, and finally pool the estimates from the different groups, we get $MSE(X)$. Now consider a more general model, $Y = Z\gamma + e$, where $C(X) \subset C(Z)$ and Z is chosen so that

$$Z = [z_{ij}'], \qquad \text{where } z_{ij}' = v_i'$$

for some vectors v_i', $i = 1, \ldots, c$. Two rows of Z are the same if and only if the corresponding rows of X are the same. Thus, the groups of observations that had the same mean value in the original model still have the same mean value in the generalized model. If there exists a lack of fit, we hope that the more general model gives a more accurate estimate of the mean value.

It turns out that there exists a most general model $Y = Z\gamma + e$ that satisfies the condition that two rows of Z are the same if and only if the corresponding rows of X are the same. We will refer to the property that rows of Z are identical if and only if the corresponding rows of X are identical as X and Z having the same *row structure*.

X was defined to have c distinct rows, therefore $r(X) \leq c$. Since Z has the same row structure as X, we also have $r(Z) \leq c$. The most general matrix Z, the one with the largest column space, will have $r(Z) = c$. We need to find Z with $C(X) \subset C(Z)$, $r(Z) = c$, and the same row structure as X. We also want to show that the column space is the same for any such Z.

Let Z be the model matrix for the model $y_{ij} = \mu_i + e_{ij}$, $i = 1, \ldots, c$, $j = 1, \ldots, N_i$. If we let $z_{ij,k}$ denote the element in the ijth row and kth column of Z, then from Chapter 4

$$Z = [z_{ij,k}], \qquad \text{where } z_{ij,k} = \delta_{ik}.$$

Z is a matrix where the kth column is 0 everywhere except that it has 1s in rows that correspond to the y_{kj}s. Since the values of $z_{ij,k}$ do not depend on j, it is clear that Z has the same row structure as X. Since the c columns of Z are linearly independent, we have $r(Z) = c$, and it is not difficult to see that $X = M_Z X$, where M_Z is the perpendicular projection operator onto $C(Z)$; so we have $C(X) \subset C(Z)$.

In fact, because of the form of Z and M_Z, it is clear that any matrix Z_1 with the same row structure as X must have $Z_1 = M_Z Z_1$ and $C(Z_1) \subset C(Z)$. If $r(Z_1) = c$, then it follows that $C(Z_1) = C(Z)$ and the column space of the most general model $Y = Z\gamma + e$ does not depend on the specific choice of Z.

If one is willing to assume that the lack of fit is not due to omitting some variable that, if included, would change the row structure of the model (i.e., if one is willing to assume that the row structure of the true model is the same as the row structure of X), then the true model can be written $Y = W\delta + e$ with $C(X) \subset C(W) \subset C(Z)$. It is easily seen that the lack-of-fit test statistic based on X and Z has a noncentral F distribution and if $Y = X\beta + e$ is the true model, the test statistic has a central F distribution.

The computations for this lack-of-fit test are quite simple. With the choice of Z indicated, $C(Z)$ is just the column space for a one-way ANOVA.

$$SSPE \equiv SSE(Z) = Y'(I - M_Z)Y = \sum_{i=1}^{c} \sum_{j=1}^{N_i} (y_{ij} - \bar{y}_{i\cdot})^2.$$

With $M_Z Y = (\bar{y}_1, \ldots \bar{y}_1, \bar{y}_2, \ldots \bar{y}_2, \ldots, \bar{y}_c, \ldots \bar{y}_c)'$, $\hat{y}_i = x_i'\hat{\beta}$ and $MY = (\hat{y}_1, \ldots \hat{y}_1, \hat{y}_2, \ldots \hat{y}_2, \ldots, \hat{y}_c, \ldots \hat{y}_c)'$, the sum of squares for lack of fit is

$$\begin{aligned} SSLF \equiv Y'(M_Z - M)Y &= [(M_Z - M)Y]'[(M_Z - M)Y] \\ &= \sum_{i=1}^{c} \sum_{j=1}^{N_i} (\bar{y}_{i\cdot} - \hat{y}_i)^2 = \sum_{i=1}^{c} N_i (\bar{y}_{i\cdot} - \hat{y}_i)^2. \end{aligned}$$

Exercise 6.11 Show that if M is the perpendicular projection operator onto $C(X)$ with

$$X = \begin{bmatrix} w_1' \\ \vdots \\ w_n' \end{bmatrix} \text{ and } M = \begin{bmatrix} T_1' \\ \vdots \\ T_n' \end{bmatrix},$$

then $w_i = w_j$ if and only if $T_i = T_j$.

Exercise 6.12 Discuss the application of the traditional lack-of-fit test to the problem where $Y = X\beta + e$ is a simple linear regression model.

As we have seen, in the traditional method of testing for lack of fit, the row structure of the model matrix X completely determines the choice of Z. Now, suppose that none of the rows of X are identical. It is still possible to have lack of fit, but the traditional method no longer applies.

6.7.2 *Near Replicate Lack of Fit Tests*

Another set of methods for testing lack of fit is based on mimicking the traditional lack-of-fit test. With these methods, rows of the model matrix that are nearly replicates are identified. One way of identifying near replicates is to use a hierarchical clustering algorithm (see Gnanadesikan 1977) to identify rows of the model matrix that are near one another. Tests for lack of fit using near replicates are reviewed by Neill and Johnson (1984). The theory behind such tests is beautifully explained in Christensen (1989, 1991). (OK, so I'm biased in favor of this particular author.) Miller et al. (1998, 1999) provide a theoretical basis for choosing near replicate clusters. Pimentel et al. (2017) propose an alternative method for defining near replicates.

Christensen (1991) suggests that a very good all-purpose near replicate lack-of-fit test was introduced by Shillington (1979). Write the regression model in terms of c clusters of near replicates with the ith cluster containing N_i cases, say

$$y_{ij} = x_{ij}'\beta + e_{ij}, \tag{1}$$

$i = 1, \ldots, c$, $j = 1, \ldots, N_i$. Note that at this point we have done nothing to the model except play with the subscripts; model (1) is just the original model. Shillington's test involves finding means of the predictor variables in each cluster and fitting the model

$$y_{ij} = \bar{x}_{i\cdot}'\beta + e_{ij}. \tag{2}$$

The numerator for Shillington's test is then the numerator mean square used in comparing this model to the one-way ANOVA model

$$y_{ij} = \mu_i + e_{ij}. \tag{3}$$

However, the denominator mean square for Shillington's test is the mean squared error from fitting the model

$$y_{ij} = x_{ij}'\beta + \mu_i + e_{ij}. \tag{4}$$

It is not difficult to see that if model (1) holds, then Shillington's test statistic has a central F distribution with the appropriate degrees of freedom. Christensen (1989, 1991) gives details and explains why this should be a good all-purpose test—even though it is not the optimal test for either of the alternatives developed by Christensen. The near replicate lack-of-fit test proposed in Christensen (1989) is the test of model (1) against model (4). The test proposed in Christensen (1991) uses the same numerator as Shillington's test, but a denominator sum of squares that is the SSE from model (1) minus the numerator sum of squares from Shillington's test. Both of Christensen's tests are optimal for certain types of lack of fit. If the clusters consist of exact replicates, then all of these tests reduce to the traditional test.

Example 6.7.1 Using the data of Example 6.2.1 we illustrate the near replicate lack-of-fit tests. Near replicates were chosen visually by plotting x_1 versus x_2. The near replicates are presented below.

Near Replicate Clusters for Steam Data

Obs. no.	x_1	x_2	Near rep.	Obs. no.	x_1	x_2	Near rep.
1	35.3	20	2	14	39.1	19	11
2	29.7	20	2	15	46.8	23	12
3	30.8	23	9	16	48.5	20	3
4	58.8	20	4	17	59.3	22	13
5	61.4	21	5	18	70.0	22	7
6	71.3	22	7	19	70.0	11	1
7	74.4	11	1	20	74.5	23	8
8	76.7	23	8	21	72.1	20	14
9	70.7	21	10	22	58.1	21	5
10	57.5	20	4	23	44.6	20	3
11	46.4	20	3	24	33.4	20	2
12	28.9	21	6	25	28.6	22	15
13	28.1	21	6				

Fitting models (1) through (4) gives the following results:

Model	(1)	(2)	(3)	(4)
dfE	22	22	10	9
SSE	46.50	50.75	12.136	7.5

Shillington's test is

$$F_S = \frac{[50.75 - 12.136]/[22 - 10]}{7.5/9} = 3.8614 > 3.073 = F(0.95, 12, 9).$$

Christensen's (1989) test is

$$F_{89} = \frac{[46.50 - 7.5]/[22 - 9]}{7.5/9} = 3.6000 > 3.048 = F(0.95, 13, 9).$$

Christensen's (1991) test is

$$F_{91} = \frac{[50.75 - 12.136]/[22 - 10]}{[46.5 - (50.75 - 12.136)]/[22 - (22 - 10)]} = 4.0804$$
$$> 2.913 = F(0.95, 12, 10).$$

All three tests indicate a lack of fit.

In this example, all of the tests behaved similarly. Christensen (1991) shows that the tests can be quite different and that for the single most interesting type of lack of fit, the 91 test will typically be more powerful than Shillington's test, which is typically better than the 89 test.

Christensen's 89 test is similar in spirit to "nonparametric" lack-of-fit tests based on spanning (or basis) functions used to approximate general regression models, see Subsection 6.2.2. In particular, it is similar to adding Haar wavelets as additional predictor variables to model (1) except that Haar wavelets amount to adding indicator variables for a predetermined partition of the space of predictor variables while the near replicate methods use the observed predictors to suggest where indicator variables are needed. See Subsection 4 and *ALM-III*, Section 1.8 for additional discussion of these nonparametric approaches.

6.7.3 Partitioning Methods

Another way to use X in determining a more general matrix Z is to partition the data. Write

$$X = \begin{bmatrix} X_1 \\ X_2 \end{bmatrix}, \quad Y = \begin{bmatrix} Y_1 \\ Y_2 \end{bmatrix}.$$

The model $Y = Z\gamma + e$ can be chosen with

$$Z = \begin{bmatrix} X_1 & 0 \\ 0 & X_2 \end{bmatrix}.$$

Clearly, $C(X) \subset C(Z)$. We again refer to the difference $SSE(X) - SSE(Z)$ as the *SSLF* and, in something of an abuse of the concept "pure," we continue to call $SSE(Z)$ the *SSPE*.

Exercise 6.13 Let M_i be the perpendicular projection operator onto $C(X_i)$, $i = 1, 2$. Show that the perpendicular projection operator onto $C(Z)$ is

$$M_Z = \begin{bmatrix} M_1 & 0 \\ 0 & M_2 \end{bmatrix}.$$

Show that $SSE(Z) = SSE(X_1) + SSE(X_2)$, where $SSE(X_i)$ is the sum of squares for error from fitting $Y_i = X_i\beta_i + e_i$, $i = 1, 2$.

If there is no lack of fit for $Y = X\beta + e$, since $C(X) \subset C(Z)$, the test statistic will have a central F distribution. Suppose that there is lack of fit and the true model is, say, $Y = W\delta + e$. It is unlikely that W will have the property that $C(X) \subset C(W) \subset C(Z)$, which would ensure that the test statistic has a noncentral F distribution. In general, if there is lack of fit, the test statistic has a doubly noncentral F distribution. (A doubly noncentral F is the ratio of two independent noncentral chi-squareds divided by their degrees of freedom.) The idea behind the lack-of-fit test based on partitioning the data is the hope that X_1 and X_2 will be chosen so that the combined fit of $Y_1 = X_1\beta_1 + e_1$ and $Y_2 = X_2\beta_2 + e_2$ will be qualitatively better than the fit of $Y = X\beta + e$. Thus, it is hoped that the noncentrality parameter of the numerator chi-squared will be larger than the noncentrality parameter of the denominator chi-squared.

Example 6.7.2 Let $Y = X\beta + e$ be the simple linear regression $y_i = \beta_0 + \beta_1 x_i + e_i$, $i = 1, \ldots, 2r$, with $x_1 \leq x_2 \leq \cdots \leq x_{2r}$. Suppose that the lack of fit is due to the true model being $y_i = \beta_0 + \beta_1 x_i + \beta_2 x_i^2 + e_i$, so the true curve is a parabola. Clearly, one can approximate a parabola better with two lines than with one line. The combined fit of $y_i = \eta_0 + \eta_1 x_i + e_i$, $i = 1, \ldots, r$, and $y_i = \tau_0 + \tau_1 x_i + e_i$, $i = r+1, \ldots, 2r$, should be better than the unpartitioned fit.

Example 6.7.3 We now test the model used in Example 6.2.1 for lack of fit using the partitioning method. The difficulty with the partitioning method lies in finding some reasonable way to partition the data. Fortunately for me, I constructed this example, so I know where the lack of fit is and I know a reasonable way to partition the data. (The example was constructed just like Example 12.4.4. The construction is explained in Chapter 12.) I partitioned the data based on the variable x_1. Any case that had a value of x_1 less than 24 went into one group. The remaining cases went into the other group. This provided a group of 12 cases with small x_1 values and a group of 13 cases with large x_1 values. The sum of squares for error for the small group was $SSE(S) = 2.925$ with 9 degrees of freedom and the sum of squares for error for the large group was $SSE(L) = 13.857$ with 10 degrees of freedom. Using the error from Example 6.2.1, we get

$$\begin{aligned}
SSPE &= 13.857 + 2.925 = 16.782,\\
dfPE &= 10 + 9 = 19,\\
MSPE &= 0.883,
\end{aligned}$$

$$\begin{aligned}
SSLF &= 46.50 - 16.78 = 29.72,\\
dfLF &= 22 - 19 = 3,\\
MSLF &= 9.91,\\
F &= 9.91/0.883 = 11.22.
\end{aligned}$$

This has 3 degrees of freedom in the numerator, 19 degrees of freedom in the denominator, is highly significant, and indicates a definite lack of fit. But remember, I knew that the lack of fit was related to x_1, so I could pick an effective partition.

Recall from Example 6.4.1 that the R^2 for Example 6.2.1 is 0.964, which indicates a very good predictive model. In spite of the high R^2, we are still able to establish a lack of fit using both the partitioning method and the near replicate method.

The partitioning method can easily be extended, cf. Atwood and Ryan (1977). For example, one could select three partitions of the data and write

$$X = \begin{bmatrix} X_1 \\ X_2 \\ X_3 \end{bmatrix}, \quad Y = \begin{bmatrix} Y_1 \\ Y_2 \\ Y_3 \end{bmatrix}, \quad Z = \begin{bmatrix} X_1 & 0 & 0 \\ 0 & X_2 & 0 \\ 0 & 0 & X_3 \end{bmatrix}.$$

The lack-of-fit test would proceed as before. Note that the partitioning method is actually a generalization of the traditional method. If the partition of the data consists of the different sets of identical rows of the model matrix, then the partitioning method gives the traditional lack-of-fit test. The partitioning method can also be used to give a near replicate lack-of-fit test with the partitions corresponding to the clusters of near replicates. As mentioned earlier, it is not clear in general how to choose an appropriate partition.

Utts (1982) presented a particular partition to be used in what she called the Rainbow Test (for lack of fit). She suggests selecting a set of rows from X that are centrally located to serve as X_1, and placing each row of X not in X_1 into a separate set that consists only of that row. With this partitioning, each of the separate sets determined by a single row corresponds to p columns of Z that are zero except for the entries in that row. These p columns are redundant. Eliminating unnecessary columns allows the Z matrix to be rewritten as

$$Z = \begin{bmatrix} X_1 & 0 \\ 0 & I \end{bmatrix}.$$

From Exercise 6.13, it is immediately seen that $SSE(Z) = SSE(X_1)$; thus, the Rainbow Test amounts to testing $Y = X\beta + e$ against $Y_1 = X_1\beta + e_1$, see also Section 9.3. To select the matrix X_1, Utts suggests looking at the diagonal elements of $M = [m_{ij}]$. The smallest values of the m_{ii}s are the most centrally located data points, cf. Section 12.1. The author's experience indicates that the Rainbow Test works best when one is quite selective about the points included in the central partition.

Example 6.7.4 Continued. First consider the Rainbow Test using half of the data set. The variables x_1, x_2, and an intercept were fitted to the 12 cases that had the smallest m_{ii} values. This gave a $SSE = 16.65$ with 9 degrees of freedom. The Rainbow Test mean square for lack of fit, mean square for pure error, and F statistic are

$$MSLF = (46.50 - 16.65)/(22 - 9) = 2.296,$$

$$MSPE = 16.65/9 = 1.850,$$

$$F = 2.296/1.850 = 1.24.$$

The F statistic is nowhere near being significant. Now consider taking the quarter of the data with the smallest m_{ii} values. These 6 data points provide a $SSE = 0.862$ with 3 degrees of freedom.

$$MSLF = (46.50 - 0.862)/(22 - 3) = 2.402,$$

$$MSPE = 0.862/3 = 0.288,$$

$$F = 2.402/0.288 = 8.35.$$

In spite of the fact that this has only 3 degrees of freedom in the denominator, the F statistic is reasonably significant. $F(0.95, 19, 3)$ is approximately 8.67 and $F(0.90, 19, 3)$ is about 5.19.

6.7.4 Nonparametric Methods

As discussed in Subsection 6.2.2, one approach to nonparametric regression is to fit very complicated linear models using "basis" functions. One particular application of this approach to nonparametric regression is the fitting of moderate (as opposed to low or high) order polynomials. *ALM-III*, Chapter 1 provides more details of this general approach to nonparametric regression and in particular Section 1.8 discusses testing lack of fit. Fundamentally, the idea is to test the original linear model $y_i = x_i'\beta + e_i$ against a larger model that incorporates nonparametric regression components, i.e.,

$y_i = x_i'\beta + \sum_{j=1}^{q} \gamma_j \phi_j(x_i) + e_i$. For high dimensional problems, the larger model may need to involve generalized additive functions.

Exercise 6.14 Test the model $y_{ij} = \beta_0 + \beta_1 x_i + \beta_2 x_i^2 + e_{ij}$ for lack of fit using the data:

x_i	1.00	2.00	0.00	−3.00	2.50
y_{ij}	3.41	22.26	−1.74	79.47	37.96
	2.12	14.91	1.32	80.04	44.23
	6.26	23.41	−2.55	81.63	
		18.39			

Exercise 6.15 Using the following data, test the model $y_{ij} = \beta_0 + \beta_1 x_{i1} + \beta_2 x_{i2} + e_{ij}$ for lack of fit. Explain and justify your method.

X_1	X_2	Y	X_1	X_2	Y
31	9.0	122.41	61	2.2	70.08
43	8.0	115.12	36	4.7	66.42
50	2.8	64.90	52	9.4	150.15
38	5.0	64.91	38	1.5	38.15
38	5.1	74.52	41	1.0	45.67
51	4.6	75.02	41	5.0	68.66
41	7.2	101.36	52	4.5	76.15
57	4.0	74.45	29	2.7	36.20
46	2.5	56.22			

6.8 Polynomial Regression and One-Way ANOVA

Polynomial regression is the special case of fitting a model

$$y_i = \beta_0 + \beta_1 x_i + \beta_2 x_i^2 + \cdots + \beta_{p-1} x_i^{p-1} + e_i,$$

i.e.,

$$Y = \begin{bmatrix} 1 & x_1 & x_1^2 & \cdots & x_1^{p-1} \\ 1 & x_2 & x_2^2 & \cdots & x_2^{p-1} \\ \vdots & \vdots & \vdots & & \vdots \\ 1 & x_n & x_n^2 & \cdots & x_n^{p-1} \end{bmatrix} \begin{bmatrix} \beta_0 \\ \beta_1 \\ \vdots \\ \beta_{p-1} \end{bmatrix} + e.$$

All of the standard multiple regression results hold, but there are some additional issues to consider. For instance, one should think very hard about whether it makes sense to test $H_0 : \beta_j = 0$ for any j other than $j = p-1$. Frequently, the model

$$y_i = \beta_0 + \beta_1 x_i + \cdots + \beta_{j-1} x_i^{j-1} + \beta_{j+1} x_i^{j+1} + \cdots + \beta_{p-1} x_i^{p-1} + e_i$$

is not very meaningful. Typically, it only makes sense to test the coefficient of the highest order term in the polynomial. One would only test $\beta_j = 0$ if it had already been decided that $\beta_{j+1} = \cdots = \beta_{p-1} = 0$.

Sometimes, polynomial regression models are fitted using orthogonal polynomials. This is a procedure that allows one to perform all the appropriate tests on the β_js without having to fit more than one regression model. The technique uses the Gram–Schmidt algorithm to orthogonalize the columns of the model matrix and then fits a model to the orthogonalized columns. Since Gram–Schmidt orthogonalizes vectors sequentially, the matrix with orthogonal columns can be written $T = XP$, where P is a nonsingular upper triangular matrix. The model $Y = X\beta + e$ is equivalent to $Y = T\gamma + e$ with $\gamma = P^{-1}\beta$. P^{-1} is also an upper triangular matrix, so γ_j is a linear function of $\beta_j, \beta_{j+1}, \ldots, \beta_{p-1}$. The test of $H_0 : \gamma_{p-1} = 0$ is equivalent to the test of $H_0 : \beta_{p-1} = 0$. If $\beta_{j+1} = \beta_{j+2} = \cdots = \beta_{p-1} = 0$, then the test of $H_0 : \gamma_j = 0$ is equivalent to the test of $H_0 : \beta_j = 0$. In other words, the test of $H_0 : \gamma_j = 0$ is equivalent to the test of $H_0 : \beta_j = 0$ in the model $y_i = \beta_0 + \cdots + \beta_j x_i^j + e_i$. However, because the columns of T are orthogonal, the sum of squares for testing $H_0 : \gamma_j = 0$ depends only on the column of T associated with γ_j. It is not necessary to do any additional model fitting to obtain the test.

An algebraic expression for the orthogonal polynomials being fitted is available in the row vector

$$[1, x, x^2, \ldots, x^{p-1}]P. \tag{1}$$

The $p - 1$ different polynomials that are contained in this row vector are orthogonal only in that the coefficients of the polynomials were determined so that XP has columns that are orthogonal. As discussed above, the test of $\gamma_j = 0$ is the same as the test of $\beta_j = 0$ when $\beta_{j+1} = \cdots = \beta_{p-1} = 0$. The test of $\gamma_j = 0$ is based on the $(j + 1)$st column of the matrix T. β_j is the coefficient of the $(j + 1)$st column of X, i.e., the jth degree term in the polynomial. By analogy, *the $(j + 1)$st column of T is called the jth degree orthogonal polynomial.*

Polynomial regression has some particularly interesting relationships with the problem of estimating pure error and with one-way ANOVA problems. It is clear that for all values of p, the row structure of the model matrices for the polynomial regression models is the same, i.e., if $x_i = x_{i'}$, then $x_i^k = x_{i'}^k$; so the i and i' rows of X are the same regardless of the order of the polynomial. Suppose there are q distinct values of x_i in the model matrix. The most general polynomial that can be fitted must give a rank q model matrix; thus the most general model must be

$$y_i = \beta_0 + \beta_1 x_i + \cdots + \beta_{q-1} x_i^{q-1} + e_i.$$

It also follows from the previous section that the column space of this model is exactly the same as the column space for fitting a one-way ANOVA with q treatments.

Using double subscript notation with $i = 1, \ldots, q$, $j = 1, \ldots, N_i$, the models

$$y_{ij} = \beta_0 + \beta_1 x_i + \cdots + \beta_{q-1} x_i^{q-1} + e_{ij} \tag{2}$$

and

$$y_{ij} = \mu + \alpha_i + e_{ij}$$

are equivalent. Since both β_0 and μ are parameters corresponding to a column of 1s, the tests of $H_0 : \beta_1 = \cdots = \beta_{q-1} = 0$ and $H_0 : \alpha_1 = \cdots = \alpha_q$ are identical. Both tests look at the orthogonal complement of J_n with respect to $C(X)$, where $C(X)$ is the column space for either of the models. Using the ideas of Section 3.6, one way to break this space up into $q-1$ orthogonal one-degree-of-freedom hypotheses is to look at the orthogonal polynomials for $i = 1, \ldots, q-1$. As seen in Section 4.2, any vector in $C(X)$ that is orthogonal to J determines a contrast in the α_is. In particular, each orthogonal polynomial corresponds to a contrast in the α_is.

Finding a set of $q-1$ orthogonal contrasts amounts to finding an orthogonal basis for $C(M_\alpha)$. If we write $T = [T_0, \ldots, T_{q-1}]$, then $T_1, \ldots, T_{q-1}$ is an orthogonal basis for $C(M_\alpha)$. Given these vectors in $C(M_\alpha)$, we can use Proposition 4.2.3 to read off the corresponding contrasts. Moreover, the test for dropping, say, T_j from the model is the test of $H_0 : \gamma_j = 0$, which is just the test that the corresponding contrast is zero. Note that testing this contrast is not of interest unless $\beta_{j+1} = \cdots = \beta_{q-1} = 0$ or, equivalently, if $\gamma_{j+1} = \cdots = \gamma_{q-1} = 0$ or, equivalently, if all the higher order polynomial contrasts are zero.

Definition 6.8.1 The orthogonal contrasts determined by the orthogonal polynomials are called the *polynomial contrasts*. The contrast corresponding to the first degree orthogonal polynomial is called the *linear contrast*. The contrasts for higher degree orthogonal polynomials are called the *quadratic*, *cubic*, *quartic*, etc., contrasts.

Using Proposition 4.2.3, if we identify an orthogonal polynomial as a vector $\rho \in C(M_\alpha)$, then the corresponding contrast can be read off. For example, the second column of the model matrix for (2) is $X_1 = [t_{ij}]$, where $t_{ij} = x_i$ for all i and j. If we orthogonalize this with respect to J, we get the linear orthogonal polynomial. Letting

$$\bar{x}_{\cdot} = \sum_{i=1}^{q} N_i x_i \Big/ \sum_{i=1}^{q} N_i$$

leads to the linear orthogonal polynomial

$$T_1 = [w_{ij}], \qquad \text{where } w_{ij} = x_i - \bar{x}_{\cdot}.$$

From Section 4.2, this vector corresponds to a contrast $\sum \lambda_i \alpha_i$, where $\lambda_i / N_i = x_i - \bar{x}_{\cdot}$. Solving for λ_i gives

$$\lambda_i = N_i(x_i - \bar{x}_{\cdot}).$$

The sum of squares for testing $H_0 : \beta_1 = \gamma_1 = 0$ is

$$\left[\sum_i N_i(x_i - \bar{x}_\cdot)\bar{y}_{i\cdot}\right]^2 \bigg/ \left[\sum_i N_i(x_i - \bar{x}_\cdot)^2\right].$$

And, of course, one would not do this test unless it had already been established that $\beta_2 = \cdots = \beta_{q-1} = 0$.

As with other directions in vector spaces and linear hypotheses, orthogonal polynomials and orthogonal polynomial contrasts are only of interest up to constant multiples. In applying the Gram–Schmidt theorem to obtain orthogonal polynomials, we really do not care about normalizing the columns. It is the sequential orthogonalization that is important. In the example of the linear contrast, we did not bother to normalize anything.

It is well known (see Exercise 6.16) that, if $N_i = N$ for all i and the quantitative levels x_i are equally spaced, then the orthogonal polynomial contrasts (up to constant multiples) depend only on q. For any value of q, the contrasts can be tabled; see Table 6.2 or, for example, Snedecor and Cochran (1980).

Although it is difficult to derive the tabled contrasts directly, one can verify the appropriateness of the tabled contrasts. Again we appeal to Chapter 4. Let $[J, Z]$ be the model matrix for the one-way ANOVA and let X be the model matrix for model (2). The model matrix for the orthogonal polynomial model is $T = XP$. With $C(X) = C(Z)$, we can write $T = ZB$ for some matrix B. Writing the $q \times q$ matrix B as $B = [b_0, \ldots, b_{q-1}]$ with $b_k = (b_{1k}, \ldots, b_{qk})'$, we will show that for $k \geq 1$ and $N_i = N$, the kth degree orthogonal polynomial contrast is $b_k'\alpha$, where $\alpha = (\alpha_1, \ldots, \alpha_q)'$. To see this, note that the kth degree orthogonal polynomial is $T_k = Zb_k$. The first column of X is a column of 1s and T is a successive orthogonalization of the columns of X; so $J_n = Zb_0$ and for $k \geq 1$, $Zb_k \perp J_n$. It follows that $b_0 = J_q$ and, from Chapter 4, for $k \geq 1$, $Zb_k \in C(M_\alpha)$. Thus, for $k \geq 1$, Zb_k determines a contrast $(Zb_k)'Z\alpha = \sum_{i=1}^q \lambda_i\alpha_i$. However, $(Zb_k)'Z\alpha = b_k'Z'Z\alpha = b_k'\text{Diag}(N_i)\alpha$. The contrast coefficients $\lambda_1, \ldots, \lambda_q$ satisfy $b_{ik}N_i = \lambda_i$. When $N_i = N$, the contrast is $\sum_i \lambda_i\alpha_i = N\sum_i b_{ik}\alpha_i = Nb_k'\alpha$. Orthogonal polynomial contrasts are defined only up to constant multiples, so the kth degree orthogonal polynomial contrast is also $b_k'\alpha$.

Given a set of contrast vectors $b_1, \ldots, b_{q-1}$, we can check whether these are the orthogonal polynomial contrasts. Simply compute the corresponding matrix $T = ZB$ and check whether this constitutes a successive orthogonalization of the columns of X.

Ideas similar to these will be used in Section 9.4 to justify the use of tabled contrasts in the analysis of balanced incomplete block designs. These ideas also relate to Section 7.3 on Polynomial Regression and the Balanced Two-Way ANOVA. Finally, these ideas relate to nonparametric regression as discussed in Subsection 6.2.2. There, polynomials were used as an example of a class of functions that can be used to approximate arbitrary continuous functions. Other examples mentioned were cosines and wavelets. The development given in this section for polynomials can be mimicked for any approximating class of functions.

Table 6.2 Orthogonal polynomial contrasts

q = 3			q = 4			q = 5		
L	Q	C	L	Q	C	L	Q	C
−1	1		−3	1	−1	−2	2	−1
0	−2		−1	−1	3	−1	−1	2
1	1		1	−1	−3	0	−2	0
			3	1	1	1	−1	−2
						2	2	1

q = 6			q = 7			q = 8		
L	Q	C	L	Q	C	L	Q	C
−5	5	−5	3	5	−1	−7	7	−7
−3	−1	7	2	0	1	−5	1	5
−1	−4	4	1	−3	1	−3	−3	7
1	−4	−4	0	−4	0	−1	−5	3
3	−1	−7	1	−3	−1	1	−5	−3
5	5	5	2	0	−1	3	−3	−7
			3	5	1	5	1	−5
						7	7	7

q = 9			q = 10			q = 11		
L	Q	C	L	Q	C	L	Q	C
−4	28	−14	−9	6	−42	−5	15	−30
−3	7	7	−7	2	14	−4	6	6
−2	−8	13	−5	−1	35	−3	−1	22
−1	−17	9	−3	−3	31	−2	−6	23
0	−20	0	−1	−4	12	−1	−9	14
1	−17	−9	1	−4	−12	0	−10	0
2	−8	−13	3	−3	−31	1	−9	−14
3	7	−7	5	−1	−35	2	−6	−23
4	28	14	7	2	−14	3	−1	−22
			9	6	42	4	6	−6
						5	15	30

L: linear, Q: quadratic, C: cubic

For completeness, an alternative idea of orthogonal polynomials should be mentioned. In equation (1), rather than using the matrix P that transforms the columns of X into T with orthonormal columns, one could instead choose a P_0 so that the transformed functions are orthogonal in an appropriate function space. The Legendre polynomials are such a collection. The fact that such orthogonal polynomials do not depend on the specific x_is in the data is both an advantage and a disadvantage. It is an advantage in that they are well known and do not have to be derived for each unique set of x_is. It is a disadvantage in that, although they display better numerical properties than the unadjusted polynomials, since $T_0 = XP_0$ typically does not have (exactly) orthonormal columns, these polynomials will not display the precise features exploited earlier in this section.

Exercise 6.16

(a) Find the model matrix for the orthogonal polynomial model $Y = T\gamma + e$ corresponding to the model

$$y_{ij} = \beta_0 + \beta_1 x_i + \beta_2 x_i^2 + \beta_3 x_i^3 + e_{ij},$$

$i = 1, 2, 3, 4$, $j = 1, \ldots, N$, where $x_i = a + (i - 1)t$.

Hint: First consider the case $N = 1$.

(b) For the model $y_{ij} = \mu + \alpha_i + e_{ij}$, $i = 1, 2, 3, 4$, $j = 1, \ldots, N$, and for $k = 1, 2, 3$, find the contrast $\sum \lambda_{ik}\alpha_i$ such that the test of $H_0 : \sum \lambda_{ik}\alpha_i = 0$ is the same as the test of $H_0 : \gamma_k = 0$, i.e., find the polynomial contrasts.

Exercise 6.17 Repeat Exercise 6.16 with $N = 2$ and $x_1 = 2$, $x_2 = 3$, $x_3 = 5$, $x_4 = 8$.

6.9 Additional Exercises

The first three exercises involve using *Fieller's method* for finding confidence intervals, cf. Exercise 3.7b.

Exercise 6.9.1 *Calibration.*

Consider the regression model $Y = X\beta + e$, $e \sim N(0, \sigma^2 I)$ and suppose that we are interested in a future observation, say y_0, that will be independent of Y and have mean $x_0'\beta$. In previous work with this situation, y_0 was not yet observed but the corresponding vector x_0 was known. The calibration problem reverses these assumptions. Suppose that we have observed y_0 and wish to infer what the corresponding vector x_0 might be.

A typical calibration problem might involve two methods of measuring some quantity: y, a cheap and easy method, and x, an expensive but very accurate method. Data are obtained to establish the relationship between y and x. Having done this, future measurements are made with y and the calibration relationship is used to identify what the value of y really means. For example, in sterilizing canned food, x would be a direct measure of the heat absorbed into the can, while y might be the number of bacterial spores of a certain strain that are killed by the heat treatment. (Obviously, one needs to be able to measure the number of spores in the can both before and after heating.)

Consider now the simplest calibration model, $y_i = \beta_0 + \beta_1 x_i + e_i$, e_is i.i.d. $N(0, \sigma^2)$, $i = 1, 2, 3, \ldots, n$. Suppose that y_0 is observed and that we wish to estimate the corresponding value x_0 (x_0 is viewed as a parameter here).

(a) Find the MLEs of β_0, β_1, x_0, and σ^2.

Hint: This is a matter of showing that the obvious estimates are MLEs.

(b) Suppose now that a series of observations $y_{01}, \ldots, y_{0r}$ were taken, all of which correspond to the same x_0. Find the MLEs of β_0, β_1, x_0, and σ^2.

Hint: Only the estimate of σ^2 changes form.

(c) Based on one observation y_0, find a $(1-\alpha)100\%$ confidence interval for x_0. When does such an interval exist?

Hint: Use an $F(1, n-2)$ distribution based on $(y_0 - \hat{\beta}_0 - \hat{\beta}_1 x_0)^2$.

Comment: Aitchison and Dunsmore (1975) discuss calibration in considerable detail, including a comparison of different methods.

Exercise 6.9.2 *Maximizing a Quadratic Response.*
Consider the model, $y_i = \beta_0 + \beta_1 x_i + \beta_2 x_i^2 + e_i$, e_is i.i.d. $N(0, \sigma^2)$, $i = 1, 2, 3, \ldots, n$. Let x_0 be the value at which the function $\mathrm{E}(y) = \beta_0 + \beta_1 x + \beta_2 x^2$ is maximized (or minimized).

(a) Find the maximum likelihood estimate of x_0.

(b) Find a $(1-\alpha)100\%$ confidence interval for x_0. Does such an interval always exist?

Hint: Use an $F(1, n-3)$ distribution based on $(\hat{\beta}_1 + 2\hat{\beta}_2 x_0)^2$.

Comment: The problem of finding values of the independent variables that maximize (or minimize) the expected y value is a basic problem in the field of response surface methods. See Box et al. (1978) or Christensen (2001, Chapter 8) or http://www.stat.unm.edu/~fletcher/TopicsInDesign) for an introduction to the subject. Box and Draper (1987) give a detailed treatment.

Exercise 6.9.3 *Two-Phase Linear Regression.*
Consider the problem of sterilizing canned pudding. As the pudding is sterilized by a heat treatment, it is simultaneously cooked. If you have ever cooked pudding, you know that it starts out soupy and eventually thickens. That, dear reader, is the point of this little tale. Sterilization depends on the transfer of heat to the pudding and the rate of transfer depends on whether the pudding is soupy or gelatinous. On an appropriate scale, the heating curve is linear in each phase. The question is, "Where does the line change phases?"

Suppose that we have collected data (y_i, x_i), $i = 1, \ldots, n+m$, and that we know that the line changes phases between x_n and x_{n+1}. The model $y_i = \beta_{10} + \beta_{11} x_i + e_i$, e_is i.i.d. $N(0, \sigma^2)$, $i = 1, \ldots, n$, applies to the first phase and the model $y_i = \beta_{20} + \beta_{21} x_i + e_i$, e_is i.i.d. $N(0, \sigma^2)$, $i = n+1, \ldots, n+m$, applies to the second phase. Let γ be the value of x at which the lines intersect.

(a) Find estimates of β_{10}, β_{11}, β_{20}, β_{21}, σ^2, and γ.

Hint: γ is a function of the other parameters.

(b) Find a $(1-\alpha)100\%$ confidence interval for γ. Does such an interval always exist?

Hint: Use an $F(1, n+m-4)$ distribution based on

$$\left[(\hat{\beta}_{10}+\hat{\beta}_{11}\gamma)-(\hat{\beta}_{20}+\hat{\beta}_{21}\gamma)\right]^2 .$$

Comment: Hinkley (1969) has treated the more realistic problem in which it is not known between which x_i values the intersection occurs.

Exercise 6.9.4 Consider the model $y_i = \beta_0 + \beta_1 x_i + e_i$, e_is i.i.d. $N(0, \sigma^2/w_i)$, $i = 1, 2, 3, \ldots, n$, where the w_is are known numbers. Derive algebraic formulas for $\hat{\beta}_0$, $\hat{\beta}_1$, $\mathrm{Var}(\hat{\beta}_0)$, and $\mathrm{Var}(\hat{\beta}_1)$.

Exercise 6.9.5 Consider the model $y_i = \beta_0 + \beta_1 x_i + e_i$, e_is i.i.d. $N(0, \sigma^2)$, $i = 1, 2, 3, \ldots, n$. If the x_is are restricted to be in the closed interval $[-10, 15]$, determine how to choose the x_is to minimize

(a) $\mathrm{Var}(\hat{\beta}_0)$.

(b) $\mathrm{Var}(\hat{\beta}_1)$.

(c) How would the choice of the x_is change if they were restricted to the closed interval $[-10, 10]$?

Exercise 6.9.6 Find $\mathrm{E}[y - \hat{E}(y|x)]^2$ in terms of the variances and covariances of x and y. Give a "natural" estimate of $\mathrm{E}\left[y - \hat{E}(y|x)\right]^2$.

Exercise 6.9.7 Test whether the data of Example 6.2.1 indicate that the multiple correlation coefficient is different from zero.

Exercise 6.9.8 Test whether the data of Example 6.2.1 indicate that the partial correlation coefficient $\rho_{y1\cdot 2}$ is different from zero.

Exercise 6.9.9 Show that

(a) $$\rho_{12\cdot 3} = \frac{\rho_{12} - \rho_{13}\rho_{23}}{\sqrt{1-\rho_{13}^2}\sqrt{1-\rho_{23}^2}}$$

(b) $$\rho_{12\cdot 34} = \frac{\rho_{12\cdot 4} - \rho_{13\cdot 4}\rho_{23\cdot 4}}{\sqrt{1-\rho_{13\cdot 4}^2}\sqrt{1-\rho_{23\cdot 4}^2}}.$$

Exercise 6.9.10 Show that in Section 2, $\gamma_* = \beta_*$ and $\beta_0 = \gamma_0 - (1/n)J_1^n Z\gamma_*$.

Hint: Examine the corresponding argument given in Section 1 for simple linear regression.

References

Aitchison, J., & Dunsmore, I. R. (1975). *Statistical prediction analysis*. Cambridge: Cambridge University Press.

Anderson, T. W. (2003). *An introduction to multivariate statistical analysis* (3rd ed.). New York: Wiley.

Atwood, C. L., & Ryan, T. A., Jr. (1977). A class of tests for lack of fit to a regression model. Unpublished manuscript.

Box, G. E. P., & Draper, N. R. (1987). *Empirical model-building and response surfaces*. New York: Wiley.

Box, G. E. P., Hunter, W. G., & Hunter, J. S. (1978). *Statistics for experimenters*. New York: Wiley.

Brockwell, P. J., & Davis, R. A. (1991). *Time series: Theory and methods* (2nd ed.). New York: Springer.

Christensen, R. (1989). Lack of fit tests based on near or exact replicates. *The Annals of Statistics, 17*, 673–683.

Christensen, R. (1991). Small sample characterizations of near replicate lack of fit tests. *Journal of the American Statistical Association, 86*, 752–756.

Christensen, R. (1995). Comment on Inman (1994). *The American Statistician, 49*, 400.

Christensen, R. (2001). *Advanced linear modeling: Multivariate, time series, and spatial data; nonparametric regression, and response surface maximization* (2nd ed.). New York: Springer.

Christensen, R. (2015). *Analysis of variance, design, and regression: Linear modeling for unbalanced data* (2nd ed.). Boca Raton: Chapman and Hall/CRC Pres.

Cook, R. D., & Weisberg, S. (1999). *Applied regression including computing and graphics*. New York: Wiley.

Cressie, N. (1993). *Statistics for spatial data* (Revised ed.). New York: Wiley.

Cressie, N. A. C., & Wikle, C. K. (2011). *Statistics for spatio-temporal data*. New York: Wiley.

Daniel, C., & Wood, F. S. (1980). *Fitting equations to data* (2nd ed.). New York: Wiley.

deLaubenfels, R. (2006). The victory of least squares and orthogonality in statistics. *The American Statistician, 60*, 315–321.

Doob, J. L. (1953). *Stochastic processes*. New York: Wiley.

Draper, N., & Smith, H. (1998). *Applied regression analysis* (3rd ed.). New York: Wiley.

Fisher, R. A. (1922). The goodness of fit of regression formulae, and the distribution of regression coefficients. *Journal of the Royal Statistical Society, 85*, 597–612.

Geisser, S. (1971). The inferential use of predictive distributions. In V.P. Godambe & D.A. Sprott (Eds.), *Foundations of statistical inference*. Toronto: Holt, Rinehart, and Winston.

Geisser, S. (1993). *Predictive inference: An introduction*. New York: Chapman and Hall.

Gnanadesikan, R. (1977). *Methods for statistical analysis of multivariate observations*. New York: Wiley.

Hartigan, J. (1969). Linear Bayesian methods. *Journal of the Royal Statistical Society, Series B, 31*, 446–454.

Hinkley, D. V. (1969). Inference about the intersection in two-phase regression. *Biometrika, 56*, 495–504.

Hodges, J. S. (2013). *Richly parameterized linear models: Additive, time series, and spatial models using random effects*. Boca Raton: Chapman and Hall/CRC.

Johnson, R. A., & Wichern, D. W. (2007). *Applied multivariate statistical analysis* (6th ed.). Englewood Cliffs: Prentice-Hall.

McCulloch, C. E., Searle, S. R., & Neuhaus, J. M. (2008). *Generalized, linear, and mixed models* (2nd ed.). New York: Wiley.

Miller, F. R., Neill, J. W., & Sherfey, B. W. (1998). Maximin clusters for near replicate regression lack of fit tests. *The Annals of Statistics, 26*, 1411–1433.

Miller, F. R., Neill, J. W., & Sherfey, B. W. (1999). Implementation of maximin power clustering criterion to select near replicates for regression lack-of-fit tests. *Journal of the American Statistical Association, 94*, 610–620.

Morrison, D. F. (2004). *Multivariate statistical methods* (4th ed.). Pacific Grove: Duxbury Press.
Neill, J. W., & Johnson, D. E. (1984). Testing for lack of fit in regression - a review. *Communications in Statistics, Part A - Theory and Methods, 13*, 485–511.
Pimentel, S. D., Small, D. S., & Rosenbaum, P. R. (2017). An exact test of fit for the Gaussian linear model using optimal nonbipartite matching. *Technometrics, 59*, 330–337.
Rao, C. R. (1973). *Linear statistical inference and its applications* (2nd ed.). New York: Wiley.
Ripley, B. D. (1981). *Spatial statistics*. New York: Wiley.
Shillington, E. R. (1979). Testing lack of fit in regression without replication. *Canadian Journal of Statistics, 7*, 137–146.
Shumway, R. H., & Stoffer, D. S. (2011). *Time series analysis and its applications: With R examples* (3rd ed.). New York: Springer.
Smith, A. F. M. (1986). Comment on an article by B. Efron. *The American Statistician, 40*, 10.
Snedecor, G. W., & Cochran, W. G. (1980). *Statistical methods* (7th ed.). Ames: Iowa State University Press.
Utts, J. (1982). The rainbow test for lack of fit in regression. *Communications in Statistics—Theory and Methods, 11*, 2801–2815.
Weisberg, S. (2014). *Applied linear regression* (4th ed.). New York: Wiley.

Chapter 7
Multifactor Analysis of Variance

Abstract This chapter presents the analysis of multifactor ANOVA models. The first three sections deal with the balanced two-way ANOVA model. Section 1 examines the no interaction model. Section 2 examines the model with interaction. Section 3 discusses the relationship between polynomial regression and the balanced two-way ANOVA model. Sections 4 and 5 discuss unbalanced two-way ANOVA models. Section 4 treats the special case of proportional numbers. Section 5 examines the general case. Finally, Section 6 extends the earlier results of the chapter to models with more than two factors. A review of the tensor concepts in Appendix B may aid the reader of this chapter.

7.1 Balanced Two-Way ANOVA Without Interaction

The balanced two-way ANOVA without interaction model is generally written

$$y_{ijk} = \mu + \alpha_i + \eta_j + e_{ijk}, \tag{1}$$

$i = 1, \ldots, a$, $j = 1, \ldots, b$, $k = 1, \ldots, N$.

R. Christensen, *Plane Answers to Complex Questions*, Springer Texts in Statistics,
https://doi.org/10.1007/978-3-030-32097-3_7

Example 7.1.1 Suppose $a = 3$, $b = 2$, $N = 4$. In matrix terms write

$$\begin{bmatrix} y_{111} \\ y_{112} \\ y_{113} \\ y_{114} \\ y_{121} \\ y_{122} \\ y_{123} \\ y_{124} \\ y_{211} \\ y_{212} \\ y_{213} \\ y_{214} \\ y_{221} \\ y_{222} \\ y_{223} \\ y_{224} \\ y_{311} \\ y_{312} \\ y_{313} \\ y_{314} \\ y_{321} \\ y_{322} \\ y_{323} \\ y_{324} \end{bmatrix} = \begin{bmatrix} 1 & 1 & 0 & 0 & 1 & 0 \\ 1 & 1 & 0 & 0 & 1 & 0 \\ 1 & 1 & 0 & 0 & 1 & 0 \\ 1 & 1 & 0 & 0 & 1 & 0 \\ 1 & 1 & 0 & 0 & 0 & 1 \\ 1 & 1 & 0 & 0 & 0 & 1 \\ 1 & 1 & 0 & 0 & 0 & 1 \\ 1 & 1 & 0 & 0 & 0 & 1 \\ 1 & 0 & 1 & 0 & 1 & 0 \\ 1 & 0 & 1 & 0 & 1 & 0 \\ 1 & 0 & 1 & 0 & 1 & 0 \\ 1 & 0 & 1 & 0 & 1 & 0 \\ 1 & 0 & 1 & 0 & 0 & 1 \\ 1 & 0 & 1 & 0 & 0 & 1 \\ 1 & 0 & 1 & 0 & 0 & 1 \\ 1 & 0 & 1 & 0 & 0 & 1 \\ 1 & 0 & 0 & 1 & 1 & 0 \\ 1 & 0 & 0 & 1 & 1 & 0 \\ 1 & 0 & 0 & 1 & 1 & 0 \\ 1 & 0 & 0 & 1 & 1 & 0 \\ 1 & 0 & 0 & 1 & 0 & 1 \\ 1 & 0 & 0 & 1 & 0 & 1 \\ 1 & 0 & 0 & 1 & 0 & 1 \\ 1 & 0 & 0 & 1 & 0 & 1 \end{bmatrix} \begin{bmatrix} \mu \\ \alpha_1 \\ \alpha_2 \\ \alpha_3 \\ \eta_1 \\ \eta_2 \end{bmatrix} + e.$$

In general, we can write the model as

$$Y = [X_0, X_1, \ldots, X_a, X_{a+1}, \ldots, X_{a+b}] \begin{bmatrix} \mu \\ \alpha_1 \\ \vdots \\ \alpha_a \\ \eta_1 \\ \vdots \\ \eta_b \end{bmatrix} + e.$$

Write $n = abN$ and the observation vector as $Y = [y_{ijk}]$, where the three subscripts i, j, and k denote a row of the vector. (Multiple subscripts denoting the rows and columns of matrices were introduced in Chapter 4.) The model matrix of the balanced two-way ANOVA model has

$$\begin{aligned} X_0 &= J, \\ X_r &= [t_{ijk}], \quad t_{ijk} = \delta_{ir}, \quad r = 1, \ldots, a, \\ X_{a+s} &= [t_{ijk}], \quad t_{ijk} = \delta_{js}, \quad s = 1, \ldots, b, \end{aligned}$$

where $\delta_{gh} = 1$ if $g = h$, and 0 otherwise. This is just a formal way of writing down model matrices that look like the one in Example 7.1.1. For example, an observation y_{rst} is subject to the effects of α_r and η_s. The rst row of the column X_r needs to be 1 so that α_r is added to y_{rst} and the rst row of X_{a+s} needs to be 1 so that η_s is added to y_{rst}. The rst rows of the columns X_j, $j = 1, \ldots, a$, $j \neq r$ and X_{a+j}, $j = 1, \ldots, b$, $j \neq s$ need to be 0 so that none of the α_js other than α_r, nor η_js other than η_s, are added to y_{rst}. The definition of the columns X_r and X_{a+s} given above ensures that this occurs.

The analysis of this model is based on doing two separate one-way ANOVAs. It is frequently said that this can be done because the treatments (groups) α and η are orthogonal. This is true in the sense that after fitting μ, the column space for the αs is orthogonal to the column space for the ηs. (See the discussion surrounding Proposition 3.6.3.) To investigate this further, consider a new matrix

$$Z = [Z_0, Z_1, \ldots, Z_a, Z_{a+1}, \ldots, Z_{a+b}],$$

where

$$Z_0 = X_0 = J$$

and

$$Z_r = X_r - \frac{X_r' J}{J'J} J,$$

for $r = 1, \ldots, a+b$. Here we have used Gram–Schmidt to eliminate the effect of J (the column associated with μ) from the rest of the columns of X. Since $J'J = abN$, $X_r' J = bN$ for $r = 1, \ldots, a$, and $X_{a+s}' J = aN$ for $s = 1, \ldots, b$, we have

$$Z_r = X_r - \frac{1}{a} J, \quad r = 1, \ldots, a,$$

$$Z_{a+s} = X_{a+s} - \frac{1}{b} J, \quad s = 1, \ldots, b.$$

Observe that

$$\begin{aligned} C(X) = & C(X_0, X_1, \ldots, X_a, X_{a+1}, \ldots, X_{a+b}) \\ & = C(Z_0, Z_1, \ldots, Z_a, Z_{a+1}, \ldots, Z_{a+b}) = C(Z), \end{aligned}$$

$$C(X_0, X_1, \ldots, X_a) = C(Z_0, Z_1, \ldots, Z_a),$$

$$C(X_0, X_{a+1}, \ldots, X_{a+b}) = C(Z_0, Z_{a+1}, \ldots, Z_{a+b}),$$

$$Z_0 \perp Z_r, \quad r = 1, \ldots, a+b,$$

and

$$C(Z_1, \ldots, Z_a) \perp C(Z_{a+1}, \ldots, Z_{a+b}).$$

To see the last of these, observe that for $r = 1, \ldots, a$ and $s = 1, \ldots, b$,

$$\begin{aligned}
Z'_{a+s}Z_r &= \sum_{ijk}(\delta_{js} - 1/b)(\delta_{ir} - 1/a) \\
&= \sum_{ijk}\delta_{js}\delta_{ir} - \sum_{ijk}\delta_{js}\frac{1}{a} - \sum_{ijk}\delta_{ir}\frac{1}{b} + \sum_{ijk}\frac{1}{ab} \\
&= \sum_{ij}N\delta_{js}\delta_{ir} - \sum_{j}\delta_{js}\frac{aN}{a} - \sum_{i}\delta_{ir}\frac{bN}{b} + \frac{abN}{ab} \\
&= N - aN/a - bN/b + N = 0.
\end{aligned}$$

We have decomposed $C(X)$ into three orthogonal parts, $C(Z_0)$, $C(Z_1, \ldots, Z_a)$, and $C(Z_{a+1}, \ldots, Z_{a+b})$. M, the perpendicular projection operator onto $C(X)$, can be written as the matrix sum of the perpendicular projection matrices onto these three spaces. By appealing to the one-way ANOVA, we can actually identify these projection matrices.

$C([X_0, X_1, \ldots, X_a])$ is the column space for the one-way ANOVA model

$$y_{ijk} = \mu + \alpha_i + e_{ijk}, \tag{2}$$

where the subscripts j and k are both used to indicate replications. Similarly, $C([X_0, X_{a+1}, \ldots, X_{a+b}])$ is the column space for the one-way ANOVA model

$$y_{ijk} = \mu + \eta_j + e_{ijk}, \tag{3}$$

where the subscripts i and k are both used to indicate replications. If one actually writes down the matrix $[X_0, X_{a+1}, \ldots, X_{a+b}]$, it looks a bit different from the usual form of a one-way ANOVA model matrix because the rows have been permuted out of the convenient order generally used.

Let M_α be the projection matrix used to test for no group effects in model (2). M_α is the perpendicular projection matrix onto $C(Z_1, \ldots, Z_a)$. Similarly, if M_η is the projection matrix for testing no group effects in model (3), then M_η is the perpendicular projection matrix onto $C(Z_{a+1}, \ldots, Z_{a+b})$. It follows that

$$M = \frac{1}{n}J_n^n + M_\alpha + M_\eta.$$

By comparing models, we see, for instance, that the test for $H_0 : \alpha_1 = \cdots = \alpha_a$ is based on

$$\frac{Y'M_\alpha Y/r(M_\alpha)}{Y'(I-M)Y/r(I-M)}.$$

It is easy to see that $r(M_\alpha) = a - 1$ and $r(I - M) = n - a - b + 1$. $Y'M_\alpha Y$ can be found as in Chapter 4 by appealing to the analysis of the one-way ANOVA model (2). In particular, since pairs jk identify replications,

$$M_\alpha Y = [t_{ijk}], \qquad \text{where } t_{ijk} = \bar{y}_{i\cdot\cdot} - \bar{y}_{\cdot\cdot\cdot}. \tag{4}$$

and

$$SS(\alpha) \equiv Y'M_\alpha Y = [M_\alpha Y]'[M_\alpha Y] = bN \sum_{i=1}^{a} (\bar{y}_{i\cdot\cdot} - \bar{y}_{\cdot\cdot\cdot})^2.$$

Expected mean squares can also be found by appealing to the one-way ANOVA

$$\mathrm{E}(Y'M_\alpha Y) = \sigma^2(a-1) + \beta'X'M_\alpha X\beta.$$

Substituting $\mu + \alpha_i + \eta_j$ for y_{ijk} in (4) gives

$$M_\alpha X\beta = [t_{ijk}], \qquad \text{where } t_{ijk} = \alpha_i - \bar{\alpha}_\cdot$$

and thus

$$\mathrm{E}\left[Y'M_\alpha Y/(a-1)\right] = \sigma^2 + \frac{bN}{a-1}\sum_{i=1}^{a}(\alpha_i - \bar{\alpha}_\cdot)^2.$$

Similar results hold for testing $H_0 : \eta_1 = \cdots = \eta_b$.

The SSE can be found using (4) and the facts that

$$M_\eta Y = [t_{ijk}], \qquad \text{where } t_{ijk} = \bar{y}_{\cdot j \cdot} - \bar{y}_{\cdot\cdot\cdot}$$

and

$$\frac{1}{n} J_n^n Y = [t_{ijk}], \qquad \text{where } t_{ijk} = \bar{y}_{\cdot\cdot\cdot}\,.$$

Because $(I-M)Y = Y - (1/n)J_n^n Y - M_\alpha Y - M_\eta Y$,

$$(I-M)Y = [t_{ijk}]$$

where

$$\begin{aligned} t_{ijk} &= y_{ijk} - \bar{y}_{\cdot\cdot\cdot} - (\bar{y}_{i\cdot\cdot} - \bar{y}_{\cdot\cdot\cdot}) - (\bar{y}_{\cdot j\cdot} - \bar{y}_{\cdot\cdot\cdot}) \\ &= y_{ijk} - \bar{y}_{i\cdot\cdot} - \bar{y}_{\cdot j\cdot} + \bar{y}_{\cdot\cdot\cdot}. \end{aligned}$$

Finally,

$$\begin{aligned} SSE &= Y'(I-M)Y = [(I-M)Y]'[(I-M)Y] \\ &= \sum_{i=1}^{a}\sum_{j=1}^{b}\sum_{k=1}^{N}(y_{ijk} - \bar{y}_{i\cdot\cdot} - \bar{y}_{\cdot j\cdot} + \bar{y}_{\cdot\cdot\cdot})^2. \end{aligned}$$

The analysis of variance table is given in Table 7.1.

Table 7.1 Balanced two-way analysis of variance table with no interaction

Matrix Notation

Source	df	SS
Grand Mean	1	$Y'\left(\frac{1}{n}J_n^n\right)Y$
Groups(α)	$a-1$	$Y'M_\alpha Y$
Groups(η)	$b-1$	$Y'M_\eta Y$
Error	$n-a-b+1$	$Y'(I-M)Y$
Total	$n=abN$	$Y'Y$

Source	SS	E(MS)
Grand Mean	$SSGM$	$\sigma^2+\beta'X'\left(\frac{1}{n}J_n^n\right)X\beta$
Groups(α)	$SS(\alpha)$	$\sigma^2+\beta'X'M_\alpha X\beta/(a-1)$
Groups(η)	$SS(\eta)$	$\sigma^2+\beta'X'M_\eta X\beta/(b-1)$
Error	SSE	σ^2
Total	$SSTot$	

Algebraic Notation

Source	df	SS
Grand Mean	$dfGM$	$n^{-1}y_{\cdot\cdot\cdot}^2=n\bar{y}_{\cdot\cdot\cdot}^2$
Groups(α)	$df(\alpha)$	$bN\sum_{i=1}^a\left(\bar{y}_{i\cdot\cdot}-\bar{y}_{\cdot\cdot\cdot}\right)^2$
Groups(η)	$df(\eta)$	$aN\sum_{j=1}^b\left(\bar{y}_{\cdot j\cdot}-\bar{y}_{\cdot\cdot\cdot}\right)^2$
Error	dfE	$\sum_{ijk}\left(y_{ijk}-\bar{y}_{i\cdot\cdot}-\bar{y}_{\cdot j\cdot}+\bar{y}_{\cdot\cdot\cdot}\right)^2$
Total	$dfTot$	$\sum_{ijk}y_{ijk}^2$

Source	MS	E(MS)
Grand Mean	$SSGM$	$\sigma^2+abN(\mu+\bar{\alpha}_\cdot+\bar{\eta}_\cdot)^2$
Groups(α)	$SS(\alpha)/(a-1)$	$\sigma^2+bN\sum_{i=1}^a(\alpha_i-\bar{\alpha}_\cdot)^2/(a-1)$
Groups(η)	$SS(\eta)/(b-1)$	$\sigma^2+aN\sum_{j=1}^b\left(\eta_j-\bar{\eta}_\cdot\right)^2/(b-1)$
Error	$SSE/(n-a-b+1)$	σ^2

7.1.1 Contrasts

We wish to show that estimation and testing of contrasts in a balanced two-way ANOVA is done exactly as in the one-way ANOVA: by ignoring the fact that a second type of group exists. This follows from showing that, say, a contrast in the α_is involves a constraint on $C(M_\alpha)$, and $C(M_\alpha)$ is defined by the one-way ANOVA without the η_js.

Theorem 7.1.2 *Let $\lambda'\beta$ be estimable and $\rho'X = \lambda'$. Then $\lambda'\beta$ is a contrast in the α_is if and only if $\rho'M = \rho'M_\alpha$. In this case, $\lambda'\hat{\beta} = \rho'MY = \rho'M_\alpha Y$, which is the estimate from the one-way ANOVA ignoring the η_js.*

Proof Let $\lambda'\beta = \sum_{i=1}^a c_i\alpha_i$ with $\sum_{i=1}^a c_i = 0$. Thus, $\lambda' = (0, c_1, \ldots, c_a, 0, \ldots, 0)$ and $\lambda' J_{a+b+1} = 0$. To have such a λ is to have ρ with

$$\rho' X_i = 0, \quad i = 0, a+1, a+2, \ldots, a+b,$$

which happens if and only if ρ is orthogonal to $C(Z_0, Z_{a+1}, \ldots, Z_{a+b}) = C(M - M_\alpha)$. In other words, $\rho'(M - M_\alpha) = 0$ and $\rho'M = \rho'M_\alpha$. □

One interpretation of this result is that having a contrast in the α_is equal to zero puts a constraint on $C(X)$ that requires $\mathrm{E}(Y) \in C(X)$ and $\mathrm{E}(Y) \perp M_\alpha\rho$. Clearly this constitutes a constraint on $C(M_\alpha)$, the space for the α groups. Another interpretation is that estimation or testing of a contrast in the α_is is done using M_α, which is exactly the way it is done in a one-way ANOVA ignoring the η_js.

Specifically, if we have a contrast $\sum_{i=1}^a c_i\alpha_i$, then the corresponding vector $M_\alpha\rho$ is

$$M_\alpha\rho = [t_{ijk}], \qquad \text{where } t_{ijk} = c_i/bN.$$

The estimated contrast is $\rho'M_\alpha Y = \sum_{i=1}^a c_i\bar{y}_{i\cdot\cdot}$ having a variance of $\sigma^2\rho'M_\alpha\rho = \sigma^2\sum_{i=1}^a c_i^2/bN$ and a sum of squares for testing $H_0 : \sum_{i=1}^a c_i\alpha_i = 0$ of

$$\left(\sum_{i=1}^a c_i\bar{y}_{i\cdot\cdot}\right)^2 \Big/ \left(\sum_{i=1}^a c_i^2/bN\right).$$

To get two orthogonal constraints on $C(M_\alpha)$, as in the one-way ANOVA, take ρ_1 and ρ_2 such that $\rho_1'X\beta$ and $\rho_2'X\beta$ are contrasts in the α_is and $\rho_1'M_\alpha\rho_2 = 0$. If $\rho_j'X\beta = \sum_{i=1}^a c_{ji}\alpha_i$, then, as shown for the one-way ANOVA, $\rho_1'M_\alpha\rho_2 = 0$ if and only if $\sum_{i=1}^a c_{1i}c_{2i} = 0$.

In the balanced two-way ANOVA without interaction, if N is greater than 1, we have a row structure to the model matrix with ab distinct rows. This allows estimation of pure error and lack of fit. The balanced two-way ANOVA with interaction retains the row structure and is equivalent to a large one-way ANOVA with ab groups.

Thus, the interaction model provides one parameterization of the model developed in Subsection 6.7.1 for testing lack of fit. The sum of squares for interaction is just the sum of squares for lack of fit of the no interaction model.

7.2 Balanced Two-Way ANOVA with Interaction

The balanced two-way ANOVA with interaction model is written

$$y_{ijk} = \mu + \alpha_i + \eta_j + \gamma_{ij} + e_{ijk},$$

$i = 1, \dots, a,\ j = 1, \dots, b,\ k = 1, \dots, N.$

Example 7.2.1 Suppose $a = 3$, $b = 2$, $N = 4$. In matrix terms we write

$$\begin{bmatrix} y_{111} \\ y_{112} \\ y_{113} \\ y_{114} \\ y_{121} \\ y_{122} \\ y_{123} \\ y_{124} \\ y_{211} \\ y_{212} \\ y_{213} \\ y_{214} \\ y_{221} \\ y_{222} \\ y_{223} \\ y_{224} \\ y_{311} \\ y_{312} \\ y_{313} \\ y_{314} \\ y_{321} \\ y_{322} \\ y_{323} \\ y_{324} \end{bmatrix} = \begin{bmatrix} 1&1&0&0&1&0&1&0&0&0&0&0 \\ 1&1&0&0&1&0&1&0&0&0&0&0 \\ 1&1&0&0&1&0&1&0&0&0&0&0 \\ 1&1&0&0&1&0&1&0&0&0&0&0 \\ 1&1&0&0&0&1&0&1&0&0&0&0 \\ 1&1&0&0&0&1&0&1&0&0&0&0 \\ 1&1&0&0&0&1&0&1&0&0&0&0 \\ 1&1&0&0&0&1&0&1&0&0&0&0 \\ 1&0&1&0&1&0&0&0&1&0&0&0 \\ 1&0&1&0&1&0&0&0&1&0&0&0 \\ 1&0&1&0&1&0&0&0&1&0&0&0 \\ 1&0&1&0&1&0&0&0&1&0&0&0 \\ 1&0&1&0&0&1&0&0&0&1&0&0 \\ 1&0&1&0&0&1&0&0&0&1&0&0 \\ 1&0&1&0&0&1&0&0&0&1&0&0 \\ 1&0&1&0&0&1&0&0&0&1&0&0 \\ 1&0&0&1&1&0&0&0&0&0&1&0 \\ 1&0&0&1&1&0&0&0&0&0&1&0 \\ 1&0&0&1&1&0&0&0&0&0&1&0 \\ 1&0&0&1&1&0&0&0&0&0&1&0 \\ 1&0&0&1&0&1&0&0&0&0&0&1 \\ 1&0&0&1&0&1&0&0&0&0&0&1 \\ 1&0&0&1&0&1&0&0&0&0&0&1 \\ 1&0&0&1&0&1&0&0&0&0&0&1 \end{bmatrix} \begin{bmatrix} \mu \\ \alpha_1 \\ \alpha_2 \\ \alpha_3 \\ \eta_1 \\ \eta_2 \\ \gamma_{11} \\ \gamma_{12} \\ \gamma_{21} \\ \gamma_{22} \\ \gamma_{31} \\ \gamma_{32} \end{bmatrix} + e.$$

In general, the model matrix can be written

$$X = [X_0, X_1, \dots, X_a, X_{a+1}, \dots, X_{a+b}, X_{a+b+1}, \dots, X_{a+b+ab}].$$

The columns $X_0, \ldots, X_{a+b}$ are exactly the same as for (7.1.1), the model without interaction. The key fact to notice is that

$$C(X) = C(X_{a+b+1}, \ldots, X_{a+b+ab});$$

this will be shown rigorously later. We can write an equivalent model using just $X_{a+b+1}, \ldots, X_{a+b+ab}$, say

$$y_{ijk} = \mu_{ij} + e_{ijk}.$$

This is just a one-way ANOVA model with *ab* groups and is sometimes called the *cell means model.*

We can decompose $C(X)$ into four orthogonal parts based on the identity

$$M = \frac{1}{n}J_n^n + M_\alpha + M_\eta + M_\gamma,$$

where

$$M_\gamma \equiv M - \frac{1}{n}J_n^n - M_\alpha - M_\eta.$$

M_α and M_η come from the no interaction model. Thus, as discussed earlier, M_α and M_η each come from a one-way ANOVA model. Since M also comes from a one-way ANOVA model, we can actually find M_γ. The interaction space is defined as $C(M_\gamma)$. The sum of squares for interaction is $Y'M_\gamma Y$. It is just the sum of squares left over after explaining as much as possible with μ, the α_is, and the η_js. The degrees of freedom for the interaction are

$$\begin{aligned} r(M_\gamma) &= r\left(M - \frac{1}{n}J_n^n - M_\alpha - M_\eta\right) \\ &= ab - 1 - (a-1) - (b-1) = (a-1)(b-1). \end{aligned}$$

The algebraic formula for the interaction sum of squares can be found as follows:

$$SS(\gamma) \equiv Y'M_\gamma Y = [M_\gamma Y]'[M_\gamma Y], \qquad (1)$$

where

$$M_\gamma Y = \left(M - \frac{1}{n}J_n^n - M_\alpha - M_\eta\right)Y = MY - \frac{1}{n}J_n^n Y - M_\alpha Y - M_\eta Y.$$

With M the projection operator for a one-way ANOVA, all of the terms on the right of the last equation have been characterized, so

$$\left(M - \frac{1}{n}J_n^n - M_\alpha - M_\eta\right)Y = [t_{ijk}],$$

where

$$\begin{aligned} t_{ijk} &= \bar{y}_{ij\cdot} - \bar{y}_{\cdots} - (\bar{y}_{i\cdot\cdot} - \bar{y}_{\cdots}) - (\bar{y}_{\cdot j\cdot} - \bar{y}_{\cdots}) \\ &= \bar{y}_{ij\cdot} - \bar{y}_{i\cdot\cdot} - \bar{y}_{\cdot j\cdot} + \bar{y}_{\cdots}. \end{aligned}$$

It follows immediately from (1) that

$$\begin{aligned} SS(\gamma) &= \sum_{i=1}^{a}\sum_{j=1}^{b}\sum_{k=1}^{N}[\bar{y}_{ij\cdot} - \bar{y}_{i\cdot\cdot} - \bar{y}_{\cdot j\cdot} + \bar{y}_{\cdots}]^2 \\ &= N\sum_{i=1}^{a}\sum_{j=1}^{b}[\bar{y}_{ij\cdot} - \bar{y}_{i\cdot\cdot} - \bar{y}_{\cdot j\cdot} + \bar{y}_{\cdots}]^2. \end{aligned}$$

The expected value of $Y'M_\gamma Y$ is $\sigma^2(a-1)(b-1) + \beta'X'M_\gamma X\beta$. The second term is a quadratic form in the γ_{ij}s because $\left(M - M_\alpha - M_\eta - [1/n]J_n^n\right)X_r = 0$ for $r = 0, 1, \ldots, a+b$. The algebraic form of $\beta'X'M_\gamma X\beta$ can be found by substituting $\mu + \alpha_i + \eta_j + \gamma_{ij}$ for y_{ijk} in $SS(\gamma)$. Simplification gives

$$\beta'X'M_\gamma X\beta = N\sum_{ij}[\gamma_{ij} - \bar{\gamma}_{i\cdot} - \bar{\gamma}_{\cdot j} - \bar{\gamma}_{\cdot\cdot}]^2.$$

The expected values of $Y'M_\alpha Y$, $Y'M_\eta Y$, and $Y'(1/n)J_n^n Y$ are now different from those found for the no interaction model. As above, algebraic forms for the expected values can be computed by substituting for the y_{ijk}s in the algebraic forms for the sums of squares. For instance,

$$\mathrm{E}(Y'M_\alpha Y) = \sigma^2(a-1) + \beta'X'M_\alpha X\beta = \sigma^2(a-1) + bN\sum_{i=1}^{a}(\alpha_i + \bar{\gamma}_{i\cdot} - \bar{\alpha}_\cdot - \bar{\gamma}_{\cdot\cdot})^2,$$

which depends on the γ_{ij}s and not just the α_is. This implies that the standard test is not a test that the α_is are all equal; it is a test that the $(\alpha_i + \bar{\gamma}_{i\cdot})$s are all equal. In fact, since the column space associated with the γ_{ij}s spans the entire space, i.e., $C(X_{a+b+1}, \ldots, X_{a+b+ab}) = C(X)$, all estimable functions of the parameters are functions of the γ_{ij}s. To see this, note that if $\lambda'\beta$ is not a function of the γ_{ij}s, but is estimable, then $\rho'X_i = 0, i = a+b+1, \ldots, a+b+ab$, and hence $\lambda' = \rho'X = 0$; so $\lambda'\beta$ is identically zero.

If we impose the "usual" side conditions, $\alpha_\cdot = \eta_\cdot = \gamma_{i\cdot} = \gamma_{\cdot j} = 0$, we obtain, for example,

$$\mathrm{E}(Y'M_\alpha Y) = \sigma^2(a-1) + bN\sum_{i=1}^{a}\alpha_i^2,$$

which looks nice but serves no purpose other than to hide the fact that these new α_i terms are averages over any interactions that exist.

As we did for a two-way ANOVA without interaction, we can put a single degree of freedom constraint on, say, $C(M_\alpha)$ by choosing a function $\lambda'\beta$ such that $\lambda' = \rho'X$ and $\rho'M = \rho'M_\alpha$. However, such a constraint no longer yields a contrast in the α_is. To examine $\rho'M_\alpha X\beta$, we examine the nature of $M_\alpha X\beta$. Since $M_\alpha Y$ is a vector whose rows are made up of terms like $(\bar{y}_{i\cdot\cdot} - \bar{y}_{\cdot\cdot\cdot})$, algebraic substitution gives $M_\alpha X\beta$ as a vector whose rows are terms like $(\alpha_i + \bar{\gamma}_{i\cdot} - \bar{\alpha}_\cdot - \bar{\gamma}_{\cdot\cdot})$. $\lambda'\beta = \rho'M_\alpha X\beta$ will be a contrast in these terms or, equivalently, in the $(\alpha_i + \bar{\gamma}_{i\cdot})$s. Such contrasts are generally hard to interpret. A contrast in the terms $(\alpha_i + \bar{\gamma}_{i\cdot})$ will be called a contrast in the α space. Similarly, a contrast in the terms $(\eta_j + \bar{\gamma}_{\cdot j})$ will be called a contrast in the η space.

There are two fundamental approaches to analyzing a two-way ANOVA with interaction. In both methods, the test for whether interaction adds to the two-way without interaction model is performed. If this is not significant, the interactions are tentatively assumed to be zero. If the effect of the interaction terms is significant, the easiest approach is to do the entire analysis as a one-way ANOVA. The alternative approach consists of trying to interpret the contrasts in $C(M_\alpha)$ and $C(M_\eta)$ and examining constraints in the interaction space.

7.2.1 Interaction Contrasts

We now consider how to define and test constraints on the interaction space. The hypothesis $H_0 : \lambda'\beta = 0$ puts a constraint on the interaction space if and only if $\lambda' = \rho'X$ has the property $\rho'M = \rho'\left(M - M_\alpha - M_\eta - [1/n]J_n^n\right)$. To find hypotheses that put constraints on the interaction space, it suffices to find $\rho \perp C\left(M_\alpha + M_\eta + [1/n]J_n^n\right)$ or, alternately, $\rho'X_i = 0$, $i = 0, \ldots, a+b$.

The goal of the following discussion is to characterize vectors in the interaction space, i.e., to characterize the vectors $M\rho$ that have the property $\rho'X_i = 0$ for $i = 0, \ldots, a+b$. A convenient way to do this is to characterize the vectors ρ that have two properties: (1) $M\rho = \rho$, and (2) $\rho'X_i = 0$ for $i = 0, \ldots, a+b$. The second property ensures that $M\rho$ is in the interaction space.

First we find a class of vectors that are contained in the interaction space. From this class of vectors we will get a class of orthogonal bases for the interaction space. The class of vectors and the class of orthogonal bases are found by combining a contrast in the α space with a contrast in the η space. This method leads naturally to the standard technique of examining interactions. Finally, a second class of vectors contained in the interaction space will be found. This class contains the first class as a special case. The second class is closed under linear combinations, so the second class is a vector space that contains a basis for the interaction space but which is also contained in the interaction space. It follows that the second class is precisely the interaction space.

At this point a problem arises. It is very convenient to write down ANOVA models as was done in Example 7.2.1, with indices to the right in y_{ijk} changing fastest. It is easy to see what the model looks like and that it can be written in a similar manner for any choices of a, b, and N. In the example it would be easy to find a vector that is in the interaction space, and it would be easy to see that the technique could be extended to cover any two-way ANOVA problem. Although it is easy to see how to write down models as in the example, it is awkward to develop a notation for it. It is also less than satisfying to have a proof that depends on the way in which the model is written down. Consequently, the material on finding a vector in the interaction space will be presented in three stages: an application to Example 7.2.1, a comment on how that argument can be generalized, and finally a rigorous presentation.

Example 7.2.2 In the model of Example 7.2.1, let $d' = (d_1, d_2, d_3) = (1, 2, -3)$ and $c' = (c_1, c_2) = (1, -1)$. The d_is determine a contrast in the α space and the c_js determine a contrast in the η space. Consider

$$\begin{aligned}[d' \otimes c'] &= [c_1d_1, c_2d_1, c_1d_2, c_2d_2, c_1d_3, c_2d_3] \\ &= [1, -1, 2, -2, -3, 3];\end{aligned}$$

and since $N = 4$, let

$$\begin{aligned}&\rho' \\ &= \frac{1}{4}[d' \otimes c'] \otimes J_1^4 = \frac{1}{4}[c_1d_1J_1^4, c_2d_1J_1^4, \ldots, c_2d_3J_1^4] \\ &= \frac{1}{4}[1, 1, 1, 1, -1, -1, -1, -1, 2, 2, 2, 2, -2, -2, -2, -2, -3, -3, -3, -3, 3, 3, 3, 3].\end{aligned}$$

It is easily seen that ρ is orthogonal to the first six columns of X; thus $M\rho$ is in the interaction space. However, it is also easily seen that $\rho \in C(X)$, so $M\rho = \rho$. The vector ρ is itself a vector in the interaction space.

Extending the argument to an arbitrary two-way ANOVA written in standard form, let $d' = (d_1, \ldots, d_a)$ with $\sum_{i=1}^{a} d_i = 0$ and $c' = (c_1, \ldots, c_b)$ with $\sum_{j=1}^{b} c_j = 0$. The d_is can be thought of as determining a contrast in the α space and the c_js as determining a contrast in the η space. Let $\rho' = (1/N)[d' \otimes c'] \otimes J_1^N$, i.e.,

$$\rho' = \frac{1}{N}(c_1d_1J_1^N, c_2d_1J_1^N, \ldots, c_bd_1J_1^N, c_1d_2J_1^N, \ldots, c_bd_aJ_1^N).$$

It is clear that $\rho'X_i = 0$ for $i = 0, \ldots, a+b$ and $\rho \in C(X_{a+b+1}, \ldots, X_{a+b+ab})$ so $\rho'M = \rho'$ and $\rho'X\beta = 0$ puts a constraint on the correct space.

To make the argument completely rigorous, with d and c defined as above, take

$$\rho = [\rho_{ijk}], \qquad \text{where } \rho_{ijk} = \frac{1}{N}(d_ic_j).$$

Using the characterizations of $X_0, \ldots, X_{a+b}$ from Section 1 we get

$$\rho' X_0 = \sum_i \sum_j \sum_k \frac{1}{N}(d_i c_j) = \sum_i d_i \sum_j c_j = 0;$$

for $r = 1, \ldots, a$, we get

$$\rho' X_r = \sum_i \sum_j \sum_k \frac{1}{N}(d_i c_j)\delta_{ir} = d_r \sum_j c_j = 0;$$

and for $s = 1, \ldots, b$, we get

$$\rho' X_{a+s} = \sum_i \sum_j \sum_k \frac{1}{N}(d_i c_j)\delta_{js} = c_s \sum_i d_i = 0.$$

This shows that $M\rho$ is in the interaction space.

To show that $M\rho = \rho$, we show that $\rho \in C(X)$. We need to characterize $C(X)$. The columns of X that correspond to the γ_{ij}s are the vectors $X_{a+b+1}, \ldots, X_{a+b+ab}$. Reindex these as $X_{(1,1)}, X_{(1,2)}, \ldots, X_{(1,b)}, \ldots, X_{(2,1)}, \ldots, X_{(a,b)}$. Thus $X_{(i,j)}$ is the column of X corresponding to γ_{ij}. We can then write

$$X_{(r,s)} = [t_{ijk}], \qquad \text{where } t_{ijk} = \delta_{(i,j)(r,s)}$$

and $\delta_{(i,j)(r,s)} = 1$ if $(i, j) = (r, s)$, and 0 otherwise. It is easily seen that

$$X_0 = \sum_{i=1}^{a} \sum_{j=1}^{b} X_{(i,j)},$$

$$X_r = \sum_{j=1}^{b} X_{(r,j)}, \quad r = 1, \ldots, a,$$

$$X_{a+s} = \sum_{i=1}^{a} X_{(i,s)}, \quad s = 1, \ldots, b.$$

This shows that $C(X) = C(X_{(1,1)}, X_{(1,2)}, \ldots, X_{(a,b)})$. (In the past, we have merely claimed this result.) It is also easily seen that

$$\rho = \sum_{i=1}^{a} \sum_{j=1}^{b} \frac{d_i c_j}{N} X_{(i,j)},$$

so that $\rho \in C(X)$.

We have found a class of vectors contained in the interaction space. If we find $(a-1)(b-1)$ such vectors, say ρ_r, $r = 1, \dots, (a-1)(b-1)$, where

$$\rho_r' \left(M - M_\alpha - M_\eta - [1/n]J_n^n\right) \rho_s = 0$$

for any $r \neq s$, then we will have an orthogonal basis for $C(M - M_\alpha - M_\eta - [1/n]J_n^n)$. Consider now another pair of contrasts for α and η, say $d^* = (d_1^*, \dots, d_a^*)'$ and $c^* = (c_1^*, \dots, c_b^*)'$, where one, say $c^* = (c_1^*, \dots, c_b^*)'$, is orthogonal to the corresponding contrast in the other pair. We can write

$$\rho_* = [\rho_{ijk}^*], \qquad \text{where } \rho_{ijk}^* = d_i^* c_j^* / N,$$

and we know that $\sum_{i=1}^a d_i^* = \sum_{j=1}^b c_j^* = \sum_{j=1}^b c_j c_j^* = 0$. With our choices of ρ and ρ_*,

$$\begin{aligned}
\rho' \left(M - M_\alpha - M_\eta - [1/n]J_n^n\right) \rho_* &= \rho' \rho_* \\
&= N^{-2} \sum_{i=1}^a \sum_{j=1}^b \sum_{k=1}^N d_i d_i^* c_j c_j^* \\
&= N^{-1} \sum_{i=1}^a d_i d_i^* \sum_{j=1}^b c_j c_j^* \\
&= 0.
\end{aligned}$$

Since there are $(a-1)$ orthogonal ways of choosing $d' = (d_1, \dots, d_a)$ and $(b-1)$ orthogonal ways of choosing $c' = (c_1, \dots, c_b)$, there are $(a-1)(b-1)$ orthogonal vectors ρ that can be chosen in this fashion. This provides the desired orthogonal breakdown of the interaction space.

To actually compute the estimates of these parametric functions, recall that M is the perpendicular projection operator for a one-way ANOVA. With ρ chosen as above,

$$\rho' X\beta = \sum_{i=1}^a \sum_{j=1}^b \sum_{k=1}^N \frac{d_i c_j}{N} \gamma_{ij} = \sum_{i=1}^a \sum_{j=1}^b d_i c_j \gamma_{ij}.$$

Its estimate reduces to

$$\rho' MY = \sum_{i=1}^a \sum_{j=1}^b d_i c_j \bar{y}_{ij\cdot}$$

and its variance to

$$\sigma^2 \rho' M \rho = \sigma^2 \rho' \rho = \sigma^2 \sum_{i=1}^a \sum_{j=1}^b d_i^2 c_j^2 / N.$$

A handy method for computing these is to write out the following two-way table:

	c_1	c_2	$\cdots$	c_b	h_i
d_1	$\bar{y}_{11\cdot}$	$\bar{y}_{12\cdot}$	$\cdots$	$\bar{y}_{1b\cdot}$	h_1
d_2	$\bar{y}_{21\cdot}$	$\bar{y}_{22\cdot}$	$\cdots$	$\bar{y}_{2b\cdot}$	h_2
$\vdots$	$\vdots$	$\vdots$	$\ddots$	$\vdots$	$\vdots$
d_a	$\bar{y}_{a1\cdot}$	$\bar{y}_{a2\cdot}$	$\cdots$	$\bar{y}_{ab\cdot}$	h_a
g_j	g_1	g_2	$\cdots$	g_b	

where $h_i = \sum_{j=1}^{b} c_j \bar{y}_{ij\cdot}$ and $g_j = \sum_{i=1}^{a} d_i \bar{y}_{ij\cdot}$. We can then write

$$\rho' MY = \sum_{j=1}^{b} c_j g_j = \sum_{i=1}^{a} d_i h_i.$$

Unfortunately, not all contrasts in the interaction space can be defined as illustrated here. The $(a-1)(b-1)$ orthogonal vectors that we have discussed finding form a basis for the interaction space, so any linear combination of these vectors is also in the interaction space. However, not all of these linear combinations can be written with the method based on two contrasts.

Let $Q = [q_{ij}]$ be any $a \times b$ matrix such that $J_a' Q = 0$ and $Q J_b = 0$ (i.e., $q_{i\cdot} = 0 = q_{\cdot j}$). If the model is written down in the usual manner, the vectors in the interaction space are the vectors of the form $\rho = (1/N)\text{Vec}(Q') \otimes J_N$. In general, we write the vector ρ with triple subscript notation as

$$\rho = [\rho_{ijk}], \qquad \text{where } \rho_{ijk} = q_{ij}/N \quad \text{with } q_{i\cdot} = q_{\cdot j} = 0. \tag{2}$$

First, note that linear combinations of vectors with this structure retain the structure; thus vectors of this structure form a vector space. Vectors ρ with this structure are in $C(X)$, and it is easily seen that $\rho' X_i = 0$ for $i = 0, 1, \ldots, a+b$. Thus, a vector with this structure is contained in the interaction space. Note also that the first method of finding vectors in the interaction space using a pair of contrasts yields a vector of the structure that we are currently considering, so the vector space alluded to above is both contained in the interaction space and contains a basis for the interaction space. It follows that the interaction space is precisely the set of all vectors with the form (2).

Exercise 7.1 Prove the claims of the previous paragraph. In particular, show that linear combinations of the vectors presented retain their structure, that the vectors are orthogonal to the columns of X corresponding to the grand mean and the group effects, and that the vectors based on contrasts have the same structure as the vector given above.

For estimation and tests of single-degree-of-freedom hypotheses in the interaction space, it is easily seen, with ρ taken as above, that

Table 7.2 Balanced two-way analysis of variance table with interaction

Matrix Notation

Source	df	SS
Grand Mean	1	$Y'\left(\frac{1}{n}J_n^n\right)Y$
Groups(α)	$a-1$	$Y'M_\alpha Y$
Groups(η)	$b-1$	$Y'M_\eta Y$
Interaction(γ)	$(a-1)(b-1)$	$Y'\left(M - M_\alpha - M_\eta - \frac{1}{n}J_n^n\right)Y$
Error	$n-ab$	$Y'(I-M)Y$
Total	$n=abN$	$Y'Y$

Source	SS	E(MS)
Grand Mean	$SSGM$	$\sigma^2 + \beta'X'\left(\frac{1}{n}J_n^n\right)X\beta$
Groups(α)	$SS(\alpha)$	$\sigma^2 + \beta'X'M_\alpha X\beta/(a-1)$
Groups(η)	$SS(\eta)$	$\sigma^2 + \beta'X'M_\eta X\beta/(b-1)$
Interaction(γ)	$SS(\gamma)$	$\sigma^2 + \beta'X'M_\gamma X\beta/(a-1)(b-1)$
Error	SSE	σ^2
Total	$SSTot$	

Algebraic Notation

Source	df	SS
Grand Mean	$dfGM$	$n^{-1}y_{\cdot\cdot\cdot}^2 = n\bar{y}_{\cdot\cdot\cdot}^2$
Groups(α)	$df(\alpha)$	$bN\sum_{i=1}^a (\bar{y}_{i\cdot\cdot} - \bar{y}_{\cdot\cdot\cdot})^2$
Groups(η)	$df(\eta)$	$aN\sum_{j=1}^b \left(\bar{y}_{\cdot j\cdot} - \bar{y}_{\cdot\cdot\cdot}\right)^2$
Interaction(γ)	$df(\gamma)$	$N\sum_{ij}\left(\bar{y}_{ij\cdot} - \bar{y}_{i\cdot\cdot} - \bar{y}_{\cdot j\cdot} + \bar{y}_{\cdot\cdot\cdot}\right)^2$
Error	dfE	$\sum_{ijk}\left(y_{ijk} - \bar{y}_{ij\cdot}\right)^2$
Total	$dfTot$	$\sum_{ijk} y_{ijk}^2$

Source	MS	E(MS)
Grand Mean	$SSGM$	$\sigma^2 + abN(\mu + \bar{\alpha}_\cdot + \bar{\eta}_\cdot + \bar{\gamma}_{\cdot\cdot})^2$
Groups(α)	$SS(\alpha)/(a-1)$	$\sigma^2 + \frac{bN}{a-1}\sum_i (\alpha_i + \bar{\gamma}_{i\cdot} - \bar{\alpha}_\cdot - \bar{\gamma}_{\cdot\cdot})^2$
Groups(η)	$SS(\eta)/(b-1)$	$\sigma^2 + \frac{aN}{b-1}\sum_j \left(\eta_j + \bar{\gamma}_{\cdot j} - \bar{\eta}_\cdot - \bar{\gamma}_{\cdot\cdot}\right)^2$
Interaction(γ)	$SS(\gamma)/df(\gamma)$	$\sigma^2 + \frac{N}{df(\gamma)}\sum_{ij}\left(\gamma_{ij} - \bar{\gamma}_{i\cdot} - \bar{\gamma}_{\cdot j} + \bar{\gamma}_{\cdot\cdot}\right)^2$
Error	$SSE/(n-ab)$	σ^2

$$\rho' X\beta = \sum_{i=1}^{a}\sum_{j=1}^{b} q_{ij}\gamma_{ij},$$

$$\rho' MY = \sum_{i=1}^{a}\sum_{j=1}^{b} q_{ij}\bar{y}_{ij\cdot},$$

$$\mathrm{Var}(\rho' MY) = \sigma^2 \sum_{i=1}^{a}\sum_{j=1}^{b} q_{ij}^2/N.$$

Table 7.2 gives the ANOVA table for the balanced two-way ANOVA with interaction. Note that if $N = 1$, there will be no pure error term available. In that case, it is often *assumed* that the interactions add nothing to the model, so that the mean square for interactions can be used as an estimate of error for the two-way ANOVA without interaction. See Example 12.2.4 for a graphical procedure that addresses this problem.

Exercise 7.2 Does the statement "the interactions add nothing to the model" mean that $\gamma_{11} = \gamma_{12} = \cdots = \gamma_{ab}$? If it does, justify the statement. If it does not, what does the statement mean?

Two final comments on exploratory work with interactions. If the $(a-1)(b-1)$ degree-of-freedom F test for interactions is not significant, then neither Scheffé's method nor the LSD method will allow us to claim significance for any contrast in the interactions. Bonferroni's method may give significance, but it is unlikely. Nevertheless, if our goal is to explore the data, there may be *suggestions* of possible interactions. For example, if you work with interactions long enough, you begin to think that some interaction contrasts have reasonable interpretations. If such a contrast exists that accounts for the bulk of the interaction sum of squares, and if the corresponding F test approaches significance, then it would be unwise to ignore this possible source of interaction. (As a word of warning though, recall that there always exists an interaction contrast, usually uninterpretable, that accounts for the entire sum of squares for interaction.) Second, in exploratory work it is very useful to plot the cell means. For example, one can plot the points $(i, \bar{y}_{ij\cdot})$ for each value of j, connecting the points to give b different curves, one for each j. If the α groups correspond to levels of some quantitative factor x_i, plot the points $(x_i, \bar{y}_{ij\cdot})$ for each value of j. If there are no interactions, the plots for different values of j should be (approximately) parallel. (If no interactions are present, the plots estimate plots of the points $(i, \mu + \alpha_i + \eta_j)$. These plots are parallel for all j.) Deviations from a parallel set of plots can suggest possible sources of interaction. The data are suggesting possible interaction contrasts, so if valid tests are desired, use Scheffé's method. Finally, it is equally appropriate to plot the points $(j, \bar{y}_{ij\cdot})$ for all i or a corresponding set of points using quantitative levels associated with the η groups.

7.3 Polynomial Regression and the Balanced Two-Way ANOVA

Consider first the balanced two-way ANOVA without interaction. Suppose that the ith level of the α groups corresponds to some number w_i and that the jth level of the η groups corresponds to some number z_j. We can write vectors taking powers of w_i and z_j. For $r = 1, \ldots, a-1$ and $s = 1, \ldots, b-1$, write

$$W^r = [t_{ijk}], \qquad \text{where } t_{ijk} = w_i^r,$$

$$Z^s = [t_{ijk}], \qquad \text{where } t_{ijk} = z_j^s.$$

Note that $W^0 = Z^0 = J$.

Example 7.3.1 Consider the model $y_{ijk} = \mu + \alpha_i + \eta_j + e_{ijk}$, $i = 1, 2, 3$, $j = 1, 2$, $k = 1, 2$. Suppose that the α groups are 1, 2, and 3 pounds of fertilizer and that the η groups are 5 and 7 pounds of manure. Then, if we write $Y = (y_{111}, y_{112}, y_{121}, y_{122}, y_{211}, \ldots, y_{322})'$, we have

$$W^1 = (1, 1, 1, 1, 2, 2, 2, 2, 3, 3, 3, 3)',$$

$$W^2 = (1, 1, 1, 1, 4, 4, 4, 4, 9, 9, 9, 9)',$$

$$Z^1 = (5, 5, 7, 7, 5, 5, 7, 7, 5, 5, 7, 7)'.$$

From the discussion of Section 6.8 on polynomial regression and one-way ANOVA, we have

$$C(J, W^1, \ldots, W^{a-1}) = C(X_0, X_1, \ldots, X_a),$$
$$C(J, Z^1, \ldots, Z^{b-1}) = C(X_0, X_{a+1}, \ldots, X_{a+b}),$$

and thus

$$C(J, W^1, \ldots, W^{a-1}, Z^1, \ldots, Z^{b-1}) = C(X_0, X_1, \ldots, X_{a+b}).$$

Fitting the two-way ANOVA is the same as fitting a joint polynomial in w_i and z_j. Writing the models out algebraically, the model

$$y_{ijk} = \mu + \alpha_i + \eta_j + e_{ijk},$$

$i = 1, \ldots, a$, $j = 1, \ldots, b$, $k = 1, \ldots, N$, is equivalent to

$$y_{ijk} = \beta_{0,0} + \beta_{1,0} w_i + \cdots + \beta_{a-1,0} w_i^{a-1} + \beta_{0,1} z_j + \cdots + \beta_{0,b-1} z_j^{b-1} + e_{ijk},$$

$i = 1, \ldots, a$, $j = 1, \ldots, b$, $k = 1, \ldots, N$. The correspondence between contrasts and orthogonal polynomials remains valid. In particular, if the w_is or z_js are equally spaced, the contrasts in Table 6.2 can be used.

We now show that the interaction model

$$y_{ijk} = \mu + \alpha_i + \eta_j + \gamma_{ij} + e_{ijk}$$

is equivalent to the model

$$y_{ijk} = \sum_{i=0}^{a-1}\sum_{j=0}^{b-1} \beta_{rs} w_i^r z_j^s + e_{ijk}.$$

Example 7.3.1 Continued. The model $y_{ijk} = \mu + \alpha_i + \eta_j + \gamma_{ij} + e_{ijk}$ is equivalent to $y_{ijk} = \beta_{00} + \beta_{10}w_i + \beta_{20}w_i^2 + \beta_{01}z_j + \beta_{11}w_iz_j + \beta_{21}w_i^2z_j + e_{ijk}$, where $w_1 = 1$, $w_2 = 2$, $w_3 = 3$, $z_1 = 5$, $z_2 = 7$.

The model matrix for this polynomial model can be written

$$S = [J_n^1, W^1, W^2, \ldots, W^{a-1}, Z^1, \ldots, Z^{b-1}, \\ W^1Z^1, \ldots, W^1Z^{b-1}, W^2Z^1, \ldots, W^{a-1}Z^{b-1}],$$

where for any two vectors in $\mathbf{R}^n$, say $U = (u_1, \ldots, u_n)'$ and $V = (v_1, \ldots, v_n)'$, we define VU to be the vector $(v_1u_1, v_2u_2, \ldots, v_nu_n)'$.

To establish the equivalence of the models, it is enough to notice that the row structure of $X = [X_0, X_1, \ldots, X_{a+b+ab}]$ is the same as the row structure of S and that $r(X) = ab = r(S)$. $C(X)$ is the column space of the one-way ANOVA model that has a separate effect for each distinct set of rows in S. From our discussion of pure error and lack of fit in Section 6.7, $C(S) \subset C(X)$. Since $r(S) = r(X)$, we have $C(S) = C(X)$; thus the models are equivalent.

We would like to characterize the test in the interaction space that is determined by, say, the quadratic contrast in the α space and the cubic contrast in the η space. We want to show that it is the test for W^2Z^3, i.e., that it is the test for $H_0 : \beta_{23} = 0$ in the model $y_{ijk} = \beta_{00} + \beta_{10}w_i + \beta_{20}w_i^2 + \beta_{01}z_j + \beta_{02}z_j^2 + \beta_{03}z_j^3 + \beta_{11}w_iz_j + \beta_{12}w_iz_j^2 + \beta_{13}w_iz_j^3 + \beta_{21}w_i^2z_j + \beta_{22}w_i^2z_j^2 + \beta_{23}w_i^2z_j^3 + e_{ijk}$. In general, we want to be able to identify the columns W^rZ^s, $r \geq 1$, $s \geq 1$, with vectors in the interaction space. Specifically, we would like to show that the test of W^rZ^s adding to the model based on $C([W^iZ^j : i = 0, \ldots, r, j = 0, \ldots, s])$ is precisely the test of the vector in the interaction space defined by the rth degree polynomial contrast in the α space and the sth degree polynomial contrast in the η space. Note that the test of the rth degree polynomial contrast in the α space is a test of whether the column W^r adds to the model based on $C(J, W^1, \ldots, W^r)$, and that the test of the sth degree polynomial contrast in the η space is a test of whether the column Z^s adds to the model based on $C(J, Z^1, \ldots, Z^s)$.

It is important to remember that the test for $W^r Z^s$ adding to the model is not a test for $W^r Z^s$ adding to the full model. It is a test for $W^r Z^s$ adding to the model spanned by the vectors $W^i Z^j$, $i = 0, \ldots, r$, $j = 0, \ldots, s$, where $W^0 = Z^0 = J_n^1$. As discussed above, the test of the vector in the interaction space corresponding to the quadratic contrast in the α space and the cubic contrast in the η space should be the test of whether $W^2 Z^3$ adds to a model with $C(J_n^1, W^1, W^2, Z^1, Z^2, Z^3, W^1Z^1, W^1Z^2, W^1Z^3, W^2Z^1, W^2Z^2, W^2Z^3)$. Intuitively, this is reasonable because the quadratic and cubic contrasts are being fitted after all terms of smaller order.

As observed earlier, if $\lambda' = \rho' X$, then the constraint imposed by $\lambda'\beta = 0$ is $M\rho$ and the test of $\lambda'\beta = 0$ is the test for dropping $M\rho$ from the model. Using the Gram–Schmidt algorithm, find $R_{0,0}$, $R_{1,0}$, …, $R_{a-1,0}$, $R_{0,1}$, …, $R_{0,b-1}$, $R_{1,1}$, …, $R_{a-1,b-1}$ by orthonormalizing, in order, the columns of S. Recall that the polynomial contrasts in the α_is correspond to the vectors $R_{i,0}$, $i = 1, \ldots, a-1$, and that $C(R_{1,0}, \ldots, R_{a-1,0}) = C(M_\alpha)$. Similarly, the polynomial contrasts in the η_js correspond to the vectors $R_{0,j}$ $j = 1, \ldots, b-1$ and $C(R_{0,1}, \ldots, R_{0,a-1}) = C(M_\eta)$. If the test of $\lambda'\beta = 0$ is to test whether, say, Z^s adds to the model after fitting Z^j, $j = 1, \ldots, s-1$, we must have $C(M\rho) = C(R_{0,s})$. Similar results hold for the vectors W^r.

First, we need to examine the relationship between vectors in $C(M_\alpha)$ and $C(M_\eta)$ with vectors in the interaction space. A contrast in the α space is defined by, say, $(d_1, \ldots, d_a)$, and a contrast in the η space by, say, $(c_1, \ldots, c_b)$. From Chapter 4, if we define

$$\rho_1 = [t_{ijk}], \qquad \text{where } t_{ijk} = d_i/Nb$$

and

$$\rho_2 = [t_{ijk}], \qquad \text{where } t_{ijk} = c_j/Na,$$

then $\rho_1 \in C(M_\alpha)$, $\rho_2 \in C(M_\eta)$, $\rho_1 X\beta = \sum_{i=1}^a d_i(\alpha_i + \bar{\gamma}_{i\cdot})$, and $\rho_2 X\beta = \sum_{j=1}^b c_j (\eta_j + \bar{\gamma}_{\cdot j})$. The vector $\rho_1\rho_2$ is

$$\rho_1\rho_2 = [t_{ijk}] \qquad \text{where } t_{ijk} = N^{-2}(ab)^{-1} d_i c_j.$$

This is proportional to a vector in the interaction space corresponding to $(d_1, \ldots, d_a)$ and $(c_1, \ldots, c_b)$.

From this argument, it follows that since $R_{r,0}$ is the vector (constraint) in $C(M_\alpha)$ for testing the rth degree polynomial contrast and $R_{0,s}$ is the vector in $C(M_\eta)$ for testing the sth degree polynomial contrast, then $R_{r,0}R_{0,s}$ is a vector in the interaction space. Since the polynomial contrasts are defined to be orthogonal, and since $R_{r,0}$ and $R_{0,s}$ are defined by polynomial contrasts, our discussion in the previous section about orthogonal vectors in the interaction space implies that the set $\{R_{r,0}R_{0,s}|r = 1, \ldots, a-1, s = 1, \ldots, b-1\}$ is an orthogonal basis for the interaction space. Moreover, with $R_{r,0}$ and $R_{0,s}$ orthonormal,

$$[R_{r,0}R_{0,s}]'[R_{r,0}R_{0,s}] = 1/abN.$$

We now check to see that $(abN)Y'[R_{r,0}R_{0,s}][R_{r,0}R_{0,s}]'Y/MSE$ provides a test of the correct thing, i.e., that W^rZ^s adds to a model containing all lower order terms. Since, by Gram–Schmidt, for some a_is and b_js we have $R_{r,0} = a_0W^r + a_1W^{r-1} + \cdots + a_{r-1}W^1 + a_rJ_n^1$ and $R_{0,s} = b_0Z^s + b_1Z^{s-1} + \cdots + b_{s-1}Z^1 + b_sJ_n^1$, we also have

$$R_{r,0}R_{0,s} = a_0b_0W^rZ^s + \sum_{j=1}^{s} b_jZ^{s-j}W^r + \sum_{i=1}^{r} a_iW^{r-i}Z^s + \sum_{j=1}^{s}\sum_{i=1}^{r} a_ib_jZ^{s-j}W^{r-i}.$$

Letting $R_{1,0} = R_{0,1} = J$, it follows immediately that

$$C(R_{i,0}R_{0,j} : i = 0, \ldots, r, j = 0, \ldots, s) \subset C(W^iZ^j : i = 0, \ldots, r, j = 0, \ldots, s).$$

However, the vectors listed in each of the sets are linearly independent and the number of vectors in each set is the same, so the ranks of the column spaces are the same and

$$C(R_{i,0}R_{0,j} : i = 0, \ldots, r, j = 0, \ldots, s) = C(W^iZ^j : i = 0, \ldots, r, j = 0, \ldots, s).$$

The vectors $R_{i,0}R_{0,j}$ are orthogonal, so $abN[R_{r,0}R_{0,s}][R_{r,0}R_{0,s}]'$ is the projection operator for testing if W^rZ^s adds to the model after fitting all terms of lower order. Since $R_{r,0}R_{0,s}$ was found as the vector in the interaction space corresponding to the rth orthogonal polynomial contrast in the α_is and the sth orthogonal polynomial contrast in the η_js, the technique for testing if W^rZ^s adds to the appropriate model is a straightforward test of an interaction contrast.

7.4 Two-Way ANOVA with Proportional Numbers

Consider the model

$$y_{ijk} = \mu + \alpha_i + \eta_j + e_{ijk},$$

$i = 1, \ldots, a$, $j = 1, \ldots, b$, $k = 1, \ldots, N_{ij}$. We say that the model has *proportional numbers* if, for $i, i' = 1, \ldots, a$ and $j, j' = 1, \ldots, b$,

$$N_{ij}/N_{ij'} = N_{i'j}/N_{i'j'}. \tag{1}$$

The special case of $N_{ij} = N$ for all i and j is the balanced two-way ANOVA.

The analysis with proportional numbers is in the same spirit as the analysis with balance presented in Section 1. After fitting the mean, μ, the column space for the α groups and the column space for the η groups are orthogonal. Before showing this we need a result on the N_{ij}s.

Lemma 7.4.1 *The N_{ij}s are proportional if and only if for any $r = 1, \ldots, a$, and $s = 1, \ldots, b$,*

$$N_{rs} = N_{r\cdot}N_{\cdot s}/N_{\cdot\cdot}\,. \tag{2}$$

Proof If the numbers are proportional,

$$N_{ij}N_{rs} = N_{rj}N_{is}\,.$$

Summing over i and j yields

$$N_{rs}N_{\cdot\cdot} = N_{rs}\sum_{i=1}^{a}\sum_{j=1}^{b}N_{ij} = \sum_{i=1}^{a}\sum_{j=1}^{b}N_{rj}N_{is} = N_{r\cdot}N_{\cdot s}.$$

Dividing by $N_{\cdot\cdot}$ gives the result. If (2) holds, substitution establishes (1). □

We now show that the α and η spaces are orthogonal after correcting for μ if and only if the model has proportional numbers. As in Section 1, we can write the model matrix as $X = [X_0, X_1, \ldots, X_{a+b}]$, where $X_0 = J_n^1$,

$$\begin{aligned} X_r &= [t_{ijk}], \quad t_{ijk} = \delta_{ir}, \quad r = 1, \ldots, a; \\ X_{a+s} &= [u_{ijk}], \quad u_{ijk} = \delta_{js}, \quad s = 1, \ldots, b. \end{aligned}$$

Orthogonalizing with respect to J_n^1 gives

$$Z_r = X_r - \frac{N_{r\cdot}}{N_{\cdot\cdot}}J, \quad r = 1, \ldots, a;$$

$$Z_{a+s} = X_{a+s} - \frac{N_{\cdot s}}{N_{\cdot\cdot}}J, \quad s = 1, \ldots, b.$$

It is easily seen that for $r = 1, \ldots, a$, $s = 1, \ldots, b$,

$$Z'_{a+s}Z_r = N_{rs} - N_{\cdot s}\frac{N_{r\cdot}}{N_{\cdot\cdot}} - N_{r\cdot}\frac{N_{\cdot s}}{N_{\cdot\cdot}} + N_{\cdot\cdot}\frac{N_{r\cdot}}{N_{\cdot\cdot}}\frac{N_{\cdot s}}{N_{\cdot\cdot}} = N_{rs} - N_{r\cdot}N_{\cdot s}/N_{\cdot\cdot}\,.$$

If follows immediately that $Z'_{a+s}Z_r = 0$ for all r and s if and only if $N_{rs} = N_{r\cdot}N_{\cdot s}/N_{\cdot\cdot}$ for all r and s.

Exercise 7.3 Find the ANOVA table for the two-way ANOVA without interaction model when there are proportional numbers. Find the least squares estimate of a contrast in the α_is. Find the variance of the contrast and give a definition of orthogonal contrasts that depends only on the contrast coefficients and the N_{ij}s. If the α_is correspond to levels of a quantitative factor, say x_is, find the linear contrast.

The analysis when interaction is included is similar. It is based on repeated use of the one-way ANOVA.

Exercise 7.4 Using proportional numbers, find the ANOVA table for the two-way ANOVA with interaction model.

7.5 Two-Way ANOVA with Unequal Numbers: General Case

Without balance or proportional numbers, there is no simplification of the model, so, typically, $R(\alpha|\mu,\eta) \neq R(\alpha|\mu)$ and $R(\eta|\mu,\alpha) \neq R(\eta|\mu)$. We are forced to analyze the model on the basis of the general theory alone.

7.5.1 Without Interaction

First consider the two-way ANOVA without interaction

$$y_{ijk} = \mu + \alpha_i + \eta_j + e_{ijk},$$

$i = 1,\ldots,a$, $j = 1,\ldots,b$, $k = 1,\ldots,N_{ij}$. Note that we have not excluded the possibility that $N_{ij} = 0$.

One approach to analyzing a two-way ANOVA is by model selection. Consider $R(\alpha|\mu,\eta)$ and $R(\eta|\mu,\alpha)$. If both of these are large, the model is taken as

$$y_{ijk} = \mu + \alpha_i + \eta_j + e_{ijk}.$$

If, say, $R(\alpha|\mu,\eta)$ is large and $R(\eta|\mu,\alpha)$ is not, then the model

$$y_{ijk} = \mu + \alpha_i + e_{ijk}$$

should be appropriate; however, a further test of α based on $R(\alpha|\mu)$ may give contradictory results. If $R(\alpha|\mu)$ is small and $R(\alpha,\eta|\mu)$ is large, we have a problem: no model seems to fit. The groups, α and η, are together having an effect, but neither seems to be helping individually. A model with μ and α is inappropriate because

$R(\alpha|\mu)$ is small. A model with μ and η is inappropriate because if μ and η are in the model and $R(\alpha|\mu, \eta)$ is large, we need α also. However, the model with μ, α, and η is inappropriate because $R(\eta|\mu, \alpha)$ is small. Finally, the model with μ alone is inappropriate because $R(\alpha, \eta|\mu)$ is large. Thus, every model that we can consider is inappropriate by some criterion. If $R(\alpha, \eta|\mu)$ had been small, the best choice probably would be a model with only μ; however, some would argue that all of μ, α, and η should be included on the basis of $R(\alpha|\mu, \eta)$ being large.

Fortunately, it is difficult for these situations to arise. Suppose $R(\alpha|\mu, \eta) = 8$ and $R(\alpha|\mu) = 6$, both with 2 degrees of freedom. Let $R(\eta|\mu, \alpha) = 10$ with 4 degrees of freedom, and $MSE = 1$ with 30 degrees of freedom. The 0.05 test for α after μ and η is

$$[8/2]/1 = 4 > 3.32 = F(0.95, 2, 30).$$

The test for α after μ is

$$[6/2]/1 = 3 < 3.32 = F(0.95, 2, 30).$$

The test for η after μ and α is

$$[10/4]/1 = 2.5 < 2.69 = F(0.95, 4, 30).$$

The test for α and η after μ is based on $R(\alpha, \eta|\mu) = R(\alpha|\mu) + R(\eta|\mu, \alpha) = 6 + 10 = 16$ with $2 + 4 = 6$ degrees of freedom. The test is

$$[16/6]/1 = 2.67 > 2.42 = F(0.95, 6, 30).$$

Although the tests are contradictory, the key point is that the P values for all four tests are about 0.05. For the first and last tests, the P values are just below 0.05; for the second and third tests, the P values are just above 0.05. The real information is not which tests are rejected and which are not rejected; the valuable information is that all four P values are approximately 0.05. All of the sums of squares should be considered to be either significantly large or not significantly large.

Because of the inflexible nature of hypothesis tests, we have chosen to discuss sums of squares that are either large or small, without giving a precise definition of what it means to be either large or small. The essence of any definition of large and small should be that the sum of two large quantities should be large and the sum of two small quantities should be small. For example, since $R(\alpha, \eta|\mu) = R(\alpha|\mu) + R(\eta|\mu, \alpha)$, it should be impossible to have $R(\alpha|\mu)$ small and $R(\eta|\mu, \alpha)$ small, but $R(\alpha, \eta|\mu)$ large. This consistency can be achieved by the following definition that exchanges one undefined term for another: *Large means significant or nearly significant.*

With this approach to the terms large and small, the contradiction alluded to above does not exist. The contradiction was based on the fact that with $R(\alpha|\mu, \eta)$ large, $R(\eta|\mu, \alpha)$ small, $R(\alpha|\mu)$ small, and $R(\alpha, \eta|\mu)$ large, no model seemed to fit.

Table 7.3 Suggested model selections: two-way analysis of variance without interaction (unequal numbers)

Table Entries Are Models as Numbered Below

		$R(\alpha\mid\mu,\eta)$	L		S	
$R(\eta\mid\mu,\alpha)$	$R(\alpha\mid\mu)$	$R(\eta\mid\mu)$	L	S	L	S
L	L		1	1	3*	I
	S		1	4,1*	3	4,1*
S	L		2*	2	1,2,3*	2
	S		I	4,1*	3	4

L indicates that the sum of squares is large.
S indicates that the sum of squares is small.
I indicates that the sum of squares is impossible.
* indicates that the model choice is debatable.

Models:
1 $y_{ijk} = \mu + \alpha_i + \eta_j + e_{ijk}$
2 $y_{ijk} = \mu + \alpha_i + e_{ijk}$
3 $y_{ijk} = \mu + \eta_j + e_{ijk}$
4 $y_{ijk} = \mu + e_{ijk}$

However, this situation is impossible. If $R(\eta|\mu, \alpha)$ is small and $R(\alpha|\mu)$ is small, then $R(\alpha, \eta|\mu)$ must be small; and the model with μ alone fits. (Note that since $R(\alpha|\mu, \eta)$ is large but $R(\alpha, \eta|\mu)$ is small, we must have $R(\eta|\mu)$ small; otherwise, two large quantities would add up to a small quantity.)

I find the argument of the previous paragraph convincing. Having built a better mousetrap, I expect the world to beat a path to my door. Unfortunately, I suspect that when the world comes, it will come carrying tar and feathers. It is my impression that in this situation most statisticians would prefer to use the model with all of μ, α, and η. In any case, the results are sufficiently unclear that further data collection would probably be worthwhile.

Table 7.3 contains some suggested model choices based on various sums of squares. Many of the suggestions are potentially controversial; these are indicated by asterisks.

The example that has been discussed throughout this section is the case in the fourth row and second column of Table 7.3. The entry in the second row and fourth column is similar.

The entry in the second row and second column is the case where each effect is important after fitting the other effect, but neither is important on its own. My inclination is to choose the model based on the results of examining $R(\alpha, \eta|\mu)$. On the other hand, it is sometimes argued that since the full model gives no basis for dropping either α or η individually, the issue of dropping them both should not be addressed.

The entry in the third row and third column is very interesting. Each effect is important on its own, but neither effect is important after fitting the other effect. The corresponding models (2 and 3 in the table) are not hierarchical, i.e., neither column space is contained in the other, so there is no way of testing which is better. From a testing standpoint, I can see no basis for choosing the full model; but since both effects have been shown to be important, many argue that both effects belong in the model. One particular argument is that the structure of the model matrix is hiding what would otherwise be a significant effect. As will be seen in Chapter 13, with collinearity in the model matrix that is quite possible.

The entries in the third row, first column and first row, third column are similar. Both effects are important by themselves, but one of the effects is not important after fitting the other. Clearly, the effect that is always important must be in the model. The model that contains only the effect for which both sums of squares are large is an adequate substitute for the full model. On the other hand, the arguments in favor of the full model from the previous paragraph apply equally well to this situation.

Great care needs to be used in all the situations where the choice of model is unclear. With unequal numbers, the possibility of collinearity in the model matrix (see Chapter 13) must be dealt with. If collinearity exists, it will affect the conclusions that can be made about the models. Of course, when the conclusions to be drawn are questionable, additional data collection would seem to be appropriate.

An alternative to selecting a model based on F tests, one that I have grown to like and to use in other books, *is to select a model based on the C_p statistic or one of the other model selection criteria that are discussed in Chapter* 14.

The discussion of model selection has been predicated on the assumption that there is no interaction. Some of the more bizarre situations that come up in model selection are more likely to arise if there is an interaction that is being ignored. A test for interaction should be made whenever possible. Of course, just because the test gives no evidence of interaction does not mean that interaction does not exist, or even that it is small enough so that it will not affect model selection.

Finally, it should be recalled that model selection is not the only possible goal. One may accept the full model and only seek to interpret it. For the purpose of interpreting the full model, $R(\alpha|\mu)$ and $R(\eta|\mu)$ are not very enlightening. In terms of the full model, the hypotheses that can be tested with these sums of squares are complicated functions of both the αs and the ηs. The exact nature of these hypotheses under the full model can be obtained from the formulae given below for the model with interaction.

7.5.2 Interaction

We now consider the two-way model with interaction. The model can be written

$$y_{ijk} = \mu + \alpha_i + \eta_j + \gamma_{ij} + e_{ijk}, \tag{1}$$

$i = 1, \ldots, a$, $j = 1, \ldots, b$, $k = 1, \ldots, N_{ij}$. However, for most purposes, we do not recommend using this parameterization of the model. The full rank *cell means* parameterization

$$y_{ijk} = \mu_{ij} + e_{ijk} \tag{2}$$

is much easier to work with. The interaction model has the column space of a one-way ANOVA with unequal numbers.

Model (1) lends itself to two distinct orthogonal breakdowns of the sum of squares for the model. These are

$$R(\mu),\ R(\alpha|\mu),\ R(\eta|\mu,\alpha),\ R(\gamma|\mu,\alpha,\eta)$$

and

$$R(\mu),\ R(\eta|\mu),\ R(\alpha|\mu,\eta),\ R(\gamma|\mu,\alpha,\eta).$$

If $R(\gamma|\mu,\alpha,\eta)$ is small, one can work with the reduced model. If $R(\gamma|\mu,\alpha,\eta)$ is large, the full model must be retained. Just as with the balanced model, the F tests for α and η test hypotheses that involve the interactions. Using the parameterization of model (2), the hypothesis associated with the test using $R(\alpha|\mu)$ is that for all i and i',

$$\sum_{j=1}^{b} N_{ij}\mu_{ij}/N_{i\cdot} = \sum_{j=1}^{b} N_{i'j}\mu_{i'j}/N_{i'\cdot}\,.$$

The hypothesis associated with the test using $R(\alpha|\mu,\eta)$ is that for all i,

$$\sum_{j=1}^{b} N_{ij}\mu_{ij} - \sum_{i'=1}^{a}\sum_{j=1}^{b} N_{ij}N_{i'j}\mu_{i'j}/N_{\cdot j} = 0.$$

Since $\mu_{ij} = \mu + \alpha_i + \eta_j + \gamma_{ij}$, the formulae above can be readily changed to involve the alternative parameterization. Similarly, by dropping the γ_{ij}, one can get the hypotheses associated with the sums of squares in the no interaction model. Neither of these hypotheses appears to be very interesting. The first of them is the hypothesis of equality of the weighted averages, taken over j, of the μ_{ij}s. It is unclear why one should be interested in weighted averages of the μ_{ij}s when the weights are the sample sizes. For many purposes, a more reasonable hypothesis would seem to be that the simple averages, taken over j, are all equal. The second hypothesis above is almost uninterpretable. In terms of model selection, these tests for "main effects" do not make much sense when there is interaction. The reduced models corresponding to such tests are not easy to find, so it is not surprising that the parametric hypotheses corresponding to these tests are not very interpretable.

Of course testing either $\alpha_1 = \cdots = \alpha_a$ or $\eta_1 = \cdots = \eta_b$ makes no sense in an interaction model because neither constraint puts a restriction on $C(X)$. What the previous paragraph discussed was what hypotheses are being tested when using certain sums of squares.

A better idea seems to be to choose tests based directly on model (2). For example, the hypothesis of equality of the simple averages $\bar{\mu}_{i\cdot}$ (i.e., $\bar{\mu}_{i\cdot} = \bar{\mu}_{i'\cdot}$ for $i, i' = 1, \ldots, a$) is easily tested using the fact that the interaction model is really just a one-way ANOVA with unequal numbers.

At this point we should say a word about computational techniques, or rather explain why we will not discuss computational techniques. The difficulty in computing sums of squares for a two-way ANOVA with interaction and unequal numbers lies in computing $R(\alpha|\mu, \eta)$ and $R(\eta|\mu, \alpha)$. The sums of squares that are needed for the analysis are $R(\gamma|\mu, \alpha, \eta)$, $R(\alpha|\mu, \eta)$, $R(\eta|\mu, \alpha)$, $R(\alpha|\mu)$, $R(\eta|\mu)$, and *SSE*. *SSE* is the error computed as in a one-way ANOVA. $R(\alpha|\mu)$ and $R(\eta|\mu)$ are computed as in a one-way ANOVA, and the interaction sum of squares can be written as $R(\gamma|\mu, \alpha, \eta) = R(\gamma|\mu) - R(\eta|\mu, \alpha) - R(\alpha|\mu)$, where $R(\gamma|\mu)$ is computed as in a one-way ANOVA. Thus, only $R(\alpha|\mu, \eta)$ and $R(\eta|\mu, \alpha)$ present difficulties.

There are formulae available for computing $R(\alpha|\mu, \eta)$ and $R(\eta|\mu, \alpha)$, but the formulae are both nasty and of little use. With the advent of high speed computing, the very considerable difficulties in obtaining these sums of squares have vanished. Moreover, the formulae add very little to one's understanding of the method of analysis. The key idea in the analysis is that for, say $R(\eta|\mu, \alpha)$, the η effects are being fitted after the α effects. The general theory of linear models provides methods for finding the sums of squares and there is little simplification available in the special case.

Of course, there is more to an analysis than just testing group effects. As mentioned above, if there is evidence of interactions, probably the simplest approach is to analyze the μ_{ij} model. If there are no interactions, then one is interested in testing contrasts in the main effects. Just as in the one-way ANOVA, it is easily seen that all estimable functions of the group effects will be contrasts; however, there is no assurance that all contrasts will be estimable. To test a contrast, say $\alpha_1 - \alpha_2 = 0$, the simplest method is to fit a model that does not distinguish between α_1 and α_2, see Example 3.2.0 and Section 3.4 and Christensen (2015) contains many similar examples. In the two-way ANOVA with unequal numbers, the model that does not distinguish between α_1 and α_2 may or may not have a different column space from that of the unrestricted model.

Example 7.5.1 Scheffé (1959) (cf. Bailey 1953) reports on an experiment in which infant female rats were given to foster mothers for nursing, and the weights, in grams, of the infants were measured after 28 days. The two factors in the experiment were the genotypes of the infants and the genotypes of the foster mothers. In the experiment, an entire litter was given to a single foster mother. The variability within each litter was negligible compared to the variability between different litters, so the analysis was performed on the litter averages. Table 7.4 contains the data.

We use the two-way ANOVA model

$$y_{ijk} = G + L_i + M_j + [LM]_{ij} + e_{ijk}, \tag{3}$$

Table 7.4 Infant rats' weight gain with foster mothers

Genotype of Litter	Genotype of Foster Mother: A	F	I	J
A	61.5	55.0	52.5	42.0
	68.2	42.0	61.8	54.0
	64.0	60.2	49.5	61.0
	65.0		52.7	48.2
	59.7			39.6
F	60.3	50.8	56.5	51.3
	51.7	64.7	59.0	40.5
	49.3	61.7	47.2	
	48.0	64.0	53.0	
		62.0		
I	37.0	56.3	39.7	50.0
	36.3	69.8	46.0	43.8
	68.0	67.0	61.3	54.5
			55.3	
			55.7	
J	59.0	59.5	45.2	44.8
	57.4	52.8	57.0	51.5
	54.0	56.0	61.4	53.0
	47.0			42.0
				54.0

where L indicates the effect of the litter genotype and M indicates the effect of the foster mother genotype. Tables 7.5 and 7.6 contain, respectively, the sums of squares for error for a variety of submodels and the corresponding reductions in sums of squares for error and F tests. In discussing these data Christensen (2015, Section 14.1) adds a column of C_p statistics to Table 7.5 and eliminates Table 7.6.

Some percentiles of the F distribution that are of interest in evaluating the statistics of Table 7.6 are $F(0.9, 9, 45) = 1.78$, $F(0.99, 3, 45) = 4.25$, and $F(0.95, 6, 45) = 2.23$. Clearly, the litter–mother interaction and the main effect for litters can be dropped from model (3). However, the main effect for mothers is important. The smallest model that fits the data is

$$y_{ijk} = G + M_j + e_{ijk}.$$

Table 7.5 Sums of squares error for fitting models to the data of Table 7.4

Model	Model	SSE	df
$G+L+M+LM$	$[LM]$	2441	45
$G+L+M$	$[L][M]$	3265	54
$G+L$	$[L]$	4040	57
$G+M$	$[M]$	3329	57
G	$[G]$	4100	60

Table 7.6 F tests for fitting models to the data of Table 7.4

Reduction in SSE		df	MS	F^*
$R(LM\|L,M,G)$	$= 3265 - 2441 = 824$	9	91.6	1.688
$R(M\|L,G)$	$= 4040 - 3265 = 775$	3	258.3	4.762
$R(L\|G)$	$= 4100 - 4040 = 60$	3	20.0	0.369
$R(L\|M,G)$	$= 3329 - 3265 = 64$	3	21.3	0.393
$R(M\|G)$	$= 4100 - 3329 = 771$	3	257.	4.738
$R(L,M\|G)$	$= 4100 - 3265 = 835$	6	139.2	2.565

*All F statistics calculated using $MSE([LM]) = 2441/45 = 54.244$ in the denominator

Table 7.7 Mean values for foster mother genotypes

Foster Mother	Parameter	$N_{\cdot j}$	Estimate
A	$G+M_1$	16	55.400
F	$G+M_2$	14	58.700
I	$G+M_3$	16	53.363
J	$G+M_4$	15	48.680

This is just a one-way ANOVA model and can be analyzed as such. By analogy with balanced two-factor ANOVAs, tests of contrasts might be best performed using $MSE([LM])$ rather than $MSE([M]) = 3329/57 = 58.40$.

The mean values for the foster mother genotypes are reported in Table 7.7, along with the number of observations for each mean. It would be appropriate to continue the analysis by comparing all pairs of means. This can be done with either Scheffé's method, Bonferroni's method, or the LSD method. The LSD method with $\alpha = 0.05$ establishes that genotype J is distinct from genotypes A and F. (Genotypes F and I are almost distinct.) Bonferroni's method with $\alpha = 0.06$ establishes that J is distinct from F and that J is almost distinct from A.

Exercise 7.5 Analyze the following data as a two-factor ANOVA where the subscripts i and j indicate the two factors.

		y_{ijk}s		
	i	1	2	3
j	1	0.620	1.228	0.615
		1.342	3.762	2.245
		0.669	2.219	2.077
		0.687	4.207	3.357
		0.155		
		2.000		
	2	1.182	3.080	2.240
		1.068	2.741	0.330
		2.545	2.522	3.453
		2.233	1.647	1.527
		2.664	1.999	0.809
		1.002	2.939	1.942
		2.506		
		4.285		
		1.696		

The dependent variable is a mathematics ineptitude score. The first factor (i) identifies economics majors, anthropology majors, and sociology majors, respectively. The second factor (j) identifies whether the student's high school background was rural (1) or urban (2).

Exercise 7.6 Analyze the following data as a two-factor ANOVA where the subscripts i and j indicate the two factors.

		y_{ijk}s		
	i	1	2	3
j	1	1.620	2.228	2.999
		1.669	3.219	1.615
		1.155	4.080	
		2.182		
		3.545		
	2	1.342	3.762	2.939
		0.687	4.207	2.245
		2.000	2.741	1.527
		1.068		0.809
		2.233		1.942
		2.664		
		1.002		

The dependent variable is again a mathematics ineptitude score and the levels of the first factor identify the same three majors as in Exercise 7.5. In these data, the second factor identifies whether the student is lefthanded (1) or righthanded (2).

7.5.3 Characterizing the Interaction Space

In a two-way ANOVA with interaction and unbalanced numbers, the cell-means parameterization is

$$y_{ijk} = \mu_{ij} + \varepsilon_{ijk}, \qquad i = 1, \ldots, a, \quad j = 1, \ldots, b, \quad k = 1, \ldots, N_{ij},$$

which is really just a one-way ANOVA with unequal numbers and the pair of subscripts ij identifying the ab different groups. The traditional parameterization is $\mu_{ij} = \mu + \alpha_i + \eta_j + \gamma_{ij}$. Write the cell-means model in matrix form as $Y = X\tilde{\mu} + e$. X has ab columns and the rs column has the form

$$X_{rs} = [t_{ijk}], \quad \text{with} \quad t_{ijk} = \delta_{(i,j)(r,s)}$$

where for any two symbols g and h, the kronecker delta has

$$\delta_{gh} = \begin{cases} 1 & \text{if } g = h \\ 0 & \text{if } g \neq h. \end{cases}$$

Note that

$$C(X) = \left\{ v | v = [v_{ijk}], \text{ with } v_{ijk} = \mu_{ij} \text{ for some } \mu_{ij} \right\}.$$

Similar to the balanced case, we will see that the interaction space is the set of vectors

$$T = [t_{ijk}], \quad \text{with} \quad t_{ijk} = q_{ij}/N_{ij} \qquad \text{where} \quad q_{i\cdot} = 0 = q_{\cdot j}, \quad \text{for all } i, j.$$

An interaction contrast is $T'X\tilde{\mu} = \sum_{ijk} q_{ij}\mu_{ij}/N_{ij} = \sum_{ij} q_{ij}\mu_{ij} = \sum_{ij} q_{ij}\gamma_{ij}$. Clearly, $T \in C(X)$, so $T'MY = T'Y$ where M is the perpendicular projection operator onto $C(X)$. It follows that the least squares estimate of $T'X\tilde{\mu}$ is $T'Y = \sum_{ij} q_{ij}\bar{y}_{ij\cdot}$. These are essentially the same results as for balanced ANOVA.

To see that vectors T characterize the interaction space, write the corresponding main effects model

$$Y = J\mu + X_\alpha \alpha + X_\eta \eta + e.$$

The matrix X_α has columns

$$X_r = [t_{ijk}], \quad t_{ijk} = \delta_{ir} \qquad r = 1, \ldots, a$$

The matrix X_η has columns

$$X_{a+s} = [t_{ijk}], \quad t_{ijk} = \delta_{js} \qquad s = 1, \ldots, b.$$

Note that the ab columns of X are generated by multiplying elementwise the a columns of X_α and the b columns of X_η because $\delta_{ir}\delta_{js} = \delta_{(i,j)(r,s)}$. By definition, the interaction space is $C(J, X_\alpha, X_\eta)^\perp_{C(X)}$. Thus, the characterization of the interaction space results from vectors of the form T spanning a space of sufficient rank $[(a-1)(b-1)$ when $N_{ij} > 0$ for all $ij]$, having $T \in C(X)$, and $X_h'T = 0$, $h = 1, \ldots, a+b$. That $X_h'T = 0$ follows from the definitions and arithmetic. Note that to have orthogonal interaction contrasts we need the T vectors corresponding to the interaction contrasts to be orthogonal.

Note also that all interaction contrasts are contrasts in the μ_{ij} one-way model, but not all μ_{ij} contrasts are interaction contrasts. As in Chapter 4, an arbitrary element of the μ_{ij} contrast space is

$$S = [s_{ijk}], \quad \text{with} \quad s_{ijk} = s_{ij}/N_{ij} \quad \text{where} \quad s_{\cdot\cdot} = 0.$$

In addition to the interaction space being a subset of the contrast space, there is an $a-1$ dimensional subspace of the contrast space consisting of vectors

$$T_\alpha = [t_{ijk}], \quad \text{with} \quad t_{ijk} = c_i/bN_{ij} \quad \text{where} \quad c_\cdot = 0$$

and a $b-1$ dimensional subspace of vectors

$$T_\eta = [t_{ijk}], \quad \text{with} \quad t_{ijk} = d_j/aN_{ij} \quad \text{where} \quad d_\cdot = 0.$$

These define contrasts in the interaction model of

$$T_\alpha' X\tilde{\mu} = \sum_{ij} c_i \mu_{ij}/b = \sum_i c_i \bar{\mu}_{i\cdot} = \sum_i c_i(\alpha_i + \bar{\gamma}_{i\cdot})$$

with estimate $T_\alpha' Y = \sum_i c_i \tilde{y}_i$ where $\tilde{y}_i \equiv \sum_j \bar{y}_{ij\cdot}/b$ which need not equal $\bar{y}_{i\cdot\cdot}$ and

$$T_\eta' X\tilde{\mu} = \sum_{ij} d_j \mu_{ij}/a = \sum_j d_j \bar{\mu}_{\cdot j} = \sum_j d_j(\eta_j + \bar{\gamma}_{\cdot j})$$

with estimate $T_\eta' Y = \sum_j d_j \tilde{y}_j$ where $\tilde{y}_j \equiv \sum_i \bar{y}_{ij\cdot}/a$ which need not equal $\bar{y}_{i\cdot\cdot}$. Under proportional numbers, these three spaces are orthogonal but in general they merely intersect in the zero vector.

For the main effects model, the vectors T_α and T_η typically are not in $C(J, X_\alpha, X_\eta)$, so although, for example, $T_\alpha'[J\mu + X_\alpha\alpha + X_\eta\eta] = \sum_i c_i\alpha_i$, the estimate is *not* $T_\alpha' Y$, it is $T_\alpha' M_0 Y$ where M_0 is the ppo onto $C(J, X_\alpha, X_\eta)$.

7.6 Three or More Way Analyses

7.6.1 Balanced Analyses

With balanced or proportional numbers, the analyses for more general models follow the same patterns as those of the two-way ANOVA. (Proportional numbers can be defined in terms of complete independence in higher dimensional tables, see Fienberg 1980 or Christensen 1997.) Consider, for example, the model

$$y_{ijkm} = \mu + \alpha_i + \eta_j + \gamma_k + (\alpha\eta)_{ij} + (\alpha\gamma)_{ik} + (\eta\gamma)_{jk} + (\alpha\eta\gamma)_{ijk} + e_{ijkm},$$

$i = 1, \dots, a$, $j = 1, \dots, b$, $k = 1, \dots, c$, $m = 1, \dots, N$. The sums of squares for the main effects of α, η, and γ are based on the one-way ANOVA ignoring all other effects, e.g.,

$$SS(\eta) = acN \sum_{j=1}^{b} (\bar{y}_{\cdot j \cdot\cdot} - \bar{y}_{\cdot\cdot\cdot\cdot})^2. \quad (1)$$

The sums of squares for the two-way interactions, $\alpha\eta$, $\alpha\gamma$, and $\eta\gamma$, are obtained as in the two-way ANOVA by ignoring the third effect, e.g.,

$$SS(\alpha\gamma) = bN \sum_{i=1}^{a} \sum_{k=1}^{c} (\bar{y}_{i \cdot k \cdot} - \bar{y}_{i \cdot\cdot\cdot} - \bar{y}_{\cdot\cdot k \cdot} + \bar{y}_{\cdot\cdot\cdot\cdot})^2. \quad (2)$$

The sum of squares for the three-way interaction $\alpha\eta\gamma$ is found by subtracting all of the other sums of squares (including the grand mean's) from the sum of squares for the full model. Note that the full model is equivalent to the one-way ANOVA model

$$y_{ijkm} = \mu_{ijk} + e_{ijkm}.$$

Sums of squares and their associated projection operators are defined from reduced models. For example, M_η is the perpendicular projection operator for fitting the ηs after μ in the model

$$y_{ijkm} = \mu + \eta_j + e_{ijkm}.$$

The subscripts i, k, and m are used to indicate the replications in this one-way ANOVA. $SS(\eta)$ is defined by

$$SS(\eta) = Y' M_\eta Y.$$

The algebraic formula for $SS(\eta)$ was given in (1). Similarly, $M_{\alpha\gamma}$ is the projection operator for fitting the $(\alpha\gamma)$s after μ, the αs, and the γs in the model

$$y_{ijkm} = \mu + \alpha_i + \gamma_k + (\alpha\gamma)_{ik} + e_{ijkm}.$$

In this model, the subscripts j and m are used to indicate replication. The sum of squares $SS(\alpha\gamma)$ is

$$SS(\alpha\gamma) = Y'M_{\alpha\gamma}Y.$$

The algebraic formula for $SS(\alpha\gamma)$ was given in (2). Because all of the projection operators (except for the three-factor interaction) are defined on the basis of reduced models that have previously been discussed, the sums of squares take on the familiar forms indicated above.

The one new aspect of the model that we are considering is the inclusion of the three-factor interaction. As mentioned above, the sum of squares for the three-factor interaction is just the sum of squares that is left after fitting everything else, i.e.,

$$SS(\alpha\eta\gamma) \equiv R\left[(\alpha\eta\gamma)|\mu, \alpha, \eta, \gamma, (\alpha\eta), (\alpha\gamma), (\eta\gamma)\right].$$

The space for the three-factor interaction is the orthogonal complement (with respect to the space for the full model) of the space for the model that includes all factors except the three-factor interaction. Thus, the space for the three-factor interaction is orthogonal to everything else. (This is true even when the numbers are not balanced.)

In order to ensure a nice analysis, we need to show that the spaces associated with all of the projection operators are orthogonal and that the projection operators add up to the perpendicular projection operator onto the space for the full model. First, we show that $C(M_\mu, M_\alpha, M_\eta, M_\gamma, M_{\alpha\eta}, M_{\alpha\gamma}, M_{\eta\gamma})$ is the column space for the model

$$y_{ijkm} = \mu + \alpha_i + \eta_j + \gamma_k + (\alpha\eta)_{ij} + (\alpha\gamma)_{ik} + (\eta\gamma)_{jk} + e_{ijkm},$$

and that all the projection operators are orthogonal. That the column spaces are the same follows from the fact that the column space of the model without the three-factor interaction is precisely the column space obtained by combining the column spaces of all of the two-factor with interaction models. Combining the spaces of the projection operators is precisely combining the column spaces of all the two-factor with interaction models. That the spaces associated with all of the projection operators are orthogonal follows easily from the fact that all of the spaces come from reduced models. For the reduced models, characterizations have been given for the various spaces. For example,

$$C(M_\eta) = \{v|v = [v_{ijkm}], \text{ where } v_{ijkm} = d_j \text{ for some } d_1, \ldots, d_b \text{ with } d_\cdot = 0\}.$$

Similarly,

$$C(M_{\alpha\gamma}) = \{w|w = [w_{ijkm}], \text{ where } w_{ijkm} = r_{ik} \text{ for some } r_{ik} \text{ with } r_{i\cdot} = r_{\cdot k} = 0 \text{ for } i = 1, \ldots, a,\ k = 1, \ldots, c\}.$$

With these characterizations, it is a simple matter to show that the projection operators define orthogonal spaces.

Let M denote the perpendicular projection operator for the full model. The projection operator onto the interaction space, $M_{\alpha\eta\gamma}$, has been defined as

$$M_{\alpha\eta\gamma} \equiv M - [M_\mu + M_\alpha + M_\eta + M_\gamma + M_{\alpha\eta} + M_{\alpha\gamma} + M_{\eta\gamma}];$$

thus,

$$M = M_\mu + M_\alpha + M_\eta + M_\gamma + M_{\alpha\eta} + M_{\alpha\gamma} + M_{\eta\gamma} + M_{\alpha\eta\gamma},$$

where the spaces of all the projection operators on the right side of the equation are orthogonal to each other.

Exercise 7.7 Show that $C(M_\eta) \perp C(M_{\alpha\gamma})$ and that $C(M_{\eta\gamma}) \perp C(M_{\alpha\gamma})$. Give an explicit characterization of a typical vector in $C(M_{\alpha\eta\gamma})$ and show that your characterization is correct.

If the α, η, and γ effects correspond to quantitative levels of some factor, the three-way ANOVA corresponds to a polynomial in three variables. The main effects and the two-way interactions can be dealt with as before. The three-way interaction can be broken down into contrasts such as the linear-by-linear-by-quadratic.

7.6.2 Unbalanced Analyses

For unequal numbers, the analysis can be performed by comparing models.

Example 7.6.1 Table 7.8 is derived from Scheffé (1959) and gives the moisture content (in grams) for samples of a food product made with three kinds of salt (A), three amounts of salt (B), and two additives (C). The amounts of salt, as measured in moles, are equally spaced. The two numbers listed for some group combinations are replications. We wish to analyze these data.

We will consider these data as a three-factor ANOVA. From the structure of the replications, the ANOVA has unequal numbers. The general model for a three-factor ANOVA with replications is

$$y_{ijkm} = G + A_i + B_j + C_k + [AB]_{ij} + [AC]_{ik} + [BC]_{jk} + [ABC]_{ijk} + e_{ijkm}.$$

Our first priority is to find out which interactions are important. Table 7.9 contains the sum of squares for error and the degrees of freedom for error for all models that include all of the main effects. Each model is identified in the table by the highest

Table 7.8 Moisture content of a food product

A (salt)		1			2			3		
B (amount salt)		1	2	3	1	2	3	1	2	3
C (additive)	1	8	17	22	7	26	34	10	24	39
			13	20	10	24		9		36
	2	5	11	16	3	17	32	5	16	33
		4	10	15	5	19	29	4		34

Table 7.9 Statistics for fitting models to the data of Table 7.8

Model	*SSE*	*df*	*F**
[*ABC*]	32.50	14	—
[*AB*][*AC*][*BC*]	39.40	18	0.743
[*AB*][*AC*]	45.18	20	0.910
[*AB*][*BC*]	40.46	20	0.572
[*AC*][*BC*]	333.2	22	16.19
[*AB*][*C*]	45.75	22	0.713
[*AC*][*B*]	346.8	24	13.54
[*BC*][*A*]	339.8	24	13.24
[*A*][*B*][*C*]	351.1	26	11.44

*The F statistics are for testing each model against the model with a three-factor interaction, i.e., [*ABC*]. The denominator of each F statistic is $MSE([ABC]) = 32.50/14 = 2.3214$

order terms in the model (cf. Table 7.5, Section 5). Readers familiar with methods for fitting log-linear models (cf. Fienberg 1980 or Christensen 1997) will notice a correspondence between Table 7.9 and similar displays used in fitting three-dimensional contingency tables. The analogies between selecting log-linear models and selecting models for unbalanced ANOVA are pervasive. Christensen (2015, Section 16.1) gives a more expansive version of this data analysis.

All of the models have been compared to the full model using F statistics in Table 7.9. It takes neither a genius nor an F table to see that the only models that fit the data are the models that include the [*AB*] interaction. A number of other comparisons can be made among models that include [*AB*]. These are [*AB*][*AC*][*BC*] versus [*AB*][*AC*], [*AB*][*AC*][*BC*] versus [*AB*][*BC*], [*AB*][*AC*][*BC*] versus [*AB*][*C*], [*AB*][*AC*] versus [*AB*][*C*], and [*AB*][*BC*] versus [*AB*][*C*]. None of the comparisons show any lack of fit. The last two comparisons are illustrated below.

$$[AB][AC] \text{ versus } [AB][C]:$$

$$R(AC|AB, C) = 45.75 - 45.18 = 0.57,$$

$$F = (0.57/2)/2.3214 = 0.123.$$

$$[AB][BC] \text{ versus } [AB][C]:$$

$$R(BC|AB, C) = 45.75 - 40.46 = 5.29,$$

$$F = (5.29/2)/2.3214 = 1.139.$$

Note that, by analogy to the commonly accepted practice for balanced ANOVAs, all tests have been performed using $MSE([ABC])$ in the denominator, i.e., the estimate of pure error from the full model.

The smallest model that seems to fit the data adequately is $[AB][C]$. The F statistics for comparing $[AB][C]$ to the larger models are all extremely small. Writing out the model $[AB][C]$, it is

$$y_{ijkm} = G + A_i + B_j + C_k + [AB]_{ij} + e_{ijkm}.$$

We need to examine the $[AB]$ interaction. Since the levels of B are quantitative, a model that is equivalent to $[AB][C]$ is a model that includes the main effects for C but, instead of fitting an interaction in A and B, fits a separate regression equation in the levels of B for each level of A. Let x_j, $j = 1, 2, 3$, denote the levels of B. There are three levels of B, so the most general polynomial we can fit is a second-degree polynomial in x_j. Since the levels of salt were equally spaced, it does not matter much what we use for the x_js. The computations were performed using $x_1 = 1$, $x_2 = 2$, $x_3 = 3$. In particular, the model $[AB][C]$ was reparameterized as

$$y_{ijkm} = A_{i0} + A_{i1}x_j + A_{i2}x_j^2 + C_k + e_{ijkm}. \quad (3)$$

With a notation similar to that used in Table 7.9, the SSE and the dfE are reported in Table 7.10 for model (3) and three reduced models.

Note that the SSE and df reported in Table 7.10 for $[A_0][A_1][A_2][C]$ are identical to the values reported in Table 7.9 for $[AB][C]$. This, of course, must be true if the models are merely reparameterizations of one another. First we want to establish whether the quadratic effects are necessary in the regressions. To do this we test

Table 7.10 Additional statistics for fitting models to the data of Table 7.8

Model	SSE	df
$[A_0][A_1][A_2][C]$	45.75	22
$[A_0][A_1][C]$	59.98	25
$[A_0][A_1]$	262.0	26
$[A_0][C]$	3130.	28

$$[A_0][A_1][A_2][C] \text{ versus } [A_0][A_1][C]:$$

$$R(A_2|A_1, A_0, C) = 59.98 - 45.75 = 14.23,$$

$$F = (14.23/3)/2.3214 = 2.04.$$

Since $F(0.95, 3, 14) = 3.34$, there is no evidence of any nonlinear effects.

At this point it might be of interest to test whether there is any linear effect. This is done by testing $[A_0][A_1][C]$ against $[A_0][C]$. The statistics needed for this test are given in Table 7.10. Instead of actually doing the test, recall that no models in Table 7.9 fit the data unless they included the $[AB]$ interaction. If we eliminated the linear effects, we would have a model that involved none of the $[AB]$ interaction. (The model $[A_0][C]$ is identical to the ANOVA model $[A][C]$.) We already know that such models do not fit.

Finally, we have never explored the possibility that there is no main effect for C. This can be done by testing

$$[A_0][A_1][C] \text{ versus } [A_0][A_1]:$$

$$R(C|A_1, A_0) = 262.0 - 59.98 = 202,$$

$$F = (202/1)/2.3214 = 87.$$

Obviously, there is a substantial main effect for C, the type of food additive.

Our conclusion is that the model $[A_0][A_1][C]$ is the smallest model yet considered that adequately fits the data. This model indicates that there is an effect for the type of additive and a linear relationship between amount of salt and moisture content. The slope and intercept of the line may depend on the type of salt. (The intercept of the line also depends on the type of additive.) Table 7.11 contains parameter estimates and standard errors for the model. All estimates in the example use the side condition $C_1 = 0$.

Table 7.11 Parameter estimates and standard errors for the model $y_{ijkm} = A_{i0} + A_{i1}x_j + C_k + e_{ijkm}$

Parameter	Estimate	S.E.
A_{10}	3.350	1.375
A_{11}	5.85	0.5909
A_{20}	−3.789	1.237
A_{21}	13.24	0.5909
A_{30}	−4.967	1.231
A_{31}	14.25	0.5476
C_1	0.	none
C_2	−5.067	0.5522

Note that, in lieu of the F test, the test for the main effect C could be performed by looking at $t = -5.067/0.5522 = -9.176$. Moreover, we should have $t^2 = F$. The t statistic squared is 84, while the F statistic reported earlier is 87. The difference is due to the fact that the S.E. reported uses the *MSE* for the model being fitted, while in performing the F test we used the *MSE*([*ABC*]).

Are we done yet? No! The parameter estimates suggest some additional questions. Are the slopes for salts 2 and 3 the same, i.e., is $A_{21} = A_{31}$? In fact, are the entire lines for salts 2 and 3 the same, i.e., are $A_{21} = A_{31}$, $A_{20} = A_{30}$? We can fit models that incorporate these assumptions.

Model	*SSE*	*df*
$[A_0][A_1][C]$	59.98	25
$[A_0][A_1][C]$, $A_{21} = A_{31}$	63.73	26
$[A_0][A_1][C]$, $A_{21} = A_{31}$, $A_{20} = A_{30}$	66.97	27

It is a small matter to check that there is no lack of fit displayed by any of these models. The smallest model that fits the data is now $[A_0][A_1][C]$, $A_{21} = A_{31}$, $A_{20} = A_{30}$. Thus there seems to be no difference between salts 2 and 3, but salt 1 has a different regression than the other two salts. (We did not actually test whether salt 1 is different, but if salt 1 had the same slope as the other two, then there would be no interaction, and we know that interaction exists.) There is also an effect for the food additives. The parameter estimates and standard errors for the final model are given in Table 7.12.

Figure 7.1 shows the fitted values for the final model. The two lines for a given additive are shockingly close when $x = 1$, which makes me wonder if $j = 1$ is the condition of no salt being used. Scheffé does not say.

Are we done yet? Probably not. We have not even considered the validity of the assumptions. Are the errors normally distributed? Are the variances the same for every group combination? Some methods for addressing these questions are discussed in Chapter 12. Technically, we need to ask whether $C_1 = C_2$ in this new model. A quick look at the estimate and standard error for C_2 answers the question in the negative. We also have not asked whether $A_{10} = A_{20}$. Personally, I find this last question so uninteresting that I would be loath to examine it. However, a look at

Table 7.12 Parameter estimates and standard errors for the model $[A_0][A_1][C]$, $A_{21} = A_{31}$, $A_{20} = A_{30}$

Parameter	Estimate	S.E.
A_{10}	3.395	1.398
A_{11}	5.845	0.6008
A_{20}	−4.466	0.9030
A_{21}	13.81	0.4078
C_1	0.	none
C_2	−5.130	0.5602

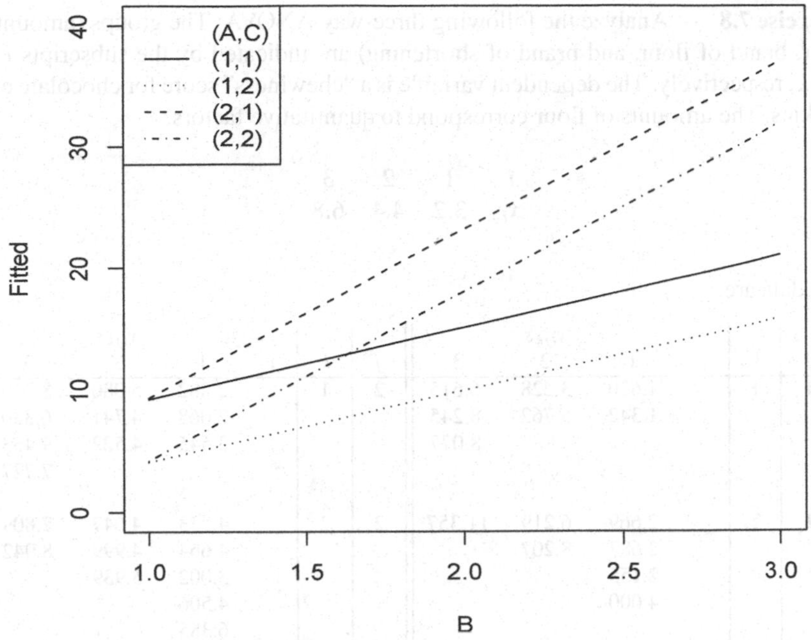

Fig. 7.1 Fitted values for moisture content data

the estimates and standard errors suggests that the answer is no. A more interesting question is whether $A_{10} + A_{11} = A_{20} + A_{21}$, i.e. that the two lines for different salts but the same additive take on the same value at $x = 1$. It is pretty clear from Figure 7.1 that there will be no evidence against this hypothesis that was suggested by the data, cf. Exercise 7.7.6. Christensen (2015, Section 16.1) considers this issue in more detail.

As mentioned in Example 7.6.1, the correspondences between log-linear models and unbalanced ANOVAs are legion. Often these correspondences have been overlooked. We have just considered a three-factor unbalanced ANOVA. What would we do with a four-factor or five-factor ANOVA? There is a considerable amount of literature in log-linear model theory about how to select models when there are a large number of factors. In particular, Benedetti and Brown (1978) and Christensen (1997) have surveyed strategies for selecting log-linear models. Those strategies can be applied to unbalanced ANOVAs with equal success.

I might also venture a personal opinion that statisticians tend not to spend enough time worrying about what high-dimensional ANOVA models actually mean. Log-linear model theorists do worry about interpreting their models. Wermuth (1976) has developed a method of searching among log-linear models that have nice interpretations, cf. also Christensen (1997). I believe that her method could be applied equally well to ANOVA models.

Exercise 7.8 Analyze the following three-way ANOVA: The groups (amount of flour, brand of flour, and brand of shortening) are indicated by the subscripts i, j, and k, respectively. The dependent variable is a "chewiness" score for chocolate chip cookies. The amounts of flour correspond to quantitative factors,

i	1	2	3
x_i	3.2	4.4	6.8

The data are:

j	k	i	1	y_{ijks} 2	3	j	k	i	1	y_{ijks} 2	3
1	1		1.620	3.228	6.615	2	1		2.282	5.080	8.240
			1.342	5.762	8.245				2.068	4.741	6.330
					8.077				3.545	4.522	9.453
											7.727
1	2		2.669	6.219	11.357	2	2		4.233	4.647	7.809
			2.687	8.207					4.664	4.999	8.942
			2.155						3.002	5.939	
			4.000						4.506		
									6.385		
									3.696		

7.7 Additional Exercises

Exercise 7.7.1 In the mid-1970s, a study on the prices of various motor oils was conducted in (what passes for) a large town in Montana. The study consisted of pricing 4 brands of oil at each of 9 stores. The data follow.

Store	Brand P	H	V	Q
1	87	95	95	82
2	96	104	106	97
3	75	87	81	70
4	81	94	91	77
5	70	85	87	65
6	85	98	97	83
7	110	123	128	112
8	83	98	95	78
9	105	120	119	98

Analyze these data.

Exercise 7.7.2 An experiment was conducted to examine thrust forces when drilling under different conditions. Data were collected for four drilling speeds and three feeds.

	Speed			
Feed	100	250	400	550
	121	98	83	58
	124	108	81	59
0.005	104	87	88	60
	124	94	90	66
	110	91	86	56
	329	291	281	265
	331	265	278	265
0.010	324	295	275	269
	338	288	276	260
	332	297	287	251
	640	569	551	487
	600	575	552	481
0.015	612	565	570	487
	620	573	546	500
	623	588	569	497

Analyze these data.

Exercise 7.7.3 Consider the model

$$y_{ijk} = \mu + \alpha_i + \eta_j + \gamma_{ij} + e_{ijk},$$

$i = 1, 2, 3, 4$, $j = 1, 2, 3$, $k = 1, \ldots, N_{ij}$, where for $i \neq 1 \neq j$, $N_{ij} = N$, and $N_{11} = 2N$. This model could arise from an experimental design having α treatments of No Treatment (NT), a_1, a_2, a_3 and η treatments of NT, b_1, b_2. This gives a total of 12 treatments: NT, a_1, a_2, a_3, b_1, a_1b_1, a_2b_1, a_3b_1 b_2, a_1b_2, a_2b_2, and a_3b_2. Since NT is a control, it might be of interest to compare all of the treatments to NT. If NT is to play such an important role in the analysis, it is reasonable to take more observations on NT than on the other treatments. Find sums of squares for testing

(a) no differences between a_1, a_2, a_3,

(b) no differences between b_1, b_2,

(c) no $\{a_1, a_2, a_3\} \times \{b_1, b_2\}$ interaction,

(d) no differences between NT and the averages of a_1, a_2, and a_3 when there is interaction,

(e) no differences between NT and the average of a_1, a_2, and a_3 when there is no interaction present,

(f) no differences between NT and the average of b_1 and b_2 when there is interaction,

(g) no differences between NT and the average of b_1 and b_2 when there is no interaction present.

Discuss the orthogonality relationships among the sums of squares. For parts (e) and (g), use the assumption of no interaction. Do not just repeat parts (d) and (f)!

Exercise 7.7.4 Consider the linear model $y_{ij} = \mu + \alpha_i + \eta_j + e_{ij}, i = 1, \ldots, a, j = 1, \ldots, b$. As in Section 1, write $X = [X_0, X_1, \ldots, X_a, X_{a+1}, \ldots, X_{a+b}]$. If we write the observations in the usual order, we can use Kronecker products to write the model matrix. Write $X = [J, X_*, X_{**}]$, where $X_* = [X_1, \ldots, X_a]$, and $X_{**} = [X_{a+1}, \ldots, X_{a+b}]$. Using Kronecker products, $X_* = [I_a \otimes J_b]$, and $X_{**} = [J_a \otimes I_b]$. In fact, with $n = ab$, $J = J_n = [J_a \otimes J_b]$. Use Kronecker products to show that $X'_*(I - [1/n]J_n^n)X_{**} = 0$. In terms of Section 1, this is the same as showing that $C(Z_1, \ldots, Z_a) \perp C(Z_{a+1}, \ldots, Z_{a+b})$. Also show that $[(1/a)J_a^a \otimes I_b]$ is the perpendicular projection operator onto $C(X_{**})$ and that $M_\eta = [(1/a)J_a^a \otimes (I_b - (1/b)J_b^b)]$.

Exercise 7.7.5 Consider the balanced two-way ANOVA with interaction model $y_{ijk} = \mu + \alpha_i + \eta_j + \gamma_{ij} + e_{ijk}$, $i = 1, \ldots, a$, $j = 1, \ldots, b$, $k = 1, \ldots, N$, with e_{ijk}s independent $N(0, \sigma^2)$. Find $\mathrm{E}[Y'(\frac{1}{n}J_n^n + M_\alpha)Y]$ in terms of μ, the α_is, the η_js, and the γ_{ij}s.

Exercise 7.7.6 For Example 7.6.1, develop a test for $H_0 : A_{10} + A_{11} = A_{20} + A_{21}$.

Exercise 7.7.7 Assuming that the reader is familiar with the analysis of two-way tables of counts and treating the N_{ij}s as a two-way table of counts, show that proportional numbers hold if and only if the N_{ij} fit the model of independence perfectly which occurs if and only if all of the odds ratios in the table equal 1.

References

Bailey, D. W. (1953). *The inheritance of maternal influences on the growth of the rat*. Ph.D. thesis, University of California.

Benedetti, J. K., & Brown, M. B. (1978). Strategies for the selection of log-linear models. *Biometrics, 34*, 680–686.

Christensen, R. (1997). *Log-linear models and logistic regression* (2nd ed.). New York: Springer.

Christensen, R. (2015). *Analysis of variance, design, and regression: Linear modeling for unbalanced data* (2nd ed.). Boca Raton: Chapman and Hall/CRC Press.

Fienberg, S. E. (1980). *The analysis of cross-classified categorical data* (2nd ed.). Cambridge: MIT Press.

Scheffé, H. (1959). *The analysis of variance*. New York: Wiley.

Wermuth, N. (1976). Model search among multiplicative models. *Biometrics, 32*, 253–264.

Chapter 8
Experimental Design Models

Abstract In this chapter we examine three models used to analyze results obtained from three specific experimental designs. The designs are the completely randomized design, the randomized complete block design, and the Latin square design. We also examine a particularly effective method of defining treatments for situations in which several factors are of interest.

The design of an experiment is extremely important. It identifies the class of linear models that can be used to analyze the results of the experiment. Perhaps more accurately, the design identifies large classes of models that are inappropriate for the analysis of the experiment. Finding appropriate models can be difficult even in well-designed experiments.

Randomization, that is, the random application of treatments to experimental units, provides a philosophical justification for inferring that the treatments actually cause the experimental results. In addition to the models considered here, Appendix G examines estimation for completely randomized designs and randomized complete block designs under an alternative error structure that is derived from the act of randomization.

Blocking is one of the most important concepts in experimental design. Blocking is used to reduce variability for the comparison of treatments. Blocking consists of grouping experimental units that will act similarly into blocks and then (randomly) applying the treatments to the units in each block. A *randomized complete block design* is one in which each block has exactly the same number of units as there are treatments. Each treatment is randomly applied once in each block. If there are more units than treatments, we can have a complete block with replications. If there are fewer units than treatments, we have an *incomplete block design*. A major issue in the subject of experimental design is how one should assign treatments to blocks in an incomplete block design. Latin square designs are complete block designs that incorporate two separate forms of blocking.

R. Christensen, *Plane Answers to Complex Questions*, Springer Texts in Statistics,
https://doi.org/10.1007/978-3-030-32097-3_8

The analysis of balanced incomplete block designs is derived in Section 9.4. Alternative methods for blocking designs are examined in the exercises for Section 11.1. For an excellent discussion of the concepts underlying the design of experiments, see Fisher (1935) or Cox (1958). For more detailed discussion of the design and analysis of experiments, there are many good books including Kempthorne (1952), Cochran and Cox (1957), John (1971), Hinkelmann and Kempthorne (2005, 2008), Casella (2008), Wu and Hamada (2009), Oehlert (2010), or [blush, blush] Christensen (1996, 2015). Cox and Reid (2000) take an interesting and different tack to the subject. My website contains design material deleted from *PA-V*, *ALM-III*, and Christensen (2015) plus some new material (http://www.stat.unm.edu/~fletcher/TopicsInDesign).

8.1 Completely Randomized Designs

The simplest experimental design is the *completely randomized design* (CRD). It involves no blocking. The experimental technique is simply to decide how many observations are to be taken on each treatment, obtain an adequate number of experimental units, and apply the treatments to the units randomly. The standard model for this design is

$$y_{ij} = \mu + \alpha_i + e_{ij},$$

$i = 1, \dots, a$, $\quad j = 1, \dots, N_i$, $\quad \mathrm{E}(e_{ij}) = 0$, $\quad \mathrm{Var}(e_{ij}) = \sigma^2$, $\quad \mathrm{Cov}(e_{ij}, e_{i'j'}) = 0$ if $(i, j) \neq (i', j')$. This is a one-way ANOVA model and is analyzed as such.

8.2 Randomized Complete Block Designs: Usual Theory

The model usually assumed for a *randomized complete block design* (RCB) is

$$y_{ij} = \mu + \alpha_i + \beta_j + e_{ij},$$

$i = 1, \dots, a, j = 1, \dots, b, \mathrm{E}(e_{ij}) = 0, \mathrm{Var}(e_{ij}) = \sigma^2, \mathrm{Cov}(e_{ij}, e_{i'j'}) = 0$ if $(i, j) \neq (i', j')$. The β_js stand for an additive effect for each block; the α_is are an additive effect for each treatment. It is assumed that any block-treatment interaction is error so that an estimate of σ^2 is available. This randomized complete block model is just a two-way ANOVA without interaction.

In this chapter, we are presenting models that are generally used for analyzing experiments conducted with certain standard experimental designs. Many people believe that these models are useful only because they are good approximations to linear models derived from the random application of the treatments to experimental units. This randomization theory leads to the conclusion that there is no valid test for block effects. On the other hand, there is nothing in the two-way ANOVA model given above to keep one from testing block effects. It is my opinion that this contradiction arises simply because the two-way ANOVA model is not very appropriate for a randomized complete block design. For example, a basic idea of blocking is that results within a block should be more alike than results in different blocks. The two-way ANOVA, with only an additive effect for blocks, is a very simplistic model. Another key idea of blocking is that it reduces the variability of treatment comparisons. It is not clear how the existence of additive block effects reduces variability in a two-way ANOVA. The question of how blocking achieves variance reduction is addressed in Exercise 8.1. Exercises 11.4–11.6 discuss alternative models for complete block designs.

Appendix G defines linear models for completely randomized designs and randomized complete block designs that are based on randomization theory. It is shown, using Theorem 10.4.5 (or Proposition 2.7.5), that least squares estimates are BLUEs for the randomization theory models.

Exercise 8.1 Using a randomized complete block design is supposed to reduce the variability of treatment comparisons. If the randomized complete block model is taken as

$$y_{ij} = \mu + \alpha_i + \beta_j + e_{ij}, \quad e_{ij}\text{s i.i.d. } N(0, \sigma^2),$$

$i = 1, \ldots, a$, $j = 1, \ldots, b$, argue that the corresponding variance for a completely randomized design should be $\sigma^2 + \sum_{j=1}^{b}(\mu_j - \bar{\mu}_\cdot)^2/b$, where $\mu_i = \mu + \beta_j$.

Hint: Figure out what population a completely randomized design would have to be sampled from.

8.3 Latin Square Designs

A *Latin square* is a design that allows for treatment effects and two different kinds of block effects. The number of treatments must equal the number of blocks of each kind. On occasion, the design is used with two kinds of treatments (each with the same number of levels) and one block effect.

Example 8.3.1 A 4×4 Latin square has four treatments, say, T_1, T_2, T_3, T_4. Consider the block effects as row effects, R_1, R_2, R_3, and R_4 and column effects C_1, C_2, C_3, and C_4. We can diagram one example of a 4×4 Latin square as

	C_1	C_2	C_3	C_4
R_1	T_1	T_2	T_3	T_4
R_2	T_2	T_3	T_4	T_1
R_3	T_3	T_4	T_1	T_2
R_4	T_4	T_1	T_2	T_3

The key idea is that each treatment occurs once in each row and once in each column.

The model for an $a \times a$ Latin square design is

$$y_{ijk} = \mu + \alpha_i + \beta_j + \gamma_k + e_{ijk},$$

$\mathrm{E}(e_{ijk}) = 0$, $\mathrm{Var}(e_{ijk}) = \sigma^2$, $\mathrm{Cov}(e_{ijk}, e_{i'j'k'}) = 0$ if $(i, j, k) \neq (i', j', k')$. The correspondence between the rows, columns, and treatments and the effects in the model is: α_i is the effect for row R_i, β_j is the effect for column C_j, and γ_k is the effect for treatment T_k. *The subscripting in this model is unusual.* The key point is that in a Latin square if you know the row and the column, the Latin square design tells you what the treatment is, so the three subscripts do not vary freely. In particular, we can specify $i = 1, \ldots, a$, $j = 1, \ldots, a$, $k \in \{1, 2, \ldots, a\}$ and $k = f(i, j)$, where for each i, $f(i, j)$ is a one-to-one function of $\{1, 2, \ldots, a\}$ onto itself, and the same is true for each j. As in the case of the randomized complete block design, this model makes no distinction between treatment effects and the two sets of block effects.

To derive the analysis of the Latin square model, we need to show that after fitting μ, the spaces for the three main effects are orthogonal. Before proceeding, note again that the terms y_{ijk} are overindexed. There are a^2 terms but a^3 possible combinations of the indices. Any two of the indices serve to identify all of the observations. For example, the mean of all a^2 observations is

$$\bar{y}_{\cdot\cdot\cdot} = \frac{1}{a^2}\sum_{i=1}^{a}\sum_{j=1}^{a} y_{ijk} = \frac{1}{a^2}\sum_{i=1}^{a}\sum_{k=1}^{a} y_{ijk} = \frac{1}{a^2}\sum_{j=1}^{a}\sum_{k=1}^{a} y_{ijk}.$$

We will use triple index notation to describe the rows of the model matrix. Write the model matrix as $X = [X_0, X_1, \ldots, X_{3a}]$, where $X_0 = J$,

$$X_r = [u_{ijk}], \quad u_{ijk} = \delta_{ir}, \quad r = 1, \ldots, a,$$

$$X_{a+s} = [u_{ijk}], \quad u_{ijk} = \delta_{js}, \quad s = 1, \ldots, a,$$

$$X_{2a+t} = [u_{ijk}], \quad u_{ijk} = \delta_{kt}, \quad t = 1, \ldots, a.$$

Example 8.3.2 The model for the 4×4 Latin square of Example 8.3.1 is

$$\begin{bmatrix} y_{111} \\ y_{122} \\ y_{133} \\ y_{144} \\ y_{212} \\ y_{223} \\ y_{234} \\ y_{241} \\ y_{313} \\ y_{324} \\ y_{331} \\ y_{342} \\ y_{414} \\ y_{421} \\ y_{432} \\ y_{443} \end{bmatrix} = \begin{bmatrix} 1&1&0&0&0&1&0&0&0&1&0&0&0 \\ 1&1&0&0&0&0&1&0&0&0&1&0&0 \\ 1&1&0&0&0&0&0&1&0&0&0&1&0 \\ 1&1&0&0&0&0&0&0&1&0&0&0&1 \\ 1&0&1&0&0&1&0&0&0&0&1&0&0 \\ 1&0&1&0&0&0&1&0&0&0&0&1&0 \\ 1&0&1&0&0&0&0&1&0&0&0&0&1 \\ 1&0&1&0&0&0&0&0&1&1&0&0&0 \\ 1&0&0&1&0&1&0&0&0&0&0&1&0 \\ 1&0&0&1&0&0&1&0&0&0&0&0&1 \\ 1&0&0&1&0&0&0&1&0&1&0&0&0 \\ 1&0&0&1&0&0&0&0&1&0&1&0&0 \\ 1&0&0&0&1&1&0&0&0&0&0&0&1 \\ 1&0&0&0&1&0&1&0&0&1&0&0&0 \\ 1&0&0&0&1&0&0&1&0&0&1&0&0 \\ 1&0&0&0&1&0&0&0&1&0&0&1&0 \end{bmatrix} \begin{bmatrix} \mu \\ \alpha_1 \\ \alpha_2 \\ \alpha_3 \\ \alpha_4 \\ \beta_1 \\ \beta_2 \\ \beta_3 \\ \beta_4 \\ \gamma_1 \\ \gamma_2 \\ \gamma_3 \\ \gamma_4 \end{bmatrix} + e.$$

Orthogonalizing columns 1 to $3a$ of X with respect to J gives the matrix Z with columns

$$Z_0 = X_0$$
$$Z_i = X_i - \frac{X_i'J}{J'J}J = X_i - \frac{a}{a^2}J, \quad i = 1, \ldots, 3a.$$

The three spaces $C(Z_1, \ldots, Z_a)$, $C(Z_{a+1}, \ldots, Z_{2a})$, and $C(Z_{2a+1}, \ldots, Z_{3a})$ are orthogonal. For example, with $r = 1, \ldots, a$ and $t = 1, \ldots, a$,

$$\begin{aligned} Z_r'Z_{2a+t} &= \sum_{i=1}^{a}\sum_{k=1}^{a}\left(\delta_{ir} - \frac{1}{a}\right)\left(\delta_{kt} - \frac{1}{a}\right) \\ &= \sum_{i=1}^{a}\sum_{k=1}^{a}\delta_{ir}\delta_{kt} - \sum_{i=1}^{a}\sum_{k=1}^{a}\delta_{ir}/a - \sum_{i=1}^{a}\sum_{k=1}^{a}\delta_{kt}/a + \sum_{i=1}^{a}\sum_{k=1}^{a}1/a^2 \\ &= 1 - 1 - 1 + 1 \\ &= 0. \end{aligned}$$

Similar computations establish the other orthogonality relationships.

Because of the orthogonality, the sum of squares for dropping, say, the α_is from the model is just the sum of squares for dropping the α_is from a one-way ANOVA model that ignores the β_j and γ_k effects. The ANOVA table is

ANOVA

Source	df	SS	E(MS)
μ	1	$a^2\bar{y}^2_{\cdot\cdot\cdot}$	$\sigma^2 + a^2(\mu + \bar{\alpha}_\cdot + \bar{\beta}_\cdot + \bar{\gamma}_\cdot)^2$
α	$a-1$	$a\sum_{i=1}^a(\bar{y}_{i\cdot\cdot} - \bar{y}_{\cdot\cdot\cdot})^2$	$\sigma^2 + \frac{a}{a-1}\sum_{i=1}^a(\alpha_i - \bar{\alpha}_\cdot)^2$
β	$a-1$	$a\sum_{j=1}^a(\bar{y}_{\cdot j\cdot} - \bar{y}_{\cdot\cdot\cdot})^2$	$\sigma^2 + \frac{a}{a-1}\sum_{j=1}^a(\beta_j - \bar{\beta}_\cdot)^2$
γ	$a-1$	$a\sum_{k=1}^a(\bar{y}_{\cdot\cdot k} - \bar{y}_{\cdot\cdot\cdot})^2$	$\sigma^2 + \frac{a}{a-1}\sum_{k=1}^a(\gamma_k - \bar{\gamma}_\cdot)^2$
Error	$(a-2)(a-1)$	By subtraction	σ^2
Total	a^2	$\sum_{i=1}^a\sum_{j=1}^a y^2_{ijk}$	

Estimation and testing in one of the treatment (block) spaces depends only on the appropriate projection operator. Since we have the usual one-way ANOVA projection operators, estimation and testing are performed in the usual way.

The Latin square model assumes that the $(\alpha\beta)$, $(\alpha\gamma)$, $(\beta\gamma)$, and $(\alpha\beta\gamma)$ interactions are nonexistent. This assumption is necessary in order to obtain an estimate of error. If an outside estimate of σ^2 is available, it might be hoped that the interactions could be examined. Unfortunately, it is impossible to tell from which interaction the degrees of freedom called "error" come from. For example, in the 4×4 Latin square of the example, the $(\alpha\beta)$ interaction can be broken up into 3 degrees of freedom for γ and 6 degrees of freedom for error. Since a similar result holds for each of the interactions, the 6 degrees of freedom for error involve all of the interactions.

Exercise 8.2 In the 4×4 Latin square of the examples, show that the 9 degrees of freedom for $(\alpha\beta)$ interaction are being divided into 3 degrees of freedom for γ and 6 degrees of freedom for error.

A *Graeco–Latin square* is a Latin square in which a second group of a treatments has been applied so that each treatment in the second group occurs once in each row, once in each column, and once with each of the treatments from the first group.

Exercise 8.3 Derive the analysis for the Graeco–Latin square given below. Use the model $y_{hijk} = \mu + \alpha_h + \beta_i + \gamma_j + \eta_k + e_{hijk}$.

	C_1	C_2	C_3	C_4	C_5
R_1	$T_1\tau_1$	$T_2\tau_3$	$T_3\tau_5$	$T_4\tau_2$	$T_5\tau_4$
R_2	$T_2\tau_2$	$T_3\tau_4$	$T_4\tau_1$	$T_5\tau_3$	$T_1\tau_5$
R_3	$T_3\tau_3$	$T_4\tau_5$	$T_5\tau_2$	$T_1\tau_4$	$T_2\tau_1$
R_4	$T_4\tau_4$	$T_5\tau_1$	$T_1\tau_3$	$T_2\tau_5$	$T_3\tau_2$
R_5	$T_5\tau_5$	$T_1\tau_2$	$T_2\tau_4$	$T_3\tau_1$	$T_4\tau_3$

Extend the analysis to an arbitrary $a \times a$ Graeco–Latin square.

8.4 Factorial Treatment Structures

For each experimental design considered in this chapter, we have assumed the existence of "a" treatments. Sometimes the treatments are chosen in such a way that the treatment space can be conveniently broken into orthogonal subspaces. One of the most common methods of doing this is to choose treatments with factorial structure.

Suppose that two or more different kinds of treatments are of interest. Each kind of treatment is called a *factor*. Each factor is of interest at some number of different levels. A very efficient way of gaining information on all of the levels of all of the factors is to use what is called a *factorial design*. In a factorial design the treatments are taken to be all possible combinations of the levels of the different factors. Since a factorial design refers only to the treatment structure, factorial designs can be used with all of the designs considered in this chapter as well as with balanced incomplete block designs (cf. Section 9.4) and split plot designs (cf. Section 11.3).

Example 8.4.1 An experiment is to be conducted examining the effects of fertilizer on potato yields. Of interest are two kinds of fertilizer, a nitrogen-based fertilizer and a phosphate-based fertilizer. The two types of fertilizer are factors. The nitrogen fertilizer is to be examined at two levels: no nitrogen fertilizer (n_0) and a single dose of nitrogen fertilizer (n_1). The phosphate fertilizer has three levels: no phosphate fertilizer (p_0), a single dose of phosphate (p_1), and a double dose of phosphate (p_2). The treatments are taken to be all six of the possible combinations:

$$n_0p_0 \quad n_0p_1 \quad n_0p_2 \quad n_1p_0 \quad n_1p_1 \quad n_1p_2.$$

The use of treatments with factorial structure has a number of advantages. One is that it allows study of the interrelationships (interactions) between the factors. In Example 8.4.1, it is possible to examine whether the levels of phosphate have different effects depending on whether or not nitrogen was applied. Another advantage is that if there are no interactions, the experimental material is used very efficiently. Suppose that there is no interaction between nitrogen and phosphate. The effect of nitrogen is the difference in yields between experimental units that have the same level of phosphate but different levels of nitrogen, that is,

$$n_0p_0 - n_1p_0,$$

$$n_0p_1 - n_1p_1,$$

and

$$n_0p_2 - n_1p_2.$$

In each case, the only difference in the pair of treatments is the difference in nitrogen. At the same time, the difference in, say, the effects of p_1 and p_2 can be examined by looking at the differences

$$n_0 p_1 - n_0 p_2$$

and

$$n_1 p_1 - n_1 p_2.$$

The same treatments are used to obtain estimates of both the nitrogen effect and the phosphate effect.

We now examine the analysis of designs having factorial treatment structure. Consider a randomized complete block design with c blocks, where the treatments are all combinations of two factors, say A and B. Suppose factor A has a levels and factor B has b levels. The total number of treatments is ab. Rewriting the model of Section 2 with τ_h denoting treatment effects and ξ_k denoting block effects, we get

$$y_{hk} = \mu + \tau_h + \xi_k + e_{hk}, \tag{1}$$

$h = 1, \ldots, ab$, $k = 1, \ldots, c$. In this model, each treatment is indicated by an index h. Since the treatments consist of all combinations of two factors, it makes sense to use two indices to identify treatments: one index for each factor. With this idea we can rewrite model (1) as

$$y_{ijk} = \mu + \tau_{ij} + \xi_k + e_{ijk}, \tag{2}$$

$i = 1, \ldots, a$, $j = 1, \ldots, b$, $k = 1, \ldots, c$.

Exercise 8.4 For model (1), $C(M_\tau)$ is given by Proposition 4.2.3. What is $C(M_\tau)$ in the notation of model (2)?

The orthogonal breakdown of the treatment space follows from a reparameterization of model (2). Model (2) can be rewritten as

$$y_{ijk} = \mu + \alpha_i + \beta_j + (\alpha\beta)_{ij} + \xi_k + e_{ijk}. \tag{3}$$

The new parameterization is simply $\tau_{ij} = \alpha_i + \beta_j + (\alpha\beta)_{ij}$. Using the ideas of Sections 7.1 and 7.2, the treatment space $C(M_\tau)$ can be broken up into three orthogonal subspaces: one for factor A, $C(M_\alpha)$, one for factor B, $C(M_\beta)$, and one for interaction, $C(M_{\alpha\beta})$. The analysis of model (3) follows along the lines of Chapter 7. Model (3) is just a balanced three-way ANOVA in which some of the interactions (namely, any interactions that involve blocks) have been thrown into the error.

In practice, it is particularly important to be able to relate contrasts in the treatments (τ_{ij}s) to contrasts in the main effects (α_is and β_js) and contrasts in the interactions ($(\alpha\beta)_{ij}$s). The relationship is demonstrated in Exercise 8.5. The relationship is illustrated in the following example.

	n_0	n_1
N	1	−1

Example 8.4.2 Consider the treatments of Example 8.4.1. There is one contrast in nitrogen,

The two orthogonal polynomial contrasts in phosphate are:

	p_0	p_1	p_2
P linear	−1	0	1
P quadratic	1	−2	1

The main effect contrasts define two orthogonal interaction contrasts: $N - P$ linear

	p_0	p_1	p_2
n_0	−1	0	1
n_1	1	0	−1

and $N - P$ quadratic

	p_0	p_1	p_2
n_0	1	−2	1
n_1	−1	2	−1

The corresponding contrasts in the six treatments are:

	n_0p_0	n_0p_1	n_0p_2	n_1p_0	n_1p_1	n_1p_2
N	1	1	1	−1	−1	−1
P linear	−1	0	1	−1	0	1
P quadratic	1	−2	1	1	−2	1
$N - P$ linear	−1	0	1	1	0	−1
$N - P$ quadradic	1	−2	1	−1	2	−1

It is easily verified that these five contrasts are orthogonal.

Exercise 8.5 Show that the contrasts in the τ_{ij}s corresponding to the contrasts $\sum \lambda_i \alpha_i$ and $\sum\sum \lambda_i \eta_j (\alpha\beta)_{ij}$ are $\sum\sum \lambda_i \tau_{ij}$ and $\sum\sum \lambda_i \eta_j \tau_{ij}$, respectively.

Hint: Any contrast in the τ_{ij}s corresponds to a vector in $C(M_\tau)$, just as any contrast in the α_is corresponds to a vector in $C(M_\alpha) \subset C(M_\tau)$. Recall that contrasts are only defined up to constant multiples; and that contrasts in the α_is also involve the interactions when interaction exists.

8.5 More on Factorial Treatment Structures

We now present a more theoretical discussion of factorial treatment structures and some interesting new models.

For two factors, say α at s levels and η at t levels, there are a total of $p \equiv st$ treatment groups. Start by considering a linear model for the data $y_{ijk} = (\alpha\eta)_{ij} + e_{ijk}$. In matrix form, the linear model $Y = X\beta + e$ has X as the indicator matrix used for a one-way ANOVA (cell means) model and $\beta = [(\alpha\eta)_{11}, \ldots, (\alpha\eta)_{st}]'$ containing a separate effect for every combination of the two factors.

More generally, if we have r factors α_i with s_i levels respectively, the factorial structure defines $p = \prod_{i=1}^{r} s_i$ groups. We again take X to be the one-way ANOVA indicator matrix and define $\beta = [\alpha_{i_1,\ldots,i_r}]$ as a vector providing a separate effect for each combination of the factors.

The idea is that factorial models are defined by specifying a linear structure for the group parameters. This consists of putting a linear constraint on β, say $\beta = U_{p\times q}\gamma$ for some known matrix U, cf. Section 3.3. For example, in the two-factor model, one such linear constraint is to force an additive main effects model for the data, $y_{ijk} = \mu + \alpha_i + \eta_j + e_{ijk}$. In matrix terms using Kronecker products, this amounts to specifying that

$$\beta = \Big([J_s \otimes J_t], [I_s \otimes J_t], [J_s \otimes I_t]\Big) \begin{bmatrix} \mu \\ \alpha \\ \eta \end{bmatrix} \equiv U\gamma, \tag{1}$$

where $\alpha = (\alpha_1, \ldots, \alpha_s)'$ and $\eta = (\eta_1, \ldots, \eta_t)'$. As in Section 3.3, the linear model for the data has been transformed to $Y = XU\gamma + e$, where the model matrix is now XU, which is a reduced model relative to the original, i.e., $C(XU) \subset C(X)$. Note that if we define a linear structure for the parameters, $\beta = U\gamma$ and a reduced structure, i.e., $\beta = U_0\gamma_0$, where $C(U_0) \subset C(U)$, then this also defines a reduced linear model for the data in that $C(XU_0) \subset C(XU)$.

I have no difficulty considering all such models to be factorial models, but McCullagh (2000) proposes a more stringent definition involving "selection invariance." The idea is that if you drop various indices from various factors, the model should somehow remain invariant. This idea can be executed in the following way: Begin with a linear structure $\beta = U\gamma$ but partition U into sets of columns $U = [U_0, U_1, \ldots, U_m]$ and then specify that

$$U_i = [V_{i1} \otimes V_{i2} \otimes \cdots \otimes V_{ir}]. \tag{2}$$

Such structures should be sufficient to satisfy the basic idea of selection invariance. However, as will be seen later, other interesting selection invariant factorial models require us to consider matrices U_i that are linear combinations of matrices with the structure (2).

An interesting aspect of factorial structures is dealing with factors that have the same levels. Such factors are called *homologous*. Example 7.5.1 involves genotypes of mothers and genotypes of litters, but the genotypes are identical for the mothers and the litters, so it provides an example of homologous factors. In fact, Christensen (2015, Section 14.4) uses that data to illustrate fitting the homologous factor models discussed below.

In the additive two-factor model $y_{ijk} = \mu + \alpha_i + \eta_j + e_{ijk}$ with $s = t$ and homologous factors, we might consider situations such as *symmetric additive effects* where $y_{ijk} = \mu + \alpha_i + \alpha_j + e_{ijk}$ or *alternating (skew symmetric) additive effects* where $y_{ijk} = \mu + \alpha_i - \alpha_j + e_{ijk}$. As illustrated in (1), we can write the additive model for the parameters in matrix form as

$$\beta = U\gamma = [J_{st}, U_1, U_2] \begin{bmatrix} \mu \\ \alpha \\ \eta \end{bmatrix}.$$

We can now specify symmetric additive effects by specifying $\alpha = \eta$ to get

$$\beta = [J_{st}, U_1, U_2] \begin{bmatrix} \mu \\ \alpha \\ \alpha \end{bmatrix} = [J_{st}, (U_1 + U_2)] \begin{bmatrix} \mu \\ \alpha \end{bmatrix},$$

thus defining the linear model $Y = X[J_{st}, (U_1 + U_2)] \begin{bmatrix} \mu \\ \alpha \end{bmatrix} + e$. Similarly, specifying alternating additive effects $\alpha = -\eta$ leads to $Y = X[J_{st}, (U_1 - U_2)] \begin{bmatrix} \mu \\ \alpha \end{bmatrix} + e$. In a 3×3 example with $\beta = [(\alpha\eta)_{11}, (\alpha\eta)_{12}, \ldots, (\alpha\eta)_{33}]'$, the additive main effects model has

$$U = [J_{st}, U_1, U_2] = \begin{bmatrix} 1 & 1 & 0 & 0 & 1 & 0 & 0 \\ 1 & 1 & 0 & 0 & 0 & 1 & 0 \\ 1 & 1 & 0 & 0 & 0 & 0 & 1 \\ 1 & 0 & 1 & 0 & 1 & 0 & 0 \\ 1 & 0 & 1 & 0 & 0 & 1 & 0 \\ 1 & 0 & 1 & 0 & 0 & 0 & 1 \\ 1 & 0 & 0 & 1 & 1 & 0 & 0 \\ 1 & 0 & 0 & 1 & 0 & 1 & 0 \\ 1 & 0 & 0 & 1 & 0 & 0 & 1 \end{bmatrix}.$$

The symmetric additive effects and alternating additive effects models have

$$[J_{st}, U_1 + U_2] = \begin{bmatrix} 1 & 2 & 0 & 0 \\ 1 & 1 & 1 & 0 \\ 1 & 1 & 0 & 1 \\ 1 & 1 & 1 & 0 \\ 1 & 0 & 2 & 0 \\ 1 & 0 & 1 & 1 \\ 1 & 1 & 0 & 1 \\ 1 & 0 & 1 & 1 \\ 1 & 0 & 0 & 2 \end{bmatrix}, \quad [J_{st}, (U_1 - U_2)] = \begin{bmatrix} 1 & 0 & 0 & 0 \\ 1 & 1 & -1 & 0 \\ 1 & 1 & 0 & -1 \\ 1 & -1 & 1 & 0 \\ 1 & 0 & 0 & 0 \\ 1 & 0 & 1 & -1 \\ 1 & -1 & 0 & 1 \\ 1 & 0 & -1 & 1 \\ 1 & 0 & 0 & 0 \end{bmatrix},$$

respectively. Given the simple structure of the original one-way ANOVA matrix X, the reduced model matrices $X[J_{st}, (U_1 + U_2)]$ and $X[J_{st}, (U_1 - U_2)]$ have structures very similar to $[J_{st}, (U_1 + U_2)]$ and $[J_{st}, (U_1 - U_2)]$. However, these linear structures for the parameters are not of the form (2), hence the need to consider linear combinations of terms like those in (2).

Models specifying such things as simple symmetry $(\alpha\eta)_{ij} = (\alpha\eta)_{ji}$ can also be specified quite easily by defining an appropriate U matrix, e.g.,

$$\begin{bmatrix} (\alpha\eta)_{11} \\ (\alpha\eta)_{12} \\ (\alpha\eta)_{13} \\ (\alpha\eta)_{21} \\ (\alpha\eta)_{22} \\ (\alpha\eta)_{23} \\ (\alpha\eta)_{31} \\ (\alpha\eta)_{32} \\ (\alpha\eta)_{33} \end{bmatrix} = \beta = U\phi = \begin{bmatrix} 1 & 0 & 0 & 0 & 0 & 0 \\ 0 & 1 & 0 & 0 & 0 & 0 \\ 0 & 0 & 1 & 0 & 0 & 0 \\ 0 & 1 & 0 & 0 & 0 & 0 \\ 0 & 0 & 0 & 1 & 0 & 0 \\ 0 & 0 & 0 & 0 & 1 & 0 \\ 0 & 0 & 1 & 0 & 0 & 0 \\ 0 & 0 & 0 & 0 & 1 & 0 \\ 0 & 0 & 0 & 0 & 0 & 1 \end{bmatrix} \begin{bmatrix} \phi_{11} \\ \phi_{12} \\ \phi_{13} \\ \phi_{22} \\ \phi_{23} \\ \phi_{33} \end{bmatrix}.$$

These also fit into the class of linear combinations of matrices with the pattern (2), e.g., the column of U associated with ϕ_{23} can be written as

$$\left(\begin{bmatrix} 0 \\ 1 \\ 0 \end{bmatrix} \otimes \begin{bmatrix} 0 \\ 0 \\ 1 \end{bmatrix} \right) + \left(\begin{bmatrix} 0 \\ 0 \\ 1 \end{bmatrix} \otimes \begin{bmatrix} 0 \\ 1 \\ 0 \end{bmatrix} \right).$$

These ideas also apply to generalized linear models. For example, in log-linear models, symmetry is sometimes an interesting model, cf. Christensen (1997, Exercise 2.7.10), and the symmetric additive effects model is a (not the) model of marginal homogeneity, cf. Christensen (1997, Exercise 10.8.6). See McCullagh (2000) for a more extensive and theoretical treatment of these ideas.

8.6 Additional Exercises

Exercise 8.6.1 A study was performed to examine the effect of two factors on increasing muscle mass in weight lifters. The first factor was dietary protein level. The levels were use of a relatively low protein diet (L) and use of a high protein diet (H). The second factor was the use of anabolic steroids. The first level consisted of no steroid use (N) and the second level involved the use of steroids (S). Subjects were chosen so that one subject was in each combination of four height groups (i) and four weight groups (j). Treatments are identified as *LN*, *LS*, *HN*, and *HS*. The dependent variable is a measure of increase in muscle mass during the treatment period. The study was replicated in two different years. The height groups and weight groups changed from the first year to the second year. The design and data are listed below. Heights are the columns of the squares and weights are the rows. Analyze the data.

Year 1

	Trt(y_{ijk})	Weight 1	2	3	4
	1	*LN*(2.7)	*LS*(4.6)	*HS*(9.3)	*HN*(0.0)
Height	2	*LS*(2.0)	*LN*(5.0)	*HN*(9.1)	*HS*(4.5)
	3	*HN*(6.4)	*HS*(10.2)	*LS*(6.1)	*LN*(2.9)
	4	*HS*(8.3)	*HN*(6.3)	*LN*(6.3)	*LS*(0.9)

Year 2

	Trt(y_{ijk})	Weight 1	2	3	4
	1	*LN*(8.6)	*LS*(3.3)	*HN*(7.6)	*HS*(9.0)
Height	2	*LS*(8.9)	*HN*(4.7)	*HS*(12.2)	*LN*(5.2)
	3	*HN*(10.0)	*HS*(7.6)	*LN*(12.3)	*LS*(5.4)
	4	*HS*(10.0)	*LN*(0.5)	*LS*(5.0)	*HN*(3.7)

Exercise 8.6.2 Show that the set of indices $i = 1, \dots, a$, $j = 1, \dots, a$, and $k = (i + j + a - 1)\,\mathrm{mod}(a)$ determines a Latin square design.

Hint: Recall that $t\,\mathrm{mod}(a)$ means t modulo a and is defined as the remainder when t is divided by a.

References

Casella, G. (2008). *Statistical design*. New York: Springer.
Christensen, R. (1996). *Analysis of variance, design, and regression: Applied statistical methods*. London: Chapman and Hall.
Christensen, R. (1997). *Log-linear models and logistic regression* (2nd ed.). New York: Springer.

Christensen, R. (2015). *Analysis of variance, design, and regression: Linear modeling for unbalanced data* (2nd ed.). Boca Raton: Chapman and Hall/CRC Press.
Cochran, W. G., & Cox, G. M. (1957). *Experimental designs* (2nd ed.). New York: Wiley.
Cox, D. R. (1958). *Planning of experiments*. New York: Wiley.
Cox, D. R., & Reid, N. (2000). *The theory of the design of experiments*. Boca Raton: Chapman and Hall/CRC.
Fisher, R. A. (1935). *The design of experiments*, (9th ed., 1971). New York: Hafner Press.
Hinkelmann, K., & Kempthorne, O. (2005). *Design and analysis of experiments: Volume 2, advanced experimental design*. Hoboken: Wiley.
Hinkelmann, K., & Kempthorne, O. (2008). *Design and analysis of experiments: Volume 1, introduction to experimental design* (2nd ed.). Hoboken: Wiley.
John, P. W. M. (1971). *Statistical design and analysis of experiments*. New York: Macmillan.
Kempthorne, O. (1952). *Design and analysis of experiments*. Huntington: Krieger.
McCullagh, P. (2000). Invariance and factorial models, with discussion. *Journal of the Royal Statistical Society, Series B, 62*, 209–238.
Oehlert, G. W. (2010). *A first course in design and analysis of experiments*. http://users.stat.umn.edu/~gary/book/fcdae.pdf.
Wu, C. F. J., & Hamada, M. S. (2009). *Experiments: Planning, analysis, and optimization* (2nd ed.). New York: Wiley.

Chapter 9
Analysis of Covariance

Abstract This chapter examines the analysis of partition models, also known as analysis of covariance, a method traditionally used for improving the analysis of designed experiments. Sections 1 and 2 present the theory of estimation and testing for general partitioned models. Sections 3 and 4 present nontraditional applications of the theory. Section 3 applies the partitioned model results to the problem of fixing up balanced ANOVA problems that have lost their balance due to the existence of some missing data. Although applying analysis of covariance to missing data problems is not a traditional experimental design application, it is an application that was used for quite some time until computational improvements made it largely unnecessary. Section 4 uses the analysis of covariance results to derive the analysis for balanced incomplete block designs. Section 5 presents Milliken and Graybill's (1970) test of a linear model versus a nonlinear alternative. I personally find the techniques of Sections 1 and 2 to be some of the most valuable tools available for deriving results in linear model theory.

Traditionally, analysis of covariance (ACOVA) has been used as a tool in the analysis of designed experiments. Suppose one or more measurements are made on a group of experimental units. In an agricultural experiment, such a measurement might be the amount of nitrogen in each plot of ground prior to the application of any treatments. In animal husbandry, the measurements might be the height and weight of animals before treatments are applied. One way to use such information is to create blocks of experimental units that have similar values of the measurements. Analysis of covariance uses a different approach. In analysis of covariance, an experimental design is chosen that does not depend on these supplemental observations. The concomitant observations come into play as regression variables that are added to the basic experimental design model.

The goal of analysis of covariance is the same as the goal of blocking. The regression variables are used to *reduce the variability of treatment comparisons*. In this traditional context, comparisons among treatments remain the primary goal of the analysis. Exercises 9.1 and 9.5 are important practical illustrations of how this is

R. Christensen, *Plane Answers to Complex Questions*, Springer Texts in Statistics,
https://doi.org/10.1007/978-3-030-32097-3_9

accomplished. Snedecor and Cochran (1980, Chapter 18) discuss the practical uses of analysis of covariance. Cox (1958, Chapter 4) discusses the proper role of concomitant observations in experimental design. *Biometrics* has devoted two entire issues to analysis of covariance: Volume 13, Number 3, 1957 and Volume 38, Number 3, 1982.

From a theoretical point of view, analysis of covariance involves the analysis of a model with a *partitioned model* matrix, say

$$Y = [X, Z]\begin{bmatrix}\beta \\ \gamma\end{bmatrix} + e, \tag{1}$$

where X is an $n \times p$ matrix, Z is an $n \times s$ matrix, $\mathrm{E}(e) = 0$, and $\mathrm{Cov}(e) = \sigma^2 I$. Analysis of covariance is a technique for analyzing model (1) based on the analysis of the reduced model

$$Y = X\delta + e. \tag{2}$$

The point is that model (2) should be a model whose analysis is relatively easy. In traditional applications, X is taken as the model matrix for a balanced analysis of variance. The Z matrix can be anything, but traditionally consists of columns of regression variables. In essence we generalize and deepen the partitioning ideas used in Subsection 6.2.1.

The practical application of general linear model theory is prohibitively difficult without a computer program to perform the worst of the calculations. There are, however, special cases: notably, simple linear regression, one-way ANOVA, and balanced multifactor ANOVA, in which the calculations are not prohibitive. Analysis of covariance allows computations of the BLUEs and the *SSE* for model (1) by performing several analyses on tractable special cases plus finding the generalized inverse of an $s \times s$ matrix. Since finding the generalized inverse of anything bigger than, say, a 3×3 matrix is difficult for hand calculations, one would typically not want more than three columns in the Z matrix for such purposes.

As mentioned earlier, in the traditional application of performing an ANOVA while adjusting for the effect of some regression variables, the primary interest is in the ANOVA. The regression variables are there only to sharpen the analysis. The inference on the ANOVA part of the model is performed after fitting the regression variables. To test whether the regression variables really help to sharpen the analysis, they should be tested after fitting the ANOVA portion of the model. The basic computation for performing these tests is finding the *SSE* for model (1). This implicitly provides a method for finding the *SSE* for submodels of model (1). Appropriate tests are performed by comparing the *SSE* for model (1) to the *SSE*s of the various submodels.

9.1 Estimation of Fixed Effects

To obtain least squares estimates, we break the estimation space of model (9.0.1) into two orthogonal parts. As usual, let M be the perpendicular projection operator onto $C(X)$. Note that $C(X, Z) = C[X, (I - M)Z]$. One way to see this is that from model (9.0.1)

$$\mathrm{E}(Y) = X\beta + Z\gamma = X\beta + MZ\gamma + (I - M)Z\gamma = [X, MZ]\begin{bmatrix}\beta\\ \gamma\end{bmatrix} + (I - M)Z\gamma.$$

Since $C(X) = C([X, MZ])$, clearly model (9.0.1) holds if and only if $\mathrm{E}(Y) \in C[X, (I - M)Z]$.

Let $\mathscr{P}$ denote the perpendicular projection matrix onto $C([X, Z]) = C([X, (I - M)Z])$. Since the two sets of column vectors in $[X, (I - M)Z]$ are orthogonal, the perpendicular projection matrix for the entire space is the sum of the perpendicular projection matrices for the subspaces $C(X)$ and $C[(I - M)Z]$, cf. Theorem B.45. Thus,

$$\mathscr{P} = M + (I - M)Z\left[Z'(I - M)Z\right]^{-} Z'(I - M)$$

and write

$$M_2 \equiv (I - M)Z\left[Z'(I - M)Z\right]^{-} Z'(I - M).$$

Least squares estimates satisfy $X\hat{\beta} + Z\hat{\gamma} = \mathscr{P}Y = MY + M_2Y$.

We now consider estimation of estimable functions of γ and β. The formulae are simpler if we incorporate the estimate

$$\hat{\gamma} \equiv \left[Z'(I - M)Z\right]^{-} Z'(I - M)Y, \tag{1}$$

which we will later show to be a least squares estimate.

First consider an estimable function of γ, say, $\xi'\gamma$. For this to be estimable, there exists a vector ρ such that $\xi'\gamma = \rho'[X\beta + Z\gamma]$. For this equality to hold for all β and γ, we must have $\rho'X = 0$ and $\rho'Z = \xi'$. The least squares estimate of $\xi'\gamma$ is

$$\begin{aligned}\rho'\mathscr{P}Y &= \rho'\{M + M_2\}Y\\ &= \rho'MY + \rho'(I - M)Z\left[Z'(I - M)Z\right]^{-} Z'(I - M)Y\\ &= 0 + \rho'Z\hat{\gamma}\\ &= \xi'\hat{\gamma}.\end{aligned}$$

The penultimate equality stems from the fact that $\rho'X = 0$ implies $\rho'M = 0$.

An arbitrary estimable function of β, say, $\lambda'\beta$, has $\lambda'\beta = \rho'[X\beta + Z\gamma]$ for some ρ. For this equality to hold for all β and γ, we must have $\rho'X = \lambda'$ and $\rho'Z = 0$. As a result, the least squares estimate is

$$\begin{aligned}\rho'\mathscr{P}Y &= \rho'\{M + M_2\}Y\\ &= \rho'MY + \rho'(I - M)Z\left[Z'(I - M)Z\right]^{-} Z'(I - M)Y\\ &= \rho'MY + \rho'(I - M)Z\hat{\gamma}\\ &= \rho'MY - \rho'MZ\hat{\gamma}\\ &= \rho'M(Y - Z\hat{\gamma})\\ &\equiv \lambda'\hat{\beta}.\end{aligned}$$

Define

$$X\hat{\beta} \equiv M(Y - Z\hat{\gamma}). \tag{2}$$

We now establish that

$$X\hat{\beta} + Z\hat{\gamma} = \mathscr{P}Y$$

so that $\hat{\gamma}$ is a least squares estimate of γ and and $X\hat{\beta}$ is a least squares estimate of $X\beta$. Write

$$\begin{aligned} X\hat{\beta} + Z\hat{\gamma} &= M(Y - Z\hat{\gamma}) + Z\hat{\gamma} \\ &= MY + (I - M)Z\hat{\gamma} \\ &= MY + M_2Y = \mathscr{P}Y. \end{aligned}$$

Often in ACOVA, the X matrix comes from a model that is simple to analyze, like one-way ANOVA or a balanced multifactor ANOVA. If the model $Y = X\beta + e$ has simple formula for computing an estimate of some function $\lambda'\beta = \rho'X\beta$, say a contrast, then that simple formula must be incorporated into $\rho'MY$. Under conditions that we will explore, $\lambda'\beta$ is also an estimable function under the ACOVA model $Y = X\beta + Z\gamma + e$ and, with the same vector ρ, the estimate is $\lambda'\hat{\beta} = \rho'M(Y - Z\hat{\gamma})$. That means that the same (simple) computational procedure that was applied to the data Y in order to estimate $\lambda'\beta$ in $Y = X\beta + e$ can also be applied to $Y - Z\hat{\gamma}$ to estimate $\lambda'\beta$ in $Y = X\beta + Z\gamma + e$, see Exercise 9.1. We now explore the conditions necessary to make this happen, along with other issues related to estimability in ACOVA models.

Often Z consists of columns of regression variables, in which case $Z'(I - M)Z$ is typically nonsingular. In this *nonsingular case*, both γ and $X\beta$ are estimable. In particular,

$$\gamma = \left[Z'(I - M)Z\right]^{-1} Z'(I - M)[X\beta + Z\gamma]$$

with estimate

$$\begin{aligned} \hat{\gamma} &= \left[Z'(I - M)Z\right]^{-1} Z'(I - M)\mathscr{P}Y \\ &= \left[Z'(I - M)Z\right]^{-1} Z'(I - M)[M + M_2]Y \\ &= \left[Z'(I - M)Z\right]^{-1} Z'(I - M)M_2Y \\ &= \left[Z'(I - M)Z\right]^{-1} Z'(I - M)Y. \end{aligned}$$

X is traditionally the model matrix for an ANOVA model, so β is usually not estimable. However, when $Z'(I - M)Z$ is nonsingular, $X\beta$ is estimable in the ACOVA model. Observe that

$$\begin{aligned} &\left\{I - Z\left[Z'(I - M)Z\right]^{-1} Z'(I - M)\right\}[X\beta + Z\gamma] \\ &= X\beta + Z\gamma - Z\left[Z'(I - M)Z\right]^{-1} Z'(I - M)Z\gamma \\ &= X\beta + Z\gamma - Z\gamma \\ &= X\beta. \end{aligned}$$

Thus, in the nonsingular case, anything that is estimable in $Y = X\beta + e$ is also estimable in the ACOVA model. In particular, if $\lambda' = \rho' X$, the estimate of $\lambda'\beta$ in $Y = X\beta + e$ is $\rho' MY$. Clearly, for the ACOVA model,

$$\rho' \left\{ I - Z\left[Z'(I-M)Z\right]^{-1} Z'(I-M) \right\} [X\beta + Z\gamma] = \rho' X\beta = \lambda'\beta$$

and the estimate is

$$\lambda'\hat{\beta} = \rho' \left\{ I - Z\left[Z'(I-M)Z\right]^{-1} Z'(I-M) \right\} \mathscr{P} Y.$$

We now show that $\lambda'\hat{\beta} = \rho' M(Y - Z\hat{\gamma})$. As mentioned earlier, the beauty of this result is that if we know how to estimate $\lambda'\beta$ in $Y = X\beta + e$ using Y, exactly the same method applied to $Y - Z\hat{\gamma}$ will give the estimate in the ACOVA model.

As discussed earlier, an estimable function $\lambda'\beta$, has $\lambda'\beta = \tilde{\rho}'[X\beta + Z\gamma]$ for some $\tilde{\rho}$ with $\tilde{\rho}' X = \lambda'$ and $\tilde{\rho}' Z = 0$. Also as before, the least squares estimate is

$$\tilde{\rho}' \mathscr{P} Y = \tilde{\rho}' M(Y - Z\hat{\gamma}).$$

In the nonsingular case, if ρ is any vector that has $\rho' X = \lambda'$, we can turn it into a vector $\tilde{\rho}$ that has both $\tilde{\rho}' X = \lambda'$ and $\tilde{\rho}' Z = 0$, simply by defining

$$\tilde{\rho}' = \rho' \left\{ I - Z\left[Z'(I-M)Z\right]^{-1} Z'(I-M) \right\}.$$

Moreover,

$$\tilde{\rho}' M(Y - Z\hat{\gamma}) = \rho' M(Y - Z\hat{\gamma}),$$

so the same estimation procedure applied to Y in $Y = X\beta + e$ gets applied to $Y - Z\hat{\gamma}$ in $Y = X\beta + Z\gamma + e$ when estimating the estimable function $\lambda'\beta$.

In general, if $\left[Z'(I-M)Z\right]$ is singular, neither γ nor $X\beta$ are estimable. The estimable functions of γ will be those that are linear functions of $(I-M)Z\gamma$. This is shown below.

Proposition 9.1.1 *$\xi'\gamma$ is estimable if and only if $\xi' = \rho'(I-M)Z$ for some vector ρ.*

Proof If $\xi'\gamma$ is estimable, there exists ρ such that $\xi'\gamma = \rho'[X, Z]\begin{bmatrix} \beta \\ \gamma \end{bmatrix}$, so $\rho'[X, Z] = (0, \xi')$ and $\rho' X = 0$. Therefore, $\xi' = \rho' Z = \rho'(I-M)Z$. Conversely, if $\xi' = \rho'(I-M)Z$ then $\rho'(I-M)[X, Z]\begin{bmatrix} \beta \\ \gamma \end{bmatrix} = \xi'\gamma$. □

Proposition 9.1.1 is phrased in terms of estimating a function of γ, but it also applies with appropriate changes to estimation of β.

Finally, if $X\beta$ and γ are estimable, that is, if $(I-M)Z$ is of full rank, it is easy to see that

$$\text{Cov}\begin{bmatrix} X\hat{\beta} \\ \hat{\gamma} \end{bmatrix} = \sigma^2 \begin{bmatrix} M + MZ\left[Z'(I-M)Z\right]^{-1} Z'M & -MZ\left[Z'(I-M)Z\right]^{-1} \\ -\left[Z'(I-M)Z\right]^{-1} Z'M & \left[Z'(I-M)Z\right]^{-1} \end{bmatrix}.$$

Exercise 9.0 In the ACOVA model (9.0.1), suppose $\lambda'\beta = \rho' X\beta$ is estimable so that

$$\text{Var}(\lambda'\hat{\beta}) = \sigma^2 \rho' \left\{ M + MZ\left[Z'(I-M)Z\right]^{-} Z'M \right\} \rho.$$

In the standard model without covariates $Y = X\beta + e$, $\text{Var}(\lambda'\hat{\beta}) = \sigma^2\rho' M\rho$, which seems to be a smaller number than the ACOVA variance. If the point of ACOVA is to sharpen up the analysis, the ACOVA variance should be smaller. Explain away this seeming contradiction. Hint: Does σ^2 mean the same thing in each model?

9.1.1 Generalized Least Squares

Consider the linear model

$$Y = X\beta + Z\gamma + e, \qquad E(e) = 0, \qquad \text{Cov}(e) = \sigma^2 V.$$

As in Section 2.7, generalized least squares estimates are BLUEs and satisfy

$$X\hat{\beta} + Z\hat{\gamma} = \mathscr{A} Y$$

where $\mathscr{A}$ is the oblique projection operator onto $C(X, Z)$ along $C[V^{-1}(X, Z)]^{\perp}$.

The matrix

$$A \equiv X[X'V^{-1}X]^{-} X'V^{-1}$$

is the oblique projection operator onto $C(X)$ along $C[V^{-1}X]^{\perp}$, i.e., if $v \in C(X)$, $Av = v$, and if $v \in C[V^{-1}X]^{\perp}$, $Av = 0$. From Exercise 2.5,

$$(I-A)'V^{-1}(I-A) = (I-A)'V^{-1} = V^{-1}(I-A). \tag{3}$$

To see that

$$\mathscr{A} = A + (I-A)Z[Z'(I-A)'V^{-1}(I-A)Z]^{-} Z'(I-A)'V^{-1} \tag{4}$$

simply write out

$$\mathscr{A} = [X, (I-A)Z]\left\{[X, (I-A)Z]'V^{-1}[X, (I-A)Z]\right\}^{-}[X, (I-A)Z]'V^{-1}$$

and simplify using (3), $(I-A)X = 0$, and the fact that a generalized inverse of a block diagonal matrix is the block diagonal of the generalized inverses.

Given (4) it is easy to see that

$$\hat{\gamma} = [Z'(I - A)'V^{-1}(I - A)Z]^- Z'(I - A)'V^{-1}(I - A)Y$$

and

$$X\hat{\beta} = A(Y - Z\hat{\gamma})$$

provide generalized least squares estimates.

9.2 Estimation of Error and Tests of Hypotheses

The estimate of the variance σ^2 is the *MSE*. We will find $SSE = Y'(I - \mathscr{P})Y$ in terms of M and Z. The error sum of squares is

$$\begin{aligned} Y'(I - \mathscr{P})Y &= Y'[I - M - M_2]Y \\ &= Y'\Big[(I - M) - (I - M)Z\left[Z'(I - M)Z\right]^- Z'(I - M)\Big]Y \qquad (1) \\ &= Y'(I - M)Y - Y'(I - M)Z\left[Z'(I - M)Z\right]^- Z'(I - M)Y. \end{aligned}$$

Using the notation $E_{AB} \equiv A'(I - M)B$, we have

$$Y'(I - \mathscr{P})Y = E_{YY} - E_{YZ}E_{ZZ}^- E_{ZY}.$$

Note the similarity of this structure to the covariance matrix of the best linear predictor's prediction error given in (6.5.1).

Example 9.2.1 Consider a balanced two-way analysis of variance with no replication or interaction and one covariate (regression variable, concomitant variable, supplemental observation) z. The analysis of covariance model can be written

$$y_{ij} = \mu + \alpha_i + \eta_j + \gamma z_{ij} + e_{ij},$$

$i = 1, \ldots, a$, $j = 1, \ldots, b$. X is the model matrix for the balanced two-way ANOVA without replication or interaction,

$$y_{ij} = \mu + \alpha_i + \eta_j + e_{ij},$$

$i = 1, \ldots, a$, $j = 1, \ldots, b$, and Z is an $ab \times 1$ matrix that contains the values of z_{ij}. The sum of squares for error in the covariate analysis is $E_{YY} - E_{YZ}^2/E_{ZZ}$, where

$$E_{YY} = \sum_{i=1}^{a}\sum_{j=1}^{b}\left(y_{ij} - \bar{y}_{i\cdot} - \bar{y}_{\cdot j} + \bar{y}_{\cdot\cdot}\right)^2,$$

$$E_{YZ} = E_{ZY} = \sum_{i=1}^{a}\sum_{j=1}^{b}\left(y_{ij} - \bar{y}_{i\cdot} - \bar{y}_{\cdot j} + \bar{y}_{\cdot\cdot}\right)\left(z_{ij} - \bar{z}_{i\cdot} - \bar{z}_{\cdot j} + \bar{z}_{\cdot\cdot}\right),$$

$$E_{ZZ} = \sum_{i=1}^{a}\sum_{j=1}^{b}\left(z_{ij} - \bar{z}_{i\cdot} - \bar{z}_{\cdot j} + \bar{z}_{\cdot\cdot}\right)^2 .$$

Tests for analysis of covariance models are found by considering the reductions in sums of squares for error due to the models. For instance, if $C(X_0) \subset C(X)$ and we want to test the reduced model

$$Y = X_0\beta_0 + Z\gamma + e$$

against the full model (9.0.1), the test statistic is

$$\frac{\left[Y'(I - \mathscr{P}_0)Y - Y'(I - \mathscr{P})Y\right]/\left[r(X,Z) - r(X_0,Z)\right]}{\left[Y'(I - \mathscr{P})Y\right]/\left[n - r(X,Z)\right]},$$

where $\mathscr{P}_0$ is the perpendicular projection operator onto $C(X_0, Z)$. We have already found $Y'(I - \mathscr{P})Y$. If M_0 is the perpendicular projection operator onto $C(X_0)$,

$$Y'(I - \mathscr{P}_0)Y = Y'(I - M_0)Y - Y'(I - M_0)Z\left[Z'(I - M_0)Z\right]^{-} Z'(I - M_0)Y.$$

In the old days, these computations were facilitated by writing an analysis of covariance table.

Example 9.2.1 Continued. The analysis of covariance table is given below in matrix notation. Recall that, for example,

$$Y'M_\alpha Y = b\sum_{i=1}^{a}\left(\bar{y}_{i\cdot} - \bar{y}_{\cdot\cdot}\right)^2,$$

$$Y'M_\alpha Z = b\sum_{i=1}^{a}\left(\bar{y}_{i\cdot} - \bar{y}_{\cdot\cdot}\right)\left(\bar{z}_{i\cdot} - \bar{z}_{\cdot\cdot}\right),$$

$$Z'M_\alpha Z = b\sum_{i=1}^{a}\left(\bar{z}_{i\cdot} - \bar{z}_{\cdot\cdot}\right)^2 .$$

ACOVA Table

Source	*df*	SS_{YY}	SS_{YZ}	SS_{ZZ}
Grand Mean	1	$Y'\frac{1}{n}J_n^nY$	$Y'\frac{1}{n}J_n^nZ$	$Z'\frac{1}{n}J_n^nZ$
Treatments (α)	$a-1$	$Y'M_\alpha Y$	$Y'M_\alpha Z$	$Z'M_\alpha Z$
Treatments (η)	$b-1$	$Y'M_\eta Y$	$Y'M_\eta Z$	$Z'M_\eta Z$
Error	$n-a-b+1$	$Y'(I-M)Y$	$Y'(I-M)Z$	$Z'(I-M)Z$

If we want to test $H_0: \eta_1 = \eta_2 = \cdots = \eta_b$, the error sum of squares under the reduced model is

$$\left[Y'(I-M)Y + Y'M_\eta Y\right] - \left[Y'(I-M)Z + Y'M_\eta Z\right] \\ \times \left[Z'(I-M)Z + Z'M_\eta Z\right]^- \left[Z'(I-M)Y + Z'M_\eta Y\right].$$

All of these terms are available from the ACOVA table. With more than one covariate, the terms in the SS_{YZ} and SS_{ZZ} columns of the table would be matrices and it would be more involved to compute $\left[Z'(I-M)Z + Z'M_\eta Z\right]^-$.

Exercise 9.1 Consider a one-way ANOVA with one covariate. The model is

$$y_{ij} = \mu + \alpha_i + \xi x_{ij} + e_{ij},$$

$i = 1, \ldots, t$, $j = 1, \ldots, N_i$. Find the BLUE of the contrast $\sum_{i=1}^t \lambda_i \alpha_i$. Find the variance of the contrast.

Exercise 9.2 Consider the problem of estimating β_p in the regression model

$$y_i = \beta_0 + \beta_1 x_{i1} + \cdots + \beta_p x_{ip} + e_i. \qquad (2)$$

Let r_i be the *ordinary residual* from fitting

$$y_i = \alpha_0 + \alpha_1 x_{i1} + \cdots + \alpha_{p-1} x_{ip-1} + e_i$$

and s_i be the residual from fitting

$$x_{ip} = \gamma_0 + \gamma_1 x_{i1} + \cdots + \gamma_{p-1} x_{ip-1} + e_i.$$

Show that the least squares estimate of β_p is $\hat{\xi}$ from fitting the model

$$r_i = \xi s_i + e_i, \quad i = 1, \ldots, n, \qquad (3)$$

that the *SSE* from models (2) and (3) are the same, and that $(\hat{\beta}_0, \ldots, \hat{\beta}_{p-1})' = \hat{\alpha} - \hat{\beta}_p \hat{\gamma}$ with $\hat{\alpha} = (\hat{\alpha}_0, \ldots, \hat{\alpha}_{p-1})'$ and $\hat{\gamma} = (\hat{\gamma}_0, \ldots, \hat{\gamma}_{p-1})'$. Discuss the usefulness of these results for computing regression estimates. (These are the key results behind the *sweep operator* that is often used in regression computations.) What happens to the results if r_i is replaced by y_i in model (3)?

Exercise 9.3 Suppose $\lambda_1'\beta$ and $\lambda_2'\gamma$ are estimable in model (9.0.1). Use the normal equations to find find least squares estimates of $\lambda_1'\beta$ and $\lambda_2'\gamma$.

Hint: Reparameterize the model as $X\beta + Z\gamma = X\delta + (I-M)Z\gamma$ and use the normal equations on the reparameterized model. Note that $X\delta = X\beta + MZ\gamma$.

Exercise 9.4 Derive the test for model (9.0.1) versus the reduced model $Y = X\beta + Z_0\gamma_0 + e$, where $C(Z_0) \subset C(Z)$. Describe how the procedure would work for testing $H_0 : \gamma_2 = 0$ in the model $y_{ij} = \mu + \alpha_i + \eta_j + \gamma_1 z_{ij1} + \gamma_2 z_{ij2} + e_{ij}$, $i = 1,\ldots,a$, $j = 1,\ldots,b$.

Exercise 9.5 An experiment was conducted with two treatments. There were four levels of the first treatment and five levels of the second treatment. Besides the data y, two covariates were measured, x_1 and x_2. The data are given below. Analyze the data with the assumption that there is no interaction between the treatments.

i	j	y_{ij}	x_{1ij}	x_{2ij}	i	j	y_{ij}	x_{1ij}	x_{2ij}
1	1	27.8	5.3	9	3	1	22.4	3.0	13
	2	27.8	5.2	11		2	21.0	4.5	12
	3	26.2	3.6	13		3	30.6	5.4	18
	4	24.8	5.2	17		4	25.4	6.6	21
	5	17.8	3.6	10		5	15.9	4.1	9
2	1	19.6	4.7	12	4	1	14.1	5.4	10
	2	28.4	5.8	17		2	29.5	6.8	18
	3	26.3	3.3	22		3	29.2	5.3	22
	4	18.3	4.1	8		4	21.5	6.2	9
	5	20.8	5.7	11		5	25.5	6.4	22

9.3 Another Adjusted Model and Missing Data

Exercise 9.2 looked at the sweep operator that is based on least squares fitting of

$$(I-M)Y = (I-M)Z\gamma + e.$$

This gives the ACOVA least squares estimates of γ and the ACOVA *SSE*. LaMotte (2014) expands on this basic idea.

Another adjusted model is to fit

$$(Y - Z\hat{\gamma}) = X\beta + e, \tag{1}$$

where $\hat{\gamma}$ is a least squares estimate from the ACOVA model. It is clear from Section 1 that a least squares fit of this model gives a least squares ACOVA estimate of $X\beta$. It is less clear that it also gives the ACOVA *SSE*. With $\hat{\gamma} = [Z'(I-M)Z]^{-}Z'(I-M)Y$, the *SSE* from the adjusted model (1) is

$$\begin{aligned}
&(Y - Z\hat{\gamma})'(I-M)(Y-Z\hat{\gamma})\\
&= Y'(I-M)Y - 2\hat{\gamma}'Z'(I-M)Y + \hat{\gamma}'Z'(I-M)Z\hat{\gamma}\\
&= Y'(I-M)Y - 2Y'(I-M)Z[Z'(I-M)Z]^{-}Z'(I-M)Y\\
&\quad + Y'(I-M)Z[Z'(I-M)Z]^{-}[Z'(I-M)Z][Z'(I-M)Z]^{-}Z'(I-M)Y\\
&= Y'(I-M)Y - Y'(I-M)Z[Z'(I-M)Z]^{-}Z'(I-M)Y,
\end{aligned}$$

A difficulty with model (1) is that you have to do the ACOVA analysis first to get $\hat{\gamma}$. Another is that, unless $X\beta$ is estimable in the ACOVA, we need to worry about ACOVA estimability rather than estimability in (1). One application of this model is in dealing with missing data.

When a few observations are missing from, say, a balanced multifactor design, the balance is lost and the analysis would seem to be quite complicated. One use of the analysis of covariance is to allow the analysis with missing data to be performed using results for the original balanced design. With modern computing power, these days we would probably just fit the unbalanced model.

Consider an original design

$$Y = X\beta + e,$$

with $Y = (y_1, \ldots, y_n)'$. For each missing observation y_i, include a covariate $z_i = (0, \ldots, 0, 1, 0, \ldots, 0)'$ with the 1 in the ith place. Set each y_i that is missing equal to zero. Before applying the adjusted model (1), consider the ACOVA model.

We wish to show that the *SSE* in this ACOVA model equals the *SSE* in the model with the missing observations deleted. The *MSE* in the covariance model will also equal the *MSE* in the model with deletions. In the covariance model, although we are artificially adding observations by setting missing observations to zero, we are also removing those degrees of freedom from the error by adding covariates.

Suppose r observations are missing. Without loss of generality, we can assume that the last r observations are missing. The $n \times r$ matrix of covariates can be written

$$Z = \begin{bmatrix} 0 \\ I_r \end{bmatrix},$$

where I_r is an $r \times r$ identity matrix and 0 is an $(n-r) \times r$ matrix of zeros. Let X be the $n \times p$ model matrix for the model with no missing observations and let X_* be the $(n-r) \times p$ model matrix for the model with the missing observations deleted. Again we can assume that

$$X = \begin{bmatrix} X_* \\ X_r \end{bmatrix},$$

where X_r is the $r \times p$ matrix whose rows are the rows of X corresponding to the missing observations. The analysis of covariance model

$$Y = X\beta + Z\gamma + e$$

can now be written as

$$Y = \begin{bmatrix} X_* & 0 \\ X_r & I_r \end{bmatrix} \begin{bmatrix} \beta \\ \gamma \end{bmatrix} + e.$$

Notice that

$$C\left(\begin{bmatrix} X_* & 0 \\ X_r & I_r \end{bmatrix}\right) = C\left(\begin{bmatrix} X_* & 0 \\ 0 & I_r \end{bmatrix}\right).$$

Let M_* be the perpendicular projection operator onto $C(X_*)$ and let $\mathscr{P}$ be the perpendicular projection operator onto

$$C\left(\begin{bmatrix} X_* & 0 \\ X_r & I_r \end{bmatrix}\right).$$

It is easy to see that

$$\mathscr{P} = \begin{bmatrix} M_* & 0 \\ 0 & I_r \end{bmatrix}.$$

Writing Y as $Y' = [Y_*', 0]$, we find that

$$Y'(I - \mathscr{P})Y = \begin{bmatrix} Y_*' & 0 \end{bmatrix} \begin{bmatrix} I - M_* & 0 \\ 0 & 0 \end{bmatrix} \begin{bmatrix} Y_* \\ 0 \end{bmatrix} = Y_*'(I - M_*)Y_*.$$

Since $Y'(I - \mathscr{P})Y$ is the *SSE* from the covariate model and $Y_*'(I - M_*)Y_*$ is the *SSE* for the model with the missing observations dropped, we are done. Note that the values we put in for the missing observations do not matter for computing the *SSE*. (The justification in Subsection 6.7.3 for Utts's Rainbow Test for Lack of Fit is similarly based on the idea of including a separate indicator for every case being deleted but there we observed the data being deleted.) Tests of hypotheses can be conducted by comparing *SSE*s for different models.

With $Y' = [Y_*', 0]$, estimation will be the same in both models. The least squares estimate of

$$\begin{bmatrix} X_*\beta \\ X_r\beta + \gamma \end{bmatrix}$$

is

$$\mathscr{P}Y = \begin{bmatrix} M_*Y_* \\ 0 \end{bmatrix}.$$

Any estimable function in the model $Y_* = X_*\beta + e_*$ is estimable in the covariate model, and the estimates are the same. The function $\rho_*'X_*\beta$ equals $\rho'(X\beta + Z\gamma)$, where $\rho' = [\rho_*', 0]$ and 0 is a $1 \times r$ matrix of zeros. Thus, $\rho'\mathscr{P}Y = \rho_*'M_*Y_*$. In particular, $X_*\hat{\beta}$ is the same in both models and if $X_r\beta$ is estimable in the X_* model, $-\hat{\gamma} = X_r\hat{\beta}$, the predicted values of $X_r\beta$ from the X_* model.

Exercise 9.6 Show that

$$\mathscr{P} = \begin{bmatrix} M_* & 0 \\ 0 & I_r \end{bmatrix}$$

and that if $X_r = P'X_*$ for some P, then γ and $X\beta$ are both estimable in the missing data ACOVA. What happens to estimability in the ACOVA model when the X model involves factors and you are missing all the observations on one level of a factor?

An alternative approach to the missing value problem is based on finding substitutes for the missing values. The substitutes are chosen so that if one acts like the substitutes are real data, the correct *SSE* is computed. (The degrees of freedom for error must be corrected.)

Setting the problem up as before, we have

$$Y = X\beta + Z\gamma + e,$$

and the estimate $\hat{\gamma}$ can be found. It is proposed to treat $Y - Z\hat{\gamma} = [Y_*', -\hat{\gamma}']'$ as the data from an experiment with model matrix X, i.e., fit the adjusted model (1). We already know that this gives the correct *SSE* and least squares estimates of $X\beta$. The variance of an estimate, say $\rho'X\hat{\beta}$, needs to be calculated as in an analysis of covariance. It is $\sigma^2(\rho'M\rho + \rho'MZ[Z'(I-M)Z]^{-1}Z'M\rho) = \sigma^2\rho_*'M_*\rho_*$, not the naive value of $\sigma^2\rho'M\rho$.

For $r = 1$ missing observation and a variety of balanced designs, formulae have been obtained for $-\hat{\gamma}$ and are available in many older statistical methods books.

Exercise 9.7 Derive $-\hat{\gamma}$ for a randomized complete block design when $r = 1$.

9.4 Balanced Incomplete Block Designs

The analysis of covariance technique can be used to develop the analysis of a balanced incomplete block design. Suppose that a design is to be set up with b blocks and t treatments, but the number of treatments that can be observed in any block is k, where $k < t$. One natural way to proceed would be to find a design where each pair of treatments occurs together in the same block a fixed number of times, say λ. Such a design is called a *balanced incomplete block (BIB) design.*

Let r be the common number of replications for each treatment. There are two well-known facts about the parameters introduced so far. First, the total number of experimental units in the design must be the number of blocks times the number of units in each block, i.e., bk, but the total number of units must also be the number of treatments times the number of times we observe each treatment, i.e., tr; thus

$$tr = bk. \tag{1}$$

Second, the number of within block comparisons between any given treatment and the other treatments is fixed. One way to count this is to multiply the number of other treatments $(t-1)$ by the number of times each occurs in a block with the given treatment (λ). Another way to count it is to multiply the number of other treatments in a block $(k-1)$ times the number of blocks that contain the given treatment (r). Therefore,

$$(t-1)\lambda = r(k-1). \tag{2}$$

The analysis that follows treats both the treatment and block effects as fixed. Previous editions of this book contained a section on the *recovery of interblock information* in which the block effects were considered random. That section now only exists on my website as part of a larger work *Topics in Experimental Design*, cf. http://www.stat.unm.edu/~fletcher/TopicsInDesign.

Example 9.4.1 An experiment was conducted to examine the effects of fertilizers on potato yields. Six treatments (A, B, C, D, E, and F) were used but blocks were chosen that contained only five experimental units. The experiment was performed using a balanced incomplete block design with six blocks. The potato yields (in pounds) along with the mean yield for each block are reported in Table 9.1.

The six treatments consist of all of the possible combinations of two factors. One factor was that a nitrogen-based fertilizer was either applied (n_1) or not applied (n_0). The other factor was that a phosphate-based fertilizer was either not applied (p_0), applied in a single dose (p_1), or applied in a double dose (p_2). In terms of the factorial structure, the six treatments are $A = n_0p_0$, $B = n_0p_1$, $C = n_0p_2$, $D = n_1p_0$, $E = n_1p_1$, and $F = n_1p_2$. From the information in Table 9.1, it is a simple matter to check that $t = 6$, $b = 6$, $k = 5$, $r = 5$, and $\lambda = 4$. After deriving the theory for balanced incomplete block designs, we will return to these data and analyze them.

Table 9.1 Potato yields in pounds for six fertilizer treatments

Block	Data					Block Means
1	*E* 583	*B* 512	*F* 661	*A* 399	*C* 525	536.0
2	*B* 439	*C* 460	*D* 424	*E* 497	*F* 592	482.4
3	*A* 334	*E* 466	*C* 492	*B* 431	*D* 355	415.6
4	*F* 570	*D* 433	*E* 514	*C* 448	*A* 344	461.8
5	*D* 402	*A* 417	*B* 420	*F* 626	*E* 615	496.0
6	*C* 450	*F* 490	*A* 268	*D* 375	*B* 347	386.0

The fixed effects balanced incomplete block model can be written as

$$y_{ij} = \mu + \beta_i + \tau_j + e_{ij}, \qquad e_{ij}\text{s i.i.d. } N(0, \sigma^2), \tag{3}$$

where $i = 1, \ldots, b$ and $j \in D_i$ or $j = 1, \ldots, t$ and $i \in A_j$. D_i is the set of indices for the treatments in block i. A_j is the set of indices for the blocks in which treatment j occurs. Note that there are k elements in each set D_i and r elements in each set A_j.

In applying the analysis of covariance, we will use the balanced one-way ANOVA determined by the grand mean and the blocks to help analyze the model with covariates. The covariates are taken as the columns of the model matrix associated with the treatments. Writing (3) in matrix terms, we get

$$Y = [X, Z] \begin{bmatrix} \beta \\ \tau \end{bmatrix} + e,$$

$$\beta' \equiv (\mu, \beta_1, \ldots, \beta_b), \qquad \tau' \equiv (\tau_1, \ldots, \tau_t).$$

Note that in performing the analysis of covariance for this model our primary interest lies in the coefficients of the covariates, i.e., the treatment effects. To perform an analysis of covariance, we need to find $Y'(I - M)Z$ and $[Z'(I - M)Z]^-$.

First, find $Z'(I - M)Z$. There are t columns in Z; write $Z = [Z_1, \ldots, Z_t]$. The rows of the mth column indicate the presence or absence of the mth treatment. Using two subscripts to denote the rows of vectors, we can write the mth column as

$$Z_m = [z_{ij,m}], \qquad \text{where } z_{ij,m} = \delta_{jm}$$

and δ_{jm} is 1 if $j = m$, and 0 otherwise. In other words, Z_m is 0 for all rows except the r rows that correspond to an observation on treatment m; those r rows are 1.

To get the $t \times t$ matrix $Z'(I - M)Z$, we find each individual element of $Z'Z$ and $Z'MZ$. This is done by finding $Z_s'Z_m$ and $Z_s'M'MZ_m$ for all values of m and s. If $m = s$, we get

$$Z_m'Z_m = \sum_i \sum_j (z_{ij,m})^2 = \sum_{j=1}^{t} \sum_{i \in A_j} \delta_{jm} = \sum_{j=1}^{t} r\delta_{jm} = r.$$

Now, if $m \neq s$, because each observation has only one treatment associated with it, either $z_{ij,s}$ or $z_{ij,m}$ equals 0; so for $s \neq m$,

$$Z_s'Z_m = \sum_i \sum_j (z_{ij,s})(z_{ij,m}) = \sum_{j=1}^{t} \sum_{i \in A_j} \delta_{js}\delta_{jm} = \sum_{j=1}^{t} r\delta_{js}\delta_{jm} = 0.$$

Thus the matrix $Z'Z$ is rI_t, where I_t is a $t \times t$ identity matrix.

Recall that in this problem, X is the model matrix for a one-way ANOVA where the groups of the ANOVA are the blocks of the BIB design and there are k observations on each group. Using two subscripts to denote each row and each column of a matrix, we can write the projection matrix as in Section 4.1,

$$M = [v_{ij,i'j'}], \qquad \text{where } v_{ij,i'j'} = \frac{1}{k}\delta_{ii'}.$$

Let

$$MZ_m = [d_{ij,m}].$$

Then

$$\begin{aligned}
d_{ij,m} &= \sum_{i'j'} v_{ij,i'j'} z_{i'j',m} \\
&= \sum_{j'=1}^{t} \sum_{i' \in A_{j'}} \frac{1}{k}\delta_{ii'}\delta_{j'm} \\
&= \sum_{j'=1}^{t} \delta_{j'm} \sum_{i' \in A_{j'}} \frac{1}{k}\delta_{ii'} \\
&= \sum_{i' \in A_m} \frac{1}{k}\delta_{ii'} \\
&= \frac{1}{k}\delta_i(A_m),
\end{aligned}$$

where $\delta_i(A_m)$ is 1 if $i \in A_m$ and 0 otherwise. In other words, if treatment m is in block i, then all k of the units in block i have $d_{ij,m} = 1/k$. If treatment m is not in block i, all k of the units in block i have $d_{ij,m} = 0$. Since treatment m is contained in exactly r blocks,

$$\begin{aligned}
Z_m' M' M Z_m &= \sum_{ij} (d_{ij,m})^2 = \sum_{i=1}^{b} \sum_{j \in D_i} k^{-2}\delta_i(A_m) \\
&= \sum_{i=1}^{b} (k/k^2)\delta_i(A_m) = \frac{r}{k}.
\end{aligned}$$

Since, for $s \neq m$, there are λ blocks in which both treatments s and m are contained,

$$\begin{aligned}
Z_s' M' M Z_m &= \sum_{ij} (d_{ij,s})(d_{ij,m}) = \sum_{i=1}^{b} \sum_{j \in D_i} (1/k^2)\delta_i(A_s)\delta_i(A_m) \\
&= \sum_{i=1}^{b} (k/k^2)\delta_i(A_s)\delta_i(A_m) = \frac{\lambda}{k}.
\end{aligned}$$

It follows that the matrix $Z'MZ$ has values r/k down the diagonal and values λ/k off the diagonal. This can be written as

$$Z'MZ = \frac{1}{k}\left[(r-\lambda)I + \lambda J_t^t\right].$$

Finally, we can now write

$$\begin{aligned} Z'(I-M)Z &= Z'Z - Z'MZ \\ &= rI - k^{-1}\left[(r-\lambda)I + \lambda J_t^t\right] \\ &= k^{-1}\left[(r(k-1)+\lambda)I - \lambda J_t^t\right]. \end{aligned}$$

This matrix can be simplified further. Define

$$W \equiv I - (1/t)J_t^t. \tag{4}$$

Note that W is a perpendicular projection operator and that equation (2) gives

$$r(k-1) + \lambda = \lambda t.$$

With these substitutions, we obtain

$$Z'(I-M)Z = (\lambda/k)\left[tI - J_t^t\right] = (\lambda t/k)\left[I - (1/t)J_t^t\right] = (\lambda t/k)W.$$

We need to find a generalized inverse of $Z'(I-M)Z$. Because W is a projection operator, it is easily seen that

$$\left[Z'(I-M)Z\right]^- = (k/\lambda t)W. \tag{5}$$

We also need to be able to find the $1 \times t$ vector $Y'(I-M)Z$. The vector $(I-M)Y$ has elements $(y_{ij} - \bar{y}_{i\cdot})$, so

$$Y'(I-M)Z_m = \sum_{ij}(y_{ij} - \bar{y}_{i\cdot})z_{ij,m} = \sum_{j=1}^{t}\delta_{jm}\sum_{i\in A_j}(y_{ij} - \bar{y}_{i\cdot}) = \sum_{i\in A_m}(y_{im} - \bar{y}_{i\cdot}).$$

Define

$$Q_m \equiv \sum_{i\in A_m}(y_{im} - \bar{y}_{i\cdot}).$$

Then

$$Y'(I-M)Z = (Q_1, \ldots, Q_t).$$

Since the β effects are for blocks, our primary interests are in estimable functions $\xi'\tau$ and in estimating σ^2. From (9.1.1) and Proposition 9.1.1, write

$$\xi' = \rho'(I - M)Z$$

to get

$$\xi'\hat{\tau} = \rho'(I - M)Z\left[Z'(I - M)Z\right]^{-} Z'(I - M)Y$$

and, from (9.2.1),

$$SSE = Y'(I - M)Y - Y'(I - M)Z\left[Z'(I - M)Z\right]^{-} Z'(I - M)Y.$$

Both of these formulae involve the term $(I - M)Z\left[Z'(I - M)Z\right]^{-}$. This term can be simplified considerably. Note that since the columns of Z are 0s and 1s, indicating the presence or absence of a treatment effect,

$$ZJ_t^1 = J_n^1.$$

Because M is defined from a one-way ANOVA,

$$0 = (I - M)J_n^1 = (I - M)ZJ_t^1. \tag{6}$$

From (5), (4), and (6), it is easily seen that

$$(I - M)Z\left[Z'(I - M)Z\right]^{-} = (k/\lambda t)(I - M)Z.$$

Using this fact, we get that the BLUE of $\xi'\tau$ is

$$\begin{aligned}
\xi'\hat{\tau} &= \rho'(I - M)Z\left[Z'(I - M)Z\right]^{-} Z'(I - M)Y \\
&= \rho'(k/\lambda t)(I - M)ZZ'(I - M)Y \\
&= (k/\lambda t)\xi'(Q_1, \ldots, Q_t)' \\
&= (k/\lambda t)\sum_{j=1}^{t} \xi_j Q_j.
\end{aligned}$$

Many computer programs, e.g., Minitab, present *adjusted treatment means* $(k/\lambda t)Q_j + \bar{y}_{\cdot\cdot}$ that can be used to estimate contrasts, because $\sum_j \xi_j \bar{y}_{\cdot\cdot} = 0$. The variance of the estimate of the contrast is

$$\begin{aligned}
\mathrm{Var}\left(\xi'\hat{\tau}\right) &= \sigma^2\rho'(I - M)Z\left[Z'(I - M)Z\right]^{-} Z'(I - M)\rho \\
&= \sigma^2(k/\lambda t)\xi'\xi.
\end{aligned}$$

From the estimate and the variance, it is a simple matter to see that

$$SS(\xi'\tau) = (k/\lambda t)\left[\sum_{j=1}^{t} \xi_j Q_j\right]^2 \Big/ \xi'\xi.$$

The error sum of squares is

$$\begin{aligned} SSE &= Y'(I-M)Y - Y'(I-M)Z\left[Z'(I-M)Z\right]^{-}Z'(I-M)Y \\ &= Y'(I-M)Y - (k/\lambda t)Y'(I-M)ZZ'(I-M)Y \\ &= \sum_{ij}(y_{ij}-\bar{y}_{i\cdot})^2 - \frac{k}{\lambda t}\sum_{j=1}^{t} Q_j^2. \end{aligned}$$

Exercise 9.8 Show that $\xi'\tau$ is estimable if and only if $\xi'\tau$ is a contrast. Hint: One direction is easy. For the other direction, show that for $\xi' = (\xi_1, \ldots, \xi_t)$,

$$\xi' = (k/\lambda t)\xi' Z'(I-M)Z.$$

Exercise 9.9 Show that if $\xi'\tau$ and $\eta'\tau$ are contrasts and that if $\xi'\eta = 0$, then $\xi'\tau = 0$ and $\eta'\tau = 0$ put orthogonal constraints on $C(X, Z)$, i.e., the treatment sum of squares can be broken down with orthogonal contrasts in the usual way. Hint: Let $\xi' = \rho_1'[X, Z]$ and $\eta' = \rho_2'[X, Z]$. Show that

$$\rho_1'(I-M)Z\left[Z'(I-M)Z\right]^{-}Z'(I-M)\rho_2 = 0.$$

Suppose that the treatments have quantitative levels, say $x_1, \ldots, x_t$, that are equally spaced. Model (3) can be reparameterized as

$$y_{ij} = \mu + \beta_i + \gamma_1 x_j + \gamma_2 x_j^2 + \cdots + \gamma_{t-1} x_j^{t-1} + e_{ij}.$$

We would like to show that the orthogonal polynomial contrasts for the balanced incomplete block design are the same as for a balanced one-way ANOVA. In other words, tabled polynomial contrasts, which are useful in balanced ANOVAs, can also be used to analyze balanced incomplete block designs. More generally, the treatments in a BIB may have a factorial structure with quantitative levels in some factor (e.g., Example 9.4.1). We would like to establish that the polynomial contrasts in the factor can be used in the usual way to analyze the data.

Because this is a balanced incomplete block design, Z is the model matrix for a balanced one-way ANOVA (without a grand mean). As in Section 7.3, define orthogonal polynomials $T = ZB$ by ignoring blocks. (Here we are not interested in J as an orthogonal polynomial, so we take B as a $t \times t-1$ matrix.) Write $B = [b_1, \ldots, b_{t-1}]$. If treatments are levels of a single quantitative factor, then the b_js are tabled orthogonal polynomial contrasts. If the treatments have factorial structure, the b_js are obtained from tabled contrasts as in the continuation of Example 9.4.1 below. The important fact is that the b_js are readily obtained. Note that $J_t'b_j = 0$ for all j, and $b_i'b_j = 0$ for $i \neq j$.

A model with treatments replaced by regression variables can be written $Y = X\beta + T\eta + e$, where $\eta = (\eta_1, \ldots, \eta_{t-1})'$. For a simple treatment structure, η_j would be the coefficient for a jth degree polynomial. For a factorial treatment structure, η_j could be the coefficient for some cross-product term. The key points are that the hypothesis $\eta_j = 0$ corresponds to some hypothesis that can be interpreted as in Section 6.7 or Section 7.3, and that the columns of T are orthogonal.

As we have seen, the model $Y = X\beta + T\eta + e$ is equivalent to the model $Y = X\delta + (I - M)T\eta + e$, where η is identical in the two models. Thus the test of $\eta_j = 0$ can be performed in the second model. In the second model, $\hat{\eta}$ is independent of $\hat{\delta}$ because of the orthogonality. If the columns of $(I - M)T$ are orthogonal, the estimates of the η_js are independent. Finally, and most importantly, we can show that the contrast in the τs that corresponds to testing $\eta_j = 0$ is simply $b_j'\tau$, where $\tau = (\tau_1, \ldots, \tau_t)'$.

To show that the columns of $(I - M)T$ are orthogonal, it suffices to show that $T'(I - M)T$ is diagonal.

$$\begin{aligned} T'(I-M)T &= B'Z'(I-M)ZB \\ &= (\lambda t/k)B'WB \\ &= (\lambda t/k)B'B. \end{aligned}$$

The last equality follows from the definition of W and the fact that $J_t'b_j = 0$ for all j. Note that $B'B$ is diagonal because $b_i'b_j = 0$ for $i \neq j$.

Finally, the contrast that corresponds to testing $\eta_j = 0$ is $\rho'(I - M)Z\tau$, where ρ is the jth column of $(I - M)T$, i.e., $\rho = (I - M)Zb_j$. This is true because $(I - M)T$ has orthogonal columns. The contrast is then

$$\begin{aligned} [(I-M)Zb_j]'(I-M)Z\tau &= b_j'Z'(I-M)Z\tau \\ &= (\lambda t/k)b_j'W\tau \\ &= (\lambda t/k)b_j'\tau \end{aligned}$$

or, equivalently, the contrast is

$$b_j'\tau.$$

We now apply these results to the analysis of the data in Example 9.4.1.

Example 9.4.1 Continued. The computation of the Q_ms is facilitated by the following table.

Treatment	n_0p_0	n_0p_1	n_0p_2	n_1p_0	n_1p_1	n_1p_2
$\sum_{i\in A_m} y_{im}$	1762.0	2149.0	2375.0	1989.0	2675.0	2939.0
$\sum_{i\in A_m} \bar{y}_{i\cdot}$	2295.4	2316.0	2281.8	2241.8	2391.8	2362.2
Q_m	−533.4	−167.0	93.2	−252.8	283.2	576.8

An analysis of variance table can be computed.

ANOVA

Source	df	SS	MS	F
Blocks (Ignoring Trts)	5	74857.77	14971.553	
Treatments (After Blks)	5	166228.98	33245.797	31.97
Error	19	19758.22	1039.906	
Total	29	260844.97		

Clearly, there are significant differences among the treatments. These can be explored further by examining contrasts. The factorial structure of the treatments suggests looking at nitrogen effects, phosphate effects, and interaction effects. With two levels of nitrogen, the only available contrast is $(1)n_0 + (-1)n_1$. Phosphate was applied at quantitative levels 0, 1, and 2. The linear contrast in phosphate is $(-1)p_0 + (0)p_1 + (1)p_2$. The quadratic contrast in phosphate is $(1)p_0 + (-2)p_1 + (1)p_2$. Combining these to obtain interaction contrasts and rewriting them as contrasts in the original six treatments gives $b_1, \ldots, b_5$.

	b_js					
Treatments	n_0p_0	n_0p_1	n_0p_2	n_1p_0	n_1p_1	n_1p_2
N	1	1	1	−1	−1	−1
P linear	−1	0	1	−1	0	1
P quadratic	1	−2	1	1	−2	1
$N - P$ linear	−1	0	1	1	0	−1
$N - P$ quadratic	1	−2	1	−1	2	−1

Source	df	SS	F
N	1	51207.20	49.24
P linear	1	110443.67	106.21
P quadratic	1	2109.76	2.03
$N - P$ linear	1	2146.30	2.06
$N - P$ quadratic	1	322.06	0.31

The conclusions to be drawn are clear. There is a substantial increase in yields due to adding the nitrogen-based fertilizer. For the dosages of phosphate used, there is a definite increasing linear relationship between amount of phosphate and yield of potatoes. There is no evidence of any interaction.

Note that the linear relationship between phosphate and yield is an approximation that holds in some neighborhood of the dosages used in the experiment. It is well known that too much fertilizer will actually kill most plants. In particular, no potato plant will survive having an entire truckload of phosphate dumped on it.

Exercise 9.10 Derive the analysis for a Latin square with one row missing. Hint: This problem is at the end of Section 9.4, not Section 9.3.

Exercise 9.11 Eighty wheat plants were grown in each of 5 different fields. Each of 6 individuals (A, B, C, D, E, and F) were asked to pick 8 "representative" plants in each field and measure the plants's heights. Measurements were taken on 6 different days. The data consist of the differences between the mean height of the 8 "representative" plants and the mean of all the heights in the field on that day. Thus the data measure the bias in selecting "representative" plants. The exact design and the data are given below. Analyze the data. (Although they are barely recognizable as such, these data are from Cochran and Cox 1957.)

	Field				
Day	1	2	3	4	5
1	E 3.50	A 0.75	C 2.28	F 1.77	D 2.28
2	D 3.78	B 1.46	A −1.06	E 1.46	F 2.76
3	F 2.32	C 2.99	B −0.28	D 1.18	E 3.39
4	C 4.13	D 4.02	E 1.81	B 1.46	A 1.50
5	A 1.38	E 1.65	F 2.64	C 2.60	B 1.50
6	B 1.22	F 2.83	D 1.57	A −1.30	C 1.97

9.5 Testing a Nonlinear Full Model

Consider testing the model

$$Y = X\beta + e, \qquad e \sim N(0, \sigma^2 I), \tag{1}$$

against a nonlinear model that involves a matrix function of $X\beta$, say

$$Y = X\beta + Z(X\beta)\gamma + e.$$

We assume that the matrix $Z(X\beta)$ has constant rank. More precisely, we need to assume that $r[(I - M)Z(X\beta)]$ and $r[X, Z(X\beta)]$ are constant functions of β. If each column of $Z(v)$ is a distinct nonlinear function of v, these conditions often hold.

Milliken and Graybill (1970) developed an exact F test for this problem. A similar result appears in the first (1965) edition of Rao (1973). When $Z(v) \equiv [v_1^2, \ldots, v_n^2]'$, the procedure gives Tukey's famous one degree of freedom for nonadditivity test, cf. Tukey (1949) or Christensen (1996, Section 10.4 or 2015, Section 7.3). Tests for Mandel's (1961, 1971) extensions of the Tukey model also fit into this class of tests. Christensen and Utts (1992) extended these tests to log-linear and logit models, and Christensen (1997, Section 7.3) examines the Tukey and Mandel models in the context of log-linear models. St. Laurent (1990) showed that Milliken and Graybill's test is equivalent to a score test and thus shares the asymptotic properties of the generalized likelihood ratio test. St. Laurent also provides references to other applications of this class of tests.

To develop the test, fit model (1) to get $\tilde{Y} \equiv MY$ and define

$$\tilde{Z} \equiv Z(\tilde{Y}).$$

Now fit the model

$$Y = X\beta + \tilde{Z}\gamma + e$$

treating $\tilde{Z}$ as a known model matrix that does not depend on Y. Let $\mathscr{P}$ be the perpendicular projection operator onto $C(X, \tilde{Z})$ and write $\mathscr{P} = M + \tilde{M}_2$ as in Section 1. The usual F test for $H_0 : \gamma = 0$ is based on

$$F = \frac{Y'\tilde{M}_2 Y / r[(I-M)\tilde{Z}]}{Y'(I-\mathscr{P})Y/[n - r(X, \tilde{Z})]} \sim F(r[(I-M)\tilde{Z}], n - r[X, \tilde{Z}]). \quad (2)$$

To show that (2) really holds under the null hypothesis of model (1), consider the distribution of Y given $\tilde{Y}$. Write

$$Y = MY + (I-M)Y = \tilde{Y} + (I-M)Y.$$

Under the null model, $\tilde{Y}$ and $(I-M)Y$ are independent and

$$(I-M)Y \sim N[0, \sigma^2(I-M)],$$

so

$$Y|\tilde{Y} \sim N[\tilde{Y}, \sigma^2(I-M)].$$

Use the general results from Section 1.3 that involve checking conditions like $VAVAV = VAV$ and $VAVBV = 0$ to establish that the F statistic has the stated $F(r[(I-M)\tilde{Z}], n - r[X, \tilde{Z}])$ distribution conditional on $\tilde{Y}$. Finally, by assumption, the degrees of freedom for the F distribution do not depend on $\tilde{Y}$, so the conditional distribution does not depend on $\tilde{Y}$ and it must also be the unconditional distribution.

Exercise 9.12 Prove that display (2) is true.

9.6 Additional Exercises

Exercise 9.6.1 Sulzberger (1953) and Williams (1959) examined the maximum compressive strength parallel to the grain (y) of 10 hoop trees and how it was affected by temperature. A covariate, the moisture content of the wood (x), was also measured. Analyze the data, which are reported below

	Temperature in Celsius									
	−20		0		20		40		60	
Tree	y	x	y	x	y	x	y	x	y	x
1	13.14	42.1	12.46	41.1	9.43	43.1	7.63	41.4	6.34	39.1
2	15.90	41.0	14.11	39.4	11.30	40.3	9.56	38.6	7.27	36.7
3	13.39	41.1	12.32	40.2	9.65	40.6	7.90	41.7	6.41	39.7
4	15.51	41.0	13.68	39.8	10.33	40.4	8.27	39.8	7.06	39.3
5	15.53	41.0	13.16	41.2	10.29	39.7	8.67	39.0	6.68	39.0
6	15.26	42.0	13.64	40.0	10.35	40.3	8.67	40.9	6.62	41.2
7	15.06	40.4	13.25	39.0	10.56	34.9	8.10	40.1	6.15	41.4
8	15.21	39.3	13.54	38.8	10.46	37.5	8.30	40.6	6.09	41.8
9	16.90	39.2	15.23	38.5	11.94	38.5	9.34	39.4	6.26	41.7
10	15.45	37.7	14.06	35.7	10.74	36.7	7.75	38.9	6.29	38.2

Exercise 9.6.2 Suppose that in Exercise 7.7.1 on motor oil pricing, the observation on store 7, brand H was lost. Treat the stores as blocks in a randomized complete block design. Plug in an estimate of the missing value and analyze the data without correcting the $MSTrts$ or any variance estimates. Compare the results of this approximate analysis to the results of a correct analysis.

Exercise 9.6.3 The missing value procedure that consists of analyzing the model $(Y - Z\hat{\gamma}) = X\beta + e$ has been shown to give the correct SSE and BLUEs; however, sums of squares explained by the model are biased upwards. For a randomized complete block design with a treatments and b blocks and the observation in the c,d cell missing, show that the correct mean square for treatments is the naive (biased) mean square treatments minus $[y_{\cdot d} - (a-1)\hat{y}_{cd}]^2/a(a-1)^2$, where $y_{\cdot d}$ is the sum of all actual observations in block d, and $\hat{y}_{cd}$ is the pseudo-observation (the nonzero element of $Z\hat{\gamma}$).

Exercise 9.6.4 State whether each design given below is a balanced incomplete block design, and if so, give the values of b, t, k, r, and λ.

(a) The experiment involves 5 treatments: A, B, C, D, and E. The experiment is laid out as follows.

Block	Treatments	Block	Treatments
1	A, B, C	6	A, B, D
2	A, B, E	7	A, C, D
3	A, D, E	8	A, C, E
4	B, C, D	9	B, C, E
5	C, D, E	10	B, D, E

(b) The following design has 9 treatments: A, B, C, D, E, F, G, H, and I.

Block	Treatments	Block	Treatments
1	B, C, D, G	6	C, D, E, I
2	A, C, E, H	7	A, D, H, I
3	A, B, F, I	8	B, E, G, I
4	A, E, F, G	9	C, F, G, H
5	B, D, F, H		

(c) The following design has 7 treatments: A, B, C, D, E, F, G.

Block	Treatments	Block	Treatments
1	C, E, F, G	5	B, C, D, G
2	A, D, F, G	6	A, C, D, E
3	A, B, E, G	7	B, D, E, F
4	A, B, C, F		

References

Christensen, R. (1996). *Analysis of variance, design, and regression: Applied statistical methods*. London: Chapman and Hall.

Christensen, R. (1997). *Log-linear models and logistic regression* (2nd ed.). New York: Springer.

Christensen, R. (2015). *Analysis of variance, design, and regression: Linear modeling for unbalanced data* (2nd ed.). Boca Raton: Chapman and Hall/CRC Press.

Christensen, R., & Utts, J. (1992). Testing for nonadditivity in log-linear and logit models. *Journal of Statistical Planning and Inference, 33*, 333–343.

Cox, D. R. (1958). *Planning of experiments*. New York: Wiley.

LaMotte, L. R. (2014). The Gram-Schmidt construction as a basis for linear models. *The American Statistician, 68*, 52–55.

Mandel, J. (1961). Nonadditivity in two-way analysis of variance. *Journal of the American Statistical Association, 56*, 878–888.

Mandel, J. (1971). A new analysis of variance model for nonadditive data. *Technometrics, 13*, 1–18.

Milliken, G. A., & Graybill, F. A. (1970). Extensions of the general linear hypothesis model. *Journal of the American Statistical Association, 65*, 797–807.

Rao, C. R. (1973). *Linear statistical inference and its applications* (2nd ed.). New York: Wiley.

Snedecor, G. W., & Cochran, W. G. (1980). *Statistical methods* (7th ed.). Ames: Iowa State University Press.

St. Laurent, R. T. (1990). The equivalence of the Milliken-Graybill procedure and the score test. *The American Statistician, 44*, 36–37.

Sulzberger, P. H. (1953). The effects of temperature on the strength of wood, plywood and glued joints. Department of supply, Report ACA-46, Aeronautical Research Consultative Committee, Australia.

Tukey, J. W. (1949). One degree of freedom for nonadditivity. *Biometrics, 5*, 232–242.

Williams, E. J. (1959). *Regression analysis*. New York: Wiley.

Chapter 10
General Gauss–Markov Models

Abstract The standard linear model assumes the data vector has a covariance matrix of $\sigma^2 I$. Sections 2.7 and 3.8 extended the theory to having a covariance matrix of $\sigma^2 V$ where V was known and positive definite. This chapter extends the theory by allowing V to be merely nonnegative definite.

A general Gauss–Markov model is a model

$$Y = X\beta + e, \quad \mathrm{E}(e) = 0, \quad \mathrm{Cov}(e) = \sigma^2 V,$$

where V is a known matrix. Linear models can be divided into four categories depending on the assumptions made about V:

(a) V is an identity matrix (the standard linear model),
(b) V is nonsingular (the generalized least squares model),
(c) V is possibly singular but $C(X) \subset C(V)$,
(d) V is possibly singular.

The categories are increasingly general. Any results for category (d) apply to all other categories. Any results for category (c) apply to categories (a) and (b). Any results for (b) apply to (a).

The majority of Chapters 1 through 9 have dealt with category (a). In Sections 2.7 and 3.8, models in category (b) were discussed. In this chapter, categories (c) and (d) are discussed. Section 1 is devoted to finding BLUEs for models in categories (c) and (d). Theorem 10.1.2 and the discussion following it give the main results for category (c). The approach is similar in spirit to Section 2.7. The model is transformed into an equivalent model that fits into category (a), and BLUEs are found for the equivalent model. Although similar in spirit to Section 2.7, the details are considerably more complicated because of the possibility that V is singular. Having found BLUEs for category (c), the extension to category (d) is very simple. The extension follows from Theorem 10.1.3. Finally, Section 1 contains some results on the uniqueness of BLUEs for category (d).

R. Christensen, *Plane Answers to Complex Questions*, Springer Texts in Statistics,
https://doi.org/10.1007/978-3-030-32097-3_10

Section 2 contains a discussion of the geometry of estimation for category (d). In particular, it points out the need for a consistency criterion and the crucial role of projection operators in linear unbiased estimation. Section 3 examines the problem of testing a model against a reduced model for category (d). Section 4 discusses the extension of least squares estimation to category (d) in light of the consistency requirement of Section 2. Section 4 also contains the very important result that least squares estimates are BLUEs if and only if $C(VX) \subset C(X)$.

Section 5 considers estimable parameters that can be known with certainty when $C(X) \not\subset C(V)$ and a relatively simple way to estimate estimable parameters that are not known with certainty. Some of the nastier parts in Sections 1 through 4 are those that provide sufficient generality to allow $C(X) \not\subset C(V)$. The simpler approach of Section 5 seems to obviate the need for much of that. Groß (2004) surveyed important results in linear models with possibly singular covariance matrices.

10.1 BLUEs with an Arbitrary Covariance Matrix

Consider the model

$$Y = X\beta + e, \quad \mathrm{E}(e) = 0, \quad \mathrm{Cov}(e) = \sigma^2 V, \tag{1}$$

where V is a known matrix. We want to find the best linear unbiased estimate of $\mathrm{E}(Y)$.

Definition 10.1.1 Let Λ' be an $r \times p$ matrix with $\Lambda' = P'X$ for some P. An estimate B_0Y is called a *best linear unbiased estimate* (*BLUE*) of $\Lambda'\beta$ if

(a) $\mathrm{E}(B_0Y) = \Lambda'\beta$ for any β, and
(b) if BY is any other linear unbiased estimate of $\Lambda'\beta$, then for any $r \times 1$ vector ξ

$$\mathrm{Var}(\xi'B_0Y) \leq \mathrm{Var}(\xi'BY).$$

Exercise 10.1 Show that A_0Y is a BLUE of $X\beta$ if and only if, for every estimable function $\lambda'\beta$ such that $\rho'X = \lambda'$, $\rho'A_0Y$ is a BLUE of $\lambda'\beta$.

Exercise 10.2 Show that if $\Lambda' = P'X$ and if A_0Y is a BLUE of $X\beta$, then $P'A_0Y$ is a BLUE of $\Lambda'\beta$.

In the case of a general covariance matrix V, it is a good idea to reconsider what the linear model (1) is really saying. The obvious thing it is saying is that $\mathrm{E}(Y) \in C(X)$. From Lemma 1.3.5, the model also says that $e \in C(V)$ or, in other notation, $Y \in C(X, V)$. If V is a nonsingular matrix, $C(V) = \mathbf{R}^n$; so the conditions on e and Y are really meaningless. When V is a singular matrix, the conditions on e and Y are extremely important.

For any matrix A, let M_A denote the perpendicular projection matrix onto $C(A)$. M without a subscript is M_X. Any property that holds with probability 1 will be said to hold almost surely (a.s.). For example, Lemma 1.3.5 indicates that $e \in C(V)$ a.s. and, adding $X\beta$ to $Y - X\beta$, the lemma gives $Y \in C(X, V)$ a.s.

If R and S are any two random vectors with $R = S$ a.s., then $\mathrm{E}(R) = \mathrm{E}(S)$ and $\mathrm{Cov}(R) = \mathrm{Cov}(S)$. In particular, if $R = S$ a.s. and R is a BLUE, then S is also a BLUE.

Results on estimation will be established by comparing the estimation problem in model (1) to the estimation problem in two other, more easily analyzed, models.

Before proceeding to the first theorem on estimation, recall the eigenvector decomposition of the symmetric, nonnegative definite matrix V. One can pick matrices E and D so that $VE = ED$. Here $D = \mathrm{Diag}(d_i)$, where the d_is are the, say m, positive eigenvalues of V (with the correct multiplicities). E can be chosen as a matrix of orthonormal columns with the ith column an eigenvector corresponding to d_i. Define $D^{1/2} \equiv \mathrm{Diag}\left(\sqrt{d_i}\right)$. Write

$$Q \equiv ED^{1/2} \quad \text{and} \quad Q^- \equiv D^{-1/2}E'.$$

Useful facts are

1. $C(V) = C(E) = C(Q)$
2. $M_V = EE' = QQ^-$
3. $I_m = Q^-Q$
4. $V = QQ'$
5. $QQ^-Q = Q$
6. $Q^-VQ^{-\prime} = I_m$
7. $Q^{-\prime}Q^- = V^-$.

Consider the linear model

$$Q^-Y = Q^-X\beta + Q^-e, \quad \mathrm{E}(Q^-e) = 0, \quad \mathrm{Cov}(Q^-e) = \sigma^2 I_m. \tag{2}$$

Models (1) and (2) are equivalent when $C(X) \subset C(V)$. Clearly, (2) can be obtained from (1) by multiplying on the left by Q^-. Moreover, with $C(X) \subset C(V)$, each of Y, $C(X)$, and e are contained in $C(V)$ a.s.; so multiplying (2) on the left by Q gives $M_V Y = M_V X\beta + M_V e$, which is model (1) a.s. Note that $Q^-Y \in \mathbf{R}^m$ and that the Gauss–Markov theorem can be used to get estimates in model (2). Moreover, if $C(X) \subset C(V)$, then $X = M_V X = QQ^-X$; so $X\beta$ is estimable in model (2).

Theorem 10.1.2 *If $C(X) \subset C(V)$, then A_0Y is a BLUE of $X\beta$ in model (1) if and only if $(A_0Q)Q^-Y$ is a BLUE of $X\beta$ in model (2).*

Proof If this theorem is to make any sense at all, we need to first show that $\mathrm{E}(A_0Y) = X\beta$ iff $\mathrm{E}(A_0QQ^-Y) = X\beta$. Recall that $Y \in C(X, V)$ a.s., so in this special case where $C(X) \subset C(V)$, we have $Y \in C(V)$ a.s. Thus, for the purposes of finding expected values and covariances, we can assume that $Y = M_V Y$. Let B be an arbitrary $n \times n$ matrix. Then $\mathrm{E}(BY) = \mathrm{E}(BM_V Y) = \mathrm{E}(BQQ^- Y)$, so $\mathrm{E}(A_0 Y) = X\beta$ iff $\mathrm{E}(A_0 QQ^- Y) = X\beta$. It is also handy to know the following fact:

$$\mathrm{Var}(\rho' BY) = \mathrm{Var}(\rho' BM_V Y) = \mathrm{Var}(\rho' BQQ^- Y).$$

Now suppose that $A_0 Y$ is a BLUE for $X\beta$ in model (1). We show that $A_0 QQ^- Y$ is a BLUE of $X\beta$ in model (2). Let $BQ^- Y$ be another unbiased estimate of $X\beta$ in model (2). Then $X\beta = \mathrm{E}(BQ^- Y)$, so $BQ^- Y$ is an unbiased estimate of $X\beta$ in model (1) and, since $A_0 Y$ is a BLUE in model (1), $\mathrm{Var}(\rho' A_0 QQ^- Y) = \mathrm{Var}(\rho' A_0 Y) \leq \mathrm{Var}(\rho' BQ^- Y)$. Thus $A_0 QQ^- Y$ is a BLUE of $X\beta$ in model (2).

Conversely, suppose that $A_0 QQ^- Y$ is a BLUE of $X\beta$ in model (2). Let BY be an unbiased estimate for $X\beta$ in model (1). Then $BQQ^- Y$ is unbiased for $X\beta$ in model (2) and $\mathrm{Var}(\rho' A_0 Y) = \mathrm{Var}(\rho' A_0 QQ^- Y) \leq \mathrm{Var}(\rho' BQQ^- Y) = \mathrm{Var}(\rho' BY)$; so $\rho' A_0 Y$ is a BLUE of $X\beta$ in model (1). □

Note that with $C(X) \subset C(V)$, $A_0 Y = A_0 M_V Y = A_0 QQ^- Y$ a.s., so Theorem 10.1.2 is really saying that $A_0 Y$ is a BLUE in model (1) if and only if $A_0 Y$ is a BLUE in model (2). The virtue of Theorem 10.1.2 is that we can actually find a BLUE for $X\beta$ in model (2). From Exercises 10.1 and 10.2, a BLUE of $X\beta = QQ^- X\beta$ from model (2) is

$$\begin{aligned} X\hat{\beta} &= QM_{Q^- X} Q^- Y \\ &= Q\left[Q^- X(X'Q^{-\prime}Q^- X)^- X'Q^{-\prime}\right] Q^- Y \\ &= M_V X(X'V^- X)^- X'V^- Y \\ &= X(X'V^- X)^- X'V^- Y. \end{aligned} \tag{3}$$

It is useful to observe that we can get a BLUE from any choice of V^- and $(X'V^- X)^-$. First, notice that $X'V^- X$ does not depend on V^-. Since $C(X) \subset C(V)$, we can write $X = VC$ for some matrix C. Then $X'V^- X = C'VV^- VC = C'VC$. Second, $Q^- X(X'Q^{-\prime}Q^- X)^- X'Q^{-\prime}$ does not depend on the choice of $(X'Q^{-\prime}Q^- X)^-$. Therefore, $X(X'V^- X)^- X'V^-$ does not depend on the choice of $(X'V^- X)^-$. Moreover, for any $Y \in C(V)$, $X(X'V^- X)^- X'V^- Y$ does not depend on the choice of V^-. To see this, write $Y = Vb$. Then $X'V^- Y = (C'V)V^-(Vb) = C'Vb$. Since $Y \in C(V)$ a.s., $X(X'V^- X)^- X'V^- Y$ is a BLUE of $X\beta$ for any choices of V^- and $(X'V^- X)^-$.

To obtain the general estimation result for arbitrary singular V, consider the linear model

$$Y_1 = X\beta + e_1, \quad \mathrm{E}(e_1) = 0, \quad \mathrm{Cov}(e_1) = \sigma^2\left(V + XUX'\right), \tag{4}$$

where U is any symmetric nonnegative definite matrix.

Theorem 10.1.3 *$A_0 Y$ is a BLUE for $X\beta$ in model (1) if and only if $A_0 Y_1$ is a BLUE for $X\beta$ in model (4).*

Proof Clearly, for any matrix B, BY is unbiased if and only if BY_1 is unbiased and both are equivalent to the condition $BX = X$. In the remainder of the proof, B will be an arbitrary matrix with $BX = X$ so that BY and BY_1 are arbitrary linear unbiased estimates of $X\beta$ in models (1) and (4), respectively. The key fact in the proof is that

$$\begin{aligned}\mathrm{Var}(\rho' BY_1) &= \sigma^2 \rho' B \left(V + XUX'\right) B'\rho \\ &= \sigma^2 \rho' BVB'\rho + \rho' BXUX'B'\rho \\ &= \mathrm{Var}(\rho' BY) + \sigma^2 \rho' XUX'\rho.\end{aligned}$$

Now suppose that $A_0 Y$ is a BLUE for $X\beta$ in model (1). Then

$$\mathrm{Var}(\rho' A_0 Y) \leq \mathrm{Var}(\rho' BY).$$

Adding $\sigma^2 \rho' XUX'\rho$ to both sides we get

$$\mathrm{Var}(\rho' A_0 Y_1) \leq \mathrm{Var}(\rho' BY_1),$$

and since BY_1 is an arbitrary linear unbiased estimate, $A_0 Y_1$ is a BLUE.

Conversely, suppose $A_0 Y_1$ is a BLUE for $X\beta$ in model (4). Then

$$\mathrm{Var}(\rho' A_0 Y_1) \leq \mathrm{Var}(\rho' BY_1),$$

or

$$\mathrm{Var}(\rho' A_0 Y) + \sigma^2 \rho' XUX'\rho \leq \mathrm{Var}(\rho' BY) + \sigma^2 \rho' XUX'\rho.$$

Subtracting $\sigma^2 \rho' XUX'\rho$ from both sides we get

$$\mathrm{Var}(\rho' A_0 Y) \leq \mathrm{Var}(\rho' BY),$$

so $A_0 Y$ is a BLUE. □

As with Theorem 10.1.2, this result is useful because a BLUE for $X\beta$ can actually be found in one of the models. Let $T = V + XUX'$. If U is chosen so that $C(X) \subset C(T)$, then Theorem 10.1.2 applies to model (4) and a BLUE for $X\beta$ is $X(X'T^- X)^- X'T^- Y$. Exercise 10.3 establishes that such matrices U exist. Since this is an application of Theorem 10.1.2, $X(X'T^- X)^- X'T^- Y$ does not depend on the choice of $(X'T^- X)^-$ and, for $Y \in C(T)$, it does not depend on the choice of T^-. Proposition 10.1.4 below shows that $C(X) \subset C(T)$ implies $C(T) = C(X, V)$; so $Y \in C(T)$ a.s., and any choice of T^- gives a BLUE.

Exercise 10.3 The BLUE of $X\beta$ can be obtained by taking $T = V + XX'$. Prove this by showing that

(a) $C(X) \subset C(T)$ if and only if $TT^- X = X$, and

(b) if $T = V + XX'$, then $TT^{-}X = X$.

Hint: Searle and Pukelsheim (1987) base a proof of (b) on

$$\begin{aligned} 0 &= (I - TT^{-})T(I - TT^{-})' \\ &= (I - TT^{-})V(I - TT^{-})' + (I - TT^{-})XX'(I - TT^{-})' \end{aligned}$$

and the fact that the last term is the sum of two nonnegative definite matrices.

Proposition 10.1.4 *If $C(X) \subset C(T)$, then $C(V) \subset C(T)$ and $C(X, V) = C(T)$.*

Proof We know that $C(X) \subset C(T)$, so $X = TG$ for some matrix G. If $v \in C(V)$, then $v = Vb_1$, and $v = Vb_1 + XUX'b_1 - XUX'b_1 = Tb_1 - TG(UX'b_1)$; so $v \in C(T)$. Thus, $C(X, V) \subset C(T)$. But clearly, by the definition of T, $C(T) \subset C(X, V)$; so we have $C(X, V) = C(T)$. □

To complete our characterization of best linear unbiased estimates, we will show that for practical purposes BLUEs of $X\beta$ are unique.

Theorem 10.1.5 *Let AY and BY be BLUEs of $X\beta$ in model (1), then* $\Pr(AY = BY) = 1$.

Proof It is enough to show that $\Pr[(A - B)Y = 0] = 1$. We need only observe that $\mathrm{E}[(A - B)Y] = 0$ and show that for any ρ, $\mathrm{Var}[\rho'(A - B)Y] = 0$.

Remembering that $\mathrm{Var}(\rho'AY) = \mathrm{Var}(\rho'BY)$, we first consider the unbiased estimate of $X\beta$, $\frac{1}{2}(A + B)Y$:

$$\mathrm{Var}(\rho'AY) \le \mathrm{Var}\left(\rho'\frac{1}{2}(A + B)Y\right)$$

but

$$\mathrm{Var}\left(\rho'\frac{1}{2}(A + B)Y\right) = \frac{1}{4}\left[\mathrm{Var}(\rho'AY) + \mathrm{Var}(\rho'BY) + 2\mathrm{Cov}(\rho'AY, \rho'BY)\right],$$

so

$$\mathrm{Var}(\rho'AY) \le \frac{1}{2}\mathrm{Var}(\rho'AY) + \frac{1}{2}\mathrm{Cov}(\rho'AY, \rho'BY).$$

Simplifying, we find that

$$\mathrm{Var}(\rho'AY) \le \mathrm{Cov}(\rho'AY, \rho'BY).$$

Now look at $\mathrm{Var}\big(\rho'(A - B)Y\big)$,

$$0 \leq \text{Var}\big(\rho'(A-B)Y\big) = \text{Var}(\rho'AY) + \text{Var}(\rho'BY) - 2\text{Cov}(\rho'AY, \rho'BY) \leq 0. \quad \square$$

Finally, the most exact characterization of a BLUE is the following:

Theorem 10.1.6 *If AY and BY are BLUEs for $X\beta$ in model (1), then $AY = BY$ for any $Y \in C(X, V)$.*

Proof It is enough to show that $AY = BY$, first when $Y \in C(X)$ and then when $Y \in C(V)$. When $Y \in C(X)$, by unbiasedness $AY = BY$.

We want to show that $AY = BY$ for all $Y \in C(V)$. Let $\mathscr{M} = \{Y \in C(V) | AY = BY\}$. It is easily seen that $\mathscr{M}$ is a vector space, and clearly $\mathscr{M} \subset C(V)$. If $C(V) \subset \mathscr{M}$, then $\mathscr{M} = C(V)$, and we are done.

From unbiasedness, $AY = X\beta + Ae$ and $BY = X\beta + Be$, so $AY = BY$ if and only if $Ae = Be$. From Theorem 10.1.5, $\Pr(Ae = Be) = 1$. We also know that $\Pr(e \in C(V)) = 1$. Therefore, $\Pr(e \in \mathscr{M}) = 1$.

Computing the covariance of e we find that

$$\sigma^2 V = \int ee' dP = \int_{e \in \mathscr{M}} ee' dP;$$

so, by Exercise 10.4, $C(V) \subset \mathscr{M}$. $\square$

Exercise 10.4 Ferguson (1967, Section 2.7, page 74) proves the following:

Lemma (3) *If S is a convex subset of $\mathbf{R}^n$, and Z is an n-dimensional random vector for which $\Pr(Z \in S) = 1$ and for which $\mathrm{E}(Z)$ exists and is finite, then $\mathrm{E}(Z) \in S$.*

Use this lemma to show that

$$C(V) = C\left[\int_{e \in \mathscr{M}} ee' dP\right] \subset \mathscr{M}.$$

After establishing Theorem 10.1.5, we only need to find any one BLUE, because for any observations that have a chance of happening, all BLUEs give the same estimates. Fortunately, we have already shown how to obtain a BLUE, so this section is finished. There is only one problem. Best linear unbiased estimates might be pure garbage. We pursue this issue in the next section.

10.2 Geometric Aspects of Estimation

The linear model

$$Y = X\beta + e, \quad E(e) = 0, \quad \mathrm{Cov}(e) = \sigma^2 V, \tag{1}$$

says two important things:

(a) $E(Y) = X\beta \in C(X)$, and
(b) $\Pr(e \in C(V)) = 1$.

Note that (b) also says something about $E(Y)$:

(b′) $X\beta = Y - e \in Y + C(V)$ a.s.

Intuitively, any reasonable estimate of $E(Y)$, say $X\hat{\beta}$, should satisfy the following definition for consistency.

Definition 10.2.1 An estimate $\widehat{X\beta}$ of $X\beta$ is called a *consistent* estimate if

(i) $\widehat{X\beta} \in C(X)$ for any Y, and
(ii) $\widehat{X\beta} \in Y + C(V)$ for any $Y \in C(X, V)$.

$\widehat{X\beta}$ is called *almost surely consistent* if conditions (i) and (ii) hold almost surely.

Note that this concept of consistency is distinct from the usual large sample idea of consistency. The idea is that a consistent estimate, in our sense, is consistent with respect to conditions (a) and (b′).

Example 10.2.2 Consider the linear model determined by

$$X = \begin{bmatrix} 1 & 0 \\ 0 & 1 \\ 0 & 0 \end{bmatrix} \quad \text{and} \quad V = \begin{bmatrix} \frac{1}{2} & 0 & \frac{1}{2} \\ 0 & 1 & 0 \\ \frac{1}{2} & 0 & \frac{1}{2} \end{bmatrix}.$$

If this model is graphed with coordinates (x, y, z), then $C(X)$ is the x, y plane and $C(V)$ is the plane determined by the y-axis and the line $[x = z, y = 0]$. (See Figure 10.1.)

Suppose that $Y = (7, 6, 2)'$. Then (see Figure 10.2) $E(Y)$ is in $C(X)$ (the x, y plane) and also in $Y + C(V)$ (which is the plane $C(V)$ in Figure 10.1, translated over until it contains Y). The intersection of $C(X)$ and $Y + C(V)$ is the line $[x = 5, z = 0]$, so any consistent estimate of $X\beta$ will be in the line $[x = 5, z = 0]$. To see this, note that $C(X)$ consists of vectors with the form $(a, b, 0)'$, and $Y + C(V)$ consists of vectors like $(7, 6, 2)' + (c, d, c)'$. The intersection is those vectors with $c = -2$, so they are of the form $(5, 6 + d, 0)'$ or $(5, b, 0)'$.

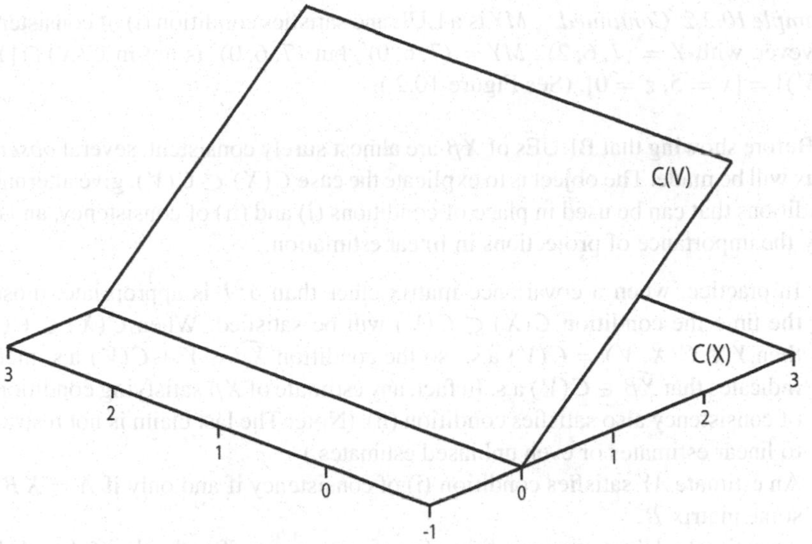

Fig. 10.1 Estimation space and singular covariance space for Example 10.2.2

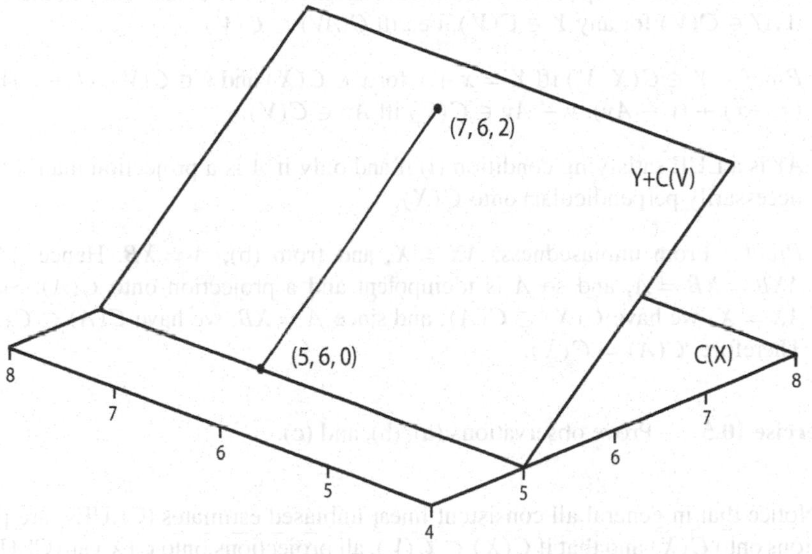

Fig. 10.2 Consistent estimation for Example 10.2.2

The problem with BLUEs of $X\beta$ is that there is no apparent reason why a BLUE should be consistent. The class of linear unbiased estimates (LUEs) is very broad. It consists of all estimates AY with $AX = X$. There are many linear unbiased estimates that are not consistent. For example, Y itself is a LUE and it satisfies condition (ii) of consistency; however, one would certainly not want to use it as an estimate.

Example 10.2.2 Continued. MY is a LUE and satisfies condition (i) of consistency; however, with $Y = (7, 6, 2)'$, $MY = (7, 6, 0)'$, but $(7, 6, 0)'$ is not in $C(X) \cap [Y + C(V)] = [x = 5, z = 0]$. (See Figure 10.2.)

Before showing that BLUEs of $X\beta$ are almost surely consistent, several *observations* will be made. The object is to explicate the case $C(X) \subset C(V)$, give alternative conditions that can be used in place of conditions (i) and (ii) of consistency, and display the importance of projections in linear estimation.

(a) In practice, when a covariance matrix other than $\sigma^2 I$ is appropriate, most of the time the condition $C(X) \subset C(V)$ will be satisfied. When $C(X) \subset C(V)$, then $Y \in C(X, V) = C(V)$ a.s., so the condition $\widehat{X\beta} \in Y + C(V)$ a.s. merely indicates that $\widehat{X\beta} \in C(V)$ a.s. In fact, any estimate of $X\beta$ satisfying condition (i) of consistency also satisfies condition (ii). (Note: The last claim is not restricted to linear estimates or even unbiased estimates.)
(b) An estimate AY satisfies condition (i) of consistency if and only if $A = XB$ for some matrix B.
(c) An estimate AY satisfies condition (ii) of consistency if and only if $(I - A)Y \in C(V)$ for any $Y \in C(X, V)$.
(d) If AY is a LUE of $X\beta$, then AY satisfies condition (ii) of consistency if and only if $AY \in C(V)$ for any $Y \in C(V)$, i.e., iff $C(AV) \subset C(V)$.

Proof $Y \in C(X, V)$ iff $Y = x + v$ for $x \in C(X)$ and $v \in C(V)$. $(I - A)Y = (x - x) + (v - Av)$. $v - Av \in C(V)$ iff $Av \in C(V)$. □

(e) AY is a LUE satisfying condition (i) if and only if A is a projection matrix (not necessarily perpendicular) onto $C(X)$.

Proof From unbiasedness, $AX = X$, and from (b), $A = XB$. Hence $AA = AXB = XB = A$, and so A is idempotent and a projection onto $C(A)$. Since $AX = X$, we have $C(X) \subset C(A)$; and since $A = XB$, we have $C(A) \subset C(X)$. Therefore, $C(A) = C(X)$. □

Exercise 10.5 Prove observations (a), (b), and (c).

Notice that in general all consistent linear unbiased estimates (CLUEs) are projections onto $C(X)$ and that if $C(X) \subset C(V)$, all projections onto $C(X)$ are CLUEs. This goes far to show the importance of projection matrices in estimation. In particular, the BLUEs that we actually found in the previous section satisfy condition (i) by observation (b). Thus, by observation (e) they are projections onto $C(X)$.

Before proving the main result, we need the following proposition:

Proposition 10.2.3 *For $T = V + XUX'$ with U nonnegative definite and $A = X(X'T^-X)^-X'T^-$ with $C(X) \subset C(T)$, $AV = VA'$.*

Proof Consider the symmetric matrix $X(X'T^-X)^-X'$, where the generalized inverses are taken as in Corollary B.41. By the choice of T, write $X = TG$.

$$X' = G'T = G'TT^-T = X'T^-T,$$
$$X(X'T^-X)^-X' = X(X'T^-X)^-X'T^-T = A(V + XUX').$$

Since A is a projection operator onto $C(X)$, $AXUX' = XUX'$; thus,

$$X(X'T^-X)^-X' - XUX' = AV.$$

The lefthand side is symmetric. □

Proposition 10.2.4 *There exists a BLUE of $X\beta$ in model (1) that is a CLUE.*

Proof $AY = X(X'T^-X)^-X'T^-Y$ is a BLUE satisfying condition (i). In Proposition 10.2.3, we proved that $AV = VA'$. By observation (d), the proposition holds. □

Theorem 10.2.5 *If AY is a BLUE of $X\beta$ in model (1), then AY is almost surely consistent.*

Proof By Proposition 10.2.4, a consistent BLUE exists. By Theorem 10.1.5, any BLUE must almost surely satisfy consistency. □

10.3 Hypothesis Testing

We consider the problem of testing the model

$$Y = X\beta + e, \quad e \sim N(0, \sigma^2 V) \tag{1}$$

against the reduced model

$$Y = X_0\gamma + e, \quad e \sim N(0, \sigma^2 V), \tag{2}$$

where $C(X_0) \subset C(X)$. In particular, we use the approach of looking at reductions in the sum of squares for error. First we need to define what we mean by the sum of squares for error.

In the remainder of this section, let $A = X(X'T^-X)^-X'T^-$, where T^- and $(X'T^-X)^-$ are chosen (for convenience) as in Corollary B.41 and as usual $T = V + XUX'$ for some nonnegative definite matrix U such that $C(X) \subset C(T)$. AY is a BLUE of $X\beta$ in model (1). We will define the sum of squares for error (*SSE*) in model (1) as

$$SSE = Y'(I - A)'T^{-}(I - A)Y. \tag{3}$$

The idea of using a quadratic form in the residuals to estimate the error is reasonable, and any quadratic form in the residuals, when normalized, gives an unbiased estimate of σ^2. The terminology *SSE* literally comes from the use of the quadratic form $Y'(I - A)'(I - A)Y$. In particular, if $V = I$, then

$$Y'(I - A)'(I - A)Y/\sigma^2 = Y'(I - A)'[\sigma^2 I]^{-1}(I - A)Y$$

has a χ^2 distribution and is independent of AY. For an arbitrary covariance matrix, an analogous procedure would be to use $Y'(I - A)'V^{-}(I - A)Y$ to get an estimate of σ^2 and develop tests. To simplify computations, we have chosen to define *SSE* as in (3). However, we will show that for $Y \in C(X, V)$, $SSE = Y'(I - A)'V^{-}(I - A)Y$ for any choice of V^{-}.

The sum of squares error in model (2) (SSE_0) is defined similarly. Let $T_0 = V + X_0U_0X_0'$ for a nonnegative definite matrix U_0 for which $C(X_0) \subset C(T_0)$. Let $A_0 = X_0(X_0'T_0^{-}X_0)^{-}X_0'T_0^{-}$, where again for convenience take T_0^{-} and $(X_0'T_0^{-}X_0)^{-}$ as in Corollary B.41.

$$SSE_0 = Y'(I - A_0)'T_0^{-}(I - A_0)Y.$$

Even though V is the same in models (1) and (2), things are complicated by the fact that T and T_0 may be different.

The test will be, for some constant K, based on

$$K\,(SSE_0 - SSE)\big/SSE,$$

which will be shown to have an F distribution. We will use Theorem 1.3.6 and Theorem 1.3.9 to obtain distribution results.

Theorem 10.3.1 *If $Y \in C(T)$, then*

(a) $Y'(I - A)'T^{-}(I - A)Y = Y'(I - A)'V^{-}(I - A)Y$ *for any* V^{-},
(b) $Y'(I - A)'T^{-}(I - A)Y = 0$ *if and only if* $(I - A)Y = 0$.

Proof The proof is given after Proposition 10.3.6. □

These results also hold when A an T are replaced by A_0 and T_0. Denote $C \equiv (I - A)'T^{-}(I - A)$ and $C_0 \equiv (I - A_0)'T_0^{-}(I - A_0)$. Thus $SSE = Y'CY$ and $SSE_0 = Y'C_0Y$.

Theorem 10.3.2

(a) $Y'CY/\sigma^2 \sim \chi^2(\text{tr}(CV), 0)$.
If $X\beta \in C(X_0, V)$, *then*
(b) $Y'(C_0 - C)Y/\sigma^2 \sim \chi^2(\text{tr}(C_0V - CV), \beta'X'C_0X\beta)$
and
(c) $Y'CY$ *and* $Y'(C_0 - C)Y$ *are independent.*

Proof The proof consists of checking the conditions in Theorems 1.3.6 and 1.3.9. There are many conditions to be checked. This is done in Lemmas 10.3.7 through 10.3.9 at the end of the section. □

The last result before stating the test establishes behavior of the distributions under the two models. Model (2) is true if and only if $X\beta \in C(X_0)$.

Theorem 10.3.3

(a) *If* $X\beta \in C(X_0)$, *then* $\Pr(Y \in C(X_0, V)) = 1$ *and* $\beta'X'C_0X\beta = 0$.
(b) *If* $X\beta \notin C(X_0)$, *then either* $X\beta \in C(X_0, V)$ *and* $\beta'X'C_0X\beta > 0$ *or* $X\beta \notin C(X_0, V)$ *and* $\Pr(Y \notin C(X_0, V)) = 1$

Proof

(a) The first part is clear, the second is because $C_0X_0 = 0$.
(b) If $X\beta \notin C(X_0)$, then either $X\beta \in C(X_0, V)$ or $X\beta \notin C(X_0, V)$. If $X\beta \in C(X_0, V)$, then by Theorem 10.3.1b, $\beta'X'C_0X\beta = 0$ if and only if $(I - A_0)X\beta = 0$ or $X\beta = A_0X\beta$. Since $X\beta \notin C(X_0)$, $\beta'X'C_0X\beta > 0$. If $X\beta \notin C(X_0, V)$, suppose $e \in C(V)$ and $Y \in C(X_0, V)$, then $X\beta = Y - e \in C(X_0, V)$, a contradiction. Therefore either $e \notin C(V)$ or $Y \notin C(X_0, V)$. Since $\Pr(e \in C(V)) = 1$, we must have $\Pr(Y \notin C(X_0, V)) = 1$. □

The test at the α level is to reject H_0 that model (2) is adequate if $Y \notin C(X_0, V)$ or if

$$\frac{(SSE_0 - SSE)/\text{tr}[(C_0 - C)V]}{SSE/\text{tr}(CV)} > F(1-\alpha, \text{tr}[(C_0 - C)V], \text{tr}(CV)).$$

This is an α level test because, under H_0, $\Pr(Y \notin C(X_0, V)) = 0$. The power of the test is at least as great as that of a noncentral F test and is always greater than α because if $\beta'X'C_0X\beta = 0$ under the alternative, Theorem 10.3.3 ensures that the test will reject with probability 1.

In the next section we consider extensions of least squares and conditions under which such extended least squares estimates are best estimates.

Proofs of Theorems 10.3.1 and 10.3.2.
Before proceeding with the proofs of the theorems, we need some background results.

Proposition 10.3.4 $A'T^{-}A = A'T^{-} = T^{-}A.$

Proof

$$A'T^{-} = T^{-}X(X'T^{-}X)^{-}X'T^{-} = T^{-}A,$$
$$A'T^{-}A = T^{-}X(X'T^{-}X)^{-}X'T^{-}X(X'T^{-}X)^{-}X'T^{-};$$

but, as discussed earlier, A does not depend on the choice of $(X'T^{-}X)^{-}$ and $(X'T^{-}X)^{-}X'T^{-}X(X'T^{-}X)^{-}$ is a generalized inverse of $X'T^{-}X$, so $A'T^{-}A = T^{-}A$. □

Corollary 10.3.5 $(I - A)'T^{-}(I - A) = (I - A)'T^{-} = T^{-}(I - A).$

In the remainder of this discussion we will let $X = TG$ and $V = TBT$. (The latter comes from Proposition 10.1.4 and symmetry.)

Proposition 10.3.6
(a) $VT^{-}(I - A) = TT^{-}(I - A)$,
(b) $VT^{-}(I - A)V = (I - A)V$.

Proof From Corollary 10.3.5 and unbiasedness, i.e., $AX = X$,

$$TT^{-}(I - A) = T(I - A)'T^{-} = V(I - A)'T^{-} = VT^{-}(I - A).$$

With $V = TBT$ and, from Proposition 10.2.3, AV symmetric, we have

$$\begin{aligned} VT^{-}(I - A)V &= TT^{-}(I - A)V = TT^{-}V(I - A)' = TT^{-}TBT(I - A)' \\ &= TBT(I - A)' = V(I - A)' = (I - A)V. \end{aligned}$$ □

Proof of Theorem 10.3.1. Recalling Proposition 10.1.4, if $Y \in C(T)$, write $Y = Xb_1 + Vb_2$.

(a) Using Proposition 10.2.3,

$$\begin{aligned} Y'(I - A)'T^{-}(I - A)Y &= (Xb_1 + Vb_2)'(I - A)'T^{-}(I - A)(Xb_1 + Vb_2) \\ &= b_2'V(I - A)'T^{-}(I - A)Vb_2 \\ &= b_2'(I - A)VT^{-}(I - A)Vb_2 \\ &= b_2'(I - A)(I - A)Vb_2 \\ &= b_2'(I - A)V(I - A)'b_2 \\ &= b_2'(I - A)VV^{-}V(I - A)'b_2 \\ &= b_2'V(I - A)'V^{-}(I - A)Vb_2 \\ &= Y'(I - A)'V^{-}(I - A)Y. \end{aligned}$$

(b) From the proof of (a),

$$Y'(I-A)'T^-(I-A)Y = b_2'(I-A)V(I-A)'b_2.$$

Recall that we can write $V = EDE'$ with $E'E = I$, $D = \text{Diag}(d_i)$, and $d_i > 0$ for all i.

$$\begin{aligned} Y'(I-A)'T^-(I-A)Y = 0 \quad \text{iff} \quad & b_2'(I-A)V(I-A)'b_2 = 0 \\ \text{iff} \quad & E'(I-A)'b_2 = 0 \\ \text{iff} \quad & (I-A)'b_2 \perp C(E) \\ \text{iff} \quad & (I-A)'b_2 \perp C(V) \\ \text{iff} \quad & V(I-A)'b_2 = 0 \\ \text{iff} \quad & (I-A)Vb_2 = 0 \\ \text{iff} \quad & (I-A)Y = 0. \end{aligned}$$

□

The following lemmas constitute the proof of Theorem 10.3.2.

Lemma 10.3.7

(a) $CVC = C$ *and* $C_0VC_0 = C_0$,
(b) $CVC_0V = CV$,
(c) $VCVCV = VCV$, $\beta'X'CVCX\beta = \beta'X'CX\beta = 0$, $VCVCX\beta = VCX\beta = 0$.

Proof

(a) Using Corollary 10.3.5 and Proposition 10.3.6, $CVC = CVT^-(I-A) = CTT^-(I-A) = (I-A)'T^-TT^-(I-A) = (I-A)'T^-(I-A) = C$.
(b) A similar argument gives the second equality: $CVC_0V = CVT_0^-(I-A_0)V = C(I-A_0)V = T^-(I-A)(I-A_0)V = T^-(I-A)V = CV$.
(c) The equalities in (c) follow from (a) and the fact that $CX = T^-(I-A)X = 0$.

□

Note that Lemma 10.3.7c leads directly to Theorem 10.3.2a. We now establish the conditions necessary for Theorem 10.3.2b.

Lemma 10.3.8 *(a)* $V(C_0-C)V(C_0-C)V = V(C_0-C)V$,
(b) $\beta'X'(C_0-C)V(C_0-C)X\beta = \beta'X'(C_0-C)X\beta = \beta'X'C_0X\beta$,
(c) if $X\beta \in C(X_0, V)$, *then* $V(C_0-C)V(C_0-C)X\beta = V(C_0-C)X\beta$.

Proof From parts (a) and (b) of Lemma 10.3.7,

$$\begin{aligned} V(C_0 - C)V(C_0 - C)V &= VC_0VC_0V - VC_0VCV - VCVC_0V + VCVCV \\ &= VC_0V - VCV - VCV + VCV \\ &= VC_0V - VCV \\ &= V(C_0 - C)V. \end{aligned}$$

To show (b), we need only show that

$$X'(C_0 - C)V(C_0 - C)X = X'(C_0 - C)X.$$

Since $CX = 0$, this is equivalent to showing $X'C_0VC_0X = X'C_0X$. The result is immediate from part (a) of Lemma 10.3.7.

To show (c), we need to show that

$$V(C_0 - C)V(C_0 - C)X\beta = V(C_0 - C)X\beta.$$

Since $CX = 0$, it is enough to show that $VC_0VC_0X\beta - VCVC_0X\beta = VC_0X\beta$. With $VC_0VC_0X\beta = VC_0X\beta$, we only need $VCVC_0X\beta = 0$. Since $X\beta \in C(T_0)$, by assumption, $(I - A_0)X\beta = T_0\gamma$ for some γ,

$$\begin{aligned} VCVC_0X\beta &= VCVT_0^-(I - A_0)X\beta = VCT_0T_0^-(I - A_0)X\beta \\ &= VCT_0T_0^-T_0\gamma = VCT_0\gamma = VC(I - A_0)X\beta \\ &= VT^-(I - A)(I - A_0)X\beta = VT^-(I - A)X\beta = 0. \end{aligned}$$

□

To establish Theorem 10.3.2c, use the following lemma:

Lemma 10.3.9

(a) $VCV(C_0 - C)V = 0$,
(b) $VCV(C_0 - C)X\beta = 0$ *if* $X\beta \in C(X_0, V)$,
(c) $V(C_0 - C)VCX\beta = 0$,
(d) $\beta'X'(C_0 - C)VCX\beta = 0$.

Proof For part (a), $VCV(C_0 - C)V = VCVC_0V - VCVCV = VCV - VCV$. Parts (c) and (d) follow because $CX = 0$; also, (b) becomes the condition $VCVC_0X\beta = 0$, as was shown in the proof of Lemma 10.3.8. □

10.4 Least Squares Consistent Estimation

Definition 10.4.1 An estimate $\tilde{\beta}$ of β is said to be *consistent* if $X\tilde{\beta}$ is a consistent estimate of $X\beta$. $\hat{\beta}$ is said to be a *least squares consistent estimate of* β if for any other consistent estimate $\tilde{\beta}$ and any $Y \in C(X, V)$,

$$(Y - X\hat{\beta})'(Y - X\hat{\beta}) \leq (Y - X\tilde{\beta})'(Y - X\tilde{\beta}).$$

How does this differ from the usual definition of least squares? In the case where $C(X) \subset C(V)$, it hardly differs at all. Any estimate $\tilde{\beta}$ will have $X\tilde{\beta} \in C(X)$, and, as we observed earlier, when $C(X) \subset C(V)$, any estimate of $X\beta$ satisfying condition (i) of consistency (Definition 10.2.1) also satisfies condition (ii). The main difference between consistent least squares and least squares is that we are restricting ourselves to consistent estimates of $X\beta$. As we saw earlier, estimates that are not consistent are just not reasonable, so we are not losing anything. (Recall Example 10.2.2, in which the least squares estimate was not consistent.) The other difference between consistent least squares estimation and regular least squares is that the current definition restricts Y to $C(X, V)$. In the case where $C(X) \subset C(V)$, this restriction would not be necessary because a least squares estimate will have to be a least squares consistent estimate. In the general case, we need to actually use condition (ii) of consistency. Condition (ii) was based on the fact that $e \in C(V)$ a.s. Since e cannot be observed, we used the related fact that $Y \in C(X, V)$ a.s., and made condition (ii) apply only when $Y \in C(X, V)$.

Theorem 10.4.2a *If AY is a CLUE and $(I - A)r \perp C(X) \cap C(V)$ for any $r \in C(X, V)$, then any $\hat{\beta}$ satisfying $AY = X\hat{\beta}$ is a least squares consistent estimate.*

Proof Let $\tilde{\beta}$ be any consistent estimate,

$$\begin{aligned}(Y - X\tilde{\beta})'(Y - X\tilde{\beta}) &= (Y - AY + AY - X\tilde{\beta})'(Y - AY + AY - X\tilde{\beta}) \\ &= Y'(I - A)'(I - A)Y + (AY - X\tilde{\beta})'(AY - X\tilde{\beta}) \\ &\quad + 2Y'(I - A)'(AY - X\tilde{\beta}).\end{aligned}$$

It is enough to show that

$$Y'(I - A)'(AY - X\tilde{\beta}) = 0 \text{ for } Y \in C(X, V).$$

Note that $AY - X\tilde{\beta} \in C(X)$. Also observe that since AY and $X\tilde{\beta}$ are consistent, $AY - X\tilde{\beta} = (Y - X\tilde{\beta}) - (I - A)Y \in C(V)$ for $Y \in C(X, V)$. Thus $AY - X\tilde{\beta} \in C(X) \cap C(V)$. For any $Y \in C(X, V)$, we have

$$(I - A)Y \perp C(X) \cap C(V);$$

so $Y'(I - A)'(AY - X\tilde{\beta}) = 0$. □

We now give a formula for a least squares consistent estimate. Choose V_0 with orthonormal columns such that $C(V_0) = C(X) \cap C(V)$. Also choose V_1 with orthonormal columns such that $C(V_1) \perp C(V_0)$ and $C(V_0, V_1) = C(V)$. It is easily seen that $MY - MV_1\hat{\gamma}$ is a CLUE, where $\hat{\gamma} = [V_1'(I - M)V_1]^{-1}V_1'(I - M)Y$. Observe that

$$C(X, V) = C(X, V_0, V_1) = C(X, (I - M)V_1).$$

To put the computations into a somewhat familiar form, consider the analysis of covariance model

$$Y = [X, (I - M)V_1]\begin{bmatrix} \beta \\ \gamma \end{bmatrix} + e, \quad \mathrm{E}(e) = 0. \tag{1}$$

We are interested in the least squares estimate of $\mathrm{E}(Y)$, so the error vector e is of no interest except that $\mathrm{E}(e) = 0$. The least squares estimate of $\mathrm{E}(Y)$ is $MY + (I - M)V_1\hat{\gamma}$, where $\hat{\gamma} = [V_1'(I - M)V_1]^{-1}V_1'(I - M)Y$. It turns out that $MY - MV_1\hat{\gamma}$ is a CLUE for $X\beta$ in model (10.1.1).

First, $MY - MV_1\hat{\gamma}$ is unbiased, because $\mathrm{E}(MY) = X\beta$ and $\mathrm{E}(\hat{\gamma}) = 0$. $\mathrm{E}(\hat{\gamma})$ is found by replacing Y with $X\beta$ in the formula for $\hat{\gamma}$, but the product of $(I - M)X\beta = 0$. Second, it is clear that for any Y

$$MY - MV_1\hat{\gamma} \in C(X).$$

Finally, it is enough to show that if $Y \in C(V)$, then $MY - MV_1\hat{\gamma} \in C(V)$. If $Y \in C(V)$, then $Y \in C(X, (I - M)V_1) = C(X, V)$. Therefore, the least squares estimate of $\mathrm{E}(Y)$ in model (1) is Y itself. We have two characterizations of the least squares estimate, and equating them gives

$$MY + (I - M)V_1\hat{\gamma} = Y$$

or

$$MY - MV_1\hat{\gamma} = Y - V_1\hat{\gamma}.$$

Now, it is clear that $V_1\hat{\gamma} \in C(V)$; so if $Y \in C(V)$, then $MY - MV_1\hat{\gamma} \in C(V)$.

Proposition 10.4.3 *If $\hat{\beta}$ is an estimate with $X\hat{\beta} = MY - MV_1\hat{\gamma}$, then $\hat{\beta}$ is a least squares consistent estimate.*

Proof By Theorem 10.4.2a, it is enough to show that

$$(Y - MY + MV_1\hat{\gamma}) \perp C(V_0)$$

for any $Y \in C(X, V)$. Let $w \in C(V_0)$; then, since $Mw = w$,

$$\begin{aligned} w'(Y - MY + MV_1\hat{\gamma}) &= w'Y - w'MY + w'MV_1\hat{\gamma} \\ &= w'Y - w'Y + w'V_1\hat{\gamma} \\ &= w'V_1\hat{\gamma} \\ &= 0. \end{aligned}$$

□

Theorem 10.4.2b *If $\hat{\beta}$ is a least squares consistent estimate, then $(Y - X\hat{\beta}) \perp C(X) \cap C(V)$ for any $Y \in C(X, V)$; and if $\tilde{\beta}$ is any other least squares consistent estimate, $X\hat{\beta} = X\tilde{\beta}$ for any $Y \in C(X, V)$.*

Proof Let A be the matrix determined by $AY = MY - MV_1\hat{\gamma}$ for all Y. We show that $AY = X\hat{\beta}$ for any $Y \in C(X, V)$ and are done.

We know, by Definition 10.4.1, that

$$(Y - X\hat{\beta})'(Y - X\hat{\beta}) = (Y - AY)'(Y - AY) \quad \text{for } Y \in C(X, V).$$

As in the proof of Theorem 10.4.2a, we also know that, for $Y \in C(X, V)$,

$$(Y - X\hat{\beta})'(Y - X\hat{\beta}) = (Y - AY)'(Y - AY) + (AY - X\hat{\beta})'(AY - X\hat{\beta}).$$

Therefore,

$$(AY - X\hat{\beta})'(AY - X\hat{\beta}) = 0;$$

hence

$$AY - X\hat{\beta} = 0$$

or

$$AY = X\hat{\beta} \quad \text{for } Y \in C(X, V). \qquad \square$$

Together, Theorems 10.4.2a and 10.4.2b give an "if and only if" condition for $\hat{\beta}$ to be a least squares CLUE.

In the future, any CLUE of $X\beta$, say AY, that satisfies $(I - A)r \perp C(X) \cap C(V)$ for any $r \in C(X, V)$ will be referred to as a least squares CLUE of $X\beta$. The most important result on least squares CLUEs is:

Theorem 10.4.4 *If in model (10.1.1), $C(VV_0) \subset C(V_0)$, then a least squares CLUE of $X\beta$ is a best CLUE of $X\beta$ (and hence a BLUE of $X\beta$).*

Proof Let AY be a least squares CLUE and BY any other CLUE. We need to show that, for any vector ρ,

$$\text{Var}(\rho' AY) \leq \text{Var}(\rho' BY).$$

We can decompose $\text{Var}(\rho' BY)$,

$$\begin{aligned}\text{Var}(\rho' BY) &= \text{Var}[\rho'(B - A)Y + \rho' AY] \\ &= \text{Var}[\rho'(B - A)Y] + \text{Var}(\rho' AY) + 2\text{Cov}[\rho'(B - A)Y, \rho' AY].\end{aligned}$$

Since variances are nonnegative, it suffices to show that

$$0 = \text{Cov}[\rho'(B - A)Y, \rho' AY] = \sigma^2 \rho'(B - A)VA'\rho.$$

First, we will establish that it is enough to show that the covariance is zero when $\rho \in C(X) \cap C(V)$. Let M_0 be the perpendicular projection matrix onto $C(X) \cap C(V)$. Then $\rho = M_0\rho + (I - M_0)\rho$ and

$$\begin{aligned}\rho'(B-A)VA'\rho &= \rho'(B-A)VA'M_0\rho + \rho'(B-A)VA'(I-M_0)\rho \\ &= \rho'M_0(B-A)VA'M_0\rho + \rho'(I-M_0)(B-A)VA'M_0\rho \\ &\quad + \rho'M_0(B-A)VA'(I-M_0)\rho \\ &\quad + \rho'(I-M_0)(B-A)VA'(I-M_0)\rho.\end{aligned}$$

It turns out that all of these terms except the first is zero. Since AY and BY are CLUEs, unbiasedness and observation (d) in Section 2 give $C(AV) \subset C(X) \cap C(V)$ and $C(BV) \subset C(X) \cap C(V)$. By orthogonality,

$$VA'(I-M_0)\rho = 0 \quad \text{and} \quad \rho'(I-M_0)(B-A)V = 0;$$

so

$$\rho'(B-A)VA'\rho = \rho'M_0(B-A)VA'M_0\rho.$$

Henceforth, assume $\rho \in C(X) \cap C(V)$.

To obtain the final result, observe that since AY is a least squares CLUE, any column of $(I-A)V$ is orthogonal to ρ; so

$$V(I-A)'\rho = 0 \quad \text{and} \quad V\rho = VA'\rho.$$

Since, by assumption, $C(VV_0) \subset C(V_0)$, we also have $V\rho \in C(X) \cap C(V)$. The covariance term is $\rho'(B-A)VA'\rho = \rho'(B-A)V\rho$. Since AY and BY are unbiased and $V\rho \in C(X)$, $(B-A)V\rho = 0$; hence

$$\rho'(B-A)VA'\rho = 0. \qquad \square$$

As mentioned in the statement of the theorem, a best CLUE is a BLUE. That occurs because all BLUEs have the same variance and there is a BLUE that is a CLUE.

As would be expected, when $C(X) \subset C(V)$, a least squares CLUE will equal MY for all $Y \in C(V)$. When $C(X) \subset C(V)$, then $C(X) = C(V_0)$; so $MV_1 = 0$ and $MY - MV_1\hat{\gamma} = MY$.

The following theorem characterizes when ordinary least squares estimates are BLUEs.

Theorem 10.4.5 *The following conditions are equivalent:*

(a) $C(VX) \subset C(X)$,
(b) $C(VV_0) \subset C(V_0)$ *and* $X'V_1 = 0$,
(c) MY *is a BLUE for* $X\beta$.

Proof Let X_1 be a matrix with orthonormal columns such that $C(X) = C(V_0, X_1)$ and $V_0'X_1 = 0$. Also let

$$V = [V_0, V_1] \begin{bmatrix} B_{11} & B_{12} \\ B_{21} & B_{22} \end{bmatrix} \begin{bmatrix} V_0' \\ V_1' \end{bmatrix};$$

so $V = V_0 B_{11} V_0' + V_1 B_{22} V_1' + V_0 B_{12} V_1' + V_1 B_{21} V_0'$. By symmetry, $B_{12} = B_{21}'$. Recall that V_0 and V_1 also have orthonormal columns.

$a \Rightarrow b$

Clearly $C(VX) \subset C(V)$; so if $C(VX) \subset C(X)$, we have $C(VX) \subset C(V_0)$. It is easily seen that

$$C(VX) \subset C(V_0) \text{ if and only if } C(VV_0) \subset C(V_0) \text{ and } C(VX_1) \subset C(V_0).$$

We show that $C(VX_1) \subset C(V_0)$ implies that $X_1'V = 0$; hence $X_1'V_1 = 0$ and $X'V_1 = 0$. First $VX_1 = V_1 B_{22} V_1' X_1 + V_0 B_{12} V_1' X_1$; we show that both terms are zero. Consider $VV_0 = V_0 B_{11} + V_1 B_{21}$; since $C(VV_0) \subset C(V_0)$, we must have $V_1 B_{21} = 0$. By symmetry, $B_{12} V_1' = 0$ and $V_0 B_{12} V_1' X_1 = 0$. To see that $VX_1 = V_1 B_{22} V_1' X_1 = 0$, observe that since $C(VX_1) \subset C(V_0)$, it must be true that $C(V_1 B_{22} V_1' X_1) \subset C(V_0)$. However, $C(V_1 B_{22} V_1' X_1) \subset C(V_1)$ but $C(V_0)$ and $C(V_1)$ are orthogonal, so $V_1 B_{22} V_1' X_1 = 0$.

$b \Rightarrow a$

If $X'V_1 = 0$, then $X_1'V_1 = 0$ and $X_1'V = 0$. Write $X = V_0 B_0 + X_1 B_1$ so $VX = VV_0 B_0 + VX_1 B_1 = VV_0 B_0$. Thus,

$$C(VX) = C(VV_0 B_0) \subset C(VV_0) \subset C(V_0) \subset C(X).$$

$b \Rightarrow c$

If $C(VV_0) \subset C(V_0)$, then $MY - MV_1\hat{\gamma}$ is a BLUE. Since $X'V_1 = 0$, $MV_1 = 0$ and MY is a BLUE.

$c \Rightarrow b$

If AY and BY are BLUEs, then $AY = BY$ for $Y \in C(X, V)$. As in Proposition 10.2.3, there exists a BLUE, say AY, such that $AV = VA'$. Since MY is a BLUE and $C(V) \subset C(X, V)$, $AV = MV$ and $MV = VM$. Finally, $VV_0 = VMV_0 = MVV_0$, so $C(VV_0) = C(MVV_0) \subset C(X)$. Since $C(VV_0) \subset C(V)$, we have $C(VV_0) \subset C(V_0)$.

From Theorem 10.4.4 we know that a least squares CLUE is a BLUE; hence $MY = MY - MV_1\hat{\gamma}$ for $Y \in C(X, V) = C(X, V_1)$. Since

$$\hat{\gamma} = \left[V_1'(I - M)V_1\right]^{-1} V_1'(I - M)Y,$$

we must have $0 = MV_1 \left[V_1'(I - M)V_1\right]^{-1} V_1'(I - M)V_1 = MV_1$. Thus $0 = X'V_1$.

□

In the proof of $a \Rightarrow b$, it was noted that $V_1 B_{21} = 0$. That means we can write $V = V_0 B_{11} V_0' + V_1 B_{22} V_1'$. For ordinary least squares estimates to be BLUEs, $C(V)$ must admit an orthogonal decomposition into a subspace contained in $C(X)$ and a subspace orthogonal to $C(X)$. Moreover, the error term e in model (10.1.1) must have $e = e_0 + e_1$, where $\mathrm{Cov}(e_0, e_1) = 0$, $\mathrm{Cov}(e_0) = V_0 B_{11} V_0'$, and $\mathrm{Cov}(e_1) = V_1 B_{22} V_1'$. Thus, with probability 1, the error can be written as the sum of two orthogonal vectors, both in $C(V)$, and one in $C(X)$. The two vectors must also be uncorrelated.

Exercise 10.6 Show that $V_1'(I - M)V_1$ is invertible.
Answer: The columns of V_1 form a basis, so $0 = V_1 b$ iff $b = 0$. Also $(I - M)V_1 b = 0$ iff $V_1 b \in C(X)$, but $V_1 b \in C(X)$ iff $b = 0$ by choice of V_1. Thus, $(I - M)V_1 b = 0$ iff $b = 0$; hence $(I - M)V_1$ has full column rank and $V_1'(I - M)V_1$ is invertible.

Exercise 10.7 Show that if $MY - MV_1\hat{\gamma}$ is a BLUE of $X\beta$, then $C(VV_0) \subset C(V_0)$.
Hint: Multiply on the right by V_0 after showing that

$$\left[M - MV_1\left[V_1'(I - M)V_1\right]^{-1} V_1'(I - M)\right]V$$
$$= V\left[M - (I - M)V_1\left[V_1'(I - M)V_1\right]^{-1} V_1'M\right].$$

We include a result that allows one to find the matrix V_1, and thus find least squares CLUEs.

Proposition 10.4.6 *$r \perp C(X) \cap C(V)$ if and only if $r \in C(I - M, I - M_V)$.*

Proof If $r \in C(I - M, I - M_V)$, then write $r = (I - M)r_1 + (I - M_V)r_2$.
Let $w \in C(X) \cap C(V)$ so that $w = M_V w = Mw$. We need to show that $w'r = 0$. Observe that

$$\begin{aligned} w'r &= w'(I - M)r_1 + w'(I - M_V)r_2 \\ &= w'M(I - M)r_1 + w'M_V(I - M_V)r_2 \\ &= 0. \end{aligned}$$

The vector space here is, say, $\mathbf{R}^n$. Let $r[C(X) \cap C(V)] = m$. From the above result, $C(I - M, I - M_V)$ is orthogonal to $C(X) \cap C(V)$. It is enough to show that the rank of $C(I - M, I - M_V)$ is $n - m$. If this is not the case, there exists a vector $w \neq 0$ such that $w \perp C(I - M, I - M_V)$ and $w \perp C(X) \cap C(V)$.
Since $w \perp C(I - M, I - M_V)$, we have $(I - M)w = 0$ or $w = Mw \in C(X)$. Similarly, $w = M_V w \in C(V)$; so $w \in C(X) \cap C(V)$, a contradiction. □

To find V_1, one could use Gram–Schmidt to first get an orthonormal basis for $C(I - M, I - M_V)$, then extend this to $\mathbf{R}^n$. The extension is a basis for $C(V_0)$. Finally, extend the basis for $C(V_0)$ to an orthonormal basis for $C(V)$. The extension is a basis for $C(V_1)$. A basis for $C(X_1)$ can be found by extending the basis for $C(V_0)$ to a basis for $C(X)$.

Exercise 10.8 Give the general form for a BLUE of $X\beta$ in model (10.1.1).
Hint: Add something to a particular BLUE.

Exercise 10.9 From inspecting Figure 10.2, give the least squares CLUE for Example 10.2.2. Do not do any matrix manipulations.

Remark Suppose that we are analyzing the model $Y = X\beta + e$, $\mathrm{E}(e) = 0$, $\mathrm{Cov}(e) = \Sigma(\theta)$, where $\Sigma(\theta)$ is some nonnegative definite matrix depending on a vector of unknown parameters θ. The special case where $\Sigma(\theta) = \sigma^2 V$ is what we have been considering so far. It is clear that if θ is known, our current theory gives BLUEs. If it happens to be the case that for any value of θ the BLUEs are identical, then the BLUEs are known even though θ may not be. This is precisely what we have been doing with $\Sigma(\theta) = \sigma^2 V$. We have found BLUEs for any value of σ^2, and they do not depend on σ^2. Another important example of this occurs when $C(\Sigma(\theta)X) \subset C(X)$ for any θ. In this case, least squares estimates are BLUEs for any θ, and least squares estimates do not depend on θ, so it does not matter that θ is unknown. The split plot design model is one in which the covariance matrix depends on two parameters, but for any value of those parameters, least squares estimates are BLUEs. Such models are examined in the next chapter.

Exercise 10.10 Show that ordinary least squares estimates are best linear unbiased estimates in the model $Y = X\beta + e$, $\mathrm{E}(e) = 0$, $\mathrm{Cov}(e) = V$ if the columns of X are eigenvectors of V.

Exercise 10.11 Use Definition B.31 and Proposition 10.4.6 to show that $M_{C(X)\cap C(V)} = M - M_W$ where $C(W) = C[M(I - M_V)]$.

Another way to find V_0 is due to Rao and Mitra (1971, pp. 118, 119). $V_0 = X\tilde{U}$ where $C(\tilde{U}) = C[X'(I - M_V)]^\perp$. Note that $(I - M_V)X\tilde{U} = 0$, so $C(X\tilde{U}) \subset C(V_0)$. To see that $C(V_0) \subset C(X\tilde{U})$, take $v \in C(V_0)$. Then $v = Xb$ for some b but also $(I - M_V)Xb = 0$, so $b \in C[X'(I - M_V)]^\perp$ and $b = \tilde{U}\gamma$ for some γ, hence $v = X\tilde{U}\gamma$. Note that $(I - M_V)$ can be replaced with any matrix B having $C(B) = C(V)^\perp$ and also that $r(V_0) = r(X) + r(V) - r(X, V)$.

10.5 Perfect Estimation and More

One of the interesting things about the linear model

$$Y = X\beta + e, \quad \mathrm{E}(e) = 0, \quad \mathrm{Cov}(e) = \sigma^2 V, \tag{1}$$

is that when $C(X) \not\subset C(V)$ you can learn things about the parameters with probability 1. We identify these estimable functions of the parameters and consider the process of estimating those parameters that are not perfectly known.

Earlier, to treat $C(X) \not\subset C(V)$, we obtained estimates and tests by replacing V in model (1) with T where $C(X) \subset C(T)$. Although we showed that the procedure works, it is probably not the first thing one would think to do. In this section, after isolating the estimable functions of $X\beta$ that can be known perfectly, we replace model (1) with a new model $\tilde{Y} = V_0\gamma + e$, $\mathrm{E}(e) = 0$, $\mathrm{Cov}(e) = \sigma^2 V$ in which $C(V_0) \subset C(V)$ and $\tilde{Y}$ is just Y minus a perfectly known component of $X\beta$. I find it far more intuitive to make adjustments to the model matrix X, something we do regularly in defining reduced and restricted models, than to adjust the covariance matrix V. Additional details are given in Christensen and Lin (2010).

Example 10.5.1 Consider a one-sample model $y_i = \mu + \varepsilon_i$, $i = 1, 2, 3$, with uncorrelated observations but in which the second and third observations have variance 0. If this model is correct, you obviously should have $\mu = y_2 = y_3$ with probability 1. The matrices for model (1) are

$$Y = \begin{bmatrix} y_1 \\ y_2 \\ y_3 \end{bmatrix}, \quad X = \begin{bmatrix} 1 \\ 1 \\ 1 \end{bmatrix}, \quad V = \begin{bmatrix} 1\,0\,0 \\ 0\,0\,0 \\ 0\,0\,0 \end{bmatrix}.$$

All of our examples in this section use this same Y vector.

The first thing to do with models having $C(X) \not\subset C(V)$ is to see whether they are even plausible for the data. In particular, Lemma 1.3.5 implies that $\Pr[(Y - X\beta) \in C(V)] = 1$ so that $Y \in C(X, V)$ a.s. This should be used as a model-checking device. If $Y \notin C(X, V)$, you clearly have the wrong model.

Example 10.5.1 Continued. In this example, if $y_2 \neq y_3$ we obviously have the wrong model. It is easily seen that

$$C(X, V) = C\left(\begin{bmatrix} 0 \\ 1 \\ 1 \end{bmatrix}, \begin{bmatrix} 1 \\ 0 \\ 0 \end{bmatrix}\right)$$

and if $y_2 \neq y_3$, $Y \notin C(X, V)$.

To identify the estimable parameters that can be known perfectly, let Q be a full rank matrix with

$$C(Q) = C(V)^{\perp}.$$

Note that if $C(X) \not\subset C(V)$, then $Q'X \neq 0$. Actually, the contrapositive is more obvious, if $Q'X = 0$ then

$$C(X) \subset C(Q)^{\perp} = \left[C(V)^{\perp}\right]^{\perp} = C(V).$$

The fact that $Q'X \neq 0$ means that the estimable function $Q'X\beta$ is nontrivial. Note also that $C(X) \not\subset C(V)$ implies that V must be singular. If V were nonsingular, then $C(V) = \mathbf{R}^n$ and $C(X)$ has to be contained in it.

We can now identify the estimable functions that are known perfectly. Because $\Pr[(Y - X\beta) \in C(V)] = 1$, clearly $\Pr[Q'(Y - X\beta) = 0] = 1$ and $Q'Y = Q'X\beta$ a.s. Therefore, whenever $C(X) \not\subset C(V)$, there are nontrivial estimable functions of β that we can learn without error. Moreover, $\text{Cov}(Q'Y) = \sigma^2 Q'VQ = 0$ and only linear functions of $Q'Y$ will have 0 covariance matrices, so only linear functions of $Q'X\beta$ will be estimated perfectly.

Exercise 10.13 Show that $\text{Cov}(B'Y) = 0$ iff $B' = B_*'Q'$ for some matrix B_*.

Hint: First, decompose B into the sum of two matrices B_0 and B_1 with $C(B_0) \subset C(V)$ and $C(B_1) \perp C(V)$. Then use a singular value decomposition of V to show that $B_0'VB_0 = 0$ iff $B_0 = 0$.

If $Q'X$ has full column rank, we can actually learn all of $X\beta$ without error. In that case, with probability 1 we can write

$$X\beta = X\left(X'QQ'X\right)^{-1}\left[X'QQ'X\right]\beta = X(X'QQ'X)^{-1}X'QQ'Y.$$

For convenience, define A so that $AY \equiv X(X'QQ'X)^{-1}X'QQ'Y$, in which case $X\beta = AY$ a.s. In particular, it is easy to see that $\text{E}[AY] = X\beta$ and $\text{Cov}[AY] = 0$. We now illustrate these matrix formulations in some simple examples to show that the matrix results give obviously correct answers.

Example 10.5.1 Continued. Using the earlier forms for X and V,

$$Q = \begin{bmatrix} 0 & 0 \\ 1 & 0 \\ 0 & 1 \end{bmatrix}.$$

It follows that

$$Q'X = \begin{bmatrix} 1 \\ 1 \end{bmatrix}$$

and $Q'Y = Q'X\beta$ reduces to

$$\begin{bmatrix} y_2 \\ y_3 \end{bmatrix} = \mu J_2 \quad \text{a.s.}$$

Obviously, for this to be true, $y_2 = y_3$ a.s. Moreover, since $Q'X$ is full rank, upon observing that $X'QQ'X = 2$ we can compute

$$\mu J_3 = X\beta = AY = [(y_2 + y_3)/2]J_3 \quad \text{a.s.}$$

Example 10.5.2 This is a two-sample problem, with the first two observations from sample one and the third from sample two. Again, observations two and three have 0 variance. Clearly, with probability 1, $\mu_1 = y_2$ and $\mu_2 = y_3$. The key matrices are

$$X = \begin{bmatrix} 1 & 0 \\ 1 & 0 \\ 0 & 1 \end{bmatrix}, \quad V = \begin{bmatrix} 1 & 0 & 0 \\ 0 & 0 & 0 \\ 0 & 0 & 0 \end{bmatrix}, \quad Q = \begin{bmatrix} 0 & 0 \\ 1 & 0 \\ 0 & 1 \end{bmatrix}.$$

Since $\beta = [\mu_1, \mu_2]'$ in the two-sample problem and

$$Q'X = \begin{bmatrix} 1 & 0 \\ 0 & 1 \end{bmatrix},$$

we have

$$\begin{bmatrix} \mu_1 \\ \mu_2 \end{bmatrix} = Q'X\beta = Q'Y = \begin{bmatrix} y_2 \\ y_3 \end{bmatrix} \quad \text{a.s.}$$

In particular, with probability 1,

$$\begin{bmatrix} \mu_1 \\ \mu_1 \\ \mu_2 \end{bmatrix} = X\beta = AY = \begin{bmatrix} y_2 \\ y_2 \\ y_3 \end{bmatrix}.$$

Example 10.5.3 This is a one-sample problem similar to Example 10.5.1 except that now the first two observations have variance 1 but the third has variance 0. The key matrices are

$$X = \begin{bmatrix} 1 \\ 1 \\ 1 \end{bmatrix}, \quad V = \begin{bmatrix} 1 & 0 & 0 \\ 0 & 1 & 0 \\ 0 & 0 & 0 \end{bmatrix}, \quad Q = \begin{bmatrix} 0 \\ 0 \\ 1 \end{bmatrix}.$$

With $Q'X = 1$ we get $Q'X\beta = \mu$ equaling $Q'Y = y_3$ with probability 1. Moreover, since $X'QQ'X = 1$ we can easily compute

$$\mu J_3 = X\beta = AY = y_3 J_3.$$

The next two examples do not have $Q'X$ with full rank, so they actually have something to estimate.

Example 10.5.4 Consider a two-sample problem similar to Example 10.5.2 except that now the first two observations have variance 1 but the third has variance 0. The key matrices are

$$X = \begin{bmatrix} 1 & 0 \\ 1 & 0 \\ 0 & 1 \end{bmatrix}, \quad V = \begin{bmatrix} 1 & 0 & 0 \\ 0 & 1 & 0 \\ 0 & 0 & 0 \end{bmatrix}, \quad Q = \begin{bmatrix} 0 \\ 0 \\ 1 \end{bmatrix}.$$

With $\beta = [\mu_1, \mu_2]'$ and

$$Q'X = \begin{bmatrix} 0 & 1 \end{bmatrix},$$

we get

$$\mu_2 = \begin{bmatrix} 0 & 1 \end{bmatrix} \begin{bmatrix} \mu_1 \\ \mu_2 \end{bmatrix} = Q'X\beta = Q'Y = y_3 \quad \text{a.s.}$$

Clearly, $y_3 = \mu_2$ a.s. but μ_1 would be estimated with the average $(y_1 + y_2)/2$. More on this later.

Example 10.5.5 Finally, in the two-sample problem we move the second observation from the first group to the second group. This time

$$X = \begin{bmatrix} 1 & 0 \\ 0 & 1 \\ 0 & 1 \end{bmatrix}, \quad V = \begin{bmatrix} 1 & 0 & 0 \\ 0 & 1 & 0 \\ 0 & 0 & 0 \end{bmatrix}, \quad Q = \begin{bmatrix} 0 \\ 0 \\ 1 \end{bmatrix},$$

with

$$Q'X = \begin{bmatrix} 0 & 1 \end{bmatrix}.$$

Again, with probability 1,

$$\mu_2 = \begin{bmatrix} 0 & 1 \end{bmatrix} \begin{bmatrix} \mu_1 \\ \mu_2 \end{bmatrix} = Q'X\beta = Q'Y = y_3,$$

so $y_3 = \mu_2$ a.s. But this time, only y_1 would be used to estimate μ_1, and y_2 is of no value for estimating the means. However, y_2 could be used to estimate an unknown variance via $(y_2 - \mu_2)^2 = (y_2 - y_3)^2$. Similar results on estimating the variance apply in all the examples except Example 10.5.4.

To get perfect estimation of *anything*, we need $C(X) \not\subset C(V)$. To get perfect estimation of *everything* we need $C(X) \cap C(V) = \{0\}$. In other words, to get perfect estimation of $X\beta$ we need $Q'X$ with full column rank and to get $Q'X$ with full column rank, we need $C(X) \cap C(V) = \{0\}$.

Proposition 10.5.6 *For X of full column rank, $Q'X$ is of full column rank if and only if $C(X) \cap C(V) = \{0\}$.*

Proof This is a special case of Lemma 10.5.7. □

Although $C(X) \cap C(V) = \{0\}$ is actually a necessary and sufficient condition for perfect estimation of $X\beta$, with the methods we have illustrated it is not obviously sufficient. For our current method, we need $Q'X$ to have full column rank, which obviously will not happen if X is not full rank. Fortunately, we can always simply choose X to have full column rank. In addition, we close this section with the mathematics needed to deal with arbitrary X.

Now consider models in which some aspect of $X\beta$ is known but some aspect is not. In particular, we know that $Q'X\beta$ is known, but how do we estimate the rest of $X\beta$? As discussed above, we must now consider the case where $C(X) \cap C(V) \neq \{0\}$. Write $\beta = \beta_0 + \beta_1$ with $\beta_0 \in C(X'Q)$ and $\beta_1 \perp C(X'Q)$. We show that $X\beta_0$ is known, so that we need only estimate $X\beta_1$ to learn all that can be learned about $X\beta$. These methods make no assumption about $r(X)$.

In fact, β_0 is known, not just $X\beta_0$. Let $P_{X'Q}$ be the ppo onto $C(X'Q)$. By the definition of β_0 as part of an (unique) orthogonal decomposition, with probability 1,

$$\beta_0 = P_{X'Q}\beta = X'Q[Q'XX'Q]^- Q'X\beta = X'Q[Q'XX'Q]^- Q'Y.$$

Since the perpendicular projection operator does not depend on the choice of generalized inverse, neither does β_0.

Now we show how to estimate $X\beta_1$. Let V_0 be such that $C(V_0) = C(X) \cap C(V)$. Proposition 10.4.6 can be used to find V_0. Note that $\beta_1 \perp C(X'Q)$ iff $Q'X\beta_1 = 0$ iff $X\beta_1 \perp C(Q)$ iff $X\beta_1 \in C(V)$ iff $X\beta_1 \in C(V_0)$ iff $X\beta_1 = V_0\gamma$ for some γ. Since $X\beta_0$ is fixed and known, it follows that $\mathrm{E}(Y - X\beta_0) = X\beta_1 \in C(V_0)$ and $\mathrm{Cov}(Y - X\beta_0) = \sigma^2 V$, so we can estimate $X\beta_1$ by fitting

$$Y - X\beta_0 = V_0\gamma + e, \quad \mathrm{E}(e) = 0, \quad \mathrm{Cov}(e) = \sigma^2 V, \tag{2}$$

and taking

$$X\hat{\beta}_1 \equiv V_0\hat{\gamma} = V_0(V_0'V^- V_0)^- V_0 V^-(Y - X\beta_0),$$

cf. Section 2. Under normality, tests are also relatively easy to construct.

Example 10.5.4 Continued. Using the earlier versions of X, V, Q, and $Q'X$, observe that

$$C(V_0) = C\left(\begin{bmatrix}1\\1\\0\end{bmatrix}\right), \quad C(X'Q) = C\left(\begin{bmatrix}0\\1\end{bmatrix}\right).$$

It follows that with $\beta = [\mu_1, \mu_2]'$,

$$\beta_0 = \begin{bmatrix} 0 \\ \mu_2 \end{bmatrix}, \quad \beta_1 = \begin{bmatrix} \mu_1 \\ 0 \end{bmatrix}.$$

Thus, since we already know that $\mu_2 = y_3$ a.s., $X\beta_0 = [0, 0, \mu_2]' = [0, 0, y_3]'$ and $X\beta_1 = [\mu_1, \mu_1, 0]'$. Finally, model (2) reduces, with probability 1, to

$$Y - X\beta_0 = \begin{bmatrix} y_1 \\ y_2 \\ y_3 - \mu_2 \end{bmatrix} = \begin{bmatrix} y_1 \\ y_2 \\ 0 \end{bmatrix} = \begin{bmatrix} 1 \\ 1 \\ 0 \end{bmatrix} \gamma + e.$$

Recalling that $X\beta_1 \equiv V_0\gamma$, it is easily seen in this example that the BLUE of $\mu_1 \equiv \gamma$ is $(y_1 + y_2)/2$.

This theory applied to Example 10.5.1 is quite degenerate, but it still works.

Example 10.5.1 Continued. Using the earlier versions of X, V, Q, and $Q'X$, observe that

$$C(V_0) = C\left(\begin{bmatrix} 0 \\ 0 \\ 0 \end{bmatrix}\right), \qquad C(X'Q) = C\left(\begin{bmatrix} 1 & 1 \end{bmatrix}\right).$$

It follows that

$$\beta_0 = \mu, \quad \beta_1 = 0.$$

Since we already know that $\mu = y_2 = y_3$ a.s.,

$$X\beta_0 = \begin{bmatrix} \mu \\ \mu \\ \mu \end{bmatrix} = \begin{bmatrix} y_2 \\ y_2 \\ y_2 \end{bmatrix} = \begin{bmatrix} y_3 \\ y_3 \\ y_3 \end{bmatrix} = \begin{bmatrix} y_2 \\ y_2 \\ y_3 \end{bmatrix} \text{ a.s.}$$

and $X\beta_1 = [0, 0, 0]'$. Finally, model (2) reduces, with probability 1, to

$$Y - X\beta_0 = \begin{bmatrix} y_1 - \mu \\ y_2 - \mu \\ y_3 - \mu \end{bmatrix} = \begin{bmatrix} y_1 - y_2 \\ 0 \\ 0 \end{bmatrix} = \begin{bmatrix} 0 \\ 0 \\ 0 \end{bmatrix} \gamma + e,$$

which provides us with one degree of freedom for estimating σ^2 using either of $(y_1 - y_i)^2$, $i = 2, 3$.

The results in the early part of this section on perfect estimation of $X\beta$ required X to be of full rank. That is never a very satisfying state of affairs. Rather than assuming X to be of full rank and considering whether $Q'X$ is also of full rank, the more general condition for estimating $X\beta$ perfectly is that $r(X) = r(Q'X)$. Moreover, with $A \equiv X(X'QQ'X)^- X'QQ'$, we always have perfect estimation of $AX\beta$ because $AX\beta$ is a linear function of $Q'X\beta = Q'Y$ a.s. but for perfect estimation of $X\beta$ we need

$$X\beta = AX\beta = AY \quad \text{a.s.}$$

for any β which requires A to be a projection operator onto $C(X)$. This added generality requires some added work.

Lemma 10.5.7

(a) $r(X) = r(Q'X)$ *iff for any* b, $Q'Xb = 0$ *implies* $Xb = 0$.
(b) $r(X) = r(Q'X)$ *iff* $C(X) \cap C(V) = \{0\}$.

Proof Proof of (a): Recall that $r(Q'X) = r(X)$ iff $r[\mathcal{N}(Q'X)] = r[\mathcal{N}(X)]$. Since the null spaces have $\mathcal{N}(X) \subset \mathcal{N}(Q'X)$, it is enough to show that $\mathcal{N}(Q'X) = \mathcal{N}(X)$ is equivalent to the condition that for any b, $Q'Xb = 0$ implies $Xb = 0$ and, in particular, it is enough to show that $\mathcal{N}(Q'X) \subset \mathcal{N}(X)$ is equivalent to the condition. But by the very definition of the null spaces, $\mathcal{N}(Q'X) \subset \mathcal{N}(X)$ is equivalent to the condition that for any b we have $Q'Xb = 0$ implies that $Xb = 0$.

Proof of (b): Note that for any b,

$$Q'Xb = 0 \quad \text{iff} \quad Xb \perp C(Q) \quad \text{iff} \quad Xb \in C(Q)^{\perp} = \left[C(V)^{\perp}\right]^{\perp} = C(V),$$

so

$$Q'Xb = 0 \quad \text{iff} \quad Xb \in C(V) \quad \text{iff} \quad Xb \in C(X) \cap C(V).$$

If follows immediately that if $C(X) \cap C(V) = \{0\}$, then $Q'Xb = 0$ implies $Xb = 0$ and $r(X) = r(Q'X)$. It also follows immediately that since $Q'Xb = 0$ is equivalent to having $Xb \in C(X) \cap C(V)$, the condition that $Q'Xb = 0$ implies $Xb = 0$ means that the only vector in $C(X) \cap C(V)$ is the 0 vector. □

Proposition 10.5.8 *If* $r(X) = r(Q'X)$, *the matrix* $A \equiv X(X'QQ'X)^{-}X'QQ'$ *is a projection operator onto* $C(X)$.

Proof By its definition we clearly have $C(A) \subset C(X)$, so it is enough to show that $AX = X$.

Let $M_{Q'X}$ be the ppo onto $C(Q'X)$. Note that for any b, $Xb - AXb \in C(X)$. Moreover, from the definitions of A and $M_{Q'X}$,

$$Q'Xb - Q'AXb = Q'Xb - M_{Q'X}Q'Xb = 0.$$

Writing

$$0 = Q'Xb - M_{Q'X}Q'Xb = Q'X\left[I - (X'QQ'X)^{-}X'QQ'X\right]b,$$

by the condition $r(X) = r(Q'X)$ and Lemma 10.5.7a, we have

$$0 = X\left[I - (X'QQ'X)^{-}X'QQ'X\right]b = Xb - AXb,$$

hence $X = AX$. □

Exercise 10.14 Show that the results in this section do not depend on the particular choice of Q.

Exercise 10.15 Let $C(V_0) = C(X) \cap C(V)$, $C(X) = C(V_0, X_1)$, $C(V) = C(V_0, V_1)$ with the columns of V_0, V_1, and X_1 being orthonormal. Show that the columns of $[V_0, V_1, X_1]$ are linearly independent.

Hint: Write $V_0 b_0 + V_1 b_1 + X_1 b_2 = 0$ and show that $b_i = 0$, $i = 0, 1, 2$. In particular, write

$$0.5V_0 b_0 + V_1 b_1 = -(0.5V_0 b_0 + X_1 b_2),$$

$0.5V_0 b_0 + V_1 b_1 \in C(V)$ and $-(0.5V_0 b_0 + X_1 b_2) \in C(X)$ so the vector is in $C(V_0) = C(X) \cap C(V)$.

References

Christensen, R., & Lin, Y. (2010). Linear models that allow perfect estimation. *Statistical Papers, 54*, 695–708.

Ferguson, T. S. (1967). *Mathematical statistics: A decision theoretic approach*. New York: Academic.

Groß, J. (2004). The general Gauss-Markov model with possibly singular dispersion matrix. *Statistical Papers, 25*, 311–336.

Rao, C. R., & Mitra, S. K. (1971). *Generalized inverse of matrices and its applications*. New York: Wiley.

Searle, S. R., & Pukelsheim, F. (1987). Estimation of the mean vector in linear models, *Technical Report BU-912-M, Biometrics Unit*. Ithaca, NY: Cornell University.

Chapter 11
Split Plot Models

Abstract This chapter introduces a cluster sampling model and then adapts that model to develop generalizations of split plot models. Split plot models are among the simplest of the mixed models considered in *ALM-III* in that they involve only two independent error terms (or, equivalently, two variance components). The chapter closes with a discussion of issues related to properly identifying the existence of two random error terms

In an experiment with at least two factors, it is sometimes convenient to apply somc of the factors to large experimental units (called *whole plots*) and then to split the large units into smaller parts to which the remaining factors are applied. The subdivisions of the whole plots are called *subplots* or *split plots*.

Split plot designs are often used when either (1) the factors applied to whole plots are not of direct interest or (2) some factors require larger experimental units than the other factors. The first case is illustrated with an experiment to evaluatc crop yields when using varying levels of a standard herbicidc and a new pesticide. If the standard herbicide is not of direct interest, but rather primary interest is in the effects of the pesticide and any possible interaction between the herbicide and the pesticide, then it is appropriate to apply herbicides as whole-plot treatments. (It will be seen later that interaction contrasts and comparisons between pesticides are subject to less error than comparisons among herbicides.) The second case that split plot designs are often used for can also be illustrated with this experiment. If the standard herbicide is applied using a tractor, but the new pesticide is applied by crop dusting, then the experimental procedure makes it necessary to use pesticides as whole-plot treatments. (Clearly, an airplane requires a larger plot of ground for spraying than does a tractor.)

It is of interest to note that a split plot design can be thought of as an (unbalanced) incomplete block design. In this approach, each whole plot is thought of as a block. Each block contains the treatments that are all combinations of the subplot factor levels with the one combination of whole-plot factor levels that was applied. As with other incomplete block designs, a split plot design is necessary when there are not

R. Christensen, *Plane Answers to Complex Questions*, Springer Texts in Statistics,
https://doi.org/10.1007/978-3-030-32097-3_11

enough blocks available that can accommodate all of the treatment combinations. In split plot designs, this means that there are not enough subplots per whole plot so that all treatment combinations could be applied at the subplot level. If, in addition, there are not enough whole plots so that each treatment combination could be applied to a whole plot, then a split plot design is an attractive option.

Mathematically, the key characteristic of a split plot model is the covariance structure. Typically, observations taken on the subplots of any particular whole plot are assumed to have a constant positive correlation. Observations taken on different whole plots are assumed to be uncorrelated.

The main purpose of this chapter is to derive the analysis for split plot models. In Section 1, we consider a special cluster sampling model. The cluster sampling model has the same covariance structure as a split plot model. In Section 2, we consider ways of generalizing the cluster sampling model that allow for an easy analysis of the data. The discussion in Section 2 is really an examination of generalized split plot models. Section 3 derives the analysis for the traditional split plot model by using the results of Section 2. Section 4 discusses the issues of identifying an appropriate error term and of subsampling.

Sections 1 and 2 are closely related to Christensen (1984) and (1987), respectively. In fact, Christensen (1987) is probably easier to read than Section 2 because it includes more introductory material and fewer of the mathematical details. Closely related work is contained in Monlezun and Blouin (1988) and Mathew and Sinha (1992). Christensen (2015) examines practical issues associated with applying this theory. A general review of methods for analyzing cluster sampling models is given in Skinner, Holt, and Smith (1989).

11.1 A Cluster Sampling Model

A commonly used technique in survey sampling is *cluster sampling* (also called *two-stage sampling*). This technique is applied when the population to be sampled consists of some kind of clusters. The sample is obtained by taking a random sample of clusters and then taking a random sample of the individuals within each of the sampled clusters. For example, suppose it was desired to sample the population of grade school students in Montana. One could take a random sample of grade schools in the state, and then for each school that was chosen take a random sample of the students in the school. One complication of this method is that students from the same school will tend to be more alike than students from different schools. In general, there will be a nonnegative correlation among the individual units within a cluster.

General Cluster Sampling Models

Suppose n observations are available from a two-stage sample with c clusters. From each cluster, m_i units are sampled and variables $y, x_1, \ldots, x_p$ are obtained. Since observations in a cluster are typically not independent, we will consider the linear model

$$Y = X\beta + e, \quad e \sim N(0, \sigma^2 V),$$

where X is $n \times p$ of rank r and (assuming the elements of Y are listed by clusters) V is the block diagonal matrix

$$V = \text{Blk diag}(V_i),$$

where V_i is an $m_i \times m_i$ *intraclass correlation* matrix

$$V_i = \begin{bmatrix} 1 & \rho & \cdots & \rho \\ \rho & 1 & \cdots & \rho \\ \vdots & \vdots & \ddots & \vdots \\ \rho & \rho & \cdots & 1 \end{bmatrix}.$$

If we let $J_{m(i)}$ be an $m_i \times 1$ vector of ones, then

$$V = (1-\rho)I + \rho \text{Blk diag}(J_{m(i)}J'_{m(i)}).$$

Now let X_1 be an $n \times c$ matrix of indicator variables for the clusters. In other words, a row of the ith column of X_1 is 1 if the row corresponds to an observation from the ith cluster and 0 otherwise. It follows that $X_1X_1' = \text{Blk diag}(J_{m(i)}J'_{m(i)})$, so

$$V = (1-\rho)I + \rho X_1X_1'. \tag{1}$$

In fact, equation (1) holds even if the elements of Y are not listed by cluster.

We can now provide an interesting condition for when ordinary least squares (OLS) estimates are best linear unbiased estimates (BLUEs) in cluster sampling models. Recall from Theorem 10.4.5 that OLS estimates are BLUEs if and only if $C(VX) \subset C(X)$. This condition holds if and only if $C(X_1X_1'X) \subset C(X)$. Since the columns of X_1 are indicator variables for the clusters, $X_1X_1'X$ takes each column of X, computes the cluster totals, and replaces each component with the corresponding cluster total. Thus, OLS estimates are BLUEs if and only if for any variable in the model, the variable formed by replacing each component with the corresponding cluster total is also, either implicitly or explicitly, contained in the model.

A Special Cluster Sampling Model

We now consider a particular cluster sampling model for which OLS estimates are BLUEs, and for which tests and confidence intervals are readily available for the most interesting parameters. Consider a model in which X can be written as $X = [X_1, X_2]$, where X_1 is again the matrix of indicator variables for the clusters. Rewriting the linear model as

$$Y = [X_1, X_2]\begin{bmatrix} \alpha \\ \gamma \end{bmatrix} + e \tag{2}$$

leads to the interpretation that the α_is are separate cluster effects. Typically, one would not be very interested in these cluster effects. One's primary interest would be in the vector γ.

It is easily seen that $C(VX) \subset C(X)$, so OLS estimates are BLUEs. The noteworthy thing about this model is that inference on the parameter vector γ can proceed just as when $\mathrm{Cov}(Y) = \sigma^2 I$. Treating (2) as an analysis of covariance model, we obtain for any estimable function $\lambda'\gamma$

$$\lambda'\hat{\gamma} = \lambda'\left[X_2'(I - M_1)X_2\right]^{-} X_2'(I - M_1)Y,$$

where M_1 is the perpendicular projection matrix onto $C(X_1)$.

The variance of $\lambda'\hat{\gamma}$ is

$$\mathrm{Var}(\lambda'\hat{\gamma}) = \sigma^2\lambda'\left[X_2'(I - M_1)X_2\right]^{-} X_2'(I - M_1)V(I - M_1)X_2\left[X_2'(I - M_1)X_2\right]^{-}\lambda. \tag{3}$$

From (1) observe that

$$V(I - M_1) = (1 - \rho)(I - M_1).$$

Substitution into (3) gives

$$\mathrm{Var}(\lambda'\hat{\gamma}) = \sigma^2(1 - \rho)\lambda'\left[X_2'(I - M_1)X_2\right]^{-}\lambda, \tag{4}$$

which, except for the term $(1 - \rho)$, is the variance from assuming $\mathrm{Cov}(Y) = \sigma^2 I$.

Exercise 11.1 Prove that equation (4) is true.

The mean square error (*MSE*) from ordinary least squares provides an independent estimate of $\sigma^2(1 - \rho)$. Let $M = X(X'X)^{-}X'$, so $MSE = Y'(I - M)Y/(n - r)$.

$$\mathrm{E}(MSE) = (n - r)^{-1}\sigma^2 \mathrm{tr}[(I - M)V].$$

Since $C(X_1) \subset C(X)$, from (1) we have

$$(I - M)V = (I - M)(1 - \rho)I = (1 - \rho)(I - M).$$

But, $\mathrm{tr}[(1 - \rho)(I - M)] = (1 - \rho)(n - r)$, so

$$\mathrm{E}(MSE) = \sigma^2(1 - \rho).$$

Theorem 11.1.1

(i) $Y'(I - M)Y/\sigma^2(1 - \rho) \sim \chi^2(n - r, 0)$.

(ii) *MSE and $X\hat{\beta}$ are independent. In particular, MSE and $\lambda'\hat{\gamma}$ are independent.*

Exercise 11.2 Prove Theorem 11.1.1.
Hint: For (i), use Theorem 1.3.6. For (ii), show that $\text{Cov}[(I-M)Y, MY] = 0$.

These results provide a basis for finding tests and confidence intervals for an estimable function $\lambda'\gamma$. We might also want to consider doing F tests. Suppose we want to test some vector of estimable restrictions on γ, say $\Lambda'\gamma = 0$. The test can be derived from $\Lambda'\hat{\gamma}$, $\text{Cov}(\Lambda'\hat{\gamma})$, and MSE, using Theorem 11.1.1 and Corollary 3.8.3. In particular,

Theorem 11.1.2

(i) $$\frac{(\Lambda'\hat{\gamma})'\left(\Lambda'\left[X_2'(I-M_1)X_2\right]^-\Lambda\right)^-(\Lambda'\hat{\gamma})/r(\Lambda)}{MSE} \sim F(r(\Lambda), n-r, \pi),$$

where $\pi = (\Lambda'\gamma)'\left(\Lambda'\left[X_2'(I-M_1)X_2\right]^-\Lambda\right)^-(\Lambda'\gamma)/2\sigma^2(1-\rho)$.

(ii) $(\Lambda'\gamma)'\left(\Lambda'\left[X_2'(I-M_1)X_2\right]^-\Lambda\right)^-(\Lambda'\gamma) = 0$ *if and only if* $\Lambda'\gamma = 0$.

An alternative to testing linear parametric functions is testing models. To test model (2) against a reduced model, say

$$Y = X_0\beta_0 + e, \quad C(X_1) \subset C(X_0) \subset C(X),$$

the test is the usual $\text{Cov}(Y) = \sigma^2 I$ test. With $M_0 = X_0(X_0'X_0)^-X_0'$, we have:

Theorem 11.1.3

(i) $$\frac{Y'(M-M_0)Y/[r(X)-r(X_0)]}{MSE} \sim F(r(X)-r(X_0), n-r, \pi),$$

where $\pi = \beta_0'X'(M-M_0)X\beta_0/2\sigma^2(1-\rho)$.
(ii) $\beta_0'X'(M-M_0)X\beta_0 = 0$ *if and only if* $\text{E}(Y) \in C(X_0)$.

Proof For part (i), see Exercise 11.3. Part (ii) follows exactly as in Theorem 3.2.1. □

Exercise 11.3 Prove Theorem 11.1.3(i).

In summary, for a model that includes separate fixed effects for each cluster, the ordinary least squares fit gives optimal estimates of all effects and valid estimates of standard errors for all effects not involving the cluster effects. If normal distributions are assumed, the usual $\text{Cov}(Y) = \sigma^2 I$ tests and confidence intervals are valid unless the cluster effects are involved. If the cluster effects are not of interest, the entire analysis can be performed with ordinary least squares. This substantially reduces the effort required to analyze the data.

The assumption that the α_is are fixed effects is necessary for the result to hold. However, if additional random cluster effects are added to the model so that there are both fixed and random cluster effects, then the basic structure of the covariance matrix remains unchanged and the optimality of ordinary least squares is retained.

Exercise 11.4 The usual model for a randomized complete block design was given in Section 8.2 as

$$y_{ij} = \mu + \alpha_i + \beta_j + e_{ij},$$

$i = 1, \ldots, a,\ j = 1, \ldots, b,\ \mathrm{Var}(e_{ij}) = \sigma^2$, and $\mathrm{Cov}(e_{ij}, e_{i'j'}) = 0$ for $(i, j) \neq (i', j')$. The β_js are considered as fixed block effects. Consider now a model

$$y_{ij} = \mu + \alpha_i + \beta_j + \eta_j + e_{ij}.$$

The η_js are independent $N(0, \sigma_2^2)$ and the e_{ij}s are independent $N(0, \sigma_1^2)$. The η_js and e_{ij}s are also independent. The block effects are now $(\beta_j + \eta_j)$. There is a fixed component and a random component with mean zero in each block effect. Use the results of this section to derive an analysis for this model. Give an ANOVA table and discuss interval estimates for contrasts in the α_is.

Exercise 11.5 An alternative model for a block design is

$$y_{ij} = \mu + \alpha_i + \beta_j + e_{ij}, \tag{5}$$

where the β_js are independent $N(0, \sigma_2^2)$ and the β_js and e_{ij}s are independent. If this model is used for a balanced incomplete block design, the BLUEs are different than they are when the β_js are assumed to be fixed. Discuss the appropriateness of the analysis based on this model in light of the results of Exercise 11.4.

Exercise 11.6 Show that when using model (5) for a randomized complete block design, the BLUE of a contrast in the α_is is the same regardless of whether the β_js are assumed random or fixed. Show that the estimates of a contrast's variance are the same.

11.2 Generalized Split Plot Models

By placing additional conditions on model (11.1.2), we can get a simple analysis of the cluster effects while retaining a simple analysis for the noncluster effects. The analysis of the cluster effects corresponds to the analysis of whole-plot treatments in a split plot model. The noncluster effects relate to effects on the subplot level.

Generalized split plot models are models obtained by imposing additional structure on the cluster sampling model (11.1.2). This additional structure involves simplifying the covariance matrix and modeling the whole-plot (cluster) effects. First, a condition on the whole plots is discussed. The condition is that the number of observations in each whole plot is the same. This condition simplifies the covariance matrix considerably. Next, the whole-plot effects are modeled by assuming a reduced model that does not allow separate effects for each whole plot. As part of the modeling process, a condition is imposed on the model matrix of the reduced model that ensures that least squares estimates are BLUEs. The problem of drawing inferences about generalized split plot models is discussed in two parts, (1) estimation and testing of estimable functions and (2) testing reduced models. A condition that allows for a simple analysis of the whole-plot effects is mentioned, a discussion of how to identify generalized split plot models is given, and finally some computational methods are presented.

The Covariance Matrix

Writing Y so that observations in each whole plot are listed contiguously, we can rewrite the covariance matrix of model (11.1.2) as

$$\sigma^2 V = \sigma^2 \left[(1-\rho)I + \rho M_1\{\text{Blk diag}(m_i I_{m(i)})\}\right],$$

where $I_{m(i)}$ is an $m_i \times m_i$ identity matrix and M_1 is the perpendicular projection matrix onto $C(X_1)$. This follows from Section 1 because $M_1 = \text{Blk diag}(m_i^{-1} J_{m(i)} J'_{m(i)})$. This characterization of V is not convenient in itself because of the Blk diag $(m_i I_{m(i)})$ term. For example, the expected value of a quadratic form, say $Y'AY$, is $\text{E}(Y'AY) = \sigma^2 \text{tr}(AV) + \beta' X' A X \beta$. The trace of AV is not easy to compute. To simplify the subsequent analysis, we impose

Condition 11.2.1 *All whole plots (clusters) are of the same size, say $m_i = m$ for all i.*

It follows that

$$V = (1-\rho)I + m\rho M_1. \tag{1}$$

This form for V will be assumed in the remainder of Section 2.

Modeling the Whole Plots

Model (11.1.2) includes X_1 which determines separate effects for each whole plot (cluster). Modeling the cluster effects consists of imposing structure on those effects. This is done by putting a constraint on $C(X_1)$. The simplest way to do this is by postulating a reduced model, say

$$Y = Z\beta_* + e, \quad Z = [X_*, X_2], \quad C(X_*) \subset C(X_1). \tag{2}$$

Partitioning β_* in conformance with Z, write

$$\beta_*' = [\delta', \gamma'].$$

Remember, this is not the same γ as in (11.1.2), but it is the coefficient for X_2, just as in (11.1.2). Define the perpendicular projection operators onto $C(Z)$ and $C(X_*)$ as M_Z and M_*, respectively.

In Section 1, it was shown that least squares estimates were BLUEs for model (11.1.2). We are now dealing with a different model, model (2), so another proof is required. To check if least squares estimates are BLUEs, we need to see whether $C(VZ) \subset C(Z)$. Since by equation (1), $V = (1-\rho)I + m\rho M_1$, we have $VZ = (1-\rho)Z + m\rho M_1 Z$. Clearly, it is enough to check whether $C(M_1 Z) \subset C(Z)$. This is true for a special case.

Proposition 11.2.2 *Let $\mathscr{M}$ and $\mathscr{N}$ be subspaces of $C(Z)$. If $C(Z) = \mathscr{M} + \mathscr{N}$, where $\mathscr{M} \subset C(X_1)$ and $\mathscr{N} \perp C(X_1)$, then $C(M_1 Z) \subset C(Z)$.*

Proof For any $v \in C(Z)$, write $v = v_1 + v_2$, where $v_1 \in \mathscr{M}$ and $v_2 \in \mathscr{N}$. $M_1 v = M_1 v_1 + M_1 v_2 = v_1$, but $v_1 \in \mathscr{M} \subset C(Z)$. □

A condition that is easy to check is

Condition 11.2.3 $C(Z) = C[X_*, (I - M_1)X_2]$ *and* $C(X_*) \subset C(X_1)$.

If Condition 11.2.3 holds, then Proposition 11.2.2 applies with $\mathscr{M} = C(X_*)$ and $\mathscr{N} = C[(I - M_1)X_2]$. If Proposition 11.2.2 applies, then least squares estimates are BLUEs. In fact, if $C(X_*) \subset C(X_1)$, all we really have to check is whether $C(M_1 X_2) \subset C(X_*)$.

Examples

Example 11.2.4A Let whole plots be denoted by the subscripts i and j, and let subplots have the subscript k. Let the dependent variable be y_{ijk} and let x_{ijk1}, x_{ijk2}, and x_{ijk3} be three covariates. The model given below is a generalized split plot model (see Exercise 11.7.):

$$\begin{aligned} y_{ijk} = \mu &+ \omega_i + \gamma_1 \bar{x}_{ij\cdot 1} + \gamma_{21} \bar{x}_{ij\cdot 2} + \eta_{ij} \\ &+ \tau_k + (\omega\tau)_{ik} + \gamma_{22}(x_{ijk2} - \bar{x}_{ij\cdot 2}) + \gamma_3(x_{ijk3} - \bar{x}_{ij\cdot 3}) \\ &+ e_{ijk}, \end{aligned} \tag{3}$$

$i = 1, \ldots, a$, $j = 1, \ldots, N_i$, $k = 1, \ldots, m$. The η_{ij}s and e_{ijk}s are all independent with $\eta_{ij} \sim N(0, \sigma_w^2)$ and $e_{ijk} \sim N(0, \sigma_s^2)$. With these assumptions

$$\sigma^2 = \sigma_w^2 + \sigma_s^2$$

and

$$\rho = \sigma_w^2/(\sigma_w^2 + \sigma_s^2).$$

The ω_is are treatment effects for a one-way ANOVA with unequal numbers in the whole plots. The whole-plot treatments can obviously be generalized to include multifactor ANOVAs with unequal numbers. The ω_is, γ_1, γ_{21}, and μ make up the δ vector. The τ_ks, $(\omega\tau)_{ik}$s, γ_{22}, and γ_3 make up the vector γ from model (2). Note that the covariate used with γ_{22} could be changed to x_{ijk2} without changing $C(Z)$ or invalidating Condition 11.2.3.

Exercise 11.7 Verify that Condition 11.2.3 holds for model (3).

It is easy to establish whether the m_is are all equal and it is easy to pick an X_* with $C(X_*) \subset C(X_1)$ because each column of X_* just has to take the same value for every subplot in a given whole plot. The difficulty in establishing Condition 11.2.3 is showing that $C(M_1 X_2) \subset C(X_*)$.

When the whole-plot model involves a balanced ANOVA, the orthogonality relationships discussed in Chapter 7 can often be used to establish Condition 11.2.3. In particular, whenever $C(J) \subset C(X_*)$, Condition 11.2.3 holds if

$$X_1'[I - (1/n)J_n^n]X_2 = 0.$$

Equivalent conditions are

$$X_1'X_2 = X_1'(1/n)J_n^n X_2,$$

and also

$$C(M_1 X_2) = C[(1/n)J_n^n X_2]$$

which implies

$$C[(I - M_1)X_2] = C\{[I - (1/n)J_n^n]X_2\}.$$

These conditions are easy to check on a computer, but they may not apply when including something as simple as a whole-plot treatment by subplot treatment interaction.

Example 11.2.4B If whole-plot treatments exist in the model and if the subplot model contains *only* balanced subplot treatments and whole-plot treatment by subplot interaction, then Condition 3 always holds. Let i identify the whole-plot treatments, let the pair ij identify the whole plots, and k identify the subplots (each of which gets a different subplot treatment). Let X_{*1} identify the whole-plot treatments, X_{21} identify the subplot treatments, and X_{22} identify the whole-plot by subplot interaction effect, i.e., $X_* = [X_{*1}, X_{*2}]$, where we don't care what X_{*2} is, and $X_2 = [X_{21}, X_{22}]$. We show that $C(M_1 X_{21}) \subset C(J)$ and $C(M_1 X_{22}) \subset C(X_{*1})$.

Write the rth column of X_{*1} as

$$X_{*1r} = [t_{ijk}], \quad t_{ijk} = \delta_{ir}.$$

Then, as for any one-way ANOVA, $M_1 Y$ just averages observations within the whole-plots, i.e.,

$$M_1 Y = [t_{ijk}] \quad t_{ijk} = \bar{y}_{ij\cdot}.$$

Most importantly, the formula for $M_1 Y$ determines what M_1 does to any vector.

Write the sth column of X_{21} as

$$X_{21s} = [t_{ijk}], \quad t_{ijk} = \delta_{ks}.$$

Then

$$M_1 X_{21s} = [t_{ijk}], \quad t_{ijk} = 1/m,$$

so $M_1 X_{21s} \in C(J)$. Similarly, write the rs column of X_{22} as

$$X_{22rs} = [t_{ijk}], \quad t_{ijk} = \delta_{(i,k)(r,s)}.$$

Then

$$M_1 X_{22rs} = [t_{ijk}], \quad t_{ijk} = (1/m)\sum_{k=1}^{m} \delta_{(i,k)(r,s)} = (1/m)\delta_{ir},$$

so $M_1 X_{22rs} \in C(X_{*1})$.

Of course this result generalizes immediately whenever the whole-plot treatments or subplot treatments have factorial treatment structure and it can serve as a tool for demonstrating Condition 11.2.3 in models with a more general X_2.

11.2.1 Estimation and Testing of Estimable Functions

We now discuss estimation and testing for model (2). Under Condition 11.2.1 and Condition 11.2.3, least squares estimates are BLUEs. Define

$$M_2 \equiv (I - M_1)X_2 \left[X_2'(I - M_1)X_2\right]^- X_2'(I - M_1).$$

From Condition 11.2.3, the perpendicular projection operator onto $C(Z)$ is

$$M_Z = M_* + M_2. \tag{4}$$

Given the perpendicular projection operator, the least squares estimates can be found in the usual way.

First, consider drawing inferences about γ. For estimable functions of γ, the estimates are exactly as in Section 1. In both cases, the estimates depend only on M_2Y. (See Proposition 9.1.1.) Since model (11.1.2) is a larger model than model (11.1.2) [i.e., $C(Z) \subset C(X)$], model (11.1.2) remains valid. It follows that all of the distributional results in Section 1 remain valid. In particular,

$$Y'(I-M)Y/\sigma^2(1-\rho) \sim \chi^2(n-r(X),0), \tag{5}$$

and $Y'(I-M)Y/[n-r(X)]$ is an unbiased estimate of $\sigma^2(1-\rho)$. In split plot models, $\sigma^2(1-\rho)$ is called the *subplot error variance.* $Y'(I-M)Y$ is called the *sum of squares for subplot error* [$SSE(s)$] and $Y'(I-M)Y/[n-r(X)]$ is the *mean square for subplot error* [$MSE(s)$].

The results in Section 1 are for functions of γ that are estimable in model (11.1.2). We now show that $\lambda'\gamma$ is estimable in (11.1.2) if and only if $\lambda'\gamma$ is estimable in (2). The argument is given for real-valued estimable functions, but it clearly applies to vector-valued estimable functions.

First, suppose that $\lambda'\gamma$ is estimable in (11.1.2). Then there exists a vector ξ such that $\lambda' = \xi'X_2$ and $\xi'X_1 = 0$. It follows immediately that $\lambda' = \xi'X_2$ and $\xi'X_* = 0$, so $\lambda'\gamma$ is estimable in model (2).

Now suppose that $\lambda'\gamma$ is estimable in model (2). There exists a vector ξ such that $\xi'X_* = 0$ and $\xi'X_2 = \lambda'$. Using equation (4) and $\xi'M_* = 0$, it is easily seen that

$$\lambda' = \xi'X_2 = \xi'M_ZX_2 = \xi'M_2X_2$$

and, since $M_2X_1 = 0$,

$$\xi'M_2X_1 = 0.$$

The vector $\xi'M_2$ satisfies the two conditions needed to show that $\lambda'\gamma$ is estimable in (11.1.2). Thus, inferences about estimable functions $\lambda'\gamma$ can be made exactly as in Section 1.

Drawing inferences about δ is trickier. The projection operator M_Z is the sum of two orthogonal projection operators M_* and M_2. Estimation of $\lambda'\gamma$ can be accomplished easily because the estimate depends on M_2Y alone. Similarly, estimable functions whose estimates depend on M_*Y alone can be handled simply. The problem lies in identifying which estimable functions have estimates that depend on M_*Y alone. Since estimable functions of γ depend on M_2Y, any estimable function with an estimate that depends on M_*Y alone must involve δ. (Of course, there exist estimable functions that depend on both M_*Y and M_2Y.) Later, a condition will be discussed that forces all estimable functions of δ to depend only on M_*Y. With this condition, we have a convenient dichotomy in that estimates of functions of δ depend on the perpendicular projection operator M_*, and estimates of functions of γ depend on the perpendicular projection operator M_2. The condition referred to is convenient, but it is not necessary for having a generalized split plot model.

As discussed in Chapter 3 the question of whether the estimate of an estimable function, say $\Lambda'\beta_*$, depends only on M_*Y is closely related to the constraint on the model imposed by the hypothesis $\Lambda'\beta_* = 0$. In particular, if $\Lambda' = P'Z$, then the constraint imposed by $\Lambda'\beta_* = 0$ is $\mathrm{E}(Y) \perp C(M_Z P)$, and $\Lambda'\hat{\beta}_*$ depends on M_*Y if and only if $C(M_Z P) \subset C(M_*) = C(X_*)$. *In the discussion that follows, $\Lambda'\beta_* = 0$ is assumed to put a constraint on $C(X_*)$.*

We seek to derive an F test for $\Lambda'\beta_* = 0$. From Corollary 3.8.3,

$$\left(\Lambda'\hat{\beta}_*\right)'\left[\mathrm{Cov}\left(\Lambda'\hat{\beta}_*\right)\right]^-\left(\Lambda'\hat{\beta}_*\right) \sim \chi^2\left(r(\Lambda), (\Lambda'\beta_*)'\left[\mathrm{Cov}\left(\Lambda'\hat{\beta}_*\right)\right]^-(\Lambda'\beta_*)/2\right).$$

We need the covariance of $\Lambda'\hat{\beta}_*$. Note that $\Lambda'\hat{\beta}_* = P'M_Z Y = P'M_* Y$. From equation (1) and the fact that $C(X_*) \subset C(X_1)$, it is easily seen that

$$M_*V = [(1-\rho) + m\rho]\, M_*; \tag{6}$$

so

$$\begin{aligned}\mathrm{Cov}\left(\Lambda'\hat{\beta}_*\right) &= \sigma^2 P'M_*VM_*P = \sigma^2\left[(1-\rho) + m\rho\right] P'M_*P \\ &= \sigma^2\left[(1-\rho) + m\rho\right] P'M_Z P \\ &= \sigma^2\left[(1-\rho) + m\rho\right] \Lambda'\left(Z'Z\right)^- \Lambda,\end{aligned}$$

which, except for the term $(1-\rho) + m\rho$, is the usual covariance of $\Lambda'\hat{\beta}_*$ from a standard linear model. We can get an F test of $\Lambda'\beta_* = 0$ if we can find an independent chi-squared estimate of $\sigma^2\left[(1-\rho) + m\rho\right]$.

Theorem 11.2.4 *Under model (2),*

$$Y'(M_1 - M_*)Y/\sigma^2\left[(1-\rho) + m\rho\right] \sim \chi^2(r(X_1) - r(X_*), 0).$$

Proof Observe that

$$M_1V = [(1-\rho) + m\rho]\, M_1. \tag{7}$$

Using equations (6) and (7), it is easy to check the conditions of Theorem 1.3.6. It remains to show that $\beta_*'Z'(M_1 - M_*)Z\beta_* = 0$. Recall that $M_Z = M_* + M_2$, and note that $M = M_1 + M_2$. It follows that $M_1 - M_* = M - M_Z$. Clearly, $(M - M_Z)Z\beta_* = 0$. □

The quadratic form $Y'(M_1 - M_*)Y$ is called the *sum of squares for whole-plot (cluster) error*. This is denoted $SSE(w)$. An unbiased estimate of $\sigma^2\left[(1-\rho) + m\rho\right]$ is available from

$$MSE(w) \equiv Y'(M_1 - M_*)Y \big/ \left[r(X_1) - r(X_*)\right].$$

To complete the derivation for the F test of $\Lambda'\beta_* = 0$, we need to show that $\Lambda'\hat{\beta}_*$ and $Y'(M_1 - M_*)Y$ are independent. It suffices to note that

$$\begin{aligned}
\text{Cov}(M_*Y, (M_1 - M_*)Y) &= \sigma^2 M_* V (M_1 - M_*) \\
&= \sigma^2 [(1-\rho) + m\rho] M_*(M_1 - M_*) \\
&= 0.
\end{aligned}$$

The F test is based on the distributional result

$$\frac{\left(\Lambda'\hat{\beta}_*\right)' \left[\Lambda'(Z'Z)^-\Lambda\right]^- \left(\Lambda'\hat{\beta}_*\right) / r(\Lambda)}{MSE(w)} \sim F(r(\Lambda), r(X_1) - r(X_*), \pi),$$

where

$$\pi = (\Lambda'\beta_*)' \left[\Lambda'(Z'Z)^-\Lambda\right]^- (\Lambda'\beta_*) / 2\sigma^2 [(1-\rho) + m\rho]$$

and $(\Lambda'\beta_*)' \left[\Lambda'(Z'Z)^-\Lambda\right]^- (\Lambda'\beta_*) = 0$ if and only if $\Lambda'\beta_* = 0$.

The argument establishing the independence of $\Lambda'\hat{\beta}_*$ and $MSE(w)$ can be extended to establish the independence of all the distinct statistics being used.

Theorem 11.2.5 *M_*Y, M_2Y, $SSE(w)$, and $SSE(s)$ are mutually independent.*

Proof Since the joint distribution of Y is multivariate normal, it suffices to use equations (6) and (7) to establish that the covariance between any pair of M_*Y, M_2Y, $(M_1 - M_*)Y$, and $(I - M)Y$ is 0. □

To summarize the results so far, if the linear model satisfies Conditions 11.2.1 and 11.2.3, then (a) least squares estimates are BLUEs, (b) inferences about estimable functions $\lambda'\gamma$ can be made in the usual way (i.e., just as if $V = I$) with the exception that the estimate of error is taken to be $MSE(s)$, and (c) inferences about estimable functions of $\Lambda'\beta_*$ that put a constraint on $C(X_*)$ can be drawn in the usual way, except that $MSE(w)$ is used as the estimate of error.

As mentioned, it is not clear what kind of estimable functions put a constraint on $C(X_*)$. Two ways of getting around this problem will be discussed. As mentioned above, one way is to place another condition on the model matrix Z of model (2), a condition that forces the estimable functions of δ to put constraints on $C(X_*)$. A second approach, that requires no additional conditions, is to abandon the idea of testing estimable functions and to look at testing models.

Inferences About δ

One of the problems with generalized split plot models is in identifying the hypotheses that put constraints on $C(X_*)$. In general, such hypotheses can involve both the δ and the γ parameters. For example, suppose that $C(X_*) \cap C(X_2)$ contains a nonzero vector ξ. Then, since $M_Z\xi = \xi \in C(X_*)$, the hypothesis $\xi'X_*\delta + \xi'X_2\gamma = 0$ puts a constraint on $C(X_*)$. However, $\xi'X_*\delta + \xi'X_2\gamma$ involves both the δ and γ parameters because, with $\xi \in C(X_*) \cap C(X_2)$, neither $\xi'X_*$ nor $\xi'X_2$ is 0. The most common example of this phenomenon occurs when X_* and X_2 are chosen so that $C(X_*)$ and $C(X_2)$ both contain a column of 1s (i.e., J_n). It follows that inferences about the grand mean, $n^{-1}J_n'Z\beta_*$, are made using $MSE(w)$.

The condition stated below ensures that any estimable function of the δs puts a constraint on $C(X_*)$. This condition is typically satisfied when Z is the model matrix for a balanced multifactor ANOVA (with some of the interactions possibly deleted).

Condition 11.2.6 *For $v \in C(X)$, if $v \perp C(X_2)$, then $v \in C(X_1)$.*
Suppose that $\lambda'\delta$ is an estimable function. Then $\lambda' = \xi' X_*$ and $\xi' X_2 = 0$ for some $\xi \in C(Z)$. Since $\xi' X_2 = 0$, Condition 11.2.7 implies that $\xi \in C(X_1)$; thus $M_2\xi = 0$. Finally, by (4),

$$M_Z \xi = M_* \xi \in C(X_*).$$

Thus, estimable functions of δ put a constraint on $C(X_*)$ and inferences about such functions are made using M_*Y and $MSE(w)$.

11.2.2 Testing Models

We will now examine the problem of testing model (2) against reduced models. To look at the complete problem, we will discuss both reduced models that put a constraint on $C(X_*)$ and reduced models that put a constraint on $C(X_2)$. For both kinds of reduced models, the tests are analogous to those developed in Section 3.2. A reduced model that puts a constraint on $C(X_*)$ can be tested by comparing the *SSE*(w) for model (2) with the *SSE*(w) for the reduced model. The difference in *SSE*(w)s is divided by the difference in the ranks of the design matrices to give a numerator mean square for the test. The denominator mean square is *MSE*(w) from model (2). The test for a reduced model that puts a constraint on $C(X_2)$ is performed in a similar fashion using *SSE*(s) and *MSE*(s). In the discussion below, specific models and notation are presented to justify these claims.

First, consider a reduced model that puts a constraint on $C(X_*)$. The reduced model is a model of the form

$$Y = Z_0\xi + e, \quad Z_0 = [X_{0*}, X_2], \quad C(X_{0*}) \subset C(X_*). \tag{8}$$

Let $M_0 \equiv Z_0(Z_0'Z_0)^- Z_0'$ and $M_{0*} \equiv X_{0*}(X_{0*}'X_{0*})^- X_{0*}'$. If the equivalent of Condition 11.2.3 holds for model (8), then

$$\frac{Y'(M_* - M_{0*})Y/[r(X_*) - r(X_{0*})]}{MSE(w)} \sim F\left(r(X_*) - r(X_{0*}), r(X_1) - r(X_*), \pi\right),$$

where

$$\pi = \beta_*' Z'(M_* - M_{0*}) Z\beta_* / 2\sigma^2[(1-\rho) + m\rho]$$

and $\beta_*' Z'(M_* - M_{0*}) Z\beta_* = 0$ if and only if $\mathrm{E}(Y) \in C(Z_0)$.

These results follow from Theorems 11.2.5 and 11.2.6 and Corollary 3.8.3 upon noticing two things: First, in Corollary 3.8.3, $A - A_0 = M_Z - M_0 = M_* - M_{0*}$. Second, from equation (6), $M_* = [(1-\rho) + m\rho]M_* V^{-1}$, with a similar result holding for $M_{0*}V^{-1}$.

The other kind of reduced model that can be treated conveniently is a reduced model that puts a constraint on $C(X_2)$. The reduced model is written as

$$Y = Z_0\xi + e, \quad Z_0 = [X_*, X_3], \quad C(X_3) \subset C(X_2). \tag{9}$$

If the equivalent of Condition 11.2.3 holds for model (9), write M_0 as before and $M_3 \equiv (I - M_1)X_3 \left[X_3'(I - M_1)X_3\right]^- X_3'(I - M_1)$. Then

$$\frac{Y'(M_2 - M_3)Y/[r(M_2) - r(M_3)]}{MSE(s)} \sim F\left(r(M_2) - r(M_3), n - r(X), \pi\right),$$

where

$$\pi = \beta_*' Z'(M_2 - M_3)Z\beta_* \big/ 2\sigma^2(1-\rho)$$

and $\beta_*' Z'(M_2 - M_3)Z\beta_* = 0$ if and only if $\mathrm{E}(Y) \in C(Z_0)$.

These results follow from Theorem 11.2.6, Corollary 3.8.3, and relation (5) upon noticing that $A - A_0 = M_Z - M_0 = M_2 - M_3$; and, since $M_2 V = (1-\rho)M_2$, we have $(1-\rho)^{-1}M_2 = M_2 V^{-1}$, and a similar result for M_3.

Identifying Generalized Split Plot Models

There are only two conditions necessary for having a generalized split plot model, Condition 11.2.1 and Condition 11.2.3. The form of generalized split plot models can be read from these conditions. Condition 11.2.1 requires that an equal number of observations be obtained within each whole plot. Condition 11.2.3 requires $C(Z) = C(X_*, (I - M_1)X_2)$, where $C(X_*) \subset C(X_1)$. Since $C(X_1)$ is the column space that allows a separate effect for each cluster, $X_*\delta$ can be anything that treats all of the observations in a given whole plot the same. The matrix X_2 can contain the columns for any ANOVA effects that are balanced within whole plots. X_2 can also contain any columns that are orthogonal to $C(X_1)$. Model (3) in Example 11.2.4a displays these characteristics.

Computations

The simplest way to actually fit generalized split plot models would seem to be to fit both models (2) and (11.1.2) using an ordinary least squares computer program. Fitting model (2) provides least squares estimates of δ and γ. Fitting model (2) also provides $Y'(I - M_Z)Y$ as the reported *SSE*. This reported *SSE* is not appropriate for any inferences, but, as seen below, it can be used to obtain the whole-plot sum of squares error. Fitting model (11.1.2) provides least squares estimates of α and γ and the reported *SSE* is $Y'(I - M)Y$. If the model is a generalized split plot model, the two estimates of γ should be identical. Since $Y(I - M)Y$ is the *SSE*(s) (sum of squares for subplot error), any conclusions about γ obtained from fitting (11.1.2) will be appropriate. To obtain *SSE*(w) (sum of squares for whole-plot error), note that

$$
\begin{aligned}
Y'(I - M_Z)Y - Y'(I - M)Y &= Y'(M - M_Z)Y \\
&= Y'(M_1 + M_2 - M_* - M_2)Y \\
&= Y'(M_1 - M_*)Y.
\end{aligned}
$$

Thus, all of the computationally intensive work can be performed on standard computer programs.

Exercise 11.8 Give detailed proofs of the test statistic's distribution for
(a) testing model (8) against model (2),
(b) testing model (9) against model (2).

Exercise 11.9 Show that model (11.1.5) for a randomized complete block design is a generalized split plot model.

11.2.3 Unbalanced Subplots

Generalized split plot models can have any structure whatsoever in the whole plots but the subplot model needs to display certain orthogonality with the whole plots and the subplots have to be balanced, i.e., the whole plots must contain the same number of observations m. If the whole plots are not balanced, a generalized split plot model does not exist. However, if you are willing to abandon the whole-plot analysis, the results of Section 1 provide a subplot analysis. And after all, the whole-plot analysis is rarely as interesting as the subplot analysis. Christensen (2015, Subsection 19.2.1) illustrates the methodology (although on data that are actually balanced).

11.3 The Split Plot Design

The traditional model for a split plot design is a special case of the model presented in Section 2. We will present the split plot model and a model equivalent to model (11.1.2). We will use the balance of the split plot design to argue that Conditions 11.2.1, 11.2.3, and 11.2.7 hold. The arguments based on the balance of the split plot model are similar to those presented in Subsection 7.6.1. Statistical inferences are based on least squares estimates and quadratic forms in corresponding perpendicular projection matrices. The balance of the split plot model dictates results that are very similar to those described in Sections 7.1, 7.2, and 7.6.

The traditional split plot model involves a randomized complete block design in the whole plots. In fact, a completely randomized design or a Latin square in the whole plots leads to an analogous analysis. Suppose that there are r blocks of t whole plots available. Within each block, a different (whole plot) treatment is applied to each whole plot. Let μ denote a grand mean, ξs denote block effects, and ωs denote whole-plot treatment effects. Let each whole plot be divided into m subplots with a different (subplot) treatment applied to each subplot. The τs denote subplot treatment effects, and the $(\omega\tau)$s denote interaction effects between the whole-plot treatments and the subplot treatments. The split plot model has two sources of error, whole plot to whole plot variation denoted by η, and subplot to subplot variation denoted by e. The split plot model is

$$y_{ijk} = \mu + \xi_i + \omega_j + \eta_{ij} + \tau_k + (\omega\tau)_{jk} + e_{ijk}, \tag{1}$$

$i = 1, \ldots, r$, $j = 1, \ldots, t$, $k = 1, \ldots, m$, η_{ij}s independent $N(0, \sigma_w^2)$, e_{ijk}s independent $N(0, \sigma_s^2)$. The η_{ij}s and e_{ijk}s are assumed to be independent. We can combine the error terms as $\varepsilon_{ijk} = \eta_{ij} + e_{ijk}$. Writing the vector of errors as $\varepsilon = [\varepsilon_{111}, \varepsilon_{112}, \ldots, \varepsilon_{rtm}]'$, we get

$$\begin{aligned} \mathrm{Cov}(\varepsilon) &= \mathrm{Blk\ diag}[\sigma_s^2 I_m + \sigma_w^2 J_m^m] \\ &= \sigma^2[(1-\rho)I + m\rho M_1], \end{aligned}$$

where $\sigma^2 = \sigma_w^2 + \sigma_s^2$ and $\rho = \sigma_w^2/(\sigma_w^2 + \sigma_s^2)$. In a split plot model, the whole plots are considered as the different combinations of i and j. There are m observations for each whole plot, so Condition 11.2.1 holds.

The model that includes separate effects α_{ij} for each whole plot can be written as

$$y_{ijk} = \alpha_{ij} + \eta_{ij} + \tau_k + (\omega\tau)_{jk} + e_{ijk}\,.$$

Combining the error terms gives

$$y_{ijk} = \alpha_{ij} + \tau_k + (\omega\tau)_{jk} + \varepsilon_{ijk}\,,$$

or, using a parameterization with interactions,

$$y_{ijk} = \mu + \xi_i + \omega_j + (\xi\omega)_{ij} + \tau_k + (\omega\tau)_{jk} + \varepsilon_{ijk}\,. \tag{2}$$

From Section 2, $C(X_1)$ is the space spanned by the columns associated with the α_{ij}s and $C(X_2)$ is the space spanned by the columns associated with the τ_ks and $(\omega\tau)_{jk}$s. $C(X)$ is the column space for model (2). Using the parameterization of model (2) and the notation and results of Subsection 7.6.1 gives

$$\begin{aligned} C(X_1) &= C(M_\mu + M_\xi + M_\omega + M_{\xi\omega}), \\ C(X_2) &= C(M_\mu + M_\omega + M_\tau + M_{\omega\tau}), \\ C(X) &= C(M_\mu + M_\xi + M_\omega + M_{\xi\omega} + M_\tau + M_{\omega\tau}). \end{aligned}$$

Recall that all of the M matrices on the right-hand sides are perpendicular projection matrices, and that all are mutually orthogonal. In particular, $M = M_\mu + M_\xi + M_\omega + M_{\xi\omega} + M_\tau + M_{\omega\tau}$ and $M_1 = M_\mu + M_\xi + M_\omega + M_{\xi\omega}$.

The split plot model (1) is a reduced model relative to model (2). The $\xi\omega$ interactions are dropped to create the split plot model. $C(X_*)$ is the space spanned by the columns associated with μ, the ξ_is, and the ω_js. Again using results from 7.6.1,

$$
\begin{aligned}
C(X_*) &= C(M_\mu + M_\xi + M_\omega),\\
C[(I - M_1)X_2] &= C(M_\tau + M_{\omega\tau}),\\
C(Z) &= C(M_\mu + M_\xi + M_\omega + M_\tau + M_{\omega\tau}).
\end{aligned}
$$

Clearly, Condition 11.2.3 applies.

In fact, even Condition 11.2.7 holds, so that estimable functions of the ξs and ωs are tested using $MSE(w)$. To check Condition 11.2.7, it suffices to show that if $v \in C(X)$ and $v \perp C(X_2)$, then $v \in C(X_1)$. If $v \in C(X)$, then $Mv = v$. If $v \perp C(X_2)$, then $(M_\mu + M_\omega + M_\tau + M_{\omega\tau})v = 0$. Thus

$$
\begin{aligned}
v = Mv &= (M_\mu + M_\xi + M_\omega + M_{\xi\omega} + M_\tau + M_{\omega\tau})v\\
&= (M_\xi + M_{\xi\omega})v.
\end{aligned}
$$

But, $v = (M_\xi + M_{\xi\omega})v \in C(X_1)$. It should be noted that with $(\omega\tau)$ interaction in the model, contrasts in the ωs and τs are not estimable. For example, the usual procedure gives estimates of contrasts in the $\omega_j + \overline{(\omega\tau)}_{j\cdot}$s. Without $(\omega\tau)$ interaction, contrasts in the ωs become estimable. In either case, the estimates are obtained using M_*.

We can now write out an ANOVA table.

Source	df	SS	E(MS)
μ	$r(M_\mu)$	$Y'M_\mu Y$	$(\sigma_s^2 + m\sigma_w^2) + \beta_*'Z'M_\mu Z\beta_*/r(M_\mu)$
ξ	$r(M_\xi)$	$Y'M_\xi Y$	$(\sigma_s^2 + m\sigma_w^2) + \beta_*'Z'M_\xi Z\beta_*/r(M_\xi)$
ω	$r(M_\omega)$	$Y'M_\omega Y$	$(\sigma_s^2 + m\sigma_w^2) + \beta_*'Z'M_\omega Z\beta_*/r(M_\omega)$
error 1	$r(M_{\xi\omega})$	$Y'M_{\xi\omega}Y$	$\sigma_s^2 + m\sigma_w^2$
τ	$r(M_\tau)$	$Y'M_\tau Y$	$\sigma_s^2 + \beta_*'Z'M_\tau Z\beta_*/r(M_\tau)$
$\omega\tau$	$r(M_{\omega\tau})$	$Y'M_{\omega\tau}Y$	$\sigma_s^2 + \beta_*'Z'M_{\omega\tau}Z\beta_*/r(M_{\omega\tau})$
error 2	$r(I - M)$	$Y'(I - M)Y$	σ_s^2
Total	n	$Y'Y$	

Note that $\sigma^2[(1-\rho) + m\rho] = \sigma_s^2 + m\sigma_w^2$ and $\sigma^2(1-\rho) = \sigma_s^2$. Algebraically, we can write the table as

Source	df	SS	$E(MS)$
μ	1	$rtm\bar{y}^2_{\cdot\cdot\cdot}$	
ξ	$r-1$	$tm\sum_i(\bar{y}_{i\cdot\cdot}-\bar{y}_{\cdot\cdot\cdot})^2$	$(\sigma_s^2+m\sigma_w^2)+A$
ω	$t-1$	$rm\sum_j(\bar{y}_{\cdot j\cdot}-\bar{y}_{\cdot\cdot\cdot})^2$	$(\sigma_s^2+m\sigma_w^2)+B$
error 1	$(r-1)(t-1)$	$m\sum_{ij}(\bar{y}_{ij\cdot}-\bar{y}_{i\cdot\cdot}-\bar{y}_{\cdot j\cdot}+\bar{y}_{\cdot\cdot\cdot})^2$	$\sigma_s^2+m\sigma_w^2$
τ	$m-1$	$rt\sum_k(\bar{y}_{\cdot\cdot k}-\bar{y}_{\cdot\cdot\cdot})^2$	σ_s^2+C
$\omega\tau$	$(t-1)(m-1)$	$r\sum_{jk}(\bar{y}_{\cdot jk}-\bar{y}_{\cdot j\cdot}-\bar{y}_{\cdot\cdot k}+\bar{y}_{\cdot\cdot\cdot})^2$	σ_s^2+D
error 2	by subtraction	by subtraction	σ_s^2
Total	n	$\sum_{ijk}y^2_{ijk}$	

$$A = tm\sum_i(\xi_i-\bar{\xi}_\cdot)^2/(r-1),$$

$$B = rm\sum_j[\omega_j+\overline{(\omega\tau)}_{j\cdot}-\bar{\omega}_\cdot-\overline{(\omega\tau)}_{\cdot\cdot}]^2/(t-1),$$

$$C = rt\sum_k[\tau_k+\overline{(\omega\tau)}_{\cdot k}-\bar{\tau}_\cdot-\overline{(\omega\tau)}_{\cdot\cdot}]^2/(m-1),$$

$$D = r\sum_{jk}[(\omega\tau)_{jk}-\overline{(\omega\tau)}_{j\cdot}-\overline{(\omega\tau)}_{\cdot k}+\overline{(\omega\tau)}_{\cdot\cdot}]^2/(t-1)(m-1).$$

Tests and confidence intervals for contrasts in the τ_ks and $(\omega\tau)_{jk}$s are based on the usual least squares estimates and use the mean square from the "error 2" line for an estimate of variance. Tests and confidence intervals in the ξ_is and ω_js also use least squares estimates, but use the mean square from the "error 1" line for an estimate of variance. Note that, even though there is no interaction in the whole-plot analysis, contrasts in the ω_js are really contrasts in the $[\omega_j+\overline{(\omega\tau)}_{j\cdot}]$s when interaction is present.

Finally, a word about missing data. If one or more whole plots are missing, the data can still be analyzed as in Section 2. If one or more subplots are missing, the data can still be analyzed as in Section 1; however, with missing subplots, some sort of ad hoc analysis for the whole-plot effects must be used.

Exercise 11.10 Consider the table of means

		τ			
		1	2	$\cdots$	m
	1	$\bar{y}_{\cdot 11}$	$\bar{y}_{\cdot 12}$	$\cdots$	$\bar{y}_{\cdot 1m}$
ω	2	$\bar{y}_{\cdot 21}$	$\bar{y}_{\cdot 22}$	$\cdots$	$\bar{y}_{\cdot 2m}$
	$\vdots$	$\vdots$	$\vdots$		$\vdots$
	t	$\bar{y}_{\cdot t1}$	$\bar{y}_{\cdot t2}$	$\cdots$	$\bar{y}_{\cdot tm}$

Let $\sum_{k=1}^{m} d_k = 0$. For any fixed j, find a confidence interval for $\sum_{k=1}^{m} d_k \mu_{jk}$, where $\mu_{jk} = \mu + \bar{\xi}_{\cdot} + \omega_j + \tau_k + (\omega\tau)_{jk}$. (Hint: The estimate of the variance comes from the "error 2" line.) Let $\sum_{j=1}^{t} c_j = 0$. Why is it not possible to find a confidence interval for $\sum_{j=1}^{t} c_j \mu_{jk}$?

11.4 Identifying the Appropriate Error

Statistics is all about drawing conclusions from data that are subject to error. One of the crucial problems in statistics is identifying and estimating the appropriate error so that valid conclusions can be made. The importance of this issue has long been recognized by the statistics community. The necessity of having a valid estimate of error is one of the main points in *The Design of Experiments*, Fisher's (1935) seminal work.

The key feature of split plot models is that they involve two separate sources of variation. The analysis involves two separate estimates of error, and a correct analysis requires that the estimates be used appropriately. If the existence of two separate sources of variability is not noticed, the estimate of error will probably be obtained by pooling the sums of squares for the two separate errors. The pooled estimate of error will generally be too small for comparing treatments applied to whole plots and too large for comparing treatments applied to subplots. The whole-plot treatments will appear more significant than they really are. The subplot treatments will appear less significant than they really are. The interactions between whole-plot and subplot treatments will also appear less significant than they really are.

The problem of identifying the appropriate error is a difficult one. In this section, some additional examples of the problem are discussed. First, the problem of subsampling is considered. The section ends with a discussion of the appropriate error for testing main effects in the presence of interactions.

11.4.1 Subsampling

One of the most commonly used, misused, and abused of models is the subsampling model.

Example 11.4.1 In an agricultural experiment, one treatment is applied to each of 6 pastures. The experiment involves 4 different treatments, so there are a total of 24 pastures in the experiment. On each pasture 10 observations are taken. The 10 observations taken on each pasture are referred to as subsamples. Each observation is subject to two kinds of variability: (1) pasture to pasture variability and (2) within pasture variability. Note, however, that in comparing the 10 observations taken on a given pasture, there is no pasture to pasture variability. The correct model for this experiment involves error terms for both kinds of variability. Typically the model is taken as

$$y_{ijk} = \mu + \omega_i + \eta_{ij} + e_{ijk}, \tag{1}$$

$i = 1, 2, 3, 4$, $j = 1, \ldots, 6$, $k = 1, \ldots, 10$, η_{ij}s i.i.d. $N(0, \sigma_w^2)$, e_{ijk}s i.i.d. $N(0, \sigma_s^2)$, and the η_{ij}s and e_{ijk}s are independent. In this model, σ_w^2 is the pasture to pasture variance and σ_s^2 is the within pasture or subsampling variance.

As will be seen below, the statistical analysis of model (1) acts like there is only 1 observation on each pasture. That 1 observation is the mean of the 10 actual observations that were taken. If the analysis acts like only 1 observation was taken on a pasture, why should an experimenter trouble to take 10 observations? Why not take just 1 observation on each pasture?

With 1 real observation on each pasture, the statistical analysis is subject to the whole weight of the within pasture variability. By taking 10 observations on a pasture and averaging them to perform the analysis, the mean of the 10 observations still has the full pasture to pasture variability, but the within pasture variability (variance) is cut by a factor of 10. The experiment is being improved by reducing the variability of the treatment estimates, but that improvement comes only in the reduction of the within pasture variability. The effects of pasture to pasture variability are not reduced by subsampling.

Rather than subsampling, it would be preferable to use more pastures. Using more pastures reduces the effects of both kinds of variability. Unfortunatcly, doing that is often not feasible. In the current example, the same reduction in within pasture variation without subsampling would require the use of 240 pastures instead of the 24 pastures that are used in the experiment with subsampling. In practice, obtaining 24 pastures for an experiment can be difficult. Obtaining 240 pastures can be well nigh impossible.

A general balanced subsampling model is

$$y_{ijk} = \mu + \omega_i + \eta_{ij} + e_{ijk}, \tag{2}$$

$$\mathrm{E}(\eta_{ij}) = \mathrm{E}(e_{ijk}) = 0, \quad \mathrm{Var}(\eta_{ij}) = \sigma_w^2, \quad \mathrm{Var}(e_{ijk}) = \sigma_s^2,$$

$i = 1, \ldots, t$, $j = 1, \ldots, r$, $k = 1, \ldots, m$ with distinct η_{ij}s and e_{ijk}s uncorrelated. By checking Conditions 11.2.1, 11.2.3, and 11.2.7, it is easily seen that this is a generalized split plot model with X_2 vacuous. An ANOVA table can be written:

Source	*df*	*SS*	E(*MS*)
ω	$t-1$	$rm\sum_i(\bar{y}_{i\cdot\cdot} - \bar{y}_{\cdot\cdot\cdot})^2$	$\sigma_s^2 + m\sigma_w^2 + A$
error 1	$t(r-1)$	$m\sum_{ij}(\bar{y}_{ij\cdot} - \bar{y}_{i\cdot\cdot})^2$	$\sigma_s^2 + m\sigma_w^2$
error 2	$rt(m-1)$	$\sum_{ijk}(y_{ijk} - \bar{y}_{ij\cdot})^2$	σ_s^2

where $A = rm\sum_i(\omega_i - \bar{\omega}_\cdot)^2/(t-1)$. The entire analysis is performed as a standard one-way ANOVA. The variance of $\bar{y}_{i\cdot\cdot}$ is $(\sigma_s^2 + m\sigma_w^2)/rm$, so the "error 1" line is used for all tests and confidence intervals.

As mentioned above, an equivalent analysis can be made with the averages of the observations in each subsample. The model based on the averages is

$$\bar{y}_{ij\cdot} = \mu + \omega_i + \xi_{ij}, \quad \mathrm{E}(\xi_{ij}) = 0, \quad \mathrm{Var}(\xi_{ij}) = [\sigma_w^2 + (\sigma_s^2/m)],$$

$i = 1,\ldots,t$, $j = 1,\ldots,r$. It is easily seen that this one-way AVOVA gives exactly the same tests and confidence intervals as those obtained from model (2).

One of the most common mistakes in statistical practice is to mistake subsampling for independent replication in an experiment. Example 11.4.1 involves 6 independent replications, i.e., the 6 pastures to which each treatment is applied. The 10 observations on a given pasture are not independent because the random effect for pastures is the same for all of them. The *incorrect* model that is often analyzed instead of model (2) is

$$y_{ij} = \mu + \omega_i + e_{ij}, \tag{3}$$

$i = 1,\ldots,t$, $j = 1,\ldots,rm$. The effect of analyzing model (3) is that the "error 1" and "error 2" lines of the ANOVA are pooled together. Since the expected mean square for "error 2" is only σ_s^2, the pooled mean square error is inappropriately small and all effects will appear to be more significant than they really are.

Subsampling can be an important tool, especially when the variability between subsamples is large. However, it is important to remember that subsampling is to be used in addition to independent replication. It does not eliminate the need for an adequate number of independent replicates. In terms of Example 11.4.1, an experimenter should first decide on a reasonable number of pastures and then address the question of how many observations to take within a pasture. If the pasture to pasture variability is large compared to the within pasture variability, subsampling will be of very limited value. If the within pasture variability is large, subsampling can be extremely worthwhile.

The existence of subsampling can usually be determined by carefully identifying the treatments and the experimental units to which the treatments are applied. In the agricultural example, identifying the subsampling structure was easy. Treatments were applied to pastures. If differences between treatments are to be examined, then differences between pastures with the same treatment must be error.

Lest the reader think that identifying subsampling is easy, let us try to confuse the issue. The analysis that has been discussed is based on the assumption that pasture to pasture variability is error, but now suppose that the experimenter has an interest in the pastures. For example, different pastures have different fertilities, so some pastures are better than others. If differences in pastures are of interest, it may be reasonable to think of the η_{ij}s as fixed effects, in which case there is no subsampling and the ANOVA table gives only one error line. The expected mean squares are:

Source	df	E(MS)
ω	3	$\sigma^2 + 20\sum_i(\omega_i + \bar{\eta}_{i\cdot} - \bar{\omega}_{\cdot} - \bar{\eta}_{\cdot\cdot})^2$
η	20	$\sigma^2 + .5\sum_{ij}(\eta_{ij} - \bar{\eta}_{i\cdot})^2$
error	216	σ^2

The mean square for η can be used to test whether there are any differences between pastures that have the same treatment. The $MSE(\omega)$ provides a test of whether the treatment effects added to their average pasture effects are different. The test from $MSE(\omega)$ may not be very interesting if there are different pasture effects. In summary, the analysis depends crucially on whether the η_{ij}s are assumed to be random or fixed. When the ω treatments are of primary importance it makes sense to treat the η effects as random.

11.4.2 Two-Way ANOVA with Interaction

The balanced two-way ANOVA with interaction model from Section 7.2 is

$$y_{ijk} = \mu + \alpha_i + \eta_j + \gamma_{ij} + e_{ijk}, \tag{4}$$

$i = 1, \ldots, a$, $j = 1, \ldots, b$, $k = 1, \ldots, N$. First, the question of subsampling in a two-way ANOVA will be addressed. The discussion of subsampling leads to an examination of independent replication, and to the question of whether the interactions should be considered fixed or random. The discussion of identifying the appropriate error begins with an example:

Example 11.4.2 We want to investigate the effects of 4 fertilizers and 6 herbicides on arable plots used for raising cane (as in sugar cane). There are a total of $4 \times 6 = 24$ treatment combinations. If each treatment combination is applied to 5 plots, then model (4) is appropriate with $a = 4$, $b = 6$, and $N = 5$.

Now consider an alternative experimental design that is easily confused with this. Suppose each treatment combination is applied to 1 plot and 5 observations are taken on each plot. There is no independent replication. The 5 observations on each plot are subsamples. Comparisons within plots do not include plot to plot variability. If model (4) is used to analyze such data, the MSE is actually the estimated subsampling variance. It is based on comparisons within plots. An analysis based on model (4) will inflate the significance of all effects.

The appropriate model for this subsampling design is

$$y_{ijk} = \mu + \alpha_i + \eta_j + \gamma_{ij} + \xi_{ij} + e_{ijk},$$

where $\xi_{ij} \sim N(0, \sigma_w^2)$ and $e_{ijk} \sim N(0, \sigma_s^2)$. Note that the indices on γ and ξ are exactly the same. It is impossible to tell interactions apart from plot to plot variability. As a result, unless the interactions can be assumed nonexistent, there is no appropriate estimate of error available in this experiment.

In the example, two extreme cases were considered, one case with no subsampling and one case with no independent replication. Of course, any combination of subsampling and independent replication can be used. In the example, the designs were clearly stated so that the subsampling structures were clear. In practice, this rarely occurs. When presented with data that have been collected, it can be very difficult to identify how the experiment was designed. (I have claimed for many years that the hardest part of any statistical consulting job is to get the clients to tell you what they actually did—as opposed to what *they* think is important.)

Designs without independent replications are actually quite common. When confronted with a two-factor ANOVA without any independent replication, the fixed interaction effects are generally assumed to be zero so that an analysis can be performed. This is precisely the assumption made in Chapter 8 in analyzing the Randomized Complete Block model. An alternate way of phrasing this idea is that any interaction effects that exist must be due to error. This idea that interaction effects can themselves be errors is important. If the interactions are errors, then model (4) needs to be changed. The standard assumption would be that the γ_{ij}s are independent $N(0, \sigma_w^2)$ random variables.

Note that the assumption of random γ_{ij}s does not imply the existence of subsampling. Subsampling is a property of the experimental design. What is being discussed is simply a choice about how to model interactions. Should they be modeled as fixed effects, or should they be modeled as random errors? One guideline is based on the repeatability of the results. If the pattern of interactions should be the same in other similar studies, then the interactions are fixed. If there is little reason to believe that the pattern of interactions would be the same in other studies, then the interactions would seem to be random.

The analysis of model (4) with random interactions is straightforward. The model is a generalized split plot model with X_2 vacuous. The mean square for interactions becomes the mean square for "error 1." The mean square error from assuming fixed interactions becomes the mean square for "error 2." The analysis for main effects uses the mean square "error 1" exclusively as the estimate of error.

Although this is not a subsampling model, it does involve two sources of variation: (1) variation due to interactions, and (2) variation due to independent replication (i.e., variation from experimental unit to experimental unit). It seems to be difficult to reduce the effects on comparisons among treatments of the variability due to interactions. The effect of variation due to experimental units can be reduced by taking additional independent replications.

11.5 Exercise: An Unusual Split Plot Analysis

Cornell (1988) considered data on the production of vinyl for automobile seat covers. Different blends involve various plasticizers, stabilizers, lubricants, pigments, fillers, drying agents, and resins. The current data involve 5 blends of vinyl.

The 5 blends represent various percentages of 3 plasticizers that together make up 40.7% of the product. The first plasticizer is restricted to be between 19.1% and 34.6% of the product. The second plasticizer is restricted to be between 0% and 10.2% of the product. The third plasticizer is restricted to be between 6.1% and 11.4% of the product. Changing these restrictions to fractions of the 40.7% of the total, we get

$$0.47 \le x_1 \le 0.85, \quad 0 \le x_2 \le 0.25, \quad 0.15 \le x_3 \le 0.28.$$

The 5 blends are

Blend	(x_1, x_2, x_3)
1	(0.85, 0.000, 0.150)
2	(0.72, 0.000, 0.280)
3	(0.60, 0.250, 0.150)
4	(0.47, 0.250, 0.280)
5	(0.66, 0.125, 0.215)

Note that the first four blends have all combinations of x_2 and x_3 at their extremes with x_1 values decreasing at about 0.13 per blend. Blend 5 is in the center of the other blends. In particular, for $i = 1, 2, 3$, the x_i value of blend 5 is the mean of the other four x_i values. Eight groups of the five different blends were prepared.

The first group of 5 blends was run with the production process set for a high rate of extrusion ($z_1 = 1$) and a low drying temperature ($z_2 = -1$). The process was then reset for low extrusion rate and high drying temperature ($z_1 = -1, z_2 = 1$), and another group of 5 blends was run. For subsequent runs of 5 blends, the process was set for $z_1 = -1, z_2 = -1$, and $z_1 = 1, z_2 = 1$ to finish the first replication. Later, the second replication was run in the order $z_1 = -1, z_2 = 1$; $z_1 = 1, z_2 = 1$; $z_1 = 1, z_2 = -1$; $z_1 = -1, z_2 = -1$. The data are presented in Table 11.1

Table 11.1 Cornell's scaled vinyl thickness values

Blend	x_1	x_2	x_3	z_1	z_2	Rep. 1	Rep. 2
1	0.85	0.000	0.150	1	−1	8	7
2	0.72	0.000	0.280	1	−1	6	5
3	0.60	0.250	0.150	1	−1	10	11
4	0.47	0.250	0.280	1	−1	4	5
5	0.66	0.125	0.215	1	−1	11	10
1	0.85	0.000	0.150	−1	1	12	10
2	0.72	0.000	0.280	−1	1	9	8
3	0.60	0.250	0.150	−1	1	13	12
4	0.47	0.250	0.280	−1	1	6	3
5	0.66	0.125	0.215	−1	1	15	11
1	0.85	0.000	0.150	−1	−1	7	8
2	0.72	0.000	0.280	−1	−1	7	6
3	0.60	0.250	0.150	−1	−1	9	10
4	0.47	0.250	0.280	−1	−1	5	4
5	0.66	0.125	0.215	−1	−1	9	7
1	0.85	0.000	0.150	1	1	12	11
2	0.72	0.000	0.280	1	1	10	9
3	0.60	0.250	0.150	1	1	14	12
4	0.47	0.250	0.280	1	1	6	5
5	0.66	0.125	0.215	1	1	13	9

Compare the RCB model for the five blend treatments

$$y_{hijk} = \gamma_h + \tau_{ijk} + \varepsilon_{hijk},$$

where the triples (i, j, k) only take on five distinct values, with the reduced regression model

$$y_{hijk} = \gamma_h + \beta_1 x_{1i} + \beta_2 x_{2j} + \beta_3 x_{3k} + \varepsilon_{hijk}.$$

Test lack of fit. Average over blends to do a whole-plot analysis. Finally, do a split plot analysis.

References

Christensen, R. (2015). *Analysis of variance, design, and regression: Linear modeling for unbalanced data* (2nd ed.). Boca Raton, FL: Chapman and Hall/CRC Pres.

Christensen, R. (1984). A note on ordinary least squares methods for two-stage sampling. *Journal of the American Statistical Association, 79*, 720–721.

Christensen, R. (1987). The analysis of two-stage sampling data by ordinary least squares. *Journal of the American Statistical Association, 82*, 492–498.
Cornell, J. A. (1988). Analyzing mixture experiments containing process variables. A split plot approach. *Journal of Quality Technology, 20*, 2–23.
Fisher, R. A. (1935). *The design of experiments*, (9th ed., 1971). New York: Hafner Press.
Mathew, T., & Sinha, B. K. (1992). Exact and optimum tests in unbalanced split-plot designs under mixed and random models. *Journal of the American Statistical Association, 87*, 192–200.
Monlezun, C. J., & Blouin, D. C. (1988). A general nested split-plot analysis of covariance. *Journal of the American Statistical Association, 83*, 818–823.
Skinner, C. J., Holt, D., & Smith, T. M. F. (1989). *Analysis of complex surveys*. New York: Wiley.

Chapter 12
Model Diagnostics

Abstract This chapter focuses on methods for evaluating the assumptions made in a standard linear model and on the use of transformations to correct such problems.

This book deals with linear model theory and as such we have largely assumed that the data are good and that the models are true. Unfortunately, good data are rare and true models are even rarer. (George Box would have us believe that true models are nonexistent.) Chapters 12–14 discuss some additional tools used by statisticians to deal with the problems presented by real data.

All models are based on assumptions. We typically assume that $\mathrm{E}(Y)$ has a linear structure, that the observations are independent, that the variance is the same for each observation, and that the observations are normally distributed. In truth, these assumptions will probably never be correct. It is our hope that if we check the assumptions and if the assumptions look plausible, then the mathematical methods presented here will work quite well.

If the assumptions are checked and found to be implausible, we need to have alternate ways of analyzing the data. In Section 2.7, Section 3.8, Chapters 10, and 11, we discussed the analysis of linear models with general covariance matrices. If an approximate covariance matrix can be found, the methods presented earlier can be used. (See also *ALM-III* and in particular the discussion of the deleterious effects of estimating covariance matrices in Chapter 4 (Christensen 2001, Section 6.5).) Another approach is to find a transformation of the data so that the assumptions seem plausible for a standard linear model in the transformed data.

The primary purpose of this chapter is to present methods of identifying when there may be trouble with the assumptions. Analysis of the residuals is the method most often used for detecting invalidity of the assumptions. Residuals are used to check for nonnormality of errors, nonindependence, lack of fit, heteroscedasticity (inequality) of variances, and outliers (unusual data). They also help identify influential observations.

R. Christensen, *Plane Answers to Complex Questions*, Springer Texts in Statistics,
https://doi.org/10.1007/978-3-030-32097-3_12

The vector of residuals is, essentially, an estimate of e in the model

$$Y = X\beta + e, \quad \mathrm{E}(e) = 0, \quad \mathrm{Cov}(e) = \sigma^2 I.$$

The residual vector is

$$\hat{e} \equiv Y - X\hat{\beta} = (I - M)Y,$$

with

$$\mathrm{E}(\hat{e}) = (I - M)X\beta = 0$$

and

$$\mathrm{Cov}(\hat{e}) = (I - M)\sigma^2 I(I - M)' = \sigma^2(I - M).$$

For many of the techniques that we will discuss, the residuals are standardized so that their variances are about 1. The *standardized residuals* are

$$r_i \equiv \hat{e}_i \Big/ \sqrt{MSE(1 - m_{ii})},$$

where m_{ii} is the ith diagonal element of M and $\hat{e} = [\hat{e}_1, \ldots, \hat{e}_n]'$.

When checking for nonnormality or heteroscedastic variances, it is important to use the standardized residuals rather than unstandardized residuals. As just seen, the ordinary residuals have heteroscedastic variances. Before they are useful in checking for equality of the variances of the observations, they need to be standardized. Moreover, methods for detecting nonnormality are often sensitive to inequality of variances, so the use of ordinary residuals can make it appear that the errors are not normal even when they are.

Once computer programs used $\hat{e}/\sqrt{MSE}$ as standardized residuals, but that seems to be a thing of the past. That method was inferior to the standardization given above because it ignores the fact that the variances of the residuals are not all equal. The standardized residuals are also sometimes called the *Studentized residuals*, but that term is also sometimes used for another quantity discussed in Section 6.

Influential observations have been mentioned. What are they? One idea is that an observation is influential if it greatly affects the fitted regression equation. Influential observations are not intrinsically good or bad, but they are always important. Typically, influential observations are outliers: data points that are, in some sense, far from the other data points being analyzed. This can happen in two ways. First, the y value associated with a particular row of the X matrix can be unlike what would be expected from examining the rest of the data. Second, a particular row of the X matrix can be unlike any of the other rows of the X matrix.

A frequently used method for analyzing residuals is to plot the (standardized) residuals against various other variables. We now consider an example that will be used throughout this chapter to illustrate the use of residual plots. In the example, we consider a model that will later be perturbed in various ways. This model and its perturbations will provide residuals that can be plotted to show characteristics of residual plots and the effects of unsatisfied assumptions.

Table 12.1 Steam data

Obs. no.	x_1	x_2	Obs. no.	x_1	x_2
1	35.3	20	14	39.1	19
2	29.7	20	15	46.8	23
3	30.8	23	16	48.5	20
4	58.8	20	17	59.3	22
5	61.4	21	18	70.0	22
6	71.3	22	19	70.0	11
7	74.4	11	20	74.5	23
8	76.7	23	21	72.1	20
9	70.7	21	22	58.1	21
10	57.5	20	23	44.6	20
11	46.4	20	24	33.4	20
12	28.9	21	25	28.6	22
13	28.1	21			

Example 12.0.1 Draper and Smith (1998) presented an example with 25 observations on a dependent variable, pounds of steam used by a company per month, and two predictor variables: x_1, the average atmospheric temperature for the month (in °F); and x_2, the number of operating days in the month. The values of x_1 and x_2 are listed in Table 12.1.

Draper and Smith's fitted equation is

$$y = 9.1266 - 0.0724x_1 + 0.2029x_2.$$

Our examples will frequently be set up so that the *true* model is

$$y_i = 9.1266 - 0.0724x_{i1} + 0.2029x_{i2} + e_i\,. \tag{1}$$

The vector $Y = (y_1, \ldots, y_{25})'$ can be obtained by generating the e_is and adding the terms on the right-hand side of the equation. Once Y is obtained, the equation $y_i = \beta_0 + \beta_1 x_{i1} + \beta_2 x_{i2} + e_i$ can be fitted by least squares and the residuals computed. By generating errors that have independent identical normal distributions, nonnormal distributions, serial correlations, or unequal variances, we can examine how residual plots should look when these conditions exist. By fitting models with incorrect mean structure, we can examine how residual plots for detecting lack of fit should look.

The material in this chapter is presented with applications to regression models in mind, but can be applied to ANOVA models with little modification. Excellent discussions of residuals and influential observations are found in Cook and Weisberg (1982) [downloadable from http://www.stat.umn.edu/rir], Atkinson (1985), and elsewhere. Cook and Weisberg (1994, 1999) and Cook (1998) give extensive discussions of regression graphics.

Exercise 12.1 Show that for a standard linear model $\hat{e}$ is the BLUP of e.

12.1 Leverage

A data point (case) that corresponds to a row of the X matrix that is "unlike" the other rows is said to have high *leverage*. In this section we will define the Mahalanobis distance and use it as a basis for identifying rows of X that are unusual. It will be shown that for regression models that include an intercept (a column of 1s), the diagonal elements of the ppo M and the Mahalanobis distances are equivalent measures. In particular, a diagonal element of the ppo is an increasing function of the corresponding Mahalanobis distance. (There are some minor technical difficulties with this claim when X is not a full rank matrix.) This equivalence justifies the use of the diagonal elements to measure the abnormality of a row of the model matrix. *The diagonal elements of the ppo are the standard tool used for measuring leverage.*

In addition to their interpretation as a measure of abnormality, it will be shown that the diagonal elements of the projection matrix can have direct implications on the fit of a regression model. A diagonal element of the projection matrix that happens to be near 1 (the maximum possible) will force the estimated regression equation to go very near the corresponding y value. Thus, cases with extremely large diagonal elements have considerable influence on the estimated regression equation. It will be shown through examples that diagonal elements that are large, but not near 1, can also have substantial influence.

High leverage points are not necessarily bad. If a case with high leverage is consistent with the remainder of the data, then the case with high leverage causes no problems. In fact, the case with high leverage can greatly reduce the variability of the least squares fit. In other words, with an essentially correct model and good data, high leverage points actually help the analysis.

On the other hand, *high leverage points are dangerous*. The regression model that one chooses is rarely the true model. Usually it is only an approximation to the truth. High leverage points can change a good approximate model into a bad approximate model. An approximate model is likely to work well only on data that are limited to some particular range of values. It is unreasonable to expect to find a model that works well in places where very little data were collected. By definition, high leverage points exist where very little data were collected, so one would not expect them to be modeled well. Ironically, just the opposite result usually occurs. The high leverage points are often fit very well, while the fit of the other data is often harmed. The model for the bulk of the data is easily distorted to accommodate the high leverage points. When high leverage points are identified, the researcher is often left to decide between a bad model for the entire data or a good model for a more limited problem.

The purpose of this discussion of cases with high leverage is to make one point. If some data were collected in unusual places, then the appropriate goal may be to find a good approximate model for the area in which the bulk of the data were collected. This is not to say that high leverage points should always be thrown out of a data set. High leverage points need to be handled with care, and the implications of excluding high leverage points from a particular data set need to be thoroughly examined. High leverage points can be the most important cases in the entire data.

We begin by defining the Mahalanobis distance and establishing its equivalence to the diagonal elements of the projection operator. This will be followed by an examination of diagonal elements that are near 1. The section closes with a series of examples.

12.1.1 Mahalanobis Distances

The *Mahalanobis distance* measures how far a random vector is from the middle of its distribution. For this purpose, we will think of the rows of the matrix X as a sample of vectors from some population. Although this contradicts our assumption that the matrix X is fixed and known, our only purpose is to arrive at a reasonable summary measure of the distance of each row from the other rows. The Mahalanobis distance provides such a measure. The notation and ideas involved in estimating Mahalanobis distances are similar to those used in estimating best linear predictors. Estimation of best linear predictors was discussed in Subsection 6.3.4. In particular, we write the ith row of X as $(1, x_i')$ so that the corresponding linear model contains an intercept.

Let x be a random vector.

Definition 12.1.1 Let $\mathrm{E}(x) = \mu$ and $\mathrm{Cov}(x) = V$. The *squared Mahalanobis distance* is

$$D^2 \equiv (x - \mu)'V^{-1}(x - \mu).$$

For a sample $x_1, \ldots, x_n$, the relative distances of the observations from the center of the distribution can be measured by the squared distances

$$D_i^2 = (x_i - \mu)'V^{-1}(x_i - \mu), \quad i = 1, \ldots, n.$$

Usually, μ and V are not available, so they must be estimated. Similar to Subsubsection 6.3.4.1, write

$$Z = \begin{bmatrix} x_1' \\ \vdots \\ x_n' \end{bmatrix},$$

estimate μ with $\bar{x}_\cdot' = (1/n)J_1^n Z$, and estimate V with

$$S = \frac{1}{n-1}\left[\sum_{i=1}^{n}(x_i - \bar{x}_\cdot)(x_i - \bar{x}_\cdot)'\right] = \frac{1}{n-1}Z'\left(I - \frac{1}{n}J_n^n\right)Z.$$

Definition 12.1.2 The *estimated squared Mahalanobis distance* for the ith case in a sample of vectors $x_1, \ldots, x_n$ is

$$\hat{D}_i^2 \equiv (x_i - \bar{x}_\cdot)' S^{-1} (x_i - \bar{x}_\cdot).$$

Note that the values $\hat{D}_i^2$ are precisely the diagonal elements of

$$(n-1)\left(I - \frac{1}{n}J_n^n\right) Z \left[Z'\left(I - \frac{1}{n}J_n^n\right) Z\right]^{-1} Z'\left(I - \frac{1}{n}J_n^n\right).$$

Our interest in these definitions is that for a regression model $Y = X\beta + e$, the distance of the ith row of X from the other rows can be measured by the estimated squared Mahalanobis distance $\hat{D}_i^2$. In this context we think of the rows of X as a sample from some population. As mentioned earlier, when the model has an intercept, the diagonal elements of M and the estimated squared Mahalanobis distances are equivalent measures. When an intercept is included in the model, X can be written as $X = [J, Z]$. Since the rows of J are identical, the matrix S, defined for the entire matrix X, is singular. Thus, S^{-1} does not exist and the estimated Mahalanobis distances are not defined. Instead, we measure the relative distances of the rows of Z from their center.

Theorem 12.1.3 *Consider the linear model $Y = X\beta + e$, where X is a full rank matrix and $X = [J, Z]$. Then,*

$$m_{ii} = \frac{1}{n} + \frac{\hat{D}^2}{n-1}.$$

Proof The theorem follows immediately from the fact that

$$M = \frac{1}{n}J_n^n + \left(I - \frac{1}{n}J_n^n\right) Z \left[Z'\left(I - \frac{1}{n}J_n^n\right) Z\right]^{-1} Z'\left(I - \frac{1}{n}J_n^n\right)$$

(cf. Sections 6.2 and 9.1). The inverse in the second term of the right-hand side exists because X, and therefore $\left(I - \frac{1}{n}J_n^n\right) Z$, are full rank matrices. □

From this theorem, it is clear that the rows with the largest squared Mahalanobis distances are precisely the rows with the largest diagonal elements of the perpendicular projection matrix. For identifying high leverage cases in a regression model with an intercept, using the diagonal elements of M is equivalent to using the squared Mahalanobis distances.

Of course, for regression models that do not include an intercept, using the squared Mahalanobis distances is not equivalent to using the diagonal elements of the projection matrix. For such models it would probably be wise to examine both measures of leverage. The diagonal elements of the projection matrix are either given by or easily obtained from many computer programs. The information in the squared Mahalanobis distances can be obtained by the artifice of adding an intercept to the model and obtaining the diagonal elements of the projection matrix for the augmented model.

For linear models in which X is not of full rank, similar definitions could be made using a generalized inverse of the (estimated) covariance matrix rather than the inverse.

Exercise 12.2 Show that for a regression model that does not contain an intercept, the diagonal elements of the perpendicular projection operator are equivalent to the estimated squared Mahalanobis distances computed with the assumption that $\mu = 0$.

12.1.2 Diagonal Elements of the Projection Operator

Having established that the diagonal elements of M are a reasonable measure of how unusual a row of X is, we are left with the problem of calibrating the measure. How big does m_{ii} have to be before we need to worry about it? The following proposition indirectly establishes several facts that allow us to provide some guidelines.

Proposition 12.1.4 *For any i*

$$m_{ii}(1 - m_{ii}) = \sum_{j \neq i} m_{ij}^2.$$

Proof Because M is a symmetric idempotent matrix,

$$m_{ii} = \sum_{j=1}^{n} m_{ij} m_{ji} = \sum_{j=1}^{n} m_{ij}^2 = m_{ii}^2 + \sum_{j \neq i} m_{ij}^2.$$

Subtracting gives

$$m_{ii}(1 - m_{ii}) = m_{ii} - m_{ii}^2 = \sum_{j \neq i} m_{ij}^2. \qquad \square$$

The term on the right-hand side of Proposition 12.1.4 is a sum of squared terms. This must be nonnegative, so $m_{ii}(1 - m_{ii}) \geq 0$. It follows immediately that the m_{ii}s must lie between 0 and 1.

Since the largest value that an m_{ii} can take is 1, any value near 1 indicates a point with extremely high leverage. Other values, considerably less than 1, can also indicate high leverage. Because $\mathrm{tr}(M) = r(M)$, the average value of the m_{ii}s is p/n. Any m_{ii} value that is substantially larger than p/n indicates a point with high leverage. Some useful but imprecise terminology is set in the following definition.

Definition 12.1.5 Any case that corresponds to a row of the model matrix that is unlike the other rows is called an *outlier in the design space* (or estimation space). Any case corresponding to an m_{ii} substantially larger than p/n is called a case with *high leverage*. Any case corresponding to an m_{ii} near 1 is called a case with *extremely high leverage*.

Points with extremely high leverage have dramatic effects. If m_{ii} happens to be near 1, then $m_{ii}(1-m_{ii})$ must be near zero and the right-hand side of Proposition 12.1.4 must be near zero. Since the right-hand side is a sum of squared terms, for all $j \neq i$ the terms m_{ij} must be near zero. As will be shown, this causes a point with extremely high leverage to dominate the fitting process.

Let ρ_i be a vector of zeros with a 1 in the ith row so that $x_i' = \rho_i' X$. The mean for the ith case is $\rho_i' X\beta = x_i'\beta$, which is estimated by $x_i'\hat{\beta} = \rho_i' MY = \sum_{j=1}^{n} m_{ij} y_j$. If m_{ii} is close to 1, m_{ij} is close to zero for all $j \neq i$; thus, the rough approximation $\rho_i' MY \doteq y_i$ applies. This approximation is by no means unusual. It simply says that the model fits well; however, the fact that the estimate largely ignores observations other than y_i indicates that something strange is occurring. Since m_{ii} is near 1, x_i' is far from the other rows of X. Thus, there is little information available about behavior at x_i' other than the observation y_i.

The fact that y_i is fit reasonably well has important implications for the estimated regression equation $f(x) = x'\hat{\beta}$. This function, when evaluated at x_i, must be near y_i. Regardless of whether y_i is an aberrant observation or whether a different approximate model is needed for observations taken near x_i, the estimated regression equation will adjust itself to fit y_i reasonably well. If necessary, the estimated regression equation will ignore the structure of the other data points in order to get a reasonable fit to the point with extremely high leverage. Thus, points with extremely high leverage have the potential to influence the estimated regression equation a great deal.

12.1.3 Examples

Example 12.1.6 *Simple Linear Regression.*
Consider the model $y_i = \beta_0 + \beta_1 x_i + e_i$, $i = 1, \dots, 6$, $x_1 = 1$, $x_2 = 2$, $x_3 = 3$, $x_4 = 4$, $x_5 = 5$, $x_6 = 15$. x_6 is far from the other x values, so it should be a case with high leverage. In particular, $m_{66} = 0.936$, so case six is an extremely high leverage point. (Section 6.1 gives a formula for M.)

For $i = 1, \dots, 5$, data were generated using the model

$$y_i = 2 + 3x_i + e_i,$$

with the e_is independent $N(0, 1)$ random variates. The y values actually obtained were $y_1 = 7.455$, $y_2 = 7.469$, $y_3 = 10.366$, $y_4 = 14.279$, $y_5 = 17.046$. The model was fit under three conditions: (1) with the sixth case deleted, (2) with $y_6 = 47 = 2 + 3(x_6)$, and (3) with $y_6 = 11 = 2 + 3(x_3)$. The results of fitting the simple linear regression model are summarized in the table below.

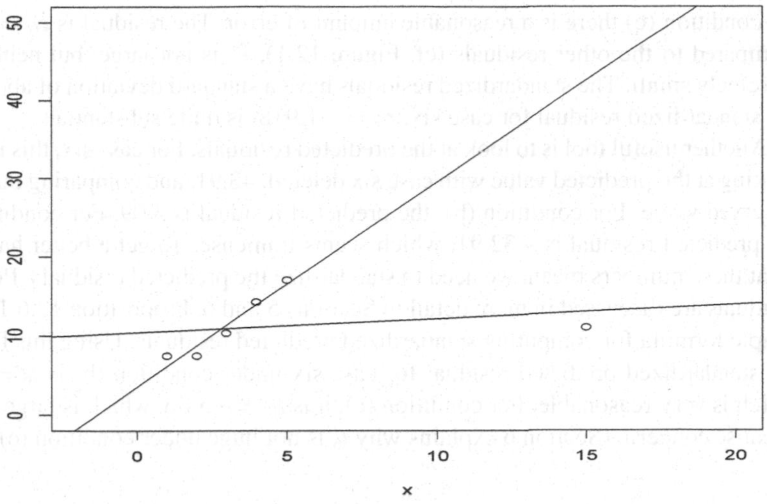

Fig. 12.1 Scatterplot and fitted lines for Example 12.1.6

Condition	y_6	$\hat{y}_6$	$\hat{\beta}_0$	$\hat{\beta}_1$	$\sqrt{MSE}$	dfE
a	deleted	43.91	3.425	2.699	1.423	3
b	47	46.80	2.753	2.937	1.293	4
c	11	13.11	10.599	0.1674	4.345	4

Figure 12.1 contains a scatterplot of the data under condition (c) and the fitted lines for conditions (a) and (c). Under conditions (a) and (b), reasonably consistent fits are obtained. In particular, the extremely high leverage case has little effect on point estimation in these situations where the data are good and the model is true. (The high leverage case could have a large effect in decreasing the size of interval estimates.)

Under condition (c), when the model is no longer valid for case six, the fit is grossly different from those obtained under conditions (a) and (b). When the extremely high leverage case is inconsistent with the rest of the data, the fit of the regression equation is dominated by that case. Note that the first five cases lead us to expect a value for y_6 in the mid-40s, but with $y_6 = 11$, the fitted value is 13.11, close to 11 and far from the mid-40s. Finally, the fit of the model, as measured by the *MSE*, is good under conditions (a) and (b), but much poorer under condition (c).

Other things being equal, under condition (b) it would be wise to include case six in the analysis. Under condition (c), it might be wiser not to try to model all the data, but rather to model only the first five cases and admit that little is known about behavior at x values substantially larger than 5. Unfortunately, in order to distinguish between conditions (b) and (c), the true model must be known, which in practice is not the case.

Finally, a word about residuals for case six. Under condition (b), the regression equation is right on the data point, $\hat{e}_6 = 0.2$, and $r_6 = 0.611$. On the other hand, for condition (c) there is a reasonable amount of error. The residual is $\hat{e}_6 = -2.11$. Compared to the other residuals (cf. Figure 12.1), $\hat{e}_6$ is not large, but neither is it extremely small. The standardized residuals have a standard deviation of about 1, so the standardized residual for case six, $r_6 = -1.918$, is quite substantial.

Another useful tool is to look at the predicted residuals. For case six, this involves looking at the predicted value with case six deleted, 43.91, and comparing that to the observed value. For condition (b), the predicted residual is 3.09. For condition (c), the predicted residual is −32.91, which seems immense. To get a better handle on what these numbers mean, we need to standardize the predicted residuals. Predicted residuals are discussed in more detail in Sections 5 and 6. Proposition 12.6.1 gives a simple formula for computing standardized predicted residuals. Using this formula, the standardized predicted residual for case six under condition (b) is $t_6 = 0.556$, which is very reasonable. For condition (c), it is $t_6 = -5.86$, which is large enough to cause concern. (Section 6 explains why t_6 is not huge under condition (c).)

Example 12.1.7 We now modify Example 12.1.6 by adding a second point far away from the bulk of the data. This is done in two ways. First, the extremely high leverage point is replicated with a second observation taken at $x = 15$. Second, a high leverage point is added that is smaller than the bulk of the data. In both cases, the y values for the first five cases remain unchanged.

With two observations at $x = 15$, $m_{66} = m_{77} = 0.48$. This is well above the value $p/n = 2/7 = 0.29$. In particular, the diagonal values for the other five cases are all less than 0.29.

To illustrate the effect of the high leverage points on the regression equation, conditions (b) and (c) of the previous example were combined. In other words, the two y values for $x = 15$ were taken as 11 and 47. The estimated regression equation becomes $\hat{y} = 6.783 + 1.5140x$. The slope of the line is about halfway between the slopes under conditions (b) and (c). More importantly, the predicted value for $x = 15$ is 29.49. The regression equation is being forced near the mean of y_6 and y_7. (The mean is 29.)

One of the salient points in this example is the effect on the root mean squared error. Under condition (b), where the high leverage point was consistent with the other data, the root mean squared error was 1.293. Under condition (c), where the high leverage point was grossly inconsistent with the other data, the root mean squared error was 4.293. In this case, with two high leverage points, one consistent with the bulk of the data and one not, the root mean squared error is 11.567. This drastic change is because, with two y values so far apart, one point almost has to be an outlier. Having an outlier in the ys at a high leverage point has a devastating effect on all estimates, especially the estimated error.

In the second illustration, x_7 was taken as −9. This was based on adding a second observation as far to the left of the bulk of the data as $x_6 = 15$ is to the right. The leverages are $m_{66} = m_{77} = 0.63$. Again, this value is well above 2/7 and is far above the other diagonal values, which are around 0.15.

To illustrate the effect of high leverage on estimation, the y values were taken as $y_6 = y_7 = 11$. The estimated regression equation was $11.0906 + 0.0906x$. The root mean squared error was 3.921. The t statistic for testing that the slope equaled zero was 0.40.

In essence, the data in the second illustration have been reduced to three points: a point $x = -9$ with a y value of 11, a point $x = 15$ with a y value of 11, and a point $x = 3$ (the mean of x_1 to x_5) with a y value of 11.523 (the mean of y_1 to y_5). Compared to the high leverage points at -9 and 15, the five points near 3 are essentially replicates.

Both of these scenarios illustrate situations where the leverages contain gaps. The first illustration has no points with leverages between 0.28 and 0.48. The second has no leverages between 0.16 and 0.63. Such *gaps in the leverages indicate that the predictor variables contain clusters of observations that are separated from each other.*

The final example of this section illustrates that high leverage points are model dependent. Since our measure of leverage is based on the perpendicular projection operator onto $C(X)$, it is not surprising that changing the model can affect the leverage of a case. The example below is a polynomial regression where a particular case does not have high leverage for fitting a line, but does have high leverage for fitting a parabola.

Example 12.1.8 *Quadratic Regression.*
Consider fitting the models

$$y_i = \beta_0 + \beta_1 x_i + e_i,$$
$$y_i = \gamma_0 + \gamma_1 x_i + \gamma_2 x_i^2 + e_i,$$

$i = 1, \ldots, 7$. The values of the x_is used were $x_1 = -10$, $x_2 = -9$, $x_3 = -8$, $x_4 = 0$, $x_5 = 8$, $x_6 = 9$, $x_7 = 10$. Note that the value of x_4 appears to be in the center of the data. For fitting a straight line, that appearance is correct. For fitting a line, the leverage of the fourth case is 0.14.

The model matrix for the quadratic model is

$$X = \begin{bmatrix} 1 & -10 & 100 \\ 1 & -9 & 81 \\ 1 & -8 & 64 \\ 1 & 0 & 0 \\ 1 & 8 & 64 \\ 1 & 9 & 81 \\ 1 & 10 & 100 \end{bmatrix}.$$

Note that the choice of the x_is makes the second column of x orthogonal to the other two columns. An orthonormal basis for $C(X)$ is easily obtained, and thus the diagonal elements of M are also easily obtained. The value of m_{44} is 0.84, which is quite large. From inspecting the third column of the model matrix, it is clear that the fourth case is unlike the rest of the data.

To make the example more specific, for $i \neq 4$ data were generated from the model

$$\begin{aligned} y_i &= 19.6 + 0.4x_i - 0.1x_i^2 + e_i \\ &= -0.1(x_i - 2)^2 + 20 + e_i, \end{aligned}$$

with the e_is independent $N(0, 1)$ random variables. The values 0, 11.5, and 19.6 were used for y_4. These values were chosen to illustrate a variety of conditions. The value 19.6 is consistent with the model given above. In particular, $19.6 = \mathrm{E}(y_4)$. The value $11.5 = \mathrm{E}(y_2 + y_6)/2$ should give a fit that is nearly linear. The value 0 is simply a convenient choice that is likely to be smaller than any other observation. The Y vector obtained was

$$Y = (6.230, 8.275, 8.580, y_4, 16.249, 14.791, 14.024)'.$$

Figure 12.2 contains a scatterplot of the data that includes all three values for y_4 as well as a plot of the true regression curve.

The linear and quadratic models were fitted with all three of the y_4 values and with the fourth case deleted from the model. For all models fitted, the coefficient of the linear term was 0.4040. As mentioned above, the second column (the linear column) of the matrix X is orthogonal to the other columns. Thus, for any value of y_4 the linear coefficient will be the same for the quadratic model and the linear model. The linear coefficient does not depend on the value of y_4 because $x_4 = 0$. Also, because $x_4 = 0$, the predicted value for y_4 is just the intercept of the line. Fitting simple linear regressions resulted in the following:

Linear Fits

y_4	$\hat{y}_4 = \hat{\beta}_0$	$\sqrt{MSE}$	dfE
deleted	11.36	1.263	4
0.0	9.74	4.836	5
11.5	11.38	1.131	5
19.6	12.54	3.594	5

As designed, the fits for y_4 deleted and $y_4 = 11.5$ are almost identical. The other values of y_4 serve merely to move the intercept up or down a bit. They do not move the line enough so that the predicted value $\hat{y}_4$ is close to the observed value y_4.

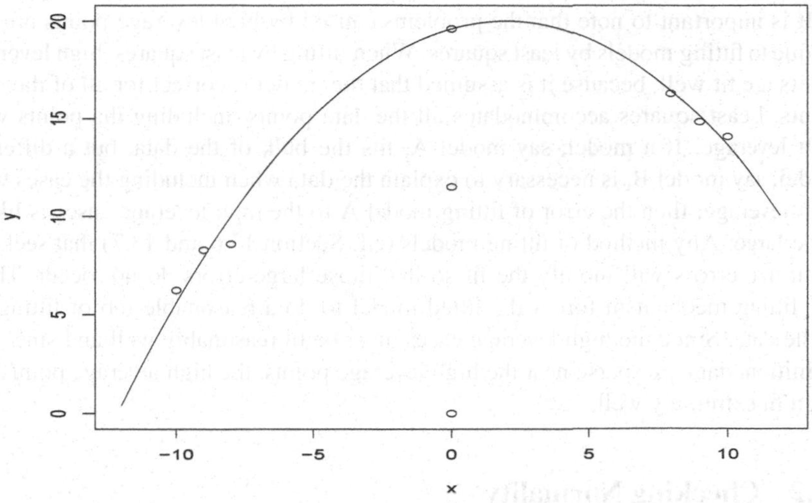

Fig. 12.2 Scatterplot for Example 12.1.8

The y_4 values of 0 and 19.6 do not fit the line well, which is reflected in the increased values for the root mean squared error. In summary, the values of y_4 do not have a great effect on the fitted lines.

The results of the quadratic fits, including the t statistic for testing $\gamma_2 = 0$, are

Quadratic Fits

y_4	$\hat{y}_4 = \hat{\gamma}_0$	$\hat{\gamma}_2$	$t(\gamma_2)$	$\sqrt{MSE}$	dfE
deleted	16.564	−0.064	−3.78	0.607	3
0.0	2.626	0.102	2.55	3.339	4
11.5	12.303	−0.013	−0.97	1.137	4
19.6	19.119	−0.094	−9.83	0.802	4

As expected, the y_4 deleted and $y_4 = 19.6$ situations are similar, and approximate the true model. The $y_4 = 11.5$ situation gives essentially a straight line; the t statistic for testing $H_0 : \gamma_2 = 0$ is very small. The true quadratic structure of all but one case is ignored in favor of the linear fit. (Note that the root mean squared error is almost identical in the linear and quadratic fits when $y_4 = 11.5$.) Finally, with $y_4 = 0$, the entire structure of the problem is turned upside down. The true model for all cases except case four is a parabola opening down. With $y_4 = 0$, the fitted parabola opens up. Although the fourth case does not have high leverage for the linear fits, the fourth case greatly affects the quadratic fits.

It is important to note that the problems caused by high leverage points are not unique to fitting models by least squares. When fitting by least squares, high leverage points are fit well, because it is assumed that the model is correct for all of the data points. Least squares accommodates all the data points, including the points with high leverage. If a model, say model A, fits the bulk of the data, but a different model, say model B, is necessary to explain the data when including the cases with high leverage, then the error of fitting model A to the high leverage cases is likely to be large. Any method of fitting models (c.f. Section 13.6 and 13.7) that seeks to minimize errors will modify the fit so that those large errors do not occur. Thus, any fitting mechanism forces the fitted model to do a reasonable job of fitting all of the data. Since the high leverage cases must be fit reasonably well and since, by definition, data are sparse near the high leverage points, the high leverage points are often fit extremely well.

12.2 Checking Normality

We give a general discussion of the problem of checking normality for a random sample and then relate it to the analysis of residuals. Suppose $v_1, \ldots, v_n$ are i.i.d. $N(\mu, \sigma^2)$ and $z_1, \ldots, z_n$ are i.i.d. $N(0, 1)$. Ordering these from smallest to largest gives the order statistics $v_{(1)} \leq \cdots \leq v_{(n)}$ and $z_{(1)} \leq \cdots \leq z_{(n)}$. The expected values of the standard normal order statistics are $\mathrm{E}[z_{(1)}], \ldots, \mathrm{E}[z_{(n)}]$. Since the v_is are normal, $[v_{(i)} - \mu]/\sigma \sim z_{(i)}$ and we should have the approximate equality, $[v_{(i)} - \mu]/\sigma \doteq \mathrm{E}[z_{(i)}]$ or $v_{(i)} \doteq \sigma\mathrm{E}[z_{(i)}] + \mu$.

Suppose now that $v_1, \ldots, v_n$ are observed and we want to see if they are a random sample from a normal distribution. If the v_is are from a normal distribution, a graph of the pairs $(\mathrm{E}[z_{(i)}], v_{(i)})$ should be an approximate straight line. If the graph is not an approximate straight line, nonnormality is indicated. These graphs are variously called *rankit plots, normal plots,* or *q-q plots.*

To make the graph, one needs the values $\mathrm{E}[z_{(i)}]$. These values, often called *rankits, normal scores,* or *theoretical (normal) quantiles*, are frequently approximated as follows. Let

$$\Phi(x) = \int_{-\infty}^{x} (2\pi)^{-1/2} \exp[-t^2/2]dt.$$

$\Phi(x)$ is the cumulative distribution function for a standard normal random variable. Let u have a uniform distribution on the interval (0, 1). Write $u \sim U(0, 1)$. It can be shown that

$$\Phi^{-1}(u) \sim N(0, 1).$$

If $z_1, \ldots, z_n$ are i.i.d. $N(0, 1)$ and $u_1, \ldots, u_n$ are i.i.d. $U(0, 1)$, then

$$z_{(i)} \sim \Phi^{-1}(u_{(i)}),$$

and

$$\mathrm{E}[z_{(i)}] = \mathrm{E}[\Phi^{-1}(u_{(i)})].$$

One reasonable approximation for $\mathrm{E}[z_{(i)}]$ is

$$\mathrm{E}[z_{(i)}] \doteq \Phi^{-1}(\mathrm{E}[u_{(i)}]) = \Phi^{-1}\left(\frac{i}{n+1}\right).$$

In practice, better approximations are available. Take

$$\mathrm{E}[z_{(i)}] \doteq \Phi^{-1}\left(\frac{i-a}{n+(1-2a)}\right). \tag{12.1}$$

For $n \geq 5$, an excellent approximation is $a = 3/8$, see Blom (1958). The R programming language defaults to $a = 3/8$ when $n \leq 10$ with $a = 0.5$ otherwise. MINITAB always uses $a = 3/8$.

To check whether $e \sim N(0, \sigma^2 I)$ in a linear model, the standardized residuals are plotted against the rankits. If the plot is not linear, nonnormality is suspected.

Example 12.2.1 For $n = 10, 25, 50$, nine random vectors, say Ei, $i = 1, \ldots, 9$, were generated so that the Eis were independent and

$$Ei \sim N(0, I).$$

The corresponding Ei—rankit plots in Figures 12.3, 12.4 and 12.5 give an idea of how straight one can reasonably expect rankit plots to be for normal data. All plots use rankits from equation (1) with $a = 3/8$.

Figure 12.3 gives rankit plots for random samples of size $n = 10$. Notice the substantial deviations from linearity in these plots even though the data are i.i.d. normal. Figure 12.4 gives rankit plots for samples of size $n = 25$ and Figure 12.5 gives plots for samples of size $n = 50$. As the sample sizes increase, the plots become more linear.

In addition, for $n = 25$ the Eis were used with (12.0.1) to generate nine Y vectors and the standardized residuals from fitting

$$y_i = \beta_0 + \beta_1 x_{i1} + \beta_2 x_{i2} + e_i \tag{12.2}$$

were computed. The Eis and the corresponding standardized residual vectors (Ris) were plotted against the approximate rankits. Two Ri—rankit plots are provided to give some idea of how the correlations among the residuals can affect the plots. Of the nine pairs of plots generated, only the two that look least normal (based on Ei as determined by the author) are displayed in Figure 12.6. The standardized residuals in plot (b) seem more normal than the original normal observations in plot (a).

Duan (1981) established that the residuals provide an asymptotically consistent estimate of the underlying error distribution. Thus for large samples, the residual—rankit plot should provide an accurate evaluation of normality. In practice, however,

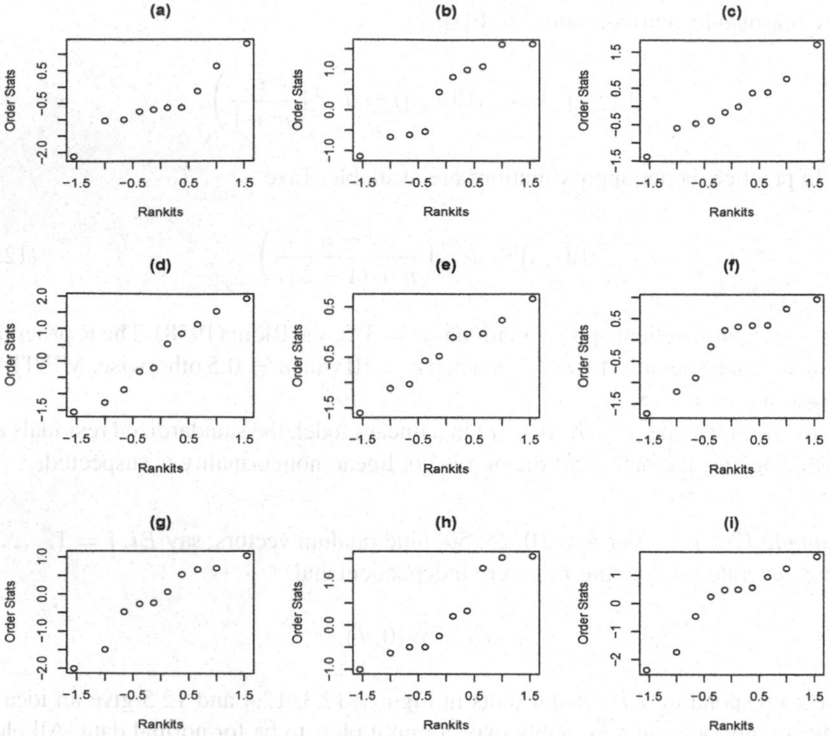

Fig. 12.3 Normal plots for normal data, $n = 10$

the real question is not whether the data are nonnormal, but whether they are sufficiently nonnormal to invalidate a normal approximation. This is a more difficult question to address. See Arnold (1981, Chapter 10) for a discussion of *asymptotic consistency* of least squares estimates.

Shapiro and Wilk (1965) developed a formal test for normality related to normal plots. Unfortunately, the test involves knowing the inverse of the covariance matrix of $z_{(1)}, \ldots, z_{(n)}$. An excellent approximation to their test was suggested by Shapiro and Francia (1972). The approximate test statistic is the square of the sample correlation coefficient computed from the pairs $(\mathrm{E}[z_{(i)}], v_{(i)})$, $i = 1, \ldots, n$. Let

$$W' = \left(\sum_{i=1}^{n} \mathrm{E}[z_{(i)}]v_{(i)}\right)^2 \Big/ \sum_{i=1}^{n} (\mathrm{E}[z_{(i)}])^2 \sum_{i=1}^{n} (v_{(i)} - \bar{v})^2.$$

(Note: $\sum_{i=1}^{n} \mathrm{E}[z_{(i)}] = 0$ by symmetry.) If W' is large, there is no evidence of nonnormality. Small values of W' are inconsistent with the hypothesis that the data are a random sample from a normal distribution. Approximate percentage points for the distribution of W' are given in Christensen (2015) and many other places.

To test for normality in linear models, the v_is are generally replaced by the r_is, i.e., the standardized residuals.

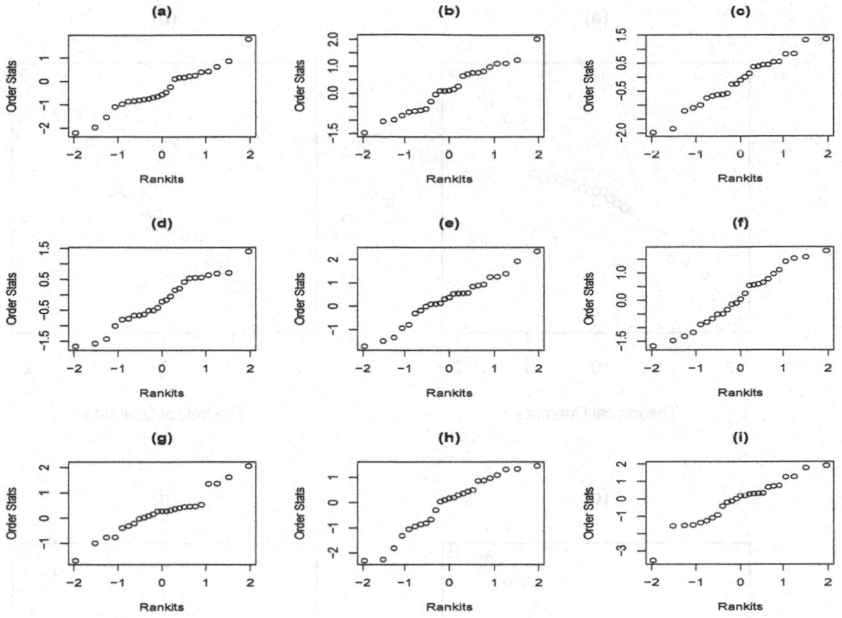

Fig. 12.4 Normal plots for normal data, $n = 25$

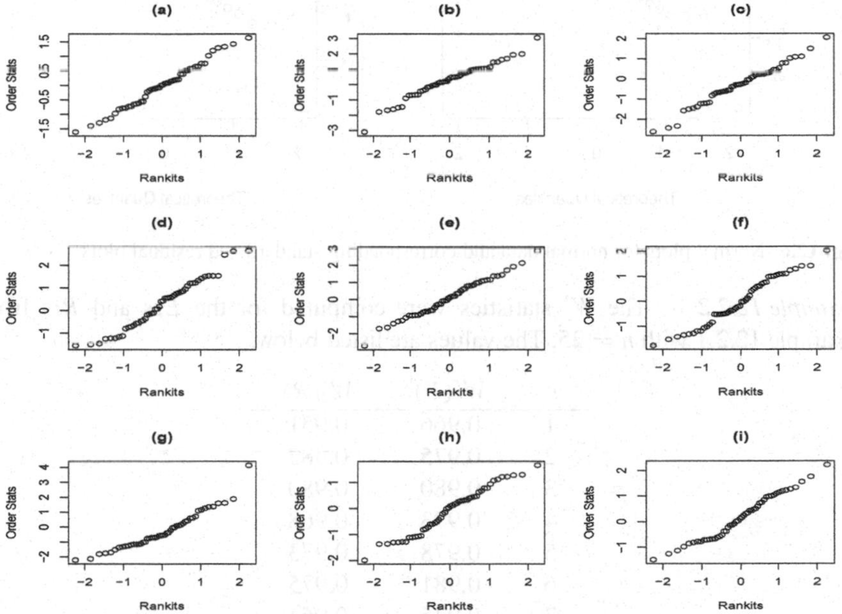

Fig. 12.5 Normal plots for normal data, $n = 50$

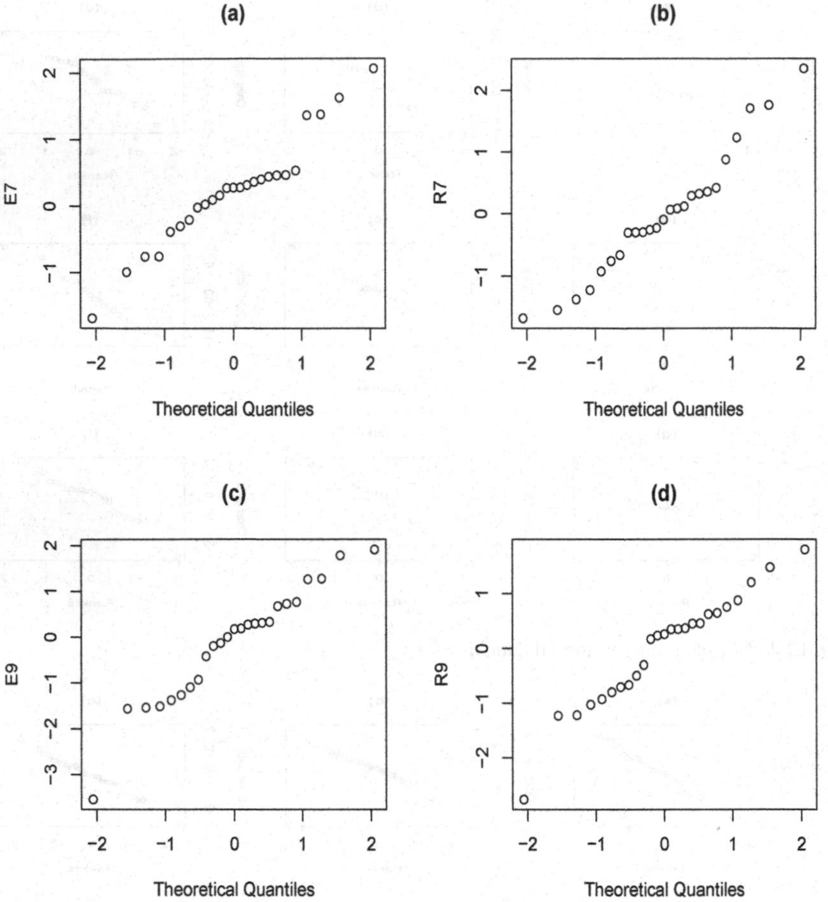

Fig. 12.6 Normal plots for normal data and corresponding standardized residual plots

Example 12.2.2 The W' statistics were computed for the Eis and Ris from Example 12.2.1 with $n = 25$. The values are listed below.

i	$W'(E)$	$W'(R)$
1	0.966	0.951
2	0.975	0.982
3	0.980	0.980
4	0.973	0.968
5	0.978	0.973
6	0.981	0.975
7	0.945	0.964
8	0.955	0.947
9	0.946	0.948

Note that, in this example (as in previous editions of the book), the W' values do not seem to be either systematically higher or lower when computed on the residuals rather than the true errors. Denoting the lower α percentile of W' from a sample of size n as $W'(\alpha, n)$, a simulation similar to that used in Christensen (2015) gives $W'(0.01, 25) = 0.874$ and $W'(0.05, 25) = 0.918$. None of the tests are rejected.

To give some indication of the power of the W' test, the example was repeated using ten samples of data generated using nonnormal errors. In one case, the errors were generated from a Cauchy distribution (a t distribution with 1 degree of freedom), and in a second case the errors were generated from a t distribution with three degrees of freedom. The results follow.

	Cauchy			$t(3)$	
i	$W'(E)$	$W'(R)$	i	$W'(E)$	$W'(R)$
1	0.491	0.553	1	0.861	0.871
2	0.539	0.561	2	0.878	0.966
3	0.903	0.909	3	0.891	0.856
4	0.822	0.783	4	0.654	0.637
5	0.575	0.644	5	0.953	0.951
6	0.354	0.442	6	0.912	0.905
7	0.502	0.748	7	0.978	0.979
8	0.753	0.792	8	0.958	0.959
9	0.921	0.952	9	0.896	0.881
10	0.276	0.293	10	0.972	0.967

With the Cauchy distribution, all but two of the tests are rejected at the 0.01 level, and one of those is rejected at the 0.05 level. With the $t(3)$ distribution, only two tests are rejected at the 0.01 level, with six of the tests based on E and five of the tests based on R rejected at 0.05.

The techniques for checking normality are applied directly to the standardized residuals. The theory assumed that the y_is were independent, but $\text{Cov}(\hat{e}) = \sigma^2(I - M)$, so both the residuals and the standardized residuals are correlated. One way to avoid this problem is to consider the $(n-p) \times 1$ vector $O'\hat{e}$, where the columns of O are an orthonormal basis for $C(I - M)$.

$$\text{Cov}(O'\hat{e}) = \sigma^2 O'(I - M)O = \sigma^2 O'O = \sigma^2 I,$$

so the procedures for checking normality can be validly applied to $O'\hat{e}$. The problem with this is that there are an infinite number of ways to pick O and the results depend on the choice of O. In fact, one can pick O so that $W' = 1$. Note that for any choice of O', $O_1 O'$ is another valid choice, where O_1 is any $n - p$ orthonormal matrix. Because O_1 is an arbitrary orthonormal matrix, $O_1 O'\hat{e}$ can be any rotation of $O'\hat{e}$. In particular, it can be one that is in exactly the same direction as the vector $a_{n-p} = (a_{n-p,1}, \ldots, a_{n-p,n-p})'$, where $a_{n,i} \equiv \text{E}[z_{(i)}]$ from a sample of size n. The sample correlation between these two vectors will be 1; thus $W' = 1$.

Exercise 12.3 Show that the sample correlation between two (mean adjusted) vectors in the same direction is 1.

Exercise 12.4 Using the model of Example 12.2.2, estimate the power of detecting a $t(3)$ with $\alpha = 0.05$ by simulation.

12.2.1 Other Applications for Normal Plots

We close this section with two variations on the use of normal rankit plots.

Example 12.2.3 Consider an ANOVA, say

$$y_{ijk} = \mu_{ij} + e_{ijk},$$

$i = 1, \ldots, a$, $j = 1, \ldots, b$, $k = 1, \ldots, N_{ij}$. Rather than using residuals to check whether the e_{ijk}s are i.i.d. $N(0, \sigma^2)$, for each pair i, j we can check whether $y_{ij1}, \ldots, y_{ijN_{ij}}$ are i.i.d. $N(\mu_{ij}, \sigma^2)$. The model assumes that for each group ij, the N_{ij} observations are a random sample from a normal population. Each group can be checked for normality individually. This leads to forming ab normal plots, one for each group. Of course, these plots will only work well if the N_{ij}s are reasonably large.

Example 12.2.4 We now present graphical methods for evaluating the interaction in a two-way ANOVA with only one observation in each cell. The model is

$$y_{ij} = \mu + \alpha_i + \eta_j + (\alpha\eta)_{ij} + e_{ij},$$

$i = 1, \ldots, a$, $j = 1, \ldots, b$. Here the e_{ij}s are assumed to be i.i.d. $N(0, \sigma^2)$. With one observation per cell, the $(\alpha\eta)_{ij}$s are confounded with the e_{ij}s; the effects of interaction cannot be separated from those of error.

As was mentioned in Section 7.2, the interactions in this model are often assumed to be nonexistent so that an analysis of the main effects can be performed. As an alternative to assuming no interaction, one can evaluate graphically an orthogonal set of $(a-1)(b-1)$ interaction contrasts, say $\lambda'_{rs}\beta$. If there are no interactions, the values $\lambda'_{rs}\hat{\beta}\Big/\sqrt{\lambda'_{rs}(X'X)^{-}\lambda_{rs}}$ are i.i.d. $N(0, \sigma^2)$. Recall that for an interaction contrast, $\lambda'\beta = \sum_{ij} q_{ij}(\alpha\eta)_{ij}$,

$$\lambda'\hat{\beta} = \sum_{ij} q_{ij}\, y_{ij}$$

and

$$\lambda'(X'X)^{-}\lambda = \sum_{ij} q_{ij}^2.$$

The graphical procedure is to order the $\lambda'_{rs}\hat{\beta}\Big/\sqrt{\lambda'_{rs}(X'X)^{-}\lambda_{rs}}$ values and form a normal rankit plot. If there are no interactions, the plot should be linear. Often there will be some estimated interactions that are near zero and some that are clearly nonzero. The near zero interactions should fall on a line, but clearly nonzero interactions will show up as deviations from the line. Contrasts that do not fit the line are identified as nonzero interaction contrasts (without having executed a formal test).

The interactions that fit on a line are used to estimate σ^2. This can be done in either of two ways. First, an estimate of the slope of the linear part of the graph can be used as an estimate of the standard deviation σ. Second, sums of squares for the contrasts that fit the line can be averaged to obtain a mean squared error.

Both methods of estimating σ^2 are open to criticism. Consider the slope estimate of σ and, in particular, assume that $(a-1)(b-1)=12$ and that there are three nonzero contrasts all yielding large positive values of $\lambda'_{rs}\hat{\beta}\Big/\sqrt{\lambda'_{rs}(X'X)^{-}\lambda_{rs}}$. In this case, the ninth largest value is plotted against the ninth largest rankit. Unfortunately, we do not know that the ninth largest value is the ninth largest observation in a random sample of size 12. If we could correct for the nonzero means of the three largest contrasts, what we observed as the ninth largest value could become anything from the ninth to the twelfth largest value. To estimate σ, we need to plot the mean adjusted statistics $\left(\lambda'_{rs}\hat{\beta}-\lambda'_{rs}\beta\right)\Big/\sqrt{\lambda'_{rs}(X'X)^{-}\lambda_{rs}}$. We know that 9 of the 12 values $\lambda'_{rs}\beta$ are zero. The ninth largest value of $\lambda'_{rs}\hat{\beta}\Big/\sqrt{\lambda'_{rs}(X'X)^{-}\lambda_{rs}}$ can be any of the order statistics of the mean adjusted values $\left(\lambda'_{rs}\hat{\beta}-\lambda'_{rs}\beta\right)\Big/\sqrt{\lambda'_{rs}(X'X)^{-}\lambda_{rs}}$ between 9 and 12. The graphical method assumes that extreme values of $\lambda'_{rs}\hat{\beta}\Big/\sqrt{\lambda'_{rs}(X'X)^{-}\lambda_{rs}}$ are also extreme values of the mean adjusted statistics. There is no justification for this assumption. If the ninth largest value were really the largest of the 12 mean adjusted statistics, then plotting the ninth largest value rather than the ninth largest mean adjusted value against the ninth largest rankit typically indicates a slope that is larger than σ. Thus the graphical procedure tends to overestimate the variance. Alternatively, the ninth largest value may not seem to fit the line and so, inappropriately, be declared nonzero. These problems should be ameliorated by dropping the three clearly nonzero contrasts and replotting the remaining contrasts as if they were a sample of size 9. In fact, the replotting method will tend to have a downward bias, as discussed in the next paragraph.

The criticism of the graphical procedure was based on what happens when there are nonzero interaction contrasts. The criticism of the mean squared error procedure is based on what happens when there are no nonzero interaction contrasts. In this case, if one erroneously identifies contrasts as being nonzero, the remaining contrasts have been selected for having small absolute values of $\lambda'_{rs}\hat{\beta}\Big/\sqrt{\lambda'_{rs}(X'X)^{-}\lambda_{rs}}$ or, equivalently, for having small sums of squares. Averaging a group of sums of squares that were chosen to be small clearly underestimates σ^2. The author's inclination is to use the mean squared error criterion and try very hard to avoid erroneously identifying zero interactions as nonzero. This avoids the problem of estimating the slope of the normal plot. Simulation envelopes such as those discussed by Atkinson (1985, Chapter 4) can be very helpful in deciding which contrasts to identify as nonzero.

Some authors contend that, for visual reasons, normal plots should be replaced with plots that do not involve the sign of the contrasts, see Atkinson (1981, 1982). Rather than having a graphical procedure based on the values $\lambda'_{rs}\hat{\beta}/\sqrt{\lambda'_{rs}(X'X)^{-}\lambda_{rs}}$, the squared values, i.e., the sums of squares for the $\lambda'_{rs}\beta$ contrasts, can be used. When there are no interactions,

$$\frac{SS(\lambda'_{rs}\beta)}{\sigma^2} \sim \chi^2(1).$$

The contrasts are orthogonal so, with no interactions, the values $SS(\lambda'_{rs}\beta)$ form a random sample from a $\sigma^2\chi^2(1)$ distribution. Let $w_1, \ldots, w_r$ be i.i.d $\chi^2(1)$, where $r = (a-1)(b-1)$. Compute the expected order statistics $\mathrm{E}[w_{(i)}]$ and plot the pairs $(\mathrm{E}[w_{(i)}], SS(\lambda'_{(i)}\beta))$, where $SS(\lambda'_{(i)}\beta)$ is the ith smallest of the sums of squares. With no interactions, this should be an approximate straight line through zero with slope σ^2. For nonzero contrasts, $SS(\lambda'_{rs}\beta)$ has a distribution that is σ^2 times a *noncentral* $\chi^2(1)$. Values of $SS(\lambda'_{rs}\beta)$ that are substantially above the linear portion of the graph indicate nonzero contrasts. A graphical estimate of σ^2 is available from the sums of squares that fit on a line; this has bias problems similar to that of a normal plot. The theoretical quantiles $\mathrm{E}[w_{(i)}]$ can be approximated by evaluating the inverse of the $\chi^2(1)$ cdf at $i/(n+1)$.

A corresponding method for estimating σ, based on the square roots of the sums of squares, is called a *half-normal plot*. The expected order statistics are often approximated as $\Phi^{-1}((n+i)/(2n+1))$.

These methods can be easily extended to handle other situations in which there is no estimate of error available. In fact, this graphical method was first proposed by Daniel (1959) for analyzing 2^n factorial designs. Daniel (1976) also contains a useful discussion. Christensen (1996) and http://www.stat.unm.edu/~fletcher/TopicsInDesign illustrate some of these ideas. Lenth (2015) argues that methods exist superior to these.

12.3 Checking Independence

Lack of independence occurs when $\mathrm{Cov}(e)$ is not diagonal. One reason that good methods for evaluating independence are difficult to develop is that, unlike the other assumptions involved in $e \sim N(0, \sigma^2 I)$, independence is not a property of the population in question. Independence is a property of the way that the population is sampled. As a result, there is no way to check independence without thinking hard about the method of sampling. Identifying lack of independence is closely related to identifying lack of fit. For example, consider data from a randomized complete block (RCB) experiment being analyzed with a one-way analysis of variance model that ignores blocks. If the blocks have fixed effects, the one-way model suffers from lack of fit. If the blocks effects are random with a common mean, the one-way model suffers from lack of independence. We begin with a general discussion of ideas for testing the independence assumption based upon Christensen and Bedrick (1997). This is followed by a subsection on detecting serial correlation.

A key idea in checking independence is the formation of rational subgroups. To evaluate whether a group of numbers form a random sample from a population, Shewhart (1931) proposed using control charts for means. The means being charted were to be formed from rational subgroups of the observations that were obtained under essentially identical conditions. Shewhart (1939, p. 42) suggests that a control chart is less a test for whether data form a random sample and more an operational definition of what it means to have a random sample. It is easily seen that a means chart based on rational subgroups is sensitive to lack of independence, lack of fit (nonconstant mean), inequality of variances, and nonnormality. In analyzing linear models, statisticians seek assurance that any lack of independence or other violations of the assumptions are not so bad as to invalidate their conclusions. Essentially, statisticians need an operational definition of when traditional linear model theory can be applied.

As used with linear models, rational subgroups are simply clusters of observations. They can be clustered in time, or in space, by having similar predictor variables, by being responses on the same individual, or by almost anything that the sampling scheme suggests could make observations within a cluster more alike than observations outside a cluster. To test for lack of independence, the near replicate lack-or-fit tests presented in Subsection 6.7.2 can be used. Simply replace the clusters of near replicates with clusters of rational subgroups determined by the sampling scheme. Christensen and Bedrick (1997) found that the analysis of covariance test, i.e., the Christensen (1989) test, worked well in a wide variety of situations, though the Shillington test often worked better when the clusters were very small. Of course, specialized tests for specific patterns of nonindependence can work much better than these general tests *when the specific patterns are appropriate.*

12.3.1 Serial Correlation

An interesting case of nonindependence is serial correlation. This occurs frequently when observations $y_1, y_2, \ldots, y_n$ are taken serially at equally spaced time periods. A model often used when the observations form such a time series is a symmetric *Toeplitz matrix*

$$\mathrm{Cov}(e) = \sigma^2 \begin{bmatrix} 1 & \rho_1 & \rho_2 & \cdots & \rho_{n-3} & \rho_{n-2} & \rho_{n-1} \\ \rho_1 & 1 & \rho_1 & \ddots & \ddots & \rho_{n-3} & \rho_{n-2} \\ \rho_2 & \rho_1 & 1 & \ddots & \ddots & \ddots & \rho_{n-3} \\ \vdots & \ddots & \ddots & \ddots & \ddots & \ddots & \vdots \\ \rho_{n-3} & \ddots & \ddots & \ddots & 1 & \rho_1 & \rho_2 \\ \rho_{n-2} & \rho_{n-3} & \ddots & \ddots & \rho_1 & 1 & \rho_1 \\ \rho_{n-1} & \rho_{n-2} & \rho_{n-3} & \cdots & \rho_2 & \rho_1 & 1 \end{bmatrix},$$

where $\rho_1, \rho_2, \ldots, \rho_{n-1}$ are such that $\mathrm{Cov}(e)$ is positive definite. Typically, only the first few of $\rho_1, \rho_2, \ldots, \rho_{n-1}$ will be substantially different from zero. One way of detecting serial correlation is to plot r_i versus i. If, say, ρ_1 and ρ_2 are positive and $\rho_3, \ldots, \rho_n$ are near zero, a plot of r_i versus i may have oscillations, but residuals that are adjacent should be close. If ρ_1 is negative, then ρ_2 must be positive. The plot of r_i versus i in this case may or may not show overall oscillations, but adjacent residuals should be oscillating rapidly. An effective way to detect a nonzero ρ_s is to plot (r_i, r_{i+s}) for $i = 1, \ldots, n-s$ or to compute the corresponding sample correlation coefficient from these pairs.

A special case of serial correlation is $\rho_s = \rho^s$ for some parameter ρ between -1 and 1. This AR(1), i.e., autoregressive order 1, covariance structure can be obtained by assuming (1) $e_1 \sim N(0, \sigma^2)$, and (2) for $i > 1$, $e_{i+1} = \rho e_i + v_{i+1}$, where $v_2, \ldots, v_n$ are i.i.d. $N(0, (1-\rho^2)\sigma^2)$ and v_{i+1} is independent of e_i for all i. Other models for serial correlation are discussed in Christensen et al. (2010, Section 10.3). Most are based on ARMA time series models, cf. *ALM-III*, Chapter 7.

Example 12.3.1 For ρ equal to $-0.9, -0.5, 0.5$, and 0.9, serially correlated error vectors were generated as just described. Dependent variable values y were obtained using (12.0.1) and the model (12.2.2) was fitted, giving a standardized residual vector r. For all values of ρ, the standardized residuals are plotted against their observation numbers. Within each figure, $z_1, z_2, \ldots, z_n$ are i.i.d. $N(0, 1)$ with $e_1 = z_1$ and $v_i = \sqrt{1-\rho^2} z_i$, so only ρ changes. Figures 12.7, 12.8 and 12.9 give three independent sets of plots. Note that when ρ is positive, adjacent observations remain near one another. The overall pattern tends to oscillate slowly. When ρ is negative, the observations oscillate very rapidly; adjacent observations tend to be far apart, but observations that are one apart (e.g., e_i and e_{i+2}) are fairly close.

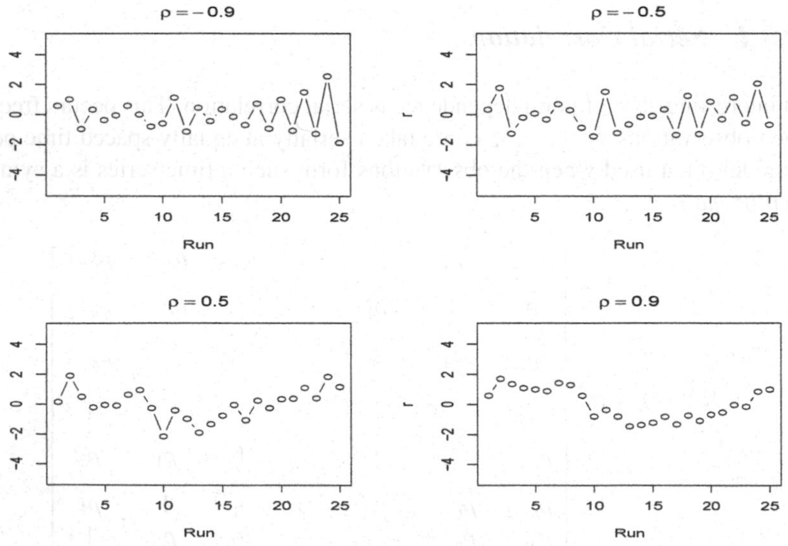

Fig. 12.7 Serial correlation standardized residual plots

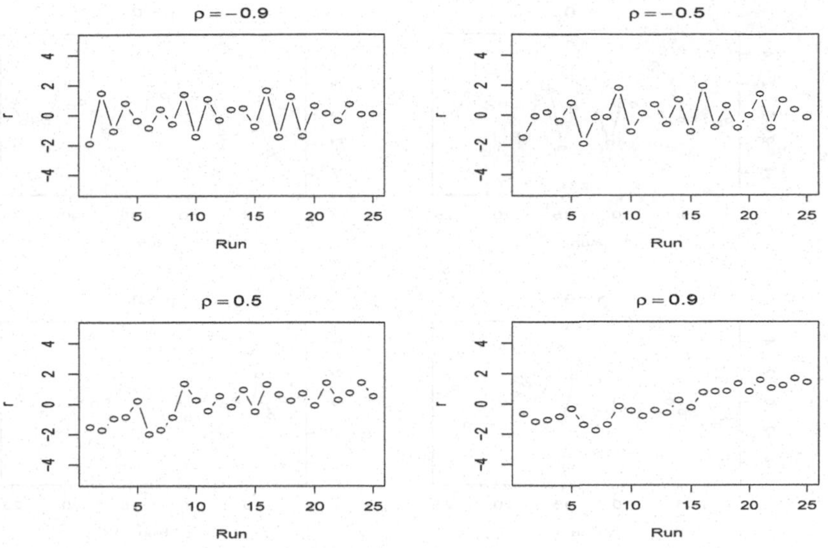

Fig. 12.8 Serial correlation standardized residual plots

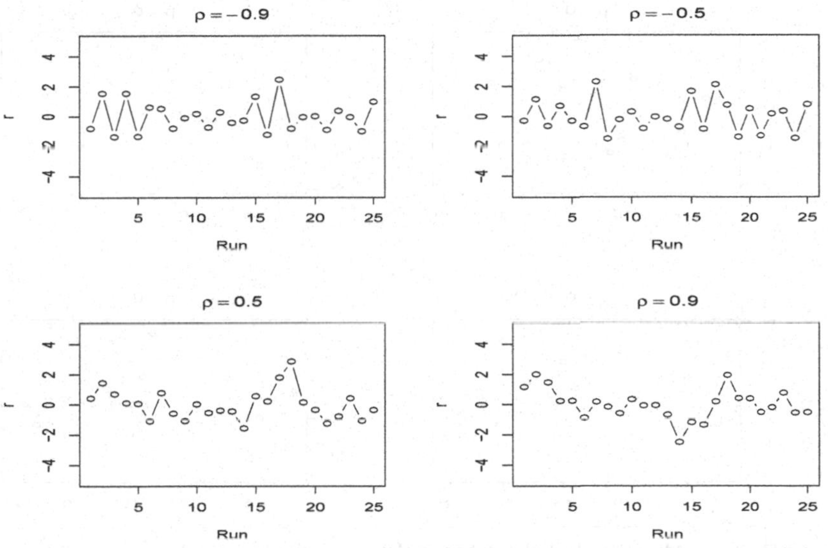

Fig. 12.9 Serial correlation standardized residual plots

Figures 12.10, 12.11, 12.12, 12.13 and 12.14 contain plots with $\rho = 0$. Figure 12.10 is in the same form as Figures 12.7, 12.8 and 12.9. The other figures use a different style. Comparing Figure 12.10 with Figures 12.7, 12.8 and 12.9, it does not seem easy to distinguish between $\rho = 0$ and moderate correlations like $\rho = \pm 0.5$.

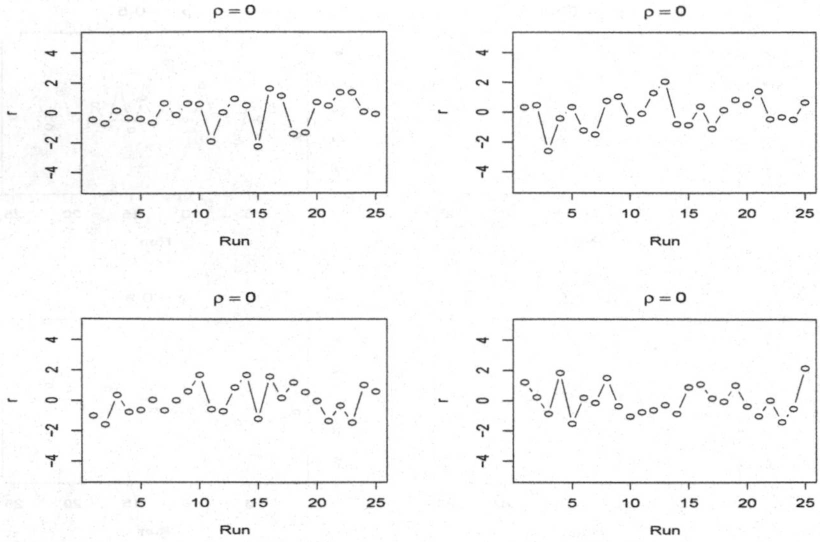

Fig. 12.10 Serial correlation standardized residual plots with uncorrelated data

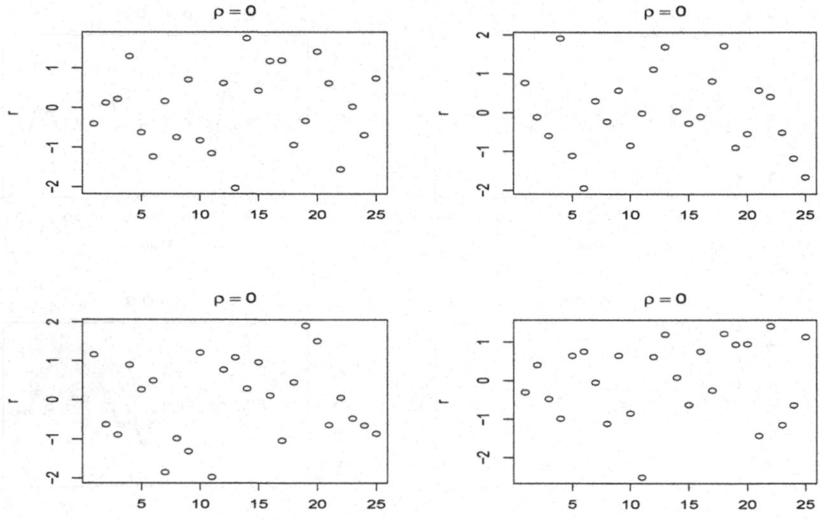

Fig. 12.11 Standardized residual plots when model assumptions are valid

Figures 12.10, 12.11, 12.12, 12.13 and 12.14 are of interest not only for illustrating a lack of serial correlation, but also as examples of what the plots in Section 4 should look like, i.e., these are standardized residual plots when all the model assumptions are valid. Note that the horizontal axis is not specified, because the residual plots should show no correlation, regardless of what they are plotted against. It is interesting to try to detect patterns in Figures 12.10, 12.11, 12.12, 12.13 and 12.14 because the

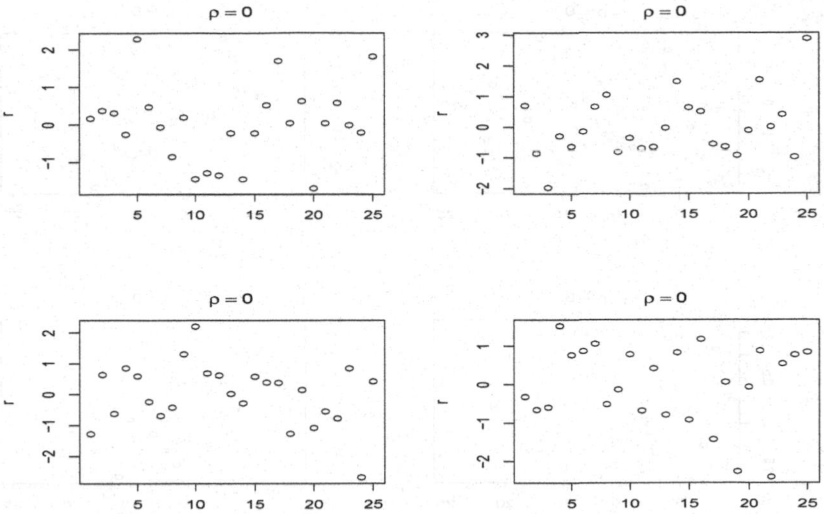

Fig. 12.12 Standardized residual plots when model assumptions are valid

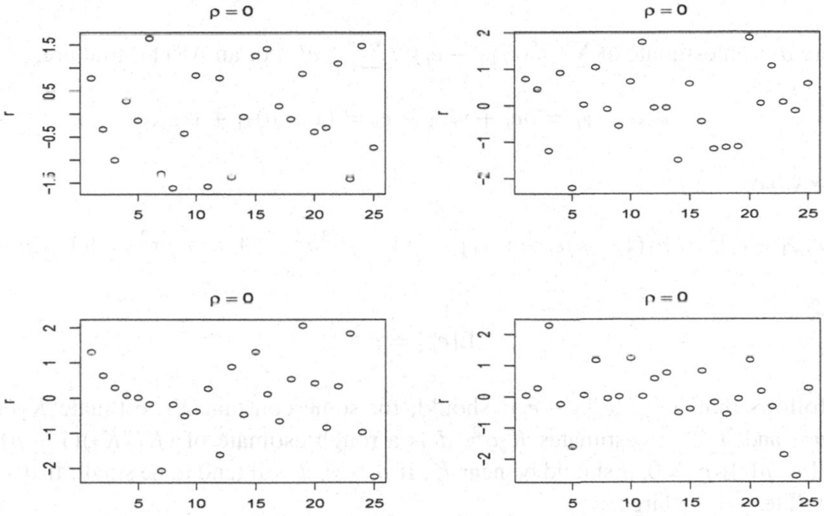

Fig. 12.13 Standardized residual plots when model assumptions are valid

human eye is good at detecting/creating patterns, even though none exist in these plots.

Durbin and Watson (1951) provided an approximate test for the hypothesis $\rho = 0$. The Durbin–Watson test statistic is

$$d = \sum_{i=1}^{n-1} (\hat{e}_{i+1} - \hat{e}_i)^2 \Big/ \sum_{i=1}^{n} \hat{e}_i^2 .$$

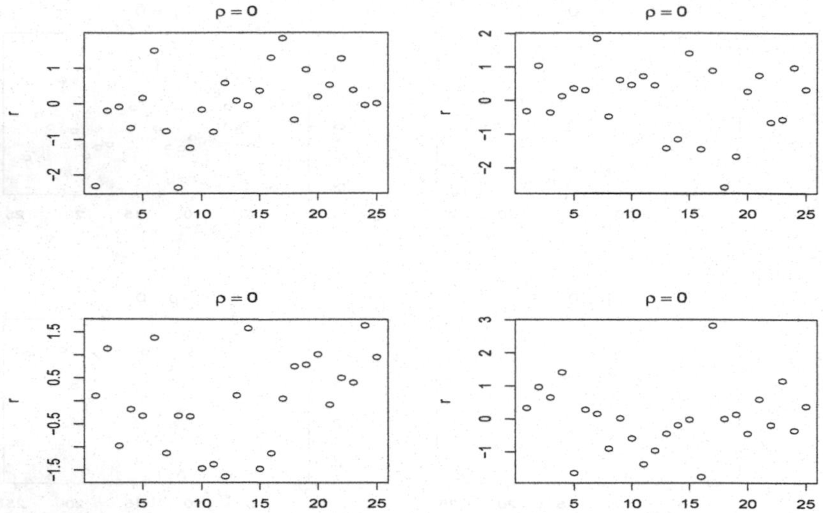

Fig. 12.14 Standardized residual plots when model assumptions are valid

Here d is an estimate of $\sum_{i=1}^{n-1}(e_{i+1} - e_i)^2 / \sum_{i=1}^{n} e_i^2$. For an AR(1) structure,

$$e_{i+1} - e_i = \rho e_i + v_{i+1} - e_i = (1 - \rho)e_i + v_{i+1},$$

so we have

$$\mathrm{E}[e_{i+1} - e_i]^2 = \mathrm{E}[(1 - \rho)e_i + v_{i+1}]^2 = (1 - \rho)^2\sigma^2 + (1 - \rho^2)\sigma^2 = 2(1 - \rho)\sigma^2,$$

and

$$\mathrm{E}[e_i^2] = \sigma^2.$$

It follows that $\sum_{i=1}^{n-1}(\hat{e}_{i+1} - \hat{e}_i)^2$ should, for some constant K_1, estimate $K_1(1 - \rho)\sigma^2$, and $\sum_{i=1}^{n} \hat{e}_i^2$ estimates $K_2\sigma^2$. d is a rough estimate of $(K_1/K_2)(1 - \rho)$ or $K[1 - \rho]$. If $\rho = 0$, d should be near K. If $\rho > 0$, d will tend to be small. If $\rho < 0$, d will tend to be large.

The exact distribution of d varies with X. For example, $\sum_{i=1}^{n-1}(\hat{e}_{i+1} - \hat{e}_i)^2$ is just a quadratic form in $\hat{e}$, say $\hat{e}A\hat{e} = Y'(I - M)A(I - M)Y$. It takes little effort to see that A is a very simple matrix. By Theorem 1.3.2,

$$\mathrm{E}(\hat{e}A\hat{e}) = \mathrm{tr}[(I - M)A(I - M)\mathrm{Cov}(Y)].$$

Thus, even the expected value of the numerator of d depends on the model matrix. Since the distribution depends on X, it is not surprising that the exact distribution of d varies with X.

Exercise 12.5 Show that d is approximately equal to $2(1 - r_a)$, where r_a is the sample (auto)correlation between the pairs $(\hat{e}_{i+1}, \hat{e}_i)$ $i = 1, \ldots, n-1$.

12.4 Heteroscedasticity and Lack of Fit

Heteroscedasticity refers to having unequal variances. In particular, an independent heteroscedastic model has

$$\text{Cov}(e) = \text{Diag}(\sigma_i^2).$$

Lack of fit refers to having an incorrect model for $\text{E}(Y)$. In Section 6.7 on testing lack of fit, we viewed this as having an insufficiently general model matrix. When lack of fit occurs, $\text{E}(e) \equiv \text{E}(Y - X\beta) \neq 0$. Both heteroscedasticity and lack of fit are diagnosed by plotting the standardized residuals against any variable of choice. The chosen variable may be case numbers, time sequence, any predictor variable included in the model, any predictor variable not included in the model, or the predicted values $\hat{Y} = MY$. If there is no lack of fit or heteroscedasticity, the residual plots should form a horizontal band. The plots in Section 3 with $\rho = 0.0$ are examples of such plots when the horizontal axis has equally spaced entries.

12.4.1 *Heteroscedasticity*

A horn-shaped pattern in a residual plot indicates that the variance of the observations is increasing or decreasing with the other variable.

Example 12.4.1 Twenty-five i.i.d. $N(0, 1)$ random variates, $z_1, \ldots, z_{25}$ were generated and y values were computed using (12.0.1) with $e_i = x_{i1}z_i/60$. The variance of the e_is increase as the x_{i1}s increase.

Figure 12.15 plots the standardized residuals $R1$ against $\hat{Y}$ and X_2. The plot of $R1$ versus $\hat{Y}$ shows something of a horn shape, but it opens to the left and is largely dependent on one large residual with a $\hat{y}$ of about 8.3. The plot against X_2 shows very little. It is difficult to detect any pattern in either plot. In (b) the two relatively small values of x_2 don't help. The top left component of Figure 12.16 plots R1 against X_1 where you can detect a pattern but, again, by no means an obvious one. The impression of a horn shape opening to the right is due almost entirely to one large residual near $x_1 = 70$. The remaining plots in Figure 12.16 as well as the plots in Figures 12.17 and 12.18 are independent replications. Often you can detect the horn shape but sometimes you cannot.

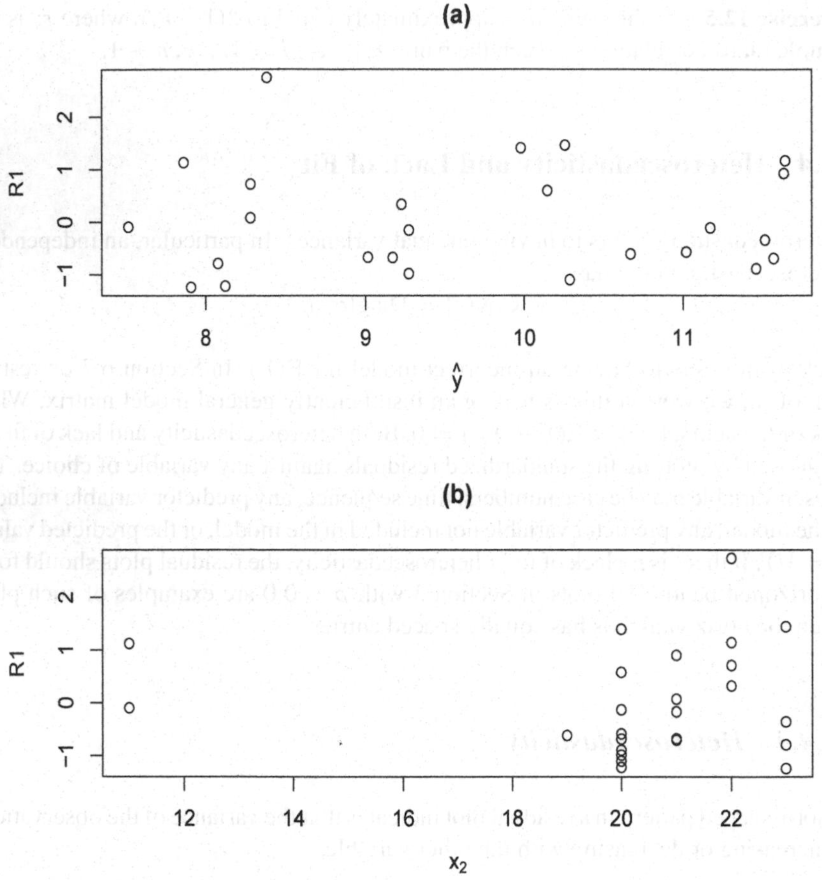

Fig. 12.15 Variance increasing with x_1

Example 12.4.2 To illustrate horn shapes that open to the left, Example 12.4.1 was repeated using $e_i = 60z_i/x_{i1}$. With these e_is, the variance decreases as x_1 increases. The plots are contained in Figures 12.19, 12.20 and 12.21. In Figure 12.19a of $R1$ versus $\hat{Y}$, we see a horn opening to the right. Note that from (12.0.1), if x_2 is held constant, y increases as x_1 decreases. In the plot, x_2 is not being held constant, but the relationship still appears. There is little to see in Figure 12.19b $R1$ versus X_2. The plot of $R1$ versus X_1 in the top left of Figure 12.20 shows a horn opening to the left. The remaining plots in Figure 12.20 as well as the plots in Figure 12.21 are independent replications.

Although plotting the residuals seems to be the standard method for examining heteroscedasticity of variances, Examples 12.4.1 and 12.4.2 indicate that residual plots are far from foolproof.

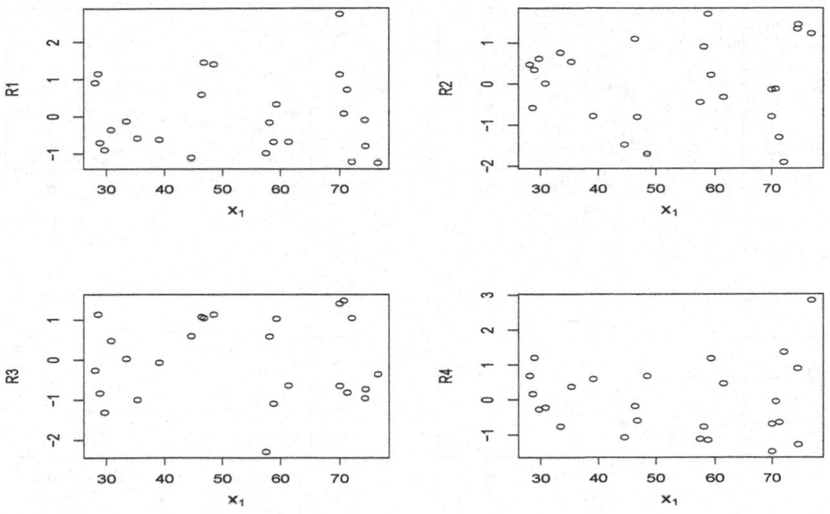

Fig. 12.16 Variance increasing with x_1

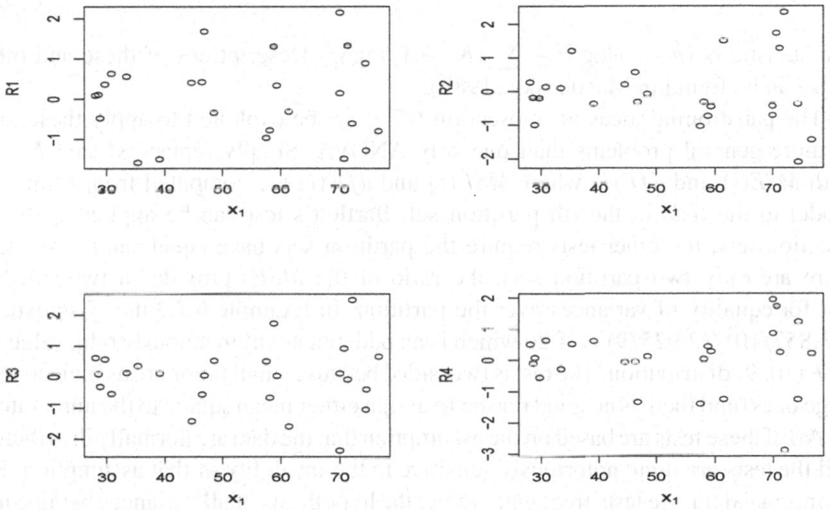

Fig. 12.17 Variance increasing with x_1

For one-way ANOVA models (and equivalent models such as two-way ANOVA with interaction), there are formal tests of heteroscedasticity available. The best known of these are Hartley's, Cochran's, and Bartlett's tests. The tests are based on the sample variances for the individual groups, say $s_1^2, \ldots, s_t^2$. Hartley's and Cochran's tests require equal sample sizes for each group; Bartlett's test does not. Hartley's test statistic is $\max_i s_i^2 / \min_i s_i^2$. Cochran's test statistic is $\max_i s_i^2 / \sum_{i=1}^t s_i^2$. Bartlett's

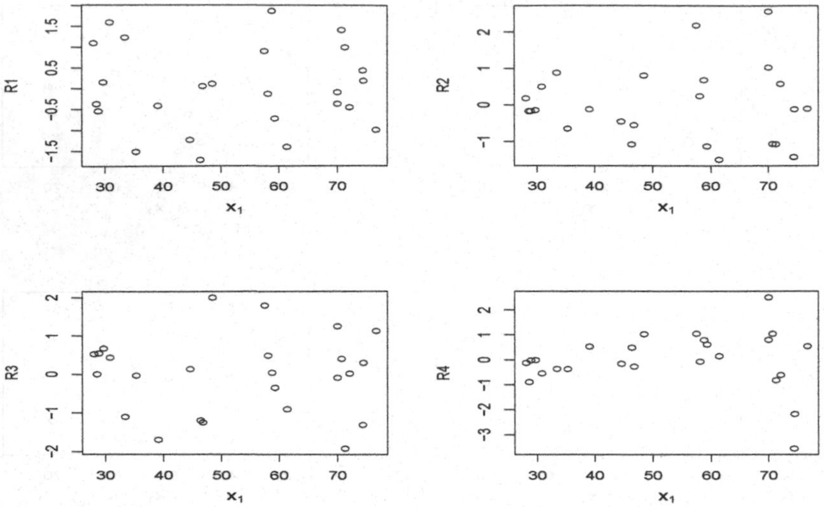

Fig. 12.18 Variance increasing with x_1

test statistic is $(n-t)\log \bar{s}^2_{\cdot} - \sum_i (N_i - 1)\log s_i^2$. Descriptions of these and other tests can be found in Mandansky (1988).

The partitioning ideas of Subsection 6.7.3 can be exploited to apply these tests to more general problems than one-way ANOVA. Simply replace s_i^2 and $N_i - 1$ with $MSE(i)$ and $dfE(i)$ where $MSE(i)$ and $dfE(i)$ are computed from fitting the model to the data in the ith partition set. Bartlett's test can be applied with any partition sets; the other tests require the partition sets have equal sample sizes. If there are only two partition sets, the ratio of the *MSE*s provides a two-sided F test for equality of variances over the partition. In Example 6.7.3 the F statistic is $(13.857/10)/(2.925/9) = 4.26$ which is an odd, but not tremendously odd, value for an $F(10, 9)$ distribution. The test is two-sided because small ratios are as indicative as large ones (and there is no good reason to assign either mean square to the numerator).

All of these tests are based on the assumption that the data are normally distributed, and the tests are quite notoriously sensitive to the invalidity of that assumption. For nonnormal data, the tests frequently reject the hypothesis of all variances being equal even when all variances are, in fact, equal. This is important because t and F tests tend not to be horribly sensitive to nonnormality. In other words, if the data are not normally distributed (and they never are), the data may be close enough to being normally distributed so that the t and F tests are approximately correct. However, the nonnormality may be enough to make Hartley's, Cochran's, and Bartlett's tests reject, so that the data analyst worries about a nonexistent problem of heteroscedasticity, cf. Box (1953).

More recently tests have been proposed that involve fitting models to functions of the residuals. The Breusch–Pagan/Cook–Weisberg test involves fitting a regression model to the squared residuals from the original fitted model. No regression

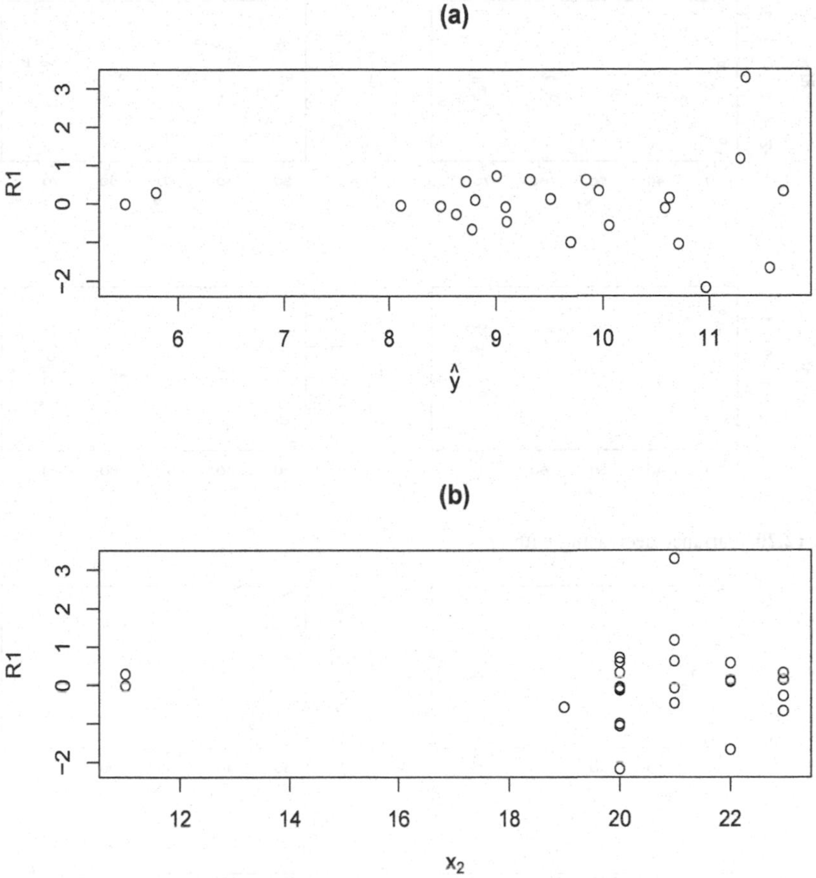

Fig. 12.19 Variance decreasing with x_1

effects suggest homoscedasticity. Significant predictors in the model for the squared residuals suggest that those predictors are related to heteroscedasticity. Rather than the squared residuals, models can also be fitted to the absolute values of the original residuals. In particular, when the original model is a one-way ANOVA, the popular Levene test is the F test for group differences in fitting an auxiliary one-way ANOVA to the absolute values of the original residuals. Again, this is generalizable to fitting arbitrary models on partition sets. It has been suggested that Levene's test is less sensitive to nonnormality than the comparable Cochran, Hartley, and Bartlett tests, cf. https://www.itl.nist.gov/div898/handbook/eda/section3/eda35a.htm.

Exercise 12.6a Show that $\mathrm{E}[\hat{e}_i^2/(1 - m_{ii})] = \sigma^2$ under a standard linear model. Justify the Breusch–Pagan/Cook–Weisberg approach when applied to the squares of the standardized residuals.

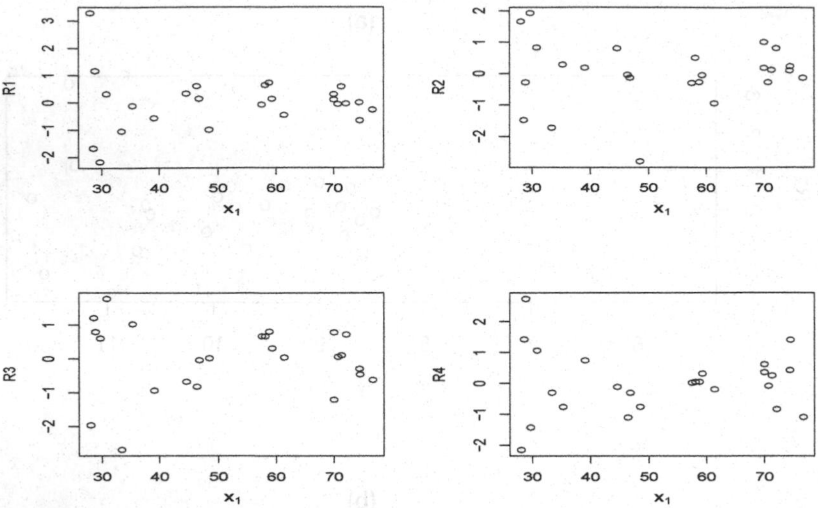

Fig. 12.20 Variance decreasing with x_1

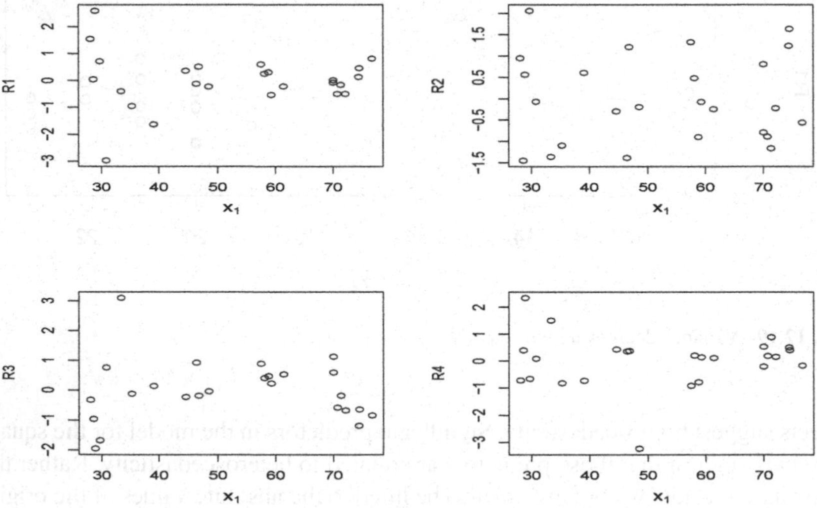

Fig. 12.21 Variance decreasing with x_1

12.4.2 Huber–White (Robust) Sandwich Estimator

Sandwich estimators get used in complicated situations to estimate the variability of point estimates. The idea is that useful point estimates are often obtained from an incorrect model but, if we suspect the model is incorrect, we still want valid estimates of their variability. It is of interest to see how sandwich estimates work for linear models.

It is often convenient to estimate β using least squares. For a standard linear model, least squares is optimal. But we want to estimate the variability of the estimates under a true heteroscedastic model. In other words, consider the problem of estimating the covariance matrix of least squares estimates when $\mathrm{Cov}(Y) \neq \sigma^2 I$ and, more specifically, under an heteroscedastic model $\mathrm{Cov}(Y) = D(\sigma_i^2)$. Our discussion follows closely that of Freedman (2006).

Assume the model

$$Y = X\beta + e, \qquad \mathrm{E}(e) = 0, \qquad \mathrm{Cov}(e) = V.$$

The generalized least squares estimates, the BLUEs and, under normality, the MLEs and the minimum variance unbiased estimates all satisfy

$$X\hat{\beta}_G = X\left[X'V^{-1}X\right]^{-1} X'V^{-1}Y$$

But frequently we do not know V.

One option, which is attractive when n is large relative to $r(X)$, is to simply continue using least squares. The least squares estimate $\hat{\beta}$ does not require knowledge of V and is still unbiased but it is inefficient in the sense that it has larger than necessary variability. However, when n is large relative to $r(X)$, even inefficient estimates can be good estimates. If we use least squares, we still need to estimate the variability of our estimates and in particular,

$$\mathrm{Cov}(\hat{\beta}) = \left[X'X\right]^{-1} X'VX\left[X'X\right]^{-1}.$$

But to estimate this covariance matrix, we still need an estimate of V.

As discussed in *ALM-III* Chapter 4, we can create a parametric model for V, say $V(\theta)$, for some s vector of parameters θ. This allows us to estimate the parameters with $\hat{\theta}$ and the covariance matrix with $\hat{V} \equiv V(\hat{\theta})$, which immediately gives an estimate for the covariance matrix for $\hat{\beta}$ of

$$\widehat{\mathrm{Cov}}(\hat{\beta}) = \left[X'X\right]^{-1} X'\hat{V}X\left[X'X\right]^{-1}.$$

But if we go to all this trouble, rather than using least squares, we might as well use the *empirical estimate* of β defined by,

$$X\tilde{\beta} = X\left[X'\hat{V}^{-1}X\right]^{-1} X'\hat{V}^{-1}Y.$$

Unfortunately, as discussed in *ALM-III* Section 4.7, the obvious way of estimating the covariance matrix of $\tilde{\beta}$ leads to underestimating the variability.

Now consider a different approach, one that is specifically *excluded* in *ALM-III*. Consider a model for $\mathrm{Cov}(Y)$ that depends on $\mathrm{E}(Y)$, say $V(X\beta)$. Having the covariance matrix depend on β (as opposed to a distinct vector θ) complicates the estimation of $X\beta$. The standard estimation methods do not apply except maximum likelihood, and the nature of the likelihood equations changes dramatically. Nonetheless, we can easily estimate the covariance matrix of least squares estimates by looking at

$$\widehat{\text{Cov}}(\hat{\beta}) = [X'X]^{-1} X'V(X\hat{\beta})X[X'X]^{-1}.$$

The sandwich estimator is something of a compromise between these two approaches. Assume a heteroscedastic, uncorrelated model $V = D(\sigma_i^2)$. Note that, if we knew β, we could get an unbiased estimate of V by simply observing that $\text{E}(y_i - x_i'\beta)^2 = \sigma_i^2$, or

$$\text{E}\left[D^2(Y - X\beta)\right] = D(\sigma_i^2).$$

This suggests using

$$\hat{V} = D^2(Y - X\hat{\beta})$$

to estimate V and gives the *Huber–White robust sandwich estimator* of the covariance of $\hat{\beta}$,

$$\widehat{\text{Cov}}(\hat{\beta}) = [X'X]^{-1} X'D^2(Y - X\hat{\beta})X[X'X]^{-1}.$$

Unfortunately, $D^2(Y - X\hat{\beta})$ is a *horrible* estimate of $D(\sigma_i^2)$. Each σ_i^2 is being estimated by $(y_i - x_i'\hat{\beta})^2$ which is a worse estimate than $(y_i - x_i'\beta)^2$ which is based on a single observation. Moreover, the model

$$Y = X\beta + e, \qquad \text{E}(e) = 0, \qquad \text{Cov}(e) = D(\sigma_i^2).$$

has $n + r(X)$ parameters. Why would anyone think that you could do a good job estimating that many parameters from n observations? Freedman (2006) repeated emphasized that Huber is not to blame for the misuse of his ideas.

The following exercise involves a situation where the sandwich estimator can be made to work. The model is a special case of the longitudinal models discussed in *ALM-III* Section 5.6.

Exercise 12.6b Consider a collection of r uncorrelated regression models

$$Y_k = X_k\beta + e_k, \qquad \text{E}(e_k) = 0, \qquad \text{Cov}(e_k) = V,$$

each with N observations. Find the least squares estimate $\hat{\beta}$ from all the data. Find an estimate of V. Find the sandwich estimator for $\text{Cov}(\hat{\beta})$. Why is this a reasonable estimate? Hint: $V = \text{E}[(Y_k - X_k\beta)(Y_k - X_k\beta)']$.

12.4.3 Lack of Fit

An additional use of residual plots is to identify lack of fit. The assumption is that $\mathrm{E}(e) = 0$, so any *systematic* pattern in the residuals (including a horn shape) can indicate lack of fit. Most commonly, one looks for a linear or quadratic trend in the residuals. Such trends indicate the existence of effects that have not been removed from the residuals, i.e., effects that have not been accounted for in the model.

Theorems 6.3.3 and 6.3.6 indicate that the residuals should be uncorrelated with any function of the predictor variables when we have the best possible model. So a nonzero correlation between the residuals and any other variable, say z, indicates that something is wrong with the model. If z were not originally in the model, then it needs to be. A quadratic relationship between the residuals and z is indicative of a nonzero correlation between the residuals and z^2, although the linear relationship might be much clearer when plotting the residuals against a standardized version of z, say $z - \bar{z}$.

For examining heteroscedasticity, the standardized residuals need to be used because the ordinary residuals are themselves heteroscedastic. For examining lack of fit, the ordinary residuals are preferred but we often use the standardized residuals for convenience.

Example 12.4.3 Data were generated using (12.0.1) and the incorrect model

$$y_i = \beta_0 + \beta_2 x_{i2} + e_i$$

was fitted. In the independently generated Figures 12.22 and 12.23, the ordinary residuals $\hat{e}$ and the standardized residuals R are plotted against x_1 in (a) and (b), respectively. The decreasing trends in the residual plots indicate that x_1 may be worth adding to the model.

Part (c) of the figures contains an *added variable plot*. To obtain it, find the ordinary residuals, say $\hat{e}(x_1)$, from fitting

$$x_{i1} = \gamma_0 + \gamma_2 x_{i2} + e_i.$$

By Exercise 9.2, a plot of $\hat{e}$ versus $\hat{e}(x_1)$ gives an exact graphical display of the effect of adding x_1 to the model.

The disadvantage of added variable plots is that it is time consuming to adjust the predictor variables under consideration for the variables already in the model. It is more convenient to plot residuals against predictor variables that have not been adjusted. As in Example 12.4.3, such plots are often informative but could, potentially, be misleading.

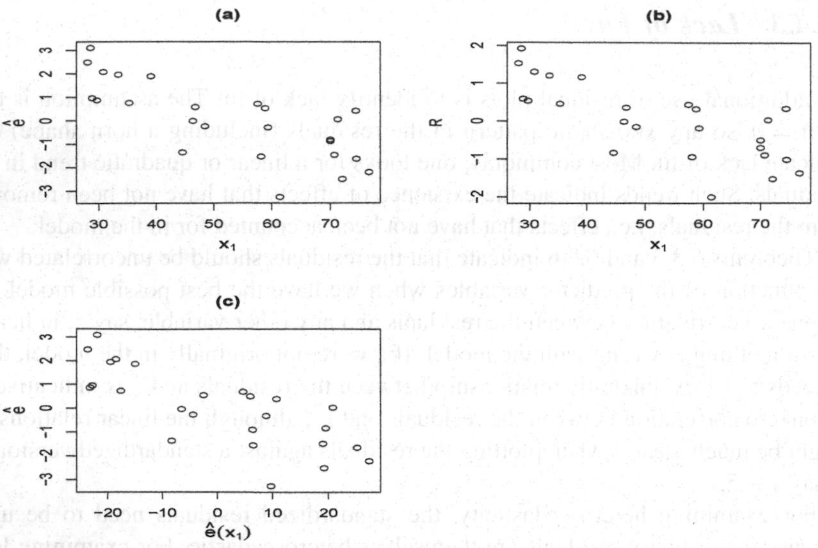

Fig. 12.22 Linear lack of fit plots

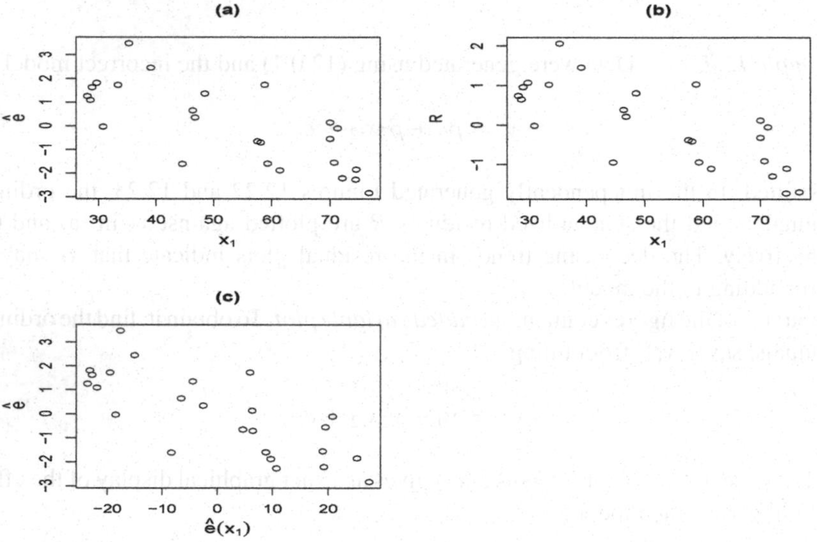

Fig. 12.23 Linear lack of fit plots

The next example in this section displays a quadratic lack of fit:

Example 12.4.4 Data were generated by adding $0.005x_1^2$ to (12.0.1). The model

$$y_i = \beta_0 + \beta_1 x_{i1} + \beta_2 x_{i2} + e_i$$

was fitted and standardized residuals R were obtained. Figures 12.24 and 12.25 are independent replications in which the standardized residuals are plotted against predicted values $\hat{y}$, against x_2, and against x_1. The quadratic trend appears clearly in the plots of residuals versus $\hat{y}$ and x_1, and even seems to be hinted at in the plot versus x_2.

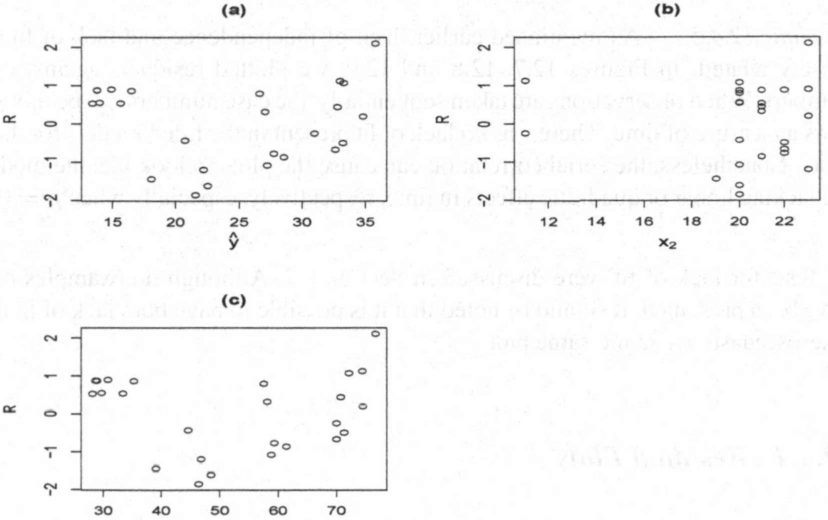

Fig. 12.24 Quadratic lack of fit plots

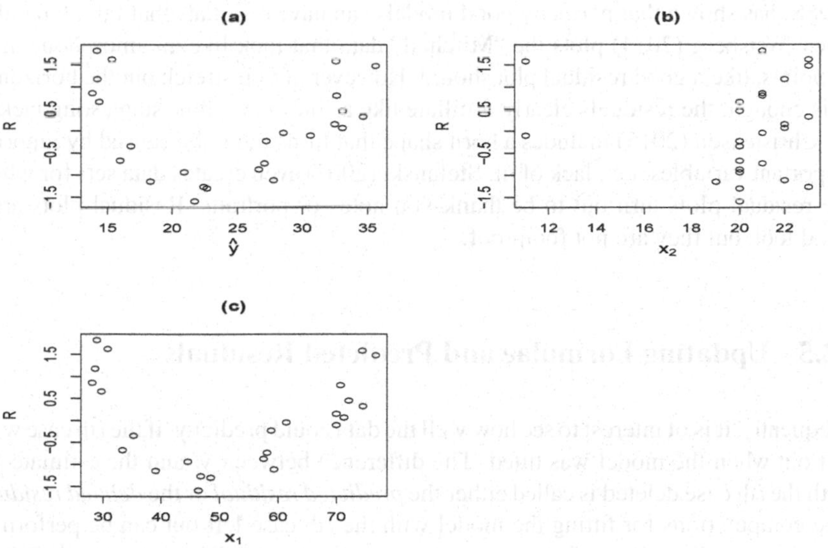

Fig. 12.25 Quadratic lack of fit plots

In Figures 12.24 and 12.25, it would not be possible to have a linear effect. The residuals are being plotted against variables x_1 and x_2 that are already included in the fitted model as well as $\hat{y}$ which is a linear combination of x_1 and x_2. Any linear effect of these variables would be eliminated by the fitting procedure. However, plotting the residuals against a variable *not* included in the model does allow the possibility of seeing a linear effect, e.g., plotting against $(x_{i1} - \bar{x}_{\cdot 1})^2$.

Example 12.4.5 As mentioned earlier, lack of independence and lack of fit are closely related. In Figures 12.7, 12.8 and 12.9, we plotted residuals against case numbers. When observations are taken sequentially, the case numbers can be thought of as a measure of time. There was *no* lack of fit present in the fitted models for these plots. Nonetheless, the serial correlation can cause the plots to look like the models are lacking linear or quadratic effects in time, respectively, especially when $\rho = 0.9$.

Tests for lack of fit were discussed in Section 6.7. Although no examples of it have been presented, it should be noted that it is possible to have both lack of fit and heteroscedasticity in the same plot.

12.4.4 Residual Plots

We have discussed what residual plots typically look like under heteroscedasticity and lack of fit. But strange things can happen, especially when the predictor variables are chosen by design, rather than being random observations from a population. Searle (1988) has shown that perfectly good models can have residuals that fall on parallel lines. Weisberg (2014) plots the "Mitchell" data that look like an amorphous mass of points, like a good residual plot should. However, if you stretch out the horizontal axis enough, the residuals clearly oscillate like a sine curve, thus suggesting lack of fit. Christensen (2015) includes a horn shape that turns out to be caused by ignoring important variables, i.e., lack of fit. Stefanski (2007) even creates data sets for which the residual plots turn out to be thank-you notes or portraits. Residual plots are a good tool, but they are not foolproof.

12.5 Updating Formulae and Predicted Residuals

Frequently, it is of interest to see how well the data could predict y_i if the ith case were left out when the model was fitted. The difference between y_i and the estimate $\hat{y}_{[i]}$ with the ith case deleted is called either the *predicted residual* or the *deleted residual*. The computations for fitting the model with the ith case left out can be performed by using simple updating formulae on the complete model. We give several of these formulae.

Let $X_{[i]}$ and $Y_{[i]}$ be X and Y with the ith row deleted. Write x_i' for the ith row of X and

$$\hat{\beta}_{[i]} \equiv (X_{[i]}'X_{[i]})^{-1}X_{[i]}'Y_{[i]}$$

for the estimate of β without the ith case. The predicted residual is defined as

$$\hat{e}_{[i]} \equiv y_i - x_i'\hat{\beta}_{[i]}.$$

The predicted residuals are useful in checking for outliers. They are also used for model selection. The Predicted REsidual Sum of Squares (PRESS) is defined as

$$\text{PRESS} \equiv \sum_{i=1}^{n} \hat{e}_{[i]}^2.$$

Models with relatively low values of the PRESS statistic should be better than models with high PRESS statistics. It is tempting to think that PRESS is a more valid measure of how well a model fits than *SSE*, because PRESS predicts values not used in fitting the model. This reasoning may seem less compelling after the updating formula for the predicted residuals has been established.

The predicted residuals can also be used to check normality, heteroscedasticity, and lack of fit in the same way that the usual residuals are used. For these purposes they should be standardized. Their variances are

$$\begin{aligned}\text{Var}(\hat{e}_{[i]}) &= \sigma^2 + \sigma^2 x_i'(X_{[i]}'X_{[i]})^{-1}x_i \\ &= \sigma^2\left[1 + x_i'(X_{[i]}'X_{[i]})^{-1}x_i\right].\end{aligned}$$

A reasonable estimate of σ^2 is $MSE_{[i]}$, the mean squared error for the model with the ith case deleted. Alternatively, σ^2 could be estimated with the regular *MSE*. If *MSE* is used, then the standardized predicted residuals are identical to the standardized residuals (see Exercise 12.6c). Standardized predicted residuals will be discussed again in Section 6. A more useful formula for $\text{Var}(\hat{e}_{[i]})$ is given in Proposition 12.5.4.

We now present a series of results that establish the updating formulae for models with one deleted case.

Proposition 12.5.1 *Let A be a $p \times p$ nonsingular matrix, and let a and b be $q \times p$ rank q matrices. Then, if all inverses exist,*

$$(A + a'b)^{-1} = A^{-1} - A^{-1}a'(I + bA^{-1}a')^{-1}bA^{-1}.$$

Proof This is a special case of Proposition B.56. □

The application of Proposition 12.5.1 is

Corollary 12.5.2

$$(X'_{[i]}X_{[i]})^{-1} = (X'X)^{-1} + [(X'X)^{-1}x_i x_i'(X'X)^{-1}]/[1 - x_i'(X'X)^{-1}x_i].$$

Proof The corollary follows from noticing that $X'_{[i]}X_{[i]} = (X'X - x_i x_i')$. □

Proposition 12.5.3 $\hat{\beta}_{[i]} = \hat{\beta} - [(X'X)^{-1}x_i\hat{e}_i]/(1 - m_{ii})$.

Proof First, note that $x_i'(X'X)^{-1}x_i = m_{ii}$ and $X'_{[i]}Y_{[i]} = X'Y - x_i y_i$. Now, from Corollary 12.5.2,

$$\begin{aligned}\hat{\beta}_{[i]} &= (X'_{[i]}X_{[i]})^{-1}X'_{[i]}Y_{[i]}\\ &= (X'_{[i]}X_{[i]})^{-1}(X'Y - x_i y_i)\\ &= \hat{\beta} - (X'X)^{-1}x_i y_i + \left[(X'X)^{-1}x_i x_i'\hat{\beta} - (X'X)^{-1}x_i x_i'(X'X)^{-1}x_i y_i\right]/(1 - m_{ii}).\end{aligned}$$

Writing $(X'X)^{-1}x_i y_i$ as $(X'X)^{-1}x_i y_i/(1 - m_{ii}) - m_{ii}(X'X)^{-1}x_i y_i/(1 - m_{ii})$, it is easily seen that

$$\begin{aligned}\hat{\beta}_{[i]} &= \hat{\beta} - [(X'X)^{-1}x_i(y_i - x_i'\hat{\beta})]/(1 - m_{ii})\\ &\quad + \left[m_{ii}(X'X)^{-1}x_i y_i - (X'X)^{-1}x_i m_{ii} y_i\right]/(1 - m_{ii})\\ &= \hat{\beta} - [(X'X)^{-1}x_i\hat{e}_i]/(1 - m_{ii}).\end{aligned}$$

□

The predicted residuals can now be written in a simple way.

Proposition 12.5.4

(a) $\hat{e}_{[i]} = \hat{e}_i/(1 - m_{ii})$.

(b) $\mathrm{Var}(\hat{e}_{[i]}) = \sigma^2/(1 - m_{ii})$.

Proof

(a)

$$\begin{aligned}\hat{e}_{[i]} &= y_i - x_i'\hat{\beta}_{[i]}\\ &= y_i - x_i'\left[\hat{\beta} - \frac{(X'X)^{-1}x_i\hat{e}_i}{1 - m_{ii}}\right]\\ &= \hat{e}_i + m_{ii}\hat{e}_i/(1 - m_{ii})\\ &= \hat{e}_i/(1 - m_{ii}).\end{aligned}$$

(b) This follows from having $\hat{e}_{[i]} = \hat{e}_i/(1 - m_{ii})$ and $\mathrm{Var}(\hat{e}_i) = \sigma^2(1 - m_{ii})$. □

The PRESS statistic can now be written as

$$\text{PRESS} = \sum_{i=1}^{n} \hat{e}_i^2/(1-m_{ii})^2.$$

The value of $\hat{e}_i^2/(1-m_{ii})^2$ will usually be large when m_{ii} is near 1. Model selection with PRESS puts a premium on having models in which observations with extremely high leverage are fitted very well. As will be discussed in Chapter 14, when fitting a model after going through a procedure to select a good model, the fitted model tends to be very optimistic in the sense of indicating much less variability than is appropriate. Model selection using the PRESS statistic tends to continue that phenomenon, cf. Picard and Cook (1984) and Picard and Berk (1990).

Later, we will also need the sum of squares for error with the ith case deleted, say $SSE_{[i]}$.

Proposition 12.5.5 $\quad SSE_{[i]} = SSE - \hat{e}_i^2/(1-m_{ii})$.

Proof By definition,

$$\begin{aligned} SSE_{[i]} &= Y_{[i]}'Y_{[i]} - Y_{[i]}'X_{[i]}(X_{[i]}'X_{[i]})^{-1}X_{[i]}'Y_{[i]} \\ &= (Y'Y - y_i^2) - Y_{[i]}'X_{[i]}\hat{\beta}_{[i]}. \end{aligned}$$

The second term can be written

$$\begin{aligned} Y_{[i]}'X_{[i]}\hat{\beta}_{[i]} &= (Y'X - y_i x_i')\left\{\hat{\beta} - [(X'X)^{-1}x_i\hat{e}_i]/(1-m_{ii})\right\} \\ &= Y'X\hat{\beta} - y_i x_i'\hat{\beta} - x_i'\hat{\beta}\hat{e}_i/(1-m_{ii}) + y_i m_{ii}\hat{e}_i/(1-m_{ii}) \\ &= Y'MY - y_i x_i'\hat{\beta} + y_i\hat{e}_i/(1-m_{ii}) - x_i'\hat{\beta}\hat{e}_i/(1-m_{ii}) \\ &\quad - y_i\hat{e}_i/(1-m_{ii}) + y_i m_{ii}\hat{e}_i/(1-m_{ii}) \\ &= Y'MY - y_i x_i'\hat{\beta} + \hat{e}_i^2/(1-m_{ii}) - y_i\hat{e}_i \\ &= Y'MY + \hat{e}_i^2/(1-m_{ii}) - y_i^2. \end{aligned}$$

Therefore,

$$\begin{aligned} SSE_{[i]} &= Y'Y - y_i^2 - [Y'MY + \hat{e}_i^2/(1-m_{ii}) - y_i^2] \\ &= Y'(I-M)Y - \hat{e}_i^2/(1-m_{ii}) \\ &= SSE - \hat{e}_i^2/(1-m_{ii}). \end{aligned}$$

□

Exercise 12.6c Show that the standardized predicted residuals with σ^2 estimated by *MSE* are the same as the standardized residuals.

12.6 Outliers and Influential Observations

Realistically, the purpose of fitting a linear model is to get a (relatively) succinct summary of the important features of the data. Rarely is the chosen linear model really correct. Usually, the linear model is no more than a rough approximation to reality.

Outliers are cases that do not seem to fit the chosen linear model. There are two kinds of outliers. Outliers may occur because the predictor variables for the case are unlike the predictor variables for the other cases. These are cases with high leverage. If we think of the linear model as being an approximation to reality, the approximation may be quite good within a certain range of the predictor variables, but poor outside that range. A few cases that fall outside the range of good approximation can greatly distort the fitted model and lead to a bad fit, even on the range of good approximation. Outliers of this kind are referred to as *outliers in the model space (or design or estimation space)*.

The other kind of outliers are those due to bizarre values of the dependent variable. These may occur because of gross measurement error, or from recording the data incorrectly. Not infrequently, data are generated from a mixture process. In other words, the data fit one pattern most of the time, but occasionally data with a different pattern are generated. Often it is appropriate to identify the different kinds of data and model them separately. If the vast majority of data fit a common pattern, there may not be enough of the rare observations for a complete analysis; but it is still important to identify such observations. In fact, these rare observations can be more important than all of the other data.

Not only is it important to be able to identify outliers, but it must also be decided whether such observations should be included when fitting the model. If they are left out, one gets an approximation to what usually happens, and one must be aware that something unusual will happen every so often.

Outliers in the design space are identified by their leverages, the m_{ii}s. Bizarre values of y_i can often be identified by their standardized residuals. Large standardized residuals indicate outliers. Typically, these are easily spotted in residual plots. However, if a case with an unusual y_i also has high leverage, the standardized residual may not be large. If there is only one bizarre value of y_i, it should be easy to identify by examining all the cases with either a large standardized residual or high leverage. With more than one bizarre value, they *may* mask each other. (I believe that careful examination of the leverages will almost always identify possible masking, cf. the comments in Section 1 on gaps in the leverage values.)

An alternative to examining the standardized residuals is to examine the *standardized predicted residuals*, also known as the *standardized deleted residuals*, the *t residuals*, and sometimes (as in the R programming language) the *Studentized residuals*. The standardized predicted residuals are

$$t_i \equiv \frac{\hat{e}_{[i]}}{\sqrt{MSE_{[i]}/(1-m_{ii})}} = \frac{y_i - x_i'\hat{\beta}_{[i]}}{\sqrt{MSE_{[i]}/(1-m_{ii})}}.$$

Since y_i, $\hat{\beta}_{[i]}$, and $MSE_{[i]}$ are independent,

$$t_i \sim t(n-p-1),$$

where $p = r(X)$. This allows a formal t test for whether the value y_i is consistent with the rest of the data. Actually, this procedure is equivalent to examining the standardized residuals, but using the $t(n-p-1)$ distribution is more convenient than using the appropriate distribution for the r_is, cf. Cook and Weisberg (1982).

When all the values $t_i, i = 1, \ldots, n$, are computed, the large values will naturally be singled out for testing. The appropriate test statistic is actually $\max_i |t_i|$. The null distribution of this statistic is quite different from a $t(n-p-1)$. Fortunately, Bonferroni's inequality provides an appropriate, actually a conservative, P value by multiplying the P value from a $t(n-p-1)$ distribution by n. Alternatively, for an α level test, use a critical value of $t(1-\alpha/2n, n-p-1)$.

If y_i corresponds to a case with extremely high leverage, the standard error for the predicted residual, $\sqrt{MSE_{[i]}/(1-m_{ii})}$, will be large, and it will be difficult to reject the t test. Recall from Example 12.1.6 that under condition (c) the y value for case six is clearly discordant. Although the absolute t value is quite large, $t_6 = -5.86$ with $m_{66} = 0.936$, it is smaller than one might expect, considering the obvious discordance of case six. In particular, the absolute t value is smaller than the critical point for the Bonferroni method with $\alpha = 0.05$. (The critical point is $t(1-0.025/6, 3) = 6.23$.) Of course, with three degrees of freedom, the power of this test is very small. A larger α level would probably be more appropriate. Using the Bonferroni method with $\alpha = 0.10$ leads to rejection.

The updating formula for t_i is

Proposition 12.6.1

$$t_i = r_i \sqrt{\frac{n-p-1}{n-p-r_i^2}}.$$

Proof Using the updating formulae of Section 5,

$$\begin{aligned}
t_i &= \hat{e}_{[i]} \Big/ \sqrt{MSE_{[i]}/(1-m_{ii})} \\
&= \hat{e}_{[i]} \sqrt{(1-m_{ii})} \Big/ \sqrt{MSE_{[i]}} \\
&= r_i \sqrt{MSE} \Big/ \sqrt{MSE_{[i]}} \\
&= r_i \sqrt{(n-p-1)/(n-p)} \sqrt{SSE/SSE_{[i]}} \\
&= r_i \sqrt{(n-p-1)/(n-p)} \sqrt{SSE/[SSE - \hat{e}_i^2/(1-m_{ii})]} \\
&= r_i \sqrt{(n-p-1)/(n-p)} \sqrt{1/[1 - r_i^2/(n-p)]} \\
&= r_i \sqrt{(n-p-1)/(n-p-r_i^2)}.
\end{aligned}$$

□

As indicated earlier, t_i really contains the same information as r_i.

A test that a given set of y values does not fit the model is easily available from general linear model theory. Suppose that the r observations $i = n - r + 1, \ldots, n$ are suspected of being outliers. The model $Y = X\beta + e$ can be written with

$$Y = \begin{bmatrix} Y_0 \\ Y_1 \end{bmatrix}, \quad X = \begin{bmatrix} X_0 \\ X_1 \end{bmatrix}, \quad e = \begin{bmatrix} e_0 \\ e_1 \end{bmatrix},$$

where Y_1, X_1, and e_1 each have r rows. If $Z = \begin{bmatrix} 0 \\ I_r \end{bmatrix}$, then the model with the possible outliers deleted

$$Y_0 = X_0\beta + e_0 \tag{12.1}$$

and the model

$$Y = X\beta + Z\gamma + e \tag{12.2}$$

are equivalent for estimating β and σ^2. A test of the reduced model $Y = X\beta + e$ against the full model $Y = X\beta + Z\gamma + e$ is rejected if

$$\frac{(SSE - SSE_0)/r}{MSE_0} > F(1 - \alpha, r, n - p - r).$$

If the test is rejected, the r observations appear to contain outliers. Note that this procedure is essentially the same as Utts's Rainbow Test for lack of fit discussed in Section 6.7. The difference is in how one identifies the cases to be eliminated.

In my opinion, the two most valuable tools for identifying outliers are the m_{ii}s and the t_is. It would be unusual to have outliers in the design space without large values of the m_{ii}s. Such outliers would have to be "far" from the other data without the Mahalanobis distance being large. For bizarre values of the y_is, it is useful to determine whether the value is so bizarre that it could not reasonably come from the model under consideration. The t_is provide a test of exactly what is needed.

Cook (1977) presented a distance measure that combines the standardized residual with the leverage to get a single measure of the influence a case has on the fit of the regression model. Cook's distance (C_i) measures the statistical distance between $\hat{\beta}$ and $\hat{\beta}_{[i]}$. It is defined as

$$C_i \equiv \frac{(\hat{\beta}_{[i]} - \hat{\beta})'(X'X)(\hat{\beta}_{[i]} - \hat{\beta})}{pMSE}.$$

Written as a function of the standardized residual and the leverage, Cook's distance is:

Proposition 12.6.2 $\quad C_i = r_i^2[m_{ii}/p(1 - m_{ii})]$.

Exercise 12.7 Prove Proposition 12.6.2.

From Proposition 12.6.2 it can be seen that Cook's distance takes the size of the standardized residual and rescales it based on the leverage of the case. For an extremely high leverage case, the squared standardized residual gets multiplied by a very large number. For low leverage cases the multiplier is very small. Another interpretation of Cook's distance is that it is a standardized version of how far the predicted values $X\hat{\beta}$ move when the ith case is deleted.

In Section 1, after establishing that the m_{ii}s were a reasonable measure of leverage, it was necessary to find guidelines for what particular values of m_{ii} meant. This can also be done for Cook's distance. Cook's distance can be calibrated in terms of confidence regions. Recall that a $(1-\alpha)100\%$ confidence region for β is

$$\left\{\beta \left| \frac{(\beta-\hat{\beta})'(X'X)(\beta-\hat{\beta})}{pMSE} < F(1-\alpha, p, n-p)\right.\right\}.$$

If $C_i \doteq F(0.75, p, n-p)$, then deleting the ith case moves the estimate of β to the edge of a 75% confidence region for β based on $\hat{\beta}$. This is a very substantial move. Since $F(0.5, p, n-p) \doteq 1$, any case with $C_i > 1$ probably has above average influence. Note that C_i does not actually have an F distribution. While many people consider such calibrations a necessity, other people, including the author, prefer simply to examine those cases with distances that are substantially larger than the other distances.

Cook's distance can be modified in an obvious way to measure the influence of any set of observations. Cook and Weisberg (1982) give a more detailed discussion of all of the topics in this section.

Updating formulae and case deletion diagnostics for linear models with general covariance matrices are discussed by Christensen, Johnson, and Pearson (1992, 1993), Christensen, Pearson, and Johnson (1992), and by Haslett and Hayes (1998) and Martin (1992). A nice review of these procedures is given by Shi and Chen (2009).

Haslett (1999) establishes two results that can greatly simplify case deletion diagnostics for correlated data. First, he shows that an analysis based on $Y_{[i]}$, $X_{[i]}$, and $V_{[ii]}$ (the covariance matrix with the ith row and column removed) is the same as an analysis based on $\tilde{Y}(i)$, X, and V where

$$\tilde{Y}(i) = \begin{bmatrix} \hat{y}_i(Y_{[i]}) \\ Y_{[i]} \end{bmatrix}$$

and $\hat{y}_i(Y_{[i]})$ is the BLUP of y_i based on $Y_{[i]}$. Second, he shows that there is a relatively simple way to find a matrix P_i such that $\tilde{Y}(i) = P_i Y$.

12.7 Transformations

If the residuals suggest nonnormality, heteroscedasticity of variances, or lack of fit, a transformation of the y_is may alleviate the problem. Cook and Weisberg (1982) and Atkinson (1985) give extensive discussions of transformations. Only a brief review is presented here.

Picking a transformation is often a matter of trial and error. Different transformations are tried until one is found for which the residuals seem reasonable. Three more systematic methods of choosing transformations will be discussed: *Box–Cox power transformations*, *variance stabilizing transformations*, and the generalized least squares approach of Grizzle et al. (1969).

Box and Cox (1964) suggested a systematic method of picking transformations. They suggested using the family of transformations

$$y^{(\lambda)} = \begin{cases} (y^{\lambda} - 1)/\lambda, & \lambda \neq 0 \\ \log y, & \lambda = 0 \end{cases}$$

and choosing λ by maximum likelihood. Convenient methods of executing this procedure are discussed in Cook and Weisberg (1982), Weisberg (2014), and Christensen (2015). They and Atkinson (1985) also discuss *constructed variable tests* for whether a Box–Cox transformation is needed. Constructed variable tests are generalizations of the famous Tukey (1949) one-degree-of-freedom test for nonadditivity.

If the distribution of the y_is is known, the commonly used *variance stabilizing transformations* can be tried (cf. Rao 1973, Section 6g, or Christensen 2015). For example,

if $y_i \sim \text{Binomial}(N_i, p_i)$,	use $\arcsin(\sqrt{y_i/N_i})$,
if $y_i \sim \text{Poisson}(\lambda_i)$,	use $\sqrt{y_i}$,
if y_i has $\sigma_i/\text{E}(y_i)$ constant,	use $\log(y_i)$.

More generally, $\arcsin(\sqrt{y_i/N_i})$ can be tried for any problem where y_i is a count between 0 and N_i or a proportion, $\sqrt{y_i}$ can be used for any problem where y_i is a count, and $\log(y_i)$ can be used if y_i is a count or amount.

The transformation $\log(y_i)$ is also frequently used because, for a linear model in $\log(y_i)$, the additive effects of the predictor variables transform to multiplicative effects on the original scale. If multiplicative effects seem reasonable, the log transformation may be appropriate.

As an alternative to the variance stabilizing transformations, there exist generalized linear models specifically designed for treating binomial and Poisson data. For Poisson data there exists a well developed theory for fitting log-linear models. One branch of the theory of log-linear models is the theory of logistic regression, which is used to analyze binomial data. As shown by Grizzle et al. (1969), generalized least squares methods can be used to fit log-linear models to Poisson data and logistic regression models to binomial data. The method involves both a transformation of the dependent variable and weights. The appropriate transformation and weights are:

Distribution of y_i	Transformation	Weights
Poisson(λ_i)	$\log y_i$	y_i
Binomial(N_i, p_i)	$\log(y_i/[N_i - y_i])$	$y_i(N_i - y_i)/N_i$

With these weights, the asymptotic variance of the transformed data is 1. Standard errors for regression coefficients are computed as usual except that no estimate of σ^2 is required (σ^2 is known to be 1). Since σ^2 is known, t tests and F tests are replaced with normal tests and chi-squared tests. In particular, if the linear model fits the data and the observations y_i are large, the *SSE* has an asymptotic chi-squared distribution with the usual degrees of freedom. If the *SSE* is too large, a lack of fit is indicated. Tests of various models can be performed by comparing the difference in the *SSE*s for the model. The difference in the *SSE*s has an asymptotic chi-squared distribution with the usual degrees of freedom. If the difference is too large, then the smaller model is deemed inadequate. Unlike the lack-of-fit test, these model comparison tests are typically valid if n is large even when the individual y_is are not. Fitting log-linear and logistic models both by generalized least squares and by maximum likelihood is discussed in detail by Christensen (1997).

Exercise 12.8 The data given below were first presented by Brownlee (1965) and have subsequently appeared in Daniel and Wood (1980), Draper and Smith (1998), and Andrews (1974), among other places. The data consist of measurements taken on 21 successive days at a plant that oxidizes ammonia into nitric acid. The dependent variable y is stack loss. It is 10 times the percentage of ammonia that is lost (in the form of unabsorbed nitric oxides) during the oxidation process. There are three predictor variables: x_1, x_2, and x_3. The first predictor variable is air flow into the plant. The second predictor variable is the cooling water temperature as it enters the countercurrent nitric oxide absorption tower. The third predictor variable is a coded version of the nitric acid concentration in the absorbing liquid. Analyze these data giving special emphasis to residual analysis and influential observations.

Obs.	x_1	x_2	x_3	y	Obs.	x_1	x_2	x_3	y
1	80	27	89	42	12	58	17	88	13
2	80	27	88	37	13	58	18	82	11
3	75	25	90	37	14	58	19	93	12
4	62	24	87	28	15	50	18	89	8
5	62	22	87	18	16	50	18	86	7
6	62	23	87	18	17	50	19	72	8
7	62	24	93	19	18	50	19	79	8
8	62	24	93	20	19	50	20	80	9
9	58	23	87	15	20	56	20	82	15
10	58	18	80	14	21	70	20	91	15
11	58	18	89	14					

Exercise 12.9 For testing whether one observation y_i is an outlier, show that the F statistic is equal to the squared standardized predicted residual.

References

Andrews, D. F. (1974). A robust method for multiple regression. *Technometrics, 16*, 523–531.
Arnold, S. F. (1981). *The theory of linear models and multivariate analysis*. New York: Wiley.
Atkinson, A. C. (1981). Two graphical displays for outlying and influential observations in regression. *Biometrika, 68*, 13–20.
Atkinson, A. C. (1982). Regression diagnostics, transformations and constructed variables (with discussion). *Journal of the Royal Statistical Society, Series B, 44*, 1–36.
Atkinson, A. C. (1985). *Plots, transformations, and regression: An introduction to graphical methods of diagnostic regression analysis*. Oxford: Oxford University Press.
Blom, G. (1958). *Statistical estimates and transformed beta variates*. New York: Wiley.
Box, G. E. P. (1953). Non-normality and tests on variances. *Biometrika, 40*, 318–335.
Box, G. E. P., & Cox, D. R. (1964). An analysis of transformations. *Journal of the Royal Statistical Society, Series B, 26*, 211–246.
Brownlee, K. A. (1965). *Statistical theory and methodology in science and engineering* (2nd ed.). New York: Wiley.
Christensen, R. (1989). Lack of fit tests based on near or exact replicates. *The Annals of Statistics, 17*, 673–683.
Christensen, R. (1996). *Analysis of variance, design, and regression: Applied statistical methods*. London: Chapman and Hall.
Christensen, R. (1997). *Log-linear models and logistic regression* (2nd ed.). New York: Springer.
Christensen, R. (2001). *Advanced linear modeling: Multivariate, time series, and spatial data; nonparametric regression, and response surface maximization* (2nd ed.). New York: Springer.
Christensen, R. (2015). *Analysis of variance, design, and regression: Linear modeling for unbalanced data* (2nd ed.). Boca Raton: Chapman and Hall/CRC Press.
Christensen, R., & Bedrick, E. J. (1997). Testing the independence assumption in linear models. *Journal of the American Statistical Association, 92*, 1006–1016.
Christensen, R., Johnson, W., & Pearson, L. M. (1992). Prediction diagnostics for spatial linear models. *Biometrika, 79*, 583–591.
Christensen, R., Johnson, W., & Pearson, L. M. (1993). Covariance function diagnostics for spatial linear models. *Mathematical Geology, 25*, 145–160.
Christensen, R., Johnson, W., Branscum, A., & Hanson, T. E. (2010). *Bayesian ideas and data analysis: An introduction for scientists and statisticians*. Boca Raton: Chapman and Hall/CRC Press.
Christensen, R., Pearson, L. M., & Johnson, W. (1992). Case deletion diagnostics for mixed models. *Technometrics, 34*, 38–45.
Cook, R. D. (1977). Detection of influential observations in linear regression. *Technometrics, 19*, 15–18.
Cook, R. D. (1998). *Regression graphics: Ideas for studying regressions through graphics*. New York: Wiley.
Cook, R. D., & Weisberg, S. (1982). *Residuals and influence in regression*. New York: Chapman and Hall.
Cook, R. D., & Weisberg, S. (1994). *An introduction to regression graphics*. New York: Wiley.
Cook, R. D., & Weisberg, S. (1999). *Applied regression including computing and graphics*. New York: Wiley.
Daniel, C. (1959). Use of half-normal plots in interpreting factorial two-level experiments. *Technometrics, 1*, 311–341.
Daniel, C. (1976). *Applications of statistics to industrial experimentation*. New York: Wiley.
Daniel, C., & Wood, F. S. (1980). *Fitting equations to data* (2nd ed.). New York: Wiley.
Draper, N., & Smith, H. (1998). *Applied regression analysis* (3rd ed.). New York: Wiley.
Duan, N. (1981). Consistency of residual distribution functions. Working Draft No. 801-1-HHS (106B-80010), Rand Corporation, Santa Monica, CA.

Durbin, J., & Watson, G. S. (1951). Testing for serial correlation in least squares regression II. *Biometrika, 38*, 159–179.

Freedman, D. A. (2006). On the so-called "Huber sandwich estimator" and "robust standard errors". *The American Statistician, 60*, 299–302.

Grizzle, J. E., Starmer, C. F., & Koch, G. G. (1969). Analysis of categorical data by linear models. *Biometrics, 25*, 489–504.

Haslett, J. (1999). A simple derivation of deletion diagnostic results for the general linear model with correlated errors. *Journal of the Royal Statistical Society, Series B, 61*, 603–609.

Haslett, J., & Hayes, K. (1998). Residuals for the linear model with general covariance structure. *Journal of the Royal Statistical Society, Series B, 60*, 201–215.

Lenth, R. V. (2015). The case against normal plots of effects (with discussion). *Journal of Quality Technology, 47*, 91–97.

Mandansky, A. (1988). *Prescriptions for working statisticians*. New York: Springer.

Martin, R. J. (1992). Leverage, influence and residuals in regression models when observations are correlated. *Communications in Statistics - Theory and Methods, 21*, 1183–1212.

Picard, R. R., & Berk, K. N. (1990). Data splitting. *The American Statistician, 44*, 140–147.

Picard, R. R., & Cook, R. D. (1984). Cross-validation of regression models. *Journal of the American Statistical Association, 79*, 575–583.

Rao, C. R. (1973). *Linear statistical inference and its applications* (2nd ed.). New York: Wiley.

Searle, S. R. (1988). Parallel lines in residual plots. *The American Statistician, 42*, 211.

Shapiro, S. S., & Francia, R. S. (1972). An approximate analysis of variance test for normality. *Journal of the American Statistical Association, 67*, 215–216.

Shapiro, S. S., & Wilk, M. B. (1965). An analysis of variance test for normality (complete samples). *Biometrika, 52*, 591–611.

Shewhart, W. A. (1931). *Economic control of quality*. New York: Van Nostrand.

Shewhart, W. A. (1939). Statistical method from the viewpoint of quality control. Graduate School of the Department of Agriculture, Washington. Reprint (1986), Dover, New York.

Shi, L., & Chen, G. (2009). Influence measures for general linear models with correlated errors. *The American Statistician, 63*, 40–42.

Stefanski, L. A. (2007). Residual (sur)realism. *The American Statistician, 61*, 163–177.

Tukey, J. W. (1949). One degree of freedom for nonadditivity. *Biometrics, 5*, 232–242.

Weisberg, S. (2014). *Applied linear regression* (4th ed.). New York: Wiley.

Chapter 13
Collinearity and Alternative Estimates

Abstract This chapter deals with problems caused by having predictor variables that are very nearly redundant. It examines estimation methods developed for dealing with those problems and then goes on to introduce a variety of alternatives to least squares estimation including robust and penalized (regularized) estimates. Penalized estimation is discussed in more detail in *ALM-III*.

Collinearity or multicollinearity refers to the problem in regression analysis of the columns of the model matrix being nearly linear dependent. Ideally, this is no problem at all. There are numerical difficulties associated with the actual computations, but there are no theoretical difficulties. If, however, one has any doubts about the accuracy of the model matrix, the analysis could be in deep trouble.

Section 1 discusses what collinearity is and what problems it can cause. Section 2 defines tolerance and variance inflation factors and relates them to our definition of collinearity. Sections 3 and 4 introduce four techniques for dealing with collinearity. These are regression in canonical form, principal component regression, generalized inverse regression, and classical ridge regression. The methods, other than classical ridge regression, are essentially the same. Classical ridge regression is a version of penalized least squares estimation, something that is discussed in more detail in *ALM-III* Chapter 2. Section 5 presents additional comments on the potential benefits of biased estimation. These results are specifically for regression in canonical form. Finally, Sections 6 and 7 present alternatives to least squares estimation. Least squares minimizes the vertical Euclidean distances between the dependent variable and the regression surface. Section 6 considers alternative methods of measuring that vertical distance. Section 7 considers minimizing the Euclidean perpendicular distance between the dependent variable and the regression surface.

The estimation methods presented in this chapter and the next typically result in biased estimates, cf. Section 2.9. When using biased estimates it might be reasonable to continue to use the original least squares residuals for estimating variability (provided the model has sufficient degrees of freedom for Error). Standard statistical methods, when applied with biased estimates, especially those from the data

R. Christensen, *Plane Answers to Complex Questions*, Springer Texts in Statistics,
https://doi.org/10.1007/978-3-030-32097-3_13

driven reduced models discussed in Chapter 14, tend to be overly optimistic in that they underestimate the appropriate variability. Data driven reduced models tend to be chosen precisely because they display small variability! For example, a prediction interval from the full model should be appropriate, if unnecessarily wide. A prediction interval from a data selected reduced model is probably unrealistically narrow, although more accurately centered. The residuals from the original (full) model should provide a more accurate estimate of the variance. A prediction interval computed from the reduced model, except with the *MSE* and *dfE* from the reduced model replaced by their corresponding values from the full model, might retain the advantages of biased estimation for β without the inappropriate downward bias in variability. But this is not a mathematical result! The mathematics becomes immensely complicated when, as in a data driven reduced model, the model matrix itself becomes a function of Y.

If the final (biased) estimate of β is a function of the least squares estimate $\hat{\beta}$, under multivariate normal errors the final estimate still has to be independent of the least squares residuals. If the final estimate is a linear function of $\hat{\beta}$, the least squares residuals are uncorrelated with it even without assuming normality. This is true for penalized (regularized) estimates like the ridge regression estimates of Section 4 (see *ALM-III* Section 2.1). Independence is also true for any estimates of estimable functions from a reduced model because they are functions of the full model least squares estimates, i.e., $X_0\hat{\beta}_0 = M_0Y = M_0MY = M_0X\hat{\beta}$, but not necessarily for data driven reduced models where X_0 is a function of Y.

In this book, the term "mean squared error" (*MSE*) has generally denoted the quantity $Y'(I-M)Y/r(I-M)$. This is a sample quantity, a function of the data. In Chapter 6, when discussing prediction, we needed a theoretical concept of the mean squared error. Fortunately, up until this point we have not needed to discuss both the sample quantity and the theoretical one at the same time. To discuss variable selection methods and techniques of dealing with collinearity, we will need both concepts simultaneously. To reduce confusion, we will refer to $Y'(I-M)Y/r(I-M)$ as the *residual mean square* (*RMS*) and $Y'(I-M)Y$ as the *residual sum of squares* (*RSS*). Since $Y'(I-M)Y = [(I-M)Y]'[(I-M)Y]$ is the sum of the squared residuals, this is a very natural nomenclature.

13.1 Defining Collinearity

In this section we define the problem of collinearity. The approach taken is to quantify the idea of having columns of the model matrix that are "nearly linearly dependent." The effects of near linear dependencies are examined. The section concludes with establishing the relationship between the definition given here and another commonly used concept of collinearity.

Suppose we have a standard regression model

$$Y = X\beta + e, \quad \mathrm{E}(e) = 0, \quad \mathrm{Cov}(e) = \sigma^2 I, \tag{1}$$

where Y is $n \times 1$, $X = [X_1, \ldots, X_p]$ is $n \times p$, β is $p \times 1$, and $r(X) = p$. Frequently, $X_1 \equiv J$, which is known without error. The essence of model (1) is that $\mathrm{E}(Y) \in C(X)$, $\mathrm{Cov}(Y) = \sigma^2 I$. Suppose that the model matrix consists of some predictor variables, say $x_1, x_2, \ldots, x_p$, that are measured with some small error. (If any variables are measured without error, e.g., $X_1 \equiv J$, adjustments to the discussion are needed.) A near linear dependence in the observed model matrix X could mean a real linear dependence in the underlying model matrix of variables measured without error. Let X_* be the underlying model matrix. If the columns of X_* are linearly dependent, there exists an infinite number of least squares estimates for the true regression coefficients. If X is nearly linearly dependent, the estimated regression coefficients may not be meaningful and may be highly variable.

The real essence of this particular problem is that $C(X)$ is too large. Generally, we hope that in some sense, $C(X)$ is close to $C(X_*)$. Regression should work well precisely when this is the case. However, when X_* has linearly dependent columns, X typically will not. Thus $r(X) > r(X_*)$. $C(X_*)$ may be close to some proper subspace of $C(X)$, but $C(X)$ has extra dimensions. By pure chance, these extra dimensions could be very good at explaining the Y vector that happened to be observed. In this case, we get an apparently good fit that has no real world significance.

The extra dimensions of $C(X)$ are due to the existence of vectors b such that $X_* b = 0$ but $Xb \neq 0$. If the errors in X are small, then Xb should be approximately zero. We would like to say that a vector w in $C(X)$ is ill-defined if there exists b such that $w = Xb$ is approximately zero. The vector w is approximately the zero vector when its length is near zero. Unfortunately, multiplying w by a scalar can increase or decrease the length of the vector arbitrarily, while not changing the direction determined within $C(X)$. To rectify this, we can restrict attention to vectors b with $b'b = 1$ (i.e., the length of b is 1), or, equivalently, we make the following:

Definition 13.1.1 A vector $w = Xb$ is said to be ε *ill-defined* if $w'w/b'b = b'X'Xb/b'b < \varepsilon$. The matrix X is ε ill-defined if any vector in $C(X)$ is ε ill-defined. We use the terms "ill-defined" and "ill-conditioned" interchangeably.

In a model with an intercept, say $X = [J, Z]$, typically we would be interested in whether the columns of Z are collinear after correcting for their mean values, i.e., evaluating whether $[I - (1/n)JJ']Z$ is ε ill-defined. In a traditional ACOVA with model matrix $[X, Z]$, when X is not subject to errors we typically would be interested in whether $(I - M)Z$ is ε ill-defined. (If individuals randomly fall into groups, rather than being placed into groups, the group indicators in X may be subject to error.)

The assumption of a real linear dependence in the X_* matrix is a strong one. We now indicate how that assumption can be weakened. Let $X = X_* + \Delta$, where the elements of Δ are uniformly small errors. Consider the vector Xb. (For simplicity, assume $b'b = 1$.) The corresponding direction in the underlying model matrix is $X_* b$.

Note that $b'X'Xb = b'X_*'X_*b + 2b'\Delta'X_*b + b'\Delta'\Delta b$. The vector $\Delta' b$ is short; so if Xb and $X_* b$ are of reasonable size, they have about the same length. Also

$$b'X_*'Xb = b'X_*'X_*b + b'X_*'\Delta b, \tag{2}$$

where $b'X_*'\Delta b$ is small; so the angle between Xb and X_*b should be near zero. (For any two vectors x and y, let θ be the angle between x and y. Then $x'y = \sqrt{x'x}\sqrt{y'y}\cos\theta$.) On the other hand, if Xb is ill-defined, X_*b will also be small, and the term $b'X_*'\Delta b$ could be a substantial part of $b'X_*'Xb$. Thus, the angle between Xb and X_*b could be substantial. In that case, the use of the direction Xb is called into question because it may be substantially different from the underlying direction X_*b. In practice, we generally cannot know if the angle between Xb and X_*b is large. Considerable care must be taken when using a direction in $C(X)$ that is ill-defined.

If $X_1 = J$, which has no measurement error, we typically orthogonalize all the other columns of X to J, i.e., we subtract their means, prior to evaluating collinearity. If the squared length of a (mean adjusted) column of X is less than ε, that direction will be ε ill-defined, regardless of what other vectors are in $C(X)$. To avoid this we typically rescale (standardize) the (mean adjusted) columns of X to have a constant length before evaluating collinearity. I think that one should not be dogmatic about the issue of whether the columns of X should be adjusted for their mean values or rescaled but see Belsley (1984) along with its excellent discussion papers and, more recently, Velilla (2018) and Christensen (2018).

The intercept term is frequently handled separately from all other variables in techniques for dealing with collinearity. The model

$$y_i = \beta_0 + \beta_1 x_{i1} + \cdots + \beta_{p-1} x_{i\,p-1} + e_i \tag{3}$$

$$Y = [J, Z]\begin{bmatrix}\beta_0\\ \beta_*\end{bmatrix} + e$$

is often rewritten as

$$y_i = \alpha + \beta_1(x_{i1} - \bar{x}_{\cdot 1}) + \ldots + \beta_{p-1}(x_{i\,p-1} - \bar{x}_{\cdot p-1}) + e_i, \tag{4}$$

or

$$Y = \left[J, \left(I - \frac{1}{n}J_n^n\right)Z\right]\begin{bmatrix}\alpha\\ \beta_*\end{bmatrix} + e,$$

where $\bar{x}_{\cdot j} = n^{-1}\sum_{i=1}^{n} x_{ij}$. It is easily seen that β_* is the same in both models, but $\beta_0 \neq \alpha$. We dispense with the concept of the underlying model matrix and define

$$X_* \equiv \left(I - \frac{1}{n}J_n^n\right)Z.$$

Because of orthogonality, $\hat{\alpha} = \bar{y}_{\cdot}$ and β_* can be estimated from the model

$$Y_* = X_*\beta_* + e, \tag{5}$$

where $Y_*' \equiv [y_1 - \bar{y}_{\cdot}, y_2 - \bar{y}_{\cdot}, \ldots, y_n - \bar{y}_{\cdot}]$.

Frequently, the scales of the x variables are also standardized. Let $q_j^2 \equiv \sum_{i=1}^{n} (x_{ij} - \bar{x}_{\cdot j})^2$. Model (5) is equivalent to

$$Y_* = X_* \text{Diag}(q_j^{-1}) \gamma_* + e, \tag{6}$$

where $\gamma_* = \text{Diag}(q_j)\beta_*$. In model (6), the model matrix is $X_*\text{Diag}(q_j^{-1})$.

Typically in techniques for dealing with collinearity, model (6) is assumed. Sometimes model (5). Rarely model (3). To retain full generality, our discussion uses model (3), but all the results apply to models (5) and (6). Note that the matrix $X_*'X_*$ is proportional to the sample covariance matrix of the Xs when $(x_{i1}, \ldots, x_{i\,p-1})$, $i = 1, \ldots, n$, is thought of as a sample of size n from a $p-1$ dimensional random vector. $\text{Diag}(q_j^{-1})X_*'X_*\text{Diag}(q_j^{-1})$ is the sample correlation matrix.

We now present the relationship between ε ill-defined matrices and other commonly used methods of identifying ill-conditioned matrices.

One of the main tools in the examination of collinearity is the examination of the eigenvalues of $X'X$. We discuss the relationship between Definition 13.1.1 and an eigen-analysis of $X'X$.

Recall that X has linearly dependent columns if and only if $X'X$ is singular, which happens if and only if $X'X$ has a zero eigenvalue. One often-used (but I think unintuitive) definition is that the columns of X are nearly linearly dependent if $X'X$ has at least one small eigenvalue. Suppose that $v_1, \ldots, v_p$ is an orthogonal set of eigenvectors for $X'X$ corresponding to the eigenvalues $\delta_1, \delta_2, \ldots, \delta_p$. Then $v_1, \ldots, v_p$ form a basis for $\mathbf{R}^p$. If $\delta_i < \varepsilon$, then $\delta_i = v_i'X'Xv_i/v_i'v_i$; so Xv_i is a direction in $C(X)$ that is ε ill-defined. Conversely, if an ε ill-defined vector exists, we show that at least one of the δ_is must be less than ε. Let $w = Xd$ be ε ill-defined. Write $d = \sum_{i=1}^{p} \alpha_i v_i$. Then $w'w = d'X'Xd = \sum_{i=1}^{p} \alpha_i^2 \delta_i$ and $d'd = \sum_{i=1}^{p} \alpha_i^2$. Since $\sum_{i=1}^{p} \alpha_i^2 \delta_i / \sum_{i=1}^{p} \alpha_i^2 < \varepsilon$, and since the δ_is are all nonnegative, at least one of the δ_is must be less than ε. We have proved:

Theorem 13.1.2 *The matrix X is ε ill-defined if and only if $X'X$ has an eigenvalue less than ε.*

The orthogonal eigenvectors of $X'X$ lead to a useful breakdown of $C(X)$ into orthogonal components. Let $\delta_1, \ldots, \delta_r$ be eigenvalues of at least ε and $\delta_{r+1}, \ldots, \delta_p$ eigenvalues less than ε. It is easily seen that $Xv_{r+1}, \ldots, Xv_p$ form an orthogonal basis for a subspace of $C(X)$ in which all vectors are ε ill-defined. The space spanned by $Xv_1, \ldots, Xv_r$ is a subspace in which none of the vectors are ε ill-defined. These two spaces are orthogonal complements with respect to $C(X)$. (Note that by taking a linear combination of a vector in each of the orthogonal subspaces one can get a vector that is ε ill-defined, but is in neither subspace.) As discussed later, the vectors Xv_j are the principal component vectors if X is centered (i.e., $X = X_*$) or centered and scaled (i.e., $X = X_*D(q_j^{-1})$).

A second commonly used method of identifying ill-conditioned matrices was presented by Belsley et al. (1980). See also Belsley (1991). They use the *condition number* (*CN*) as the basis of their definition of collinearity. If the columns of X are rescaled to have length 1, the condition number is

$$CN \equiv \sqrt{\max_i \delta_i \Big/ \min_i \delta_i}\,.$$

Large values of the condition number indicate that X is ill-conditioned.

Theorem 13.1.3

(a) *If* $CN > \sqrt{p/\varepsilon}$, *then* X *is* ε *ill-defined.*
(b) *If* X *is* ε *ill-defined, then* $CN > \sqrt{1/\varepsilon}$.

Proof See Exercise 13.2. □

If the columns of X have not been rescaled to have length 1, the *CN* is inappropriate. If X has two orthogonal columns, with one of length 10^3 and one of length 10^{-3}, the condition number is 10^6. Rescaling would make the columns of X orthonormal, which is the ideal of noncollinearity. (Note that such an X matrix is also ill-defined for any $\varepsilon > 10^{-6}$.)

Exercise 13.1 Show that any linear combination of ill-defined orthonormal eigenvectors is ill-defined. In particular, if $w = X(av_i + bv_j)$, then

$$\frac{w'w}{(av_i + bv_j)'(av_i + bv_j)} \le \max(\delta_i, \delta_j).$$

Exercise 13.2 Prove Theorem 13.1.3.

Hint: For a model matrix with columns of length 1, $\mathrm{tr}(X'X) = p$. It follows that $1 \le \max_i \delta_i \le p$.

13.2 Tolerance and Variance Inflation Factors

Regression assumes that the model matrix in $Y = X\beta + e$ has full rank. Mathematically, either the columns of X are linearly independent or they are not. In practice, computational difficulties arise if the columns of X are nearly linearly dependent, e.g., if $Xb \doteq 0$ for some b with $b'b = 1$. In this section we focus on an equivalent idea, that one column of X can be nearly reproduced by the other columns, i.e., when Proposition 2.1.6 *almost* holds.

Most computer programs for doing regression analysis fit the predictor variables sequentially. *Tolerance* is a concept used to determine whether a variable is too collinear with the previous variables to allow its entry into the model because it will present numerical difficulties. The relationship between the concepts of tolerance and ill-defined vectors is complex.

Suppose the model $Y = X\beta + e$ has an intercept, say,

$$y_i = \beta_0 + \beta_1 x_{i1} + \cdots + \beta_{p-1} x_{i\,p-1} + e_i,$$

and we are considering adding variable x_p to the model. To check the tolerance associated with adding x_p, fit

$$x_{ip} = \alpha_0 + \alpha_1 x_{i1} + \cdots + \alpha_{p-1} x_{i\,p-1} + e_i. \tag{1}$$

If the coefficient of determination $R^2(p)$ from this model is high, the column vectors of X, say $J, X_1, \ldots, X_{p-1}$, are nearly linearly dependent with $X_p \equiv [x_{1p}, \ldots, x_{np}]'$. The tolerance of x_p (relative to $x_1, \ldots, x_{p-1}$) is defined as the value

$$T_p \equiv 1 - R^2(p)$$

for fitting model (1). If the tolerance is too small, variable x_p is not used. Often, in a computer program, the user can define which values of the tolerance should be considered too small.

More generally, the tolerance of predictor variable x_j is defined as

$$T_j \equiv 1 - R^2(j)$$

where $R^2(j)$ is the coefficient of determination from regressing X_j on an intercept and all of the other variables. $R^2(j)$, being a measure of predictive ability, examines the predictive ability after correcting *all* variables for their means.

A closely related concept is that of the *variance inflation factor (VIF)* associated with the regression coefficient γ_j in the model

$$y_i = \gamma_0 + \gamma_1 x_{i1} + \cdots + \gamma_p x_{ip} + e_i.$$

The variance inflation factor is defined as

$$VIF_j \equiv 1/T_j.$$

The VIF can be viewed as the variance of $\hat{\gamma}_j$ divided by the variance that $\hat{\gamma}_j$ would have if all the predictors were uncorrelated, which is also the variance of $\hat{\eta}_j$ in the simple linear regression $y_i = \eta_0 + \eta_j x_{ij} + e_i$. Specifically, from Chapter 9 and for $j = p$ (only because we have that notation already set),

$$VIF_p = \frac{X_p'\left(I - \frac{1}{n}J_n^n\right)X_p}{X_p(I - M)X_p} = \frac{\mathrm{Var}(\hat{\gamma}_p)}{\mathrm{Var}(\hat{\eta}_p)}.$$

Recall that M is the ppo onto $C(X) = C(J, X_1, \ldots, X_{p-1})$.

While the above definition of T_j is usually appropriate, in the discussion of Section 1, where X_1 is not assumed to be the intercept, it would be more appropriate to define tolerance as

$$T_j = \text{sum of squares error/sum of squares total(uncorrected)}$$

for regressing X_j on the other variables. If it is decided to adjust all variables for their means as in model (13.1.5), this becomes the usual definition. *In the following discussion of the relationship between tolerance and ε ill-defined, we will use this alternative definition of T_j.*

In a more subtle way, T_j also adjusts for the scale of $X_j = [x_{1j}, \ldots, x_{nj}]'$. The scale adjustment comes because T_j is the ratio of two squared lengths. This issue arises again in our later analysis.

Rewrite model (1) as

$$X_p = X\alpha + e. \tag{2}$$

The new (orthogonal) dimension added to $C(X)$ by including the variable x_p in the regression is the vector of residuals from fitting model (2), i.e., $X_p - X\hat{\alpha}$. Our question is whether $X_p - X\hat{\alpha}$ is ill-defined within $C(X, X_p)$.

By the alternative definition,

$$\begin{aligned} T_p &= X_p'(I - M)X_p \big/ X_p'X_p \\ &= X_p'(I - M)X_p \big/ [X_p'(I - M)X_p + X_p'MX_p]. \end{aligned}$$

T_p is small when $X_p'(I - M)X_p$ is small relative to $X_p'MX_p$.

The residual vector, $X_p - X\hat{\alpha}$, is ε ill-defined in $C(X, X_p)$ if

$$\varepsilon > \frac{(X_p - X\hat{\alpha})'(X_p - X\hat{\alpha})}{1 + \hat{\alpha}'\hat{\alpha}} = \frac{X_p'(I - M)X_p}{1 + \hat{\alpha}'\hat{\alpha}}.$$

It is not too difficult to see, cf. Exercise 13.3a, that

$$\delta_1 \frac{X_p'(I - M)X_p}{\delta_1 + X_p'MX_p} \le \frac{X_p'(I - M)X_p}{1 + \hat{\alpha}'\hat{\alpha}} \le \delta_p \frac{X_p'(I - M)X_p}{\delta_p + X_p'MX_p}, \tag{3}$$

where δ_1 and δ_p are the smallest and largest eigenvalues of $X'X$. Note that for positive a and b, $a/(a + b) < \varepsilon$ implies $a/(\delta_p + b) < a/b < \varepsilon/(1 - \varepsilon)$; so it follows immediately that if $T_p < \varepsilon$, $X_p - X\hat{\alpha}$ is $\delta_p\varepsilon/(1 - \varepsilon)$ ill-defined. On the other hand, if $X_p - X\hat{\alpha}$ is ε ill-defined, some algebra shows that

$$T_p < \frac{\delta_1\varepsilon}{(\delta_1 + \varepsilon)X_p'X_p} + \frac{\varepsilon}{\delta_1 + \varepsilon}.$$

By picking ε small, the tolerance can be made small. This bound depends on the squared length of X_p. In practice, however, if X_p is short, X_p may not have small tolerance, but the vector of residuals may be ill-defined. If X_p is standardized so that $X_p'X_p = 1$, this problem is eliminated. Standardizing the columns of X also ensures that there are some reasonable limits on the values of δ_1 and δ_p. (One would assume in this context that X is not ill-defined.)

Exercise 13.3a Use a singular value decomposition $X'X = PD(\delta_i)P'$ to show that $v'(X'X)^{-1}v/\delta_p \le v'(X'X)^{-2}v \le v'(X'X)^{-1}v/\delta_1$, where $\delta_1 \le \cdots \le \delta_p$ and v is any vector. Use this to show inequality (3). Hints: The inequalities are reversed if you invert the numbers. Focus on rewriting $\hat{\alpha}'\hat{\alpha}$.

Exercise 13.3b Writing model (2) as $X_p = J\alpha_0 + Z\alpha_* + e$ and using the original definition of tolerance $T_p = X_p(I-M)X_p/X_p[I-(1/n)JJ']X_p$, find the relationship between tolerance and having $X_p - X\hat{\alpha}$ be ε ill-defined in $C\{[I-(1/n)JJ'](X, X_p)\}$.

13.3 Regression in Canonical Form and on Principal Components

13.3.1 Regression in Canonical Form

Regression in canonical form involves transforming the Y and β vectors so that the model matrix is particularly nice. Regression in canonical form is closely related to two procedures that have been proposed for dealing with collinearity. To transform the regression problem we need

Theorem 13.3.1 The Singular Value Decomposition.
Let X be an $n \times p$ matrix with rank s. Then X can be written as

$$X = ULV',$$

where U is $n \times s$, L is $s \times s$, V is $p \times s$, and

$$L \equiv Diag(\lambda_j).$$

The λ_js are the positive square roots of the positive eigenvalues (singular values) of $X'X$ and XX' (i.e., $\lambda_j^2 = \delta_j$). The columns of V are s orthonormal eigenvectors of $X'X$ corresponding to the positive eigenvalues with

$$X'XV = VL^2,$$

and the columns of U are s orthonormal eigenvectors of XX′ with

$$XX'U = UL^2.$$

Proof We can pick L and V as indicated in the theorem. We need to show that we can find U so that the theorem holds. (Orthonormal eigenvectors of XX' are not unique and not just any set of them will do!) If we take $U = XVL^{-1}$, then

$$XX'U = XX'XVL^{-1} = XVL^2L^{-1} = XVL = XVL^{-1}L^2 = UL^2;$$

so the columns of U are eigenvectors of XX' corresponding to the λ_i^2s. The columns of U are orthonormal because the columns of V are orthonormal and

$$U'U = L^{-1}V'X'XVL^{-1} = L^{-1}V'VL^2L^{-1} = L^{-1}I_sL = I_s.$$

Having found U we need to show that $X = ULV'$. Note that

$$U'XV = U'XVL^{-1}L = U'UL = L,$$

thus

$$UU'XVV' = ULV'.$$

From Appendix B.2 and Proposition B.51, $C(U) = C(X)$ and $C(V) = C(X')$; so by Theorem B.35, UU' is the ppo onto $C(X)$ and VV' is the ppo onto $C(X')$, hence $X = ULV'$. □

We can now derive the canonical form of a regression problem. Consider the standard linear model

$$Y = X\beta + e, \quad \mathrm{E}(e) = 0, \quad \mathrm{Cov}(e) = \sigma^2 I. \tag{1}$$

Write $X = ULV'$ as in Theorem 13.3.1 and write $U_* = [U, U_1]$, where the columns of U_* are an orthonormal basis for $\mathbf{R}^n$. Transform model (1) to

$$U_*'Y = U_*'X\beta + U_*'e. \tag{2}$$

Let $Y_* = U_*'Y$ and $e_* = U_*'e$. Then

$$\mathrm{E}(e_*) = 0, \quad \mathrm{Cov}(e_*) = \sigma^2 U_*'U_* = \sigma^2 I.$$

Using Theorem 13.3.1 again,

$$U_*'X\beta = U_*'ULV'\beta = \begin{bmatrix} U' \\ U_1' \end{bmatrix} ULV'\beta = \begin{bmatrix} L \\ 0 \end{bmatrix} V'\beta.$$

Reparameterizing by letting $\gamma \equiv V'\beta$ gives the canonical regression model

$$Y_* = \begin{bmatrix} L \\ 0 \end{bmatrix} \gamma + e_*, \quad \mathrm{E}(e_*) = 0, \quad \mathrm{Cov}(e_*) = \sigma^2 I. \tag{3}$$

Since this was obtained by a nonsingular transformation of model (1), it contains all of the information in model (1).

Estimation of parameters becomes trivial in this model:

$$\hat{\gamma} = (L^{-1}, 0)Y_* = L^{-1}U'Y,$$

$$\mathrm{Cov}(\hat{\gamma}) = \sigma^2 L^{-2},$$

$$RSS = Y_*' \begin{bmatrix} 0 & 0 \\ 0 & I_{n-p} \end{bmatrix} Y_*,$$

$$RMS = \sum_{i=s+1}^{n} y_{*i}^2 \Big/ (n-p).$$

In particular, the estimate of γ_j is $\hat{\gamma}_j = y_{*j}/\lambda_j$, and the variance of $\hat{\gamma}_j$ is σ^2/λ_j^2. The estimates $\hat{\gamma}_j$ and $\hat{\gamma}_k$ have zero covariance if $j \neq k$.

Models (1) and (3) are also equivalent to the (principal component) model

$$Y = UL\gamma + e, \tag{4}$$

so, writing $U = [u_1, \ldots, u_s]$ and $V = [v_1, \ldots, v_s]$, γ_j is the coefficient in the direction $\lambda_j u_j$. If λ_j is small, $\lambda_j u_j = Xv_j$ is ill-defined, and the variance of $\hat{\gamma}_j$, σ^2/λ_j^2, is large. Since the variance is large, it will be difficult to reject $H_0 : \gamma_j = 0$. If the data are consistent with $\gamma_j = 0$, life is great. We conclude that there is no evidence of an effect in the ill-defined direction $\lambda_j u_j$. If $H_0 : \gamma_j = 0$ is rejected, we have to weigh the evidence that the direction u_j is important in explaining Y against the evidence that the ill-defined direction u_j should not be included in $C(X)$.

Regression in canonical form can, of course, be applied to models (13.1.5) and (13.1.6), where the predictor variables have been standardized.

Exercise 13.3c Show the following.

(1) $Y'MY = \sum_{i=1}^{s} y_{*i}^2$.

(2) $Y'(I-M)Y = \sum_{i=s+1}^{n} y_{*i}^2$.

(3) $\hat{\beta}'\hat{\beta} = \sum_{i=1}^{s} y_{*i}^2/\lambda_i^2$.

(4) If $\lambda_1^2 \leq \cdots \leq \lambda_s^2$, then

$$\lambda_1^2 \frac{Y'(I-M)Y}{\lambda_1^2 + Y'MY} \leq \frac{Y'(I-M)Y}{1+\hat{\beta}'\hat{\beta}} \leq \lambda_s^2 \frac{Y'(I-M)Y}{\lambda_s^2 + Y'MY}.$$

13.3.2 Principal Component Regression

If we take as our original model a standardized version such as (13.1.5) or (13.1.6), the model matrix UL of model (4) has columns that are the principal components of the multivariate data set $(x_{i1}, \ldots, x_{i\,p-1})'$, $i = 1, \ldots, n$. See Johnson and Wichern (2007) or *ALM-III*, Chapter 14 for a discussion of principal components.

If the direction u_j is ill-defined, we may decide that the direction should not be used for estimation. Not using the direction u_j amounts to setting $\gamma_j = 0$ in model (4). If ill-defined directions are not to be used, and if ill-defined is taken to mean that $\lambda_j < \sqrt{\varepsilon}$ for some small value of ε, then we can take as our estimate of γ, $\tilde{\gamma} = (\tilde{\gamma}_1, \ldots, \tilde{\gamma}_p)'$, where

$$\tilde{\gamma}_j = \begin{cases} \hat{\gamma}_j, & \text{if } \lambda_j \geq \sqrt{\varepsilon} \\ 0, & \text{if } \lambda_j < \sqrt{\varepsilon}. \end{cases}$$

As an estimate of β in the original model (1) we can use $\tilde{\beta} = V\tilde{\gamma}$. This is reasonable in a regression model with $p = s$ because $V'\beta = \gamma$; so $V\gamma = VV'\beta = \beta$. This procedure for obtaining an estimate of β is referred to as *principal component regression (PCR)*, see also Subsection 13.4.1. Christensen (2015, Section 11.6) contains a numerical example.

Mohammad Hattab and Gabriel Huerta have brought to my attention that principal component regression is also a viable method for fitting models with $n < p$. In that case it is simpler to use the eigenvalues and vectors of XX' rather than $X'X$.

13.3.3 Generalized Inverse Regression

To deal with collinearity, Marquardt (1970) suggested using the estimate

$$\tilde{\beta} \equiv (X'X)_r^- X'Y,$$

where

$$(X'X)_r^- \equiv \sum_{j=p-r+1}^{p} v_j v_j'/\lambda_j^2,$$

and the λ_js are written so that $\lambda_1 \leq \lambda_2 \leq \cdots \leq \lambda_p$. $(X'X)_r^-$ would be the (Moore–Penrose) generalized inverse of $(X'X)$ if $r(X'X) = r$, i.e., if $0 = \lambda_1 = \cdots = \lambda_{p-r}$. Since $X = ULV'$,

$$\tilde{\beta} = (X'X)_r^- X'Y = \sum_{j=p-r+1}^{p} v_j v_j' VLU'Y/\lambda_j^2 = \sum_{j=p-r+1}^{p} v_j \hat{\gamma}_j.$$

This is essentially the same procedure as principal component regression. One method chooses r principal components and the other uses all the principal components that have sufficiently large eigenvalues. Marquardt originally suggested this as an alternative to classical ridge regression, which is the subject of the next section.

13.4 Classical Ridge Regression

Ridge regression was originally proposed by Hoerl and Kennard (1970) as a method to deal with collinearity. Now it is more commonly viewed as a form of penalized likelihood estimation, which makes it a form of Bayesian estimation. In this section, we consider the traditional view of ridge regression. *ALM-III*, Chapter 2 relates ridge regression to the more general issue of penalized estimation.

Hoerl and Kennard (1970) looked at the mean squared error, $\mathrm{E}[(\hat{\beta}-\beta)'(\hat{\beta}-\beta)]$, for estimating β with least squares. This is the expected value of a quadratic form in $(\hat{\beta}-\beta)$. $\mathrm{E}(\hat{\beta}-\beta)=0$ and $\mathrm{Cov}(\hat{\beta}-\beta)=\sigma^2(X'X)^{-1}$; so by Theorem 1.3.2

$$\mathrm{E}[(\hat{\beta}-\beta)'(\hat{\beta}-\beta)] = \mathrm{tr}[\sigma^2(X'X)^{-1}].$$

If $\lambda_1^2,\ldots,\lambda_p^2$ are the eigenvalues of $(X'X)$, we see that $\mathrm{tr}[(X'X)^{-1}]=\sum_{j=1}^{p}\lambda_j^{-2}$; so

$$\mathrm{E}[(\hat{\beta}-\beta)'(\hat{\beta}-\beta)] = \sigma^2\sum_{j=1}^{p}\lambda_j^{-2}.$$

If some of the values λ_j^2 are small, the mean squared error will be large.

Hoerl and Kennard suggested using the estimate

$$\tilde{\beta} \equiv (X'X+kI)^{-1}X'Y, \tag{1}$$

where k is some fixed scalar. The choice of k will be discussed briefly later. The consequences of using this estimate are easily studied in the canonical regression model. The canonical regression model (13.2.3) is

$$Y_* = \begin{bmatrix} L \\ 0 \end{bmatrix}\gamma + e_*.$$

The ridge regression estimate is

$$\tilde{\gamma} = (L'L+kI)^{-1}[L',0]Y_* = (L^2+kI)^{-1}L^2\hat{\gamma}. \tag{2}$$

In particular,

$$\tilde{\gamma}_j = \frac{\lambda_j^2}{\lambda_j^2+k}\hat{\gamma}_j.$$

If λ_j is small, $\tilde{\gamma}_j$ will be shrunk toward zero. If λ_j is large, $\tilde{\gamma}_j$ will change relatively little from $\hat{\gamma}_j$.

Exercise 13.4 illustrates the relationship between ridge regression performed on the canonical model and ridge regression performed on the usual model. The transformation matrix V is defined as in Section 3.

Exercise 13.4 Use equations (1) and (2) to show that
(a) $\tilde{\beta} = V\tilde{\gamma}$,
(b) $\mathrm{E}[(\tilde{\beta}-\beta)'(\tilde{\beta}-\beta)] = \mathrm{E}[(\tilde{\gamma}-\gamma)'(\tilde{\gamma}-\gamma)]$.

The estimate $\tilde{\beta}$ has expected mean square

$$\mathrm{E}[(\tilde{\beta}-\beta)'(\tilde{\beta}-\beta)] = \sigma^2\mathrm{tr}[(X'X+kI)^{-1}X'X(X'X+kI)^{-1}] \\ +\beta'\left\{(X'X+kI)^{-1}X'X-I\right\}'\left\{(X'X+kI)^{-1}X'X-I\right\}\beta.$$

Writing $X'X = VL^2V'$, $I = VV'$, and in the second term $I = (X'X+kI)^{-1}(X'X+kI)$, so that $(X'X+kI)^{-1}X'X - I = -(X'X+kI)^{-1}kI$, this can be simplified to

$$\mathrm{E}[(\tilde{\beta}-\beta)'(\tilde{\beta}-\beta)] = \sigma^2\sum_{j=1}^{p}\lambda_j^2/(\lambda_j^2+k)^2 + k^2\beta'(X'X+kI)^{-2}\beta.$$

The derivative of this with respect to k at $k=0$ can be shown to be negative. Since $k=0$ is least squares estimation, in terms of mean squared error there exists $k>0$ that gives better estimates of β than least squares. Unfortunately, the particular values of such k are not known.

Frequently, a *ridge trace* is used to determine k. A ridge trace is a simultaneous plot of the estimated regression coefficients (which are functions of k) against k. The value of k is chosen so that the regression coefficients change little for any larger values of k. In addition to providing a good overall review of ridge regression, Draper and van Nostrand (1979) provide references to criticisms that have been raised against the ridge trace.

Because the mean squared error, $\mathrm{E}[(\tilde{\beta}-\beta)'(\tilde{\beta}-\beta)]$, puts equal weight on each regression coefficient, it is often suggested that ridge regression be used only on the rescaled model (13.1.6).

The ridge regression technique admits obvious generalizations. One is to use $\tilde{\beta} = (X'X+K)^{-1}X'Y$, where $K = \mathrm{Diag}(k_j)$. The ridge estimates for canonical regression become

$$\tilde{\gamma}_j = \frac{\lambda_j^2}{\lambda_j^2+k_j}\hat{\gamma}_j.$$

The ridge regression estimate (1) can also be arrived at from a Bayesian argument. With $Y = X\beta + e$ and $e \sim N(0,\sigma^2 I)$, incorporating prior information of the form $\beta|\sigma^2 \sim N[0,(\sigma^2/k)I]$ leads to fitting a version of (2.10.3) that has

Table 13.1 Coleman report data

Obs.	x_1	x_2	x_3	x_4	x_5	y
1	3.83	28.87	7.20	26.60	6.19	37.01
2	2.89	20.10	−11.71	24.40	5.17	26.51
3	2.86	69.05	12.32	25.70	7.04	36.51
4	2.92	65.40	14.28	25.70	7.10	40.70
5	3.06	29.59	6.31	25.40	6.15	37.10
6	2.07	44.82	6.16	21.60	6.41	33.90
7	2.52	77.37	12.70	24.90	6.86	41.80
8	2.45	24.67	−0.17	25.01	5.78	33.40
9	3.13	65.01	9.85	26.60	6.51	41.01
10	2.44	9.99	−0.05	28.01	5.57	37.20
11	2.09	12.20	−12.86	23.51	5.62	23.30
12	2.52	22.55	0.92	23.60	5.34	35.20
13	2.22	14.30	4.77	24.51	5.80	34.90
14	2.67	31.79	−0.96	25.80	6.19	33.10
15	2.71	11.60	−16.04	25.20	5.62	22.70
16	3.14	68.47	10.62	25.01	6.94	39.70
17	3.54	42.64	2.66	25.01	6.33	31.80
18	2.52	16.70	−10.99	24.80	6.01	31.70
19	2.68	86.27	15.03	25.51	7.51	43.10
20	2.37	76.73	12.77	24.51	6.96	41.01

$$\begin{bmatrix} Y \\ 0 \end{bmatrix} = \begin{bmatrix} X \\ I \end{bmatrix} \beta + \begin{bmatrix} e \\ \tilde{e} \end{bmatrix}, \quad \begin{bmatrix} e \\ \tilde{e} \end{bmatrix} \sim N\left(\begin{bmatrix} 0_{n\times 1} \\ 0_{p\times 1} \end{bmatrix}, \sigma^2 \begin{bmatrix} I_n & 0 \\ 0 & (1/k)I_p \end{bmatrix} \right).$$

It is easily seen that the generalized least squares estimate of β associated with this model is (1). When k is near 0, the prior variance is large, so the prior information is very weak and the posterior mean is very close to the least squares estimate.

Exercise 13.5 Mosteller and Tukey (1977) reported the data in Table 13.1 on verbal test scores for sixth graders. They used a sample of 20 Mid-Atlantic and New England schools taken from *The Coleman Report*. The dependent variable y was the mean verbal test score for each school. The predictor variables were: x_1 = staff salaries per pupil, x_2 = percent of sixth grader's fathers employed in white collar jobs, x_3 = a composite score measuring socioeconomic status, x_4 = the mean score on a verbal test administered to teachers, and x_5 = one-half of the sixth grader's mothers' mean number of years of schooling. Evaluate the data for collinearity problems and if necessary apply an appropriate procedure.

Exercise 13.6 It could be argued that the canonical model should be standardized before applying ridge regression. Define an appropriate standardization so that under the standardization

$$\tilde{\gamma}_j = \frac{1}{1+k_j}\hat{\gamma}_j.$$

13.4.1 Ridge Applied to Principal Components

The standard regression model is

$$Y = X\beta + e.$$

Using the notation of the singular value decomposition, (ignoring complications due to centering and scaling) the principal component (PC) regression model is

$$Y = \tilde{X}\gamma + e, \qquad \tilde{X} \equiv XV = UL.$$

These are equivalent models because $C(X) = C(\tilde{X})$. In particular

$$\begin{aligned} \mathrm{E}(Y) = \tilde{X}\gamma &= UL\gamma = ULV'V\gamma \\ &= XV\gamma = X\beta. \end{aligned}$$

As a result we can transform *any* estimate of γ into an estimate of β through

$$\beta = V\gamma.$$

(If the model is not a regression, life is not so simple because β can be any vector with $\beta = V\gamma + w$ for $w \perp C(X')$.) Traditional PCR involves dropping ill-defined directions from $\tilde{X}$ but here we apply ridge regression to all of the principal components.

First we want to examine least squares and ridge regression estimates on the PCR model because the behavior of the ridge estimates is very clear there. The least squares estimates are

$$\hat{\gamma} \equiv (\tilde{X}'\tilde{X})^{-1}\tilde{X}'Y = L^{-1}U'Y. \tag{3}$$

The ridge regression estimates turn out to be

$$\hat{\gamma}_R \equiv (\tilde{X}'\tilde{X} + kI)^{-1}\tilde{X}'Y = D\left(\frac{\lambda_i^2}{\lambda_i^2 + k}\right)\hat{\gamma}. \tag{4}$$

In *ALM-III* we discuss something called the kernel trick which amounts to fitting $XX'\eta$ in place of $X\beta$. Since Proposition B.51 establishes $C(XX') = C(X)$, the kernel-trick model is equivalent to the original model. The point of the kernel trick is that in some situations it is easier to specify XX' than to specify X. In the kernel-trick model, any estimate of η can be transformed into an estimate of β via $\beta = X'\eta$. Note that if $\hat{\eta}$ is any least squares estimate, $\hat{\beta} \equiv X'\hat{\eta}$ will be the unique least squares estimate in $C(X')$, cf. Exercise 2.11.8. In applications the kernel trick is typically used with some form of biased estimation, so we also investigate ridge regression applied to the kernel-trick model.

We also want to apply the kernel trick to the PC model, thus we fit $\tilde{X}\tilde{X}'\delta$ in place of $\tilde{X}\gamma$. Moreover we want find the ridge estimate for the kernel-trick PC model. Applying ridge regression to the kernel-trick PC model gives

$$\hat{\delta}_{KR} \equiv (\tilde{X}\tilde{X}'\tilde{X}\tilde{X}' + kI)^{-1}\tilde{X}\tilde{X}'Y$$

and transforming the estimate back to the PC model eventually gives

$$\hat{\gamma}_{KR} \equiv \tilde{X}'\hat{\delta}_{KR} = \tilde{X}'(\tilde{X}\tilde{X}'\tilde{X}\tilde{X}' + kI)^{-1}\tilde{X}\tilde{X}'Y = D\left(\frac{\lambda_i^4}{\lambda_i^4 + k}\right)\hat{\gamma}. \tag{5}$$

That the two ridge PC estimates are both diagonal transformations of the least squares PC estimates elucidates their relationship. The notation for the ridge estimates suppresses their dependence on the tuning parameter k but it is now clear from (4) and (5) that if you pick any tuning parameter, say k_R, to use in the ridge estimate $\hat{\gamma}_R$, there exists a (different) tuning parameter k_{KR} that gives $\hat{\gamma}_R = \hat{\gamma}_{KR}$, and vice versa.

It is not obvious but the ridge estimates on the PC model transform directly to the ridge estimates on the original model. In particular, the least squares estimates for the original model are

$$\hat{\beta} \equiv (X'X)^{-1}X'Y = V\hat{\gamma}. \tag{6}$$

The ridge estimates are

$$\hat{\beta}_R \equiv (X'X + kI)^{-1}X'Y = V\hat{\gamma}_R. \tag{7}$$

Even the kernel-trick ridge estimates are a direct transformation of the kernel-trick PC ridge estimates,

$$\hat{\beta}_{KR} \equiv X'(XX'XX' + kI)^{-1}XX'Y = V\hat{\gamma}_{KR}. \tag{8}$$

Exercise 13.7 Prove equations (3) through (8). Hint: All but (5) and (8) follow directly from the characterizations $X = ULV'$ and $\tilde{X} = UL$. Those two involve applying Proposition B.56 to find, say, the inverse of $[kI + X(X'X)X']$ prior to using the characterizations.

13.5 More on Mean Squared Error

In Section 2.9 we showed that the biased estimates obtained from reduced models can provide better estimates than the least squares estimates obtained from the full model. We now explore more general results for the canonical regression model.

For the canonical regression model, Goldstein and Smith (1974) have shown that for $0 \le h_j \le 1$, if $\tilde{\gamma}_j$ is a general *shrinkage estimator* defined by

$$\tilde{\gamma}_j = h_j \hat{\gamma}_j$$

and if

$$\frac{\gamma_j^2}{\mathrm{Var}(\hat{\gamma}_j)} < \frac{1+h_j}{1-h_j}, \tag{1}$$

then $\tilde{\gamma}_j$ is a better estimate than $\hat{\gamma}_j$ in that

$$\mathrm{E}(\tilde{\gamma}_j - \gamma_j)^2 \le \mathrm{E}(\hat{\gamma}_j - \gamma_j)^2.$$

In particular, if

$$\gamma_j^2 < \mathrm{Var}(\hat{\gamma}_j) = \sigma^2/\lambda_j^2, \tag{2}$$

then $\tilde{\gamma}_j = 0$ is a better estimate than $\hat{\gamma}_j$. Estimating σ^2 with *RMS* and γ_j with $\hat{\gamma}_j$ leads to taking $\tilde{\gamma}_j = 0$ if the absolute t statistic for testing $H_0 : \gamma_j = 0$ is less than 1. This is only an approximation to condition (2), so taking $\tilde{\gamma}_j = 0$ for larger values of the t statistic may well be justified.

For generalized ridge regression, $h_j = \lambda_j^2/(\lambda_j^2 + k_j)$, and condition (1) becomes

$$\gamma_j^2 < \frac{\sigma^2}{\lambda_j^2}\frac{2\lambda_j^2 + k_j}{k_j} = \sigma^2\left[\frac{2}{k_j} + \frac{1}{\lambda_j^2}\right].$$

If λ_j^2 is small, almost any value of k_j will give an improvement over least squares. If λ_j^2 is large, only very small values of k_j will give an improvement.

Note that, since the $\hat{\gamma}_j$s are unbiased, the $\tilde{\gamma}_j$s will, in general, be biased estimates.

13.6 Robust Estimation and Alternative Distance Measures

Robust estimates of β have less sensitivity to outlying y_i values than least squares estimates, see, for example, Huber and Ronchetti (2009). Robust estimates work better when the distribution of the y_is has fatter tails than the normal distribution, e.g. Laplace, logistic, $t(df)$. Optimal estimates for such distributions tend to be nonlinear. In standard linear models least squares estimates are BLUEs, so they should be reasonable, if not optimal, for most errors that are i.i.d.. The robust estimates discussed here are still sensitive to high leverage x_i vectors.

Write the $n \times p$ model matrix as

$$X = \begin{bmatrix} x_1' \\ \vdots \\ x_n' \end{bmatrix}.$$

As in Section 2.2, least squares estimates minimize

$$\|Y - X\beta\|^2 = \sum_{i=1}^{n} (y_i - x_i'\beta)^2. \tag{1}$$

This is a geometric estimation criterion but we showed that for a standard linear model with $\mathrm{Cov}(Y) = \sigma^2 I$ the least squares estimates are BLUEs and if Y also has a multivariate normal distribution the least squares estimates have other optimal properties. Because the squared distance in (1) is always nonnegative, it is equivalent (but less convenient) to minimize

$$\|Y - X\beta\| = \sqrt{\sum_{i=1}^{n} (y_i - x_i'\beta)^2}.$$

In Section 2.7 we discussed generalized least squares estimates that minimized, for some positive definite matrix W,

$$\|Y - X\beta\|_W^2 \equiv (Y - X\beta)'W(Y - X\beta).$$

In particular we showed that if $\mathrm{Cov}(Y) = \sigma^2 V$, the choice $W = V^{-1}$ leads to BLUEs and if Y also has a multivariate normal distribution these generalized least squares estimates have other optimal properties. With W positive definite, it is equivalent to minimize $\|Y - X\beta\|_W$.

The simplest form of generalized least squares estimation is weighted least squares estimation in which W is a positive definite diagonal matrix, i.e., $W = D(w)$ for a vector $w = (w_1, \ldots, w_n)'$ with $w_i > 0$. In this case

$$\|Y - X\beta\|_{D(w)}^2 = (Y - X\beta)'D(w)(Y - X\beta) = \sum_{i=1}^{n} w_i(y_i - x_i'\beta)^2. \tag{2}$$

When $\mathrm{Cov}(Y) = \sigma^2 D(v_i)$, the optimal weights in (2) are $w_i = 1/v_i$.

For an arbitrary n vector v, the measures $\|v\|$ and $\|v\|_W$ provide alternative definitions for the length of a vector. A wide variety of estimates for β can be obtained by defining yet other concepts of the length of a vector. One of the most common concepts of length used in mathematics is $\mathbf{L}^p$ length in which, for $p \geq 1$,

$$\|v\|_p \equiv \left[\sum_{i=1}^{n} |v_i|^p\right]^{1/p}.$$

There is also

$$\|v\|_\infty \equiv \max_i\{|v_1|, \ldots, |v_n|\}.$$

In this book, we have used the notation

$$\|\cdot\| \equiv \|\cdot\|_2.$$

A minimum $\mathbf{L}^p$ estimate of β minimizes the distance $\|Y - X\beta\|_p$ or, equivalently, minimizes $\left(\|Y - X\beta\|_p\right)^p$. Not that I have ever seen anyone do it, but one could even estimate β by minimizing $\|Y - X\beta\|_\infty$. When $1 \leq p < 2$, minimum $\mathbf{L}^p$ estimates are robust to unusual y_i values. Taking $p > 2$ makes the estimates *more* sensitive to unusual y_i values, something statisticians rarely want. (Hence my never seeing anyone use minimum $\mathbf{L}^\infty$ estimation.) But recall that when estimating the mean of a thin tailed distribution, like the uniform, it is the most extreme observations that provide the most information.

Minimum $\mathbf{L}^p$ estimation provides an immediate analogy to finding weighted least squares estimates: just minimize $\sum_{i=1}^n w_i|y_i - x_i'\beta|^p$ for positive w_is. Unfortunately, for $p \neq 2$, $\mathbf{L}^p$ distances do not readily generalize to incorporate relationships between observations as does generalized least squares estimation. When not using a quadratic form to define length, it is not clear how to include an entire positive definite matrix that totally redefines the concept of distance. (As opposed to weighted versions that merely rescale individual variables.) One thing you could do is apply minimum $\mathbf{L}^p$ estimation to model (2.7.2).

In the search for good robust estimates, people have gone well past the use of minimum weighted $\mathbf{L}^p$ estimation with $1 \leq p < 2$. *M-estimates* involve choosing a nonnegative loss function $\mathscr{L}(y, u)$ and weights and then picking $\tilde{\beta}$ to minimize the weighted sum of the losses, i.e.,

$$\sum_{i=1}^{n} w_i \mathscr{L}(y_i, x_i'\tilde{\beta}) = \min_\beta \sum_{i=1}^{n} w_i \mathscr{L}(y_i, x_i'\beta). \tag{3}$$

Whether this gives robust estimation or not depends on the choice of loss function.

The M in M-estimation is an allusion to maximum likelihood type estimates. The loss function $\mathscr{L}(y, u) = (y - u)^2$, with equal weights, leads to least squares and thus to maximum likelihood estimates for standard linear models with multivariate normal data. In generalized linear models for binomial data, wherein y_i denotes the proportion of successes, maximum likelihood estimates of β can typically be cast as minimizing a weighted sum of losses where the weights equal the binomial sample sizes, cf. *ALM-III*, Section 13.1. Even Support Vector Machines can be cast as estimating β in $x'\beta$ by minimizing a sum of losses, cf. *ALM-III*, Section 13.5.

In linear models, loss functions typically take the form

$$\mathscr{L}(y,u) = \mathscr{L}(y-u).$$

If $\mathscr{L}(\xi)$ is differentiable everywhere, Newton–Raphson can be used to find the minimizing value. One of the more famous families of robust loss functions is *Tukey's biweight* which is typically defined by its derivative:

$$\mathbf{d}_\xi \mathscr{L}_c(\xi) \equiv \begin{cases} \xi\left(1-\frac{\xi^2}{c^2}\right) & \text{if } |\xi| < c \\ 0 & \text{if } |\xi| \geq c \end{cases}$$

for some scale factor c.

Exercise 13.8 Find the derivative of the loss function for least squares and minimum $\mathbf{L}^1$ estimation. Graph them along with the $c = 1$ biweight derivative.

13.7 Orthogonal Regression

Suppose we have bivariate data (x_i, y_i) and want to fit a line $\hat{y} = \hat{\beta}_0 + \hat{\beta}_1 x$. Rather than using least squares (that minimizes vertical distances to the line) we will do orthogonal regression that minimizes perpendicular distances to the line. We will run this line through the point $(\bar{x}_\cdot, \bar{y}_\cdot)$ so we need only worry about the slope of the line.

Transform $(x_i, y_i)'$ into $v_i \equiv (x_i - \bar{x}_\cdot, y_i - \bar{y}_\cdot)'$. We want to find a vector a, or more properly a one-dimensional column space $C(a)$, that minimizes the squared perpendicular distances between the data v_i and the regression line that consists of all multiples of the vector a. We need to take v_i and project it onto the line, that is, project it into $C(a)$. The projection is $M_a v_i$. The squared perpendicular distance between the data and the line is $\|v_i - M_a v_i\|^2 = \|(I - M_a)v_i\|^2 = v_i'(I - M_a)v_i$. It will become important later to recognize this as the squared length of the perpendicular projection of v_i onto the orthogonal complement of the regression surface vector space. In any case, we need to pick the line so as to minimize the sum of all these squared distances, i.e., so that $\sum_{i=1}^n v_i'(I - M_a)v_i$ is minimized. However, if a minimizes $\sum_{i=1}^n v_i'(I - M_a)v_i$, it maximizes $\sum_{i=1}^n v_i' M_a v_i$. Note also that

$$\begin{aligned}
\max_a \sum_{i=1}^{n} v_i' M_a v_i &= \max_a \sum_{i=1}^{n} v_i' a(a'a)^{-1} a' v_i \\
&= \max_a \frac{1}{a'a} \sum_{i=1}^{n} a' v_i v_i' a \\
&= \max_a \frac{1}{a'a} a' \left(\sum_{i=1}^{n} v_i v_i' \right) a \\
&= \max_a \frac{n-1}{a'a} a' S a,
\end{aligned}$$

where S is the sample covariance matrix of the data.

It is enough to find $\hat{a}$ such that

$$\frac{\hat{a}' S \hat{a}}{\hat{a}'\hat{a}} = \max_a \frac{a' S a}{a'a}.$$

It is well-known (*ALM-III*, Proposition 12.7.4) that this max is achieved by eigenvectors associated with the largest eigenvalue of S. In particular, if we pick a maximizing eigenvector $\hat{a}$ to be $\hat{a}' = (1, \hat{\beta})$, then $\hat{\beta}$ is the orthogonal regression slope estimate. The estimated line becomes

$$\hat{y} = \bar{y}_\cdot + \hat{\beta}(x - \bar{x}_\cdot).$$

Technically, this occurs because for the fitted values to fall on the regression line, they must themselves determine a maximizing eigenvector, i.e., $\hat{y}$ is *defined* so that

$$\begin{bmatrix} x - \bar{x}_\cdot \\ \hat{y} - \bar{y}_\cdot \end{bmatrix} \equiv (x - \bar{x}_\cdot) \begin{bmatrix} 1 \\ \hat{\beta} \end{bmatrix}.$$

Normally, we would find the eigenvector corresponding to the largest eigenvalue computationally, but our problem can be solved analytically. To find the maximizing eigenvector we need to solve the matrix equation

$$\begin{bmatrix} s_x^2 - \lambda & s_{xy} \\ s_{xy} & s_y^2 - \lambda \end{bmatrix} \begin{bmatrix} 1 \\ \hat{\beta} \end{bmatrix} = \begin{bmatrix} 0 \\ 0 \end{bmatrix}. \tag{1}$$

This simplifies to the set of equations

$$\lambda = s_x^2 + s_{xy}\hat{\beta} \tag{2}$$

and

$$s_{xy} + (s_y^2 - \lambda)\hat{\beta} = 0. \tag{3}$$

Substituting λ from (2) into (3),

$$s_{xy} + (s_y^2 - s_x^2)\hat{\beta} - s_{xy}\hat{\beta}^2 = 0. \tag{4}$$

Applying the quadratic formula gives

$$\hat{\beta} = \frac{-(s_y^2 - s_x^2) \pm \sqrt{(s_y^2 - s_x^2)^2 + 4s_{xy}^2}}{-2s_{xy}} = \frac{(s_y^2 - s_x^2) \pm \sqrt{(s_y^2 - s_x^2)^2 + 4s_{xy}^2}}{2s_{xy}}.$$

Substituting $\hat{\beta}$ back into (2), the larger of the two values of λ corresponds to

$$\hat{\beta} = \frac{(s_y^2 - s_x^2) + \sqrt{(s_y^2 - s_x^2)^2 + 4s_{xy}^2}}{2s_{xy}},$$

which is our slope estimate.

If you find the eigenvalues and eigenvectors computationally, remember that eigenvectors for a given eigenvalue (essentially) form a vector space. (Eigenvectors are not allowed to be 0.) Thus, a reported eigenvector $(a_1, a_2)'$ for the largest eigenvalue also determines the eigenvector we want, $(1, a_2/a_1)'$.

It turns out that we can also get least squares estimates by modifying the matrix equations (1). Consider

$$\begin{bmatrix} s_x^2 + [s_y^2 - s_{xy}^2/s_x^2] - \lambda & s_{xy} \\ s_{xy} & s_y^2 - \lambda \end{bmatrix} \begin{bmatrix} 1 \\ \hat{\beta} \end{bmatrix} = \begin{bmatrix} 0 \\ 0 \end{bmatrix}$$

or equivalently,

$$\begin{bmatrix} s_x^2 + [s_y^2 - s_{xy}^2/s_x^2] & s_{xy} \\ s_{xy} & s_y^2 \end{bmatrix} \begin{bmatrix} 1 \\ \hat{\beta} \end{bmatrix} = \lambda \begin{bmatrix} 1 \\ \hat{\beta} \end{bmatrix}.$$

Rather than solving this for $\hat{\beta}$, simply observe that one solution is $\lambda = s_x^2 + s_y^2$ and $(1, \hat{\beta})' = (1, s_{xy}/s_x^2)'$. Note that $[s_y^2 - s_{xy}^2/s_x^2] = [(n-2)/(n-1)]MSE$ where MSE is the mean squared error from the least squares fit of $y_i = \beta_0 + \beta_1 x_i + \varepsilon_i$.

To generalize the orthogonal regression procedure to multiple regression we need a more oblique approach. With data (x_i', y_i) consisting of p-dimensional vectors, a linear regression surface corresponds to a $(p-1)$-dimensional hyperplane so that if we specify a $(p-1)$ vector of predictor variables x, we know the fitted value $\hat{y}$ because it is the point corresponding to x on the hyperplane. By considering data $v_i' \equiv (x_i' - \bar{x}_\cdot', y_i - \bar{y}_\cdot)$, the regression surface goes through the origin and becomes a $(p-1)$-dimensional vector space. Rather than specify the $(p-1)$-dimensional space, it is easier to find the orthogonal complement which is one-dimensional. Writing the one-dimensional space as $C(a)$ for a p vector a, the regression surface will be $C(a)^\perp$. The squared distances from the v_is to the $(p-1)$-dimensional regression space are now $v_i' M_a v_i$, so we want to minimize $\sum_{i=1}^n v_i' M_a v_i$. Similar to our earlier argument,

$$\min_a \sum_{i=1}^{n} v_i' M_a v_i = \min_a \frac{n-1}{a'a} a'Sa,$$

with S being the sample covariance matrix of the complete data. However, now we obtain our fitted values differently. The minimum is achieved by choosing the eigenvector $\hat{a} = (\hat{\beta}', -1)'$ corresponding to the smallest eigenvalue. Our fitted values $\hat{y}$ now must determine vectors that are orthogonal to this eigenvector, so they satisfy

$$[\hat{\beta}' \; -1] \begin{bmatrix} x - \bar{x}. \\ \hat{y} - \bar{y}. \end{bmatrix} = 0$$

or

$$\hat{y} = \bar{y}. + \hat{\beta}'(x - \bar{x}.).$$

As illustrated earlier for $p = 2$, any eigenvector (for the smallest eigenvalue) reported by a computer program is easily rescaled to $(\hat{\beta}', -1)'$

Finally, this approach better give the same answers for simple linear regression that we got from our first procedure. It is not difficult to see that

$$\begin{bmatrix} s_x^2 - \lambda & s_{xy} \\ s_{xy} & s_y^2 - \lambda \end{bmatrix} \begin{bmatrix} \hat{\beta} \\ -1 \end{bmatrix} = \begin{bmatrix} 0 \\ 0 \end{bmatrix}$$

once again leads to equation (4) but now

$$\hat{\beta} = \frac{(s_y^2 - s_x^2) + \sqrt{(s_y^2 - s_x^2)^2 + 4s_{xy}^2}}{2s_{xy}}$$

corresponds to the smallest eigenvalue. If you think about it, with distinct eigenvalues, the eigenvector $(1, \hat{\beta})'$ corresponding to the largest eigenvalue must be orthogonal to any eigenvector for the smallest eigenvalue, and $(\hat{\beta}, -1)'$ is orthogonal to $(1, \hat{\beta})'$.

Like least squares, this procedure is a geometric justification for an estimate, not a statistical justification. In Chapter 2 we showed that least squares estimates have statistical optimality properties like being BLUEs, MVUEs, and MLEs. The ideas used here are similar to those needed for looking at the separating hyperplanes often used to motivate support vector machines, see Moguerza and Muñoz (2006) or Zhu (2008).

References

Belsley, D. A. (1984). Demeaning conditioning diagnostics through centering (with discussion). *The American Statistician, 38*, 73–77.

Belsley, D. A. (1991). *Collinearity diagnostics: Collinearity and weak data in regression*. New York: Wiley.

Belsley, D. A., Kuh, E., & Welsch, R. E. (1980). *Regression diagnostics: Identifying influential data and sources of collinearity*. New York: Wiley.

Christensen, R. (2015). *Analysis of variance, design, and regression: Linear modeling for unbalanced data* (2nd ed.). Boca Raton, FL: Chapman and Hall/CRC Press.

Christensen, R. (2018). Comment on "A note on collinearity diagnostics and centering" by Velilla (2018). *The American Statistician, 72*, 114–117.

Draper, N. R., & van Nostrand, R. C. (1979). Ridge regression and James-Stein estimation: Review and comments. *Technometrics, 21*, 451–466.

Goldstein, M., & Smith, A. F. M. (1974). Ridge-type estimators for regression analysis. *Journal of the Royal Statistical Society, Series B, 26*, 284–291.

Hoerl, A. E., & Kennard, R. (1970). Ridge regression: Biased estimation for non-orthogonal problems. *Technometrics, 12*, 55–67.

Huber, P. J., & Ronchetti, E. M. (2009). *Robust statistics* (2nd ed.). New York: Wiley.

Johnson, R. A., & Wichern, D. W. (2007). *Applied multivariate statistical analysis* (6th ed.). Englewood Cliffs, NJ: Prentice-Hall.

Marquardt, D. W. (1970). Generalized inverses, ridge regression, biased linear estimation, and nonlinear estimation. *Technometrics, 12*, 591–612.

Moguerza, J. M., & Muñoz, A. (2006). Support vector machines with applications. *Statistical Science, 21*, 322–336.

Mosteller, F., & Tukey, J. W. (1977). *Data analysis and regression*. Reading, MA: Addison-Wesley.

Velilla, S. (2018). A note on collinearity diagnostics and centering. *The American Statistician, 72*, 140–146.

Zhu, M. (2008). Kernels and ensembles: Perspectives on statistical learning. *The American Statistician, 62*, 97–109.

Chapter 14
Variable Selection

Abstract This chapter addresses the question of which predictor variables should be included in a linear model. The easiest version of the problem is, given a linear model, which variables should be excluded. To that end we examine the question of selecting the best subset of predictor variables from amongst the original variables. To do this requires us to define a "best" model and we examine several competing measures. We also examine the "greedy" algorithm for this problem known as backward elimination. The more difficult problem of deciding which variables to place into a linear model is addressed by the greedy algorithm of forward selection. (These algorithms are greedy in the sense of always wanting the best thing right now, rather than seeking a global sense of what is best.) We examine traditional forward selection as well as the modern adaptations of forward selection known as boosting, bagging, and random forests.

Suppose we have a set of variables $y, x_1, \ldots, x_s$ and observations on these variables $y_i, x_{i1}, \ldots, x_{is}$, $i = 1, \ldots, n$. We want to identify which of the predictor variables x_j are important for a regression on y. There are several methods available for doing this. Recall from Section 2.9 that models with fewer predictors, sometimes even when the models are incorrect, can provide better estimates than a full model.

Tests for the adequacy of various reduced models can be performed, assuming that the full model

$$y_i = \beta_0 + \beta_1 x_{i1} + \cdots + \beta_s x_{is} + e_i \tag{1}$$

is an adequate model for the data. This largest model will be written $Y = X\beta + e$. A candidate (reduced) model with $p < s$ predictor variables will be written

$$y_i = \gamma_0 + \gamma_1 x_{i1} + \cdots + \gamma_p x_{ip} + e_i, \tag{2}$$

or as $Y = X_0\gamma + e$.

Obviously, the most complete method of evaluating candidate models is to look at all possible regression equations involving $x_1, \ldots, x_s$. There are 2^s of these. Even

R. Christensen, *Plane Answers to Complex Questions*, Springer Texts in Statistics,
https://doi.org/10.1007/978-3-030-32097-3_14

if one has the time and money to compute all of them, it may be difficult to assimilate that much information.

A more efficient procedure than computing all possible regressions is to choose a criterion for ranking how well different candidate models fit and to compute only the best fitting models. Typically, one would want to *identify several of the best fitting models and investigate them further*. The computing effort for this *best subset regression* method is still considerable. Most computer programs put limits on how large s can be. For example, the R package `leaps` and Minitab 18 both require $s \leq 31$. Section 1 discusses criteria for choosing the best models.

An older group of methods is stepwise regression. These methods consider the efficacy of adding or deleting individual variables to a model that is currently under consideration. These methods have the flaw of considering variables only one at a time. For example, there is no reason to believe that the best two variables to add to a model are the one variable that adds most to the model followed by the one variable that adds the most to this augmented model. The flaw of stepwise procedures is also their virtue. Because computations go one variable at a time, they are relatively easy. Stepwise methods are considered in Section 2. Section 3 discusses the best subset and traditional stepwise approaches to variable selection.

Problems with s large relative to n seem to have become more important with the rise of computing power. In the early '90s I was involved in discussions of data on ceramic superconductors. Collecting the data involved destroying the object and each superconductor cost $20,000. Naturally n was small and naturally they measured everything they could on every superconductor, so s was larger than n. Even in situations where the cost per unit is not large, if it easy to measure large numbers of variables on each unit, people do so.

Cook et al. (2013, 2015) and Tarpey et al. (2015) are a few of many works that discuss asymptotics when the number of predictors s increases as a fixed percentage of the sample size, but even those problems are much more tractable than having s as large or larger than n. If s is close to n, fitting the full model (1) may be unrealistic or unwise. It seems to me that a crucial factor is having an adequate number of degrees of freedom for error. If $s + 1 = n$, a regression model will be saturated, i.e. $dfE = 0$ and $\hat{Y} = Y$. When $s + 1 > n$, the model can no longer be a regression model because $(X'X)^{-1}$ will not exist.

I am not aware of any truly good general methods for handling data with really large numbers of predictor variables. When the predictor variables are actual measurements, I suspect that the best way to handle $s > n$ is to model the covariance structure of the predictor variables with fewer parameters than there are data points and then use best linear prediction theory as the basis for analysis. That idea seems less appropriate when the predictors in the model include many nonlinear functions of some original measured predictor variables, as occurs in many nonparametric regression models, cf. *ALM-III*, Chapter 1.

Despite its obvious faults, forward selection seems to be the best available general procedure for dealing with large s. (Although I am intrigued by the potential of PCR.) Section 4 discusses some modern ideas on how to improve forward selection when s is large.

Another popular method of variable selection is to employ the *lasso (least absolute shrinkage and selection operator)*. The lasso was discussed briefly in *PA-IV* and is discussed more extensively in *ALM-III*, Chapter 2 but it is not discussed here.

14.1 All Possible Regressions and Best Subset Regression

There is little to say about the all possible regressions technique. The efficient computation of all possible regressions is due to Schatzoff et al. (1968). Their algorithm was a major advance. Further advances have made this method obsolete. It is a waste of money to compute all possible regressions. One should only compute those regressions that consist of the best subsets of the predictor variables.

The efficient computation of the best regressions is due to Furnival and Wilson (1974). "Best" is defined by ranking models on the basis of some measure of how well they fit. The most commonly used of these measures were R^2, adjusted R^2, and Mallows's C_p. In recent years *AIC* and *BIC* have become increasingly popular measures. Except for R^2, all of these criteria introduce a penalty for fitting more parameters. *Cost complexity pruning* determines the best model by using cross-validation to determine the most appropriate penalty. All of these criteria are discussed in the subsections that follow. Although nominally discussed for regression models, *all of these measures are trivially adapted to general linear models by replacing the number of columns in model matrices by their ranks.*

Although the criteria for identifying best models are traditionally used in the context of finding the best subsets among all possible regression models, they can be used to identify the best within any collection of linear or generalized linear models. For example, Christensen (2015) uses the C_p statistic to identify best unbalanced ANOVA models, Christensen (1997) used AIC to identify best ANOVA-like log-linear models for categorical data, and in the next section we mention using them on the sequences of models created by stepwise regression procedures.

14.1.1 R^2

The coefficient of determination, R^2, was discussed in Section 6.4. It is computed as

$$R^2 = \frac{SSReg}{SSTot - C}$$

and is just the ratio of the variability in y explained by the regression to the total variability of y. R^2 measures how well a regression model predicts (fits) the data as compared to just fitting a mean to the data. If one has two models with, say, p independent variables, other things being equal, the model with the higher R^2 will be the better model.

Using R^2 is not a valid way to compare models with different numbers of independent variables. With $s > p$, the R^2 for the full model

$$y_i = \beta_0 + \beta_1 x_{i1} + \cdots + \beta_s x_{is} + e_i \tag{1}$$

must be at least as large as the R^2 for the candidate model

$$y_i = \beta_0 + \beta_1 x_{i1} + \cdots + \beta_p x_{ip} + e_i \tag{2}$$

The full model has all the variables in the candidate model plus more, so

$$SSReg(1) \geq SSReg(2)$$

and

$$R^2(1) \geq R^2(2).$$

Typically, if the R^2 criterion is chosen, a program for doing best subset regression will print out the models with the highest R^2 for each possible value of p, the number of predictor variables. It is the use of the R^2 criterion in best subset regression that makes computing all possible regressions obsolete. The R^2 criterion fits all the good models one could ever want. In fact, it probably fits too many models.

14.1.2 Adjusted R^2

The adjusted R^2 is a modification of R^2 so that it can be used to compare models with different numbers of predictor variables. For a candidate model with p predictor variables plus an intercept, the adjusted R^2 is defined as

$$\text{Adj } R^2 = 1 - \frac{n-1}{n-p-1}\left(1 - R^2\right).$$

Define $s_y^2 = (SSTot - C)/(n-1)$, then s_y^2 is the sample variance of the y_is ignoring any regression structure. It is easily seen (Exercise 14.1) that

$$\text{Adj } R^2 = 1 - \frac{RMS}{s_y^2}.$$

As in the previous chapter, to avoid confusion with theoretical expected (mean) squared errors, we refer to the unbiased variance estimate as the residual mean square. The best models based on the Adj R^2 criterion are those models with the smallest residual mean squares.

As a method of identifying sets of good models, this is very attractive. The models with the smallest residual mean squares should be among the best models. However, the model with the smallest residual mean square may very well not be the best model.

Consider the question of deleting some variables from a model. In particular, consider testing model (2) against model (1). If the F statistic is greater than 1, then deleting the variables will increase the residual mean square. By the adjusted R^2 criterion, the variables should not be deleted. (See Exercise 14.2.) However, unless the F value is substantially greater than 1, the variables probably should be deleted. The Adj R^2 criterion tends to include too many variables in the model.

Exercise 14.1 Show that Adj $R^2 = 1 - (RMS/s_y^2)$.

Exercise 14.2 Consider testing the candidate regression model (2) against (1). Show that $F > 1$ if and only if the Adj R^2 for model (1) is greater than the Adj R^2 for model (2).

Some people define a *predicted* R^2 as $R^2 \equiv 1 - [PRESS/(n-1)s_y^2]$ where PRESS was discussed in Section 12.5.

14.1.3 Mallows's C_p

Suppose the full model $y_i = \beta_0 + \beta_1 x_{i1} + \cdots + \beta_s x_{is} + e_i$, i.e. $Y = X\beta + e$, is correct. In variable selection the first problem is that some of the β_js may be zero. As discussed in Section 2.9, we can even get better fitted values by eliminating variables that merely have small β_js. Rather than trying to identify which β_js are zero, Mallows suggested that the appropriate criterion for evaluating a reduced candidate model $Y = X_0\gamma + e$ is via its mean squared error for estimating $X\beta$, i.e.,

$$\mathrm{E}\left[(X_0\hat{\gamma} - X\beta)'(X_0\hat{\gamma} - X\beta)\right].$$

This is the same criterion used in Section 2.9. As mentioned earlier, to distinguish between this use of the term "mean squared error" and the estimate of the variance in the full model we write $RSS(\beta) \equiv Y'(I-M)Y$ for the residual sum of squares and $RMS(\beta) \equiv Y'(I-M)Y/r(I-M)$ for the residual mean square. The statistics $RSS(\gamma)$ and $RMS(\gamma)$ are the corresponding quantities for the model $Y = X_0\gamma + e$.

The quantity

$$(X_0\hat{\gamma} - X\beta)'(X_0\hat{\gamma} - X\beta)$$

is a quadratic form in the vector $(X_0\hat{\gamma} - X\beta)$. Writing

$$M_0 = X_0(X_0'X_0)^- X_0'$$

gives

$$(X_0\hat{\gamma} - X\beta) = M_0 Y - X\beta,$$

$$\mathrm{E}(X_0\hat{\gamma} - X\beta) = M_0X\beta - X\beta = -(I - M_0)X\beta,$$
$$\mathrm{Cov}(X_0\hat{\gamma} - X\beta) = \sigma^2 M_0.$$

From Theorem 1.3.2

$$\mathrm{E}\big[(X_0\hat{\gamma} - X\beta)'(X_0\hat{\gamma} - X\beta)\big] = \sigma^2\mathrm{tr}(M_0) + \beta'X'(I - M_0)X\beta.$$

We do not know σ^2 or β, but we can estimate the mean squared error. First note that

$$\mathrm{E}\big[Y'(I - M_0)Y\big] = \sigma^2\mathrm{tr}(I - M_0) + \beta'X'(I - M_0)X\beta;$$

so

$$\begin{aligned}\mathrm{E}\big[(X_0\hat{\gamma} - X\beta)'(X_0\hat{\gamma} - X\beta)\big] &= \sigma^2\mathrm{tr}(M_0) + \mathrm{E}\big[Y'(I - M_0)Y\big] - \sigma^2\mathrm{tr}(I - M_0) \\ &= \sigma^2\left[2\mathrm{tr}(M_0) - n\right] + \mathrm{E}\big[Y'(I - M_0)Y\big].\end{aligned}$$

With $p + 1 = \mathrm{tr}(M_0) = r(X_0)$, an unbiased estimate of the mean squared error is

$$RMS(\beta)[2(p + 1) - n] + RSS(\gamma).$$

Mallows's C_p statistic simply rescales the estimated mean squared error,

$$C_p \equiv \frac{RSS(\gamma)}{RMS(\beta)} - [n - 2(p + 1)].$$

The models with the smallest values of C_p have the smallest estimated mean squared error and should be among the best models for the data.

C_p *allows us to estimate which reduced models will give better* β *estimates through incorporating bias.* (A standard reduced model biases β towards 0. Incorporating a known offset vector Xb into a reduced model biases β towards b.)

C_p is the only measure discussed in this section that involves fitting a full model.

Exercise 14.3 For the specific candidate model (2) that drops some of the variables in (1), it is easy to see how an estimate of γ determines an estimate of β in (1); some of the $\hat{\beta}_j$s are forced to be 0. For a candidate model specified only by $C(X_0) \subset C(X)$, show how an estimate of γ determines an estimate of β.

Exercise 14.4

(a) Consider the F statistic for testing model (2) against model (1). Show that

$$C_p = (s - p)(F - 2) + s + 1.$$

(b) If $\mathrm{E}(Y) = X_1\delta$ has $r(X_1) = r$ and $C(X_0) \subset C(X_1) \subset C(X)$, show that

$$\frac{[RSS(\gamma) - RSS(\delta)]/[r-p]}{RMS(\beta)} > 2$$

if and only if C_p for the γ model is greater than C_p for the δ model.

Exercise 14.5 Give an informal argument to show that if $Y = X_0\gamma + e$ is a correct model, then the value of C_p should be around $p+1$. Provide a formal argument for this fact. Show that if $(n-s) > 3$, then $E(C_p) = (p+1) + 2(s-p)/(n-s-3)$. To do this you need to know that if $W \sim F(u, v, 0)$, then $E(W) = v/(v-2)$ for $v > 2$. For large values of n (relative to s and p), what is the approximate value of $E(C_p)$?

14.1.4 Information Criteria: AIC, BIC

Unlike the previous criteria that depend only on second moment assumptions, information criteria depend on the likelihood of the data.

As seen in Section 2.4, the log-likelihood for a standard linear model under normal theory is

$$\ell(\beta, \sigma^2) = \frac{-n}{2}\log(2\pi) - \frac{n}{2}\log[\sigma^2] - (Y - X\beta)'(Y - X\beta)/2\sigma^2.$$

The maximum likelihood estimates of β and σ^2 are the least squares $\hat{\beta}$ and $\hat{\sigma}^2 \equiv RSS/n$, so

$$\begin{aligned} -2\ell(\hat{\beta}, \hat{\sigma}^2) &= n\log(2\pi) + n\log[(\hat{\sigma}^2)] + (Y - X\hat{\beta})'(Y - X\hat{\beta})/\hat{\sigma}^2 \\ &= n\log(2\pi) + n\log(RSS/n) + RSS/(RSS/n). \\ &= n\log(2\pi) + n\log(RSS) - n\log(n) + n. \end{aligned}$$

The *Akaike Information Criterion (AIC)* is -2 times the maximum of the log-likelihood plus 2 times the number of parameters in the model. Better models have smaller AIC values.

For a candidate regression model with p predictors plus an intercept plus an unknown variance, the MLEs are $\hat{\gamma}$, $\hat{\sigma}^2_\gamma = RSS(\gamma)/n$, so

$$\begin{aligned} AIC &= -2\ell(\hat{\gamma}, \hat{\sigma}^2_\gamma) + 2(p+2) \\ &= n\log(2\pi) + n\log[RSS(\gamma)] - n\log(n) + n + 2(p+2). \\ &= \{n\log(2\pi) - n\log(n) + n + 4\} + n\log[RSS(\gamma)] + 2p. \end{aligned}$$

Everything in the first term of the last line is a constant that does not depend on the particular model, so, for comparing candidate models with intercepts, effectively

$$AIC = n\log[RSS(\gamma)] + 2p.$$

If you want to use AIC to compare models with different data distributions, the constant term needs to be included.

While it is somewhat advantageous that AIC can be computed without worrying about the existence of a largest model, it is of interest to compare how AIC and C_p work when both are applicable. Using the notation for C_p statistics, AIC picks candidate models with small values of

$$RSS(\gamma)e^{2p/n} = \exp(AIC/n),$$

whereas C_p picks models with small values of

$$RSS(\gamma) + 2pRMS(\beta).$$

Exercise 14.6 Find the AIC for a linear model in which the variance is known. Show that the C_p statistic is a simple function of the known variance AIC but where the variance is estimated from the largest available model.

Asymptotically, something similar to Exercise 14.4a kicks in for AIC where a larger model is preferred if the F statistic is above 2. Again using the notation of the previous subsection, suppose s and p are fixed but n is large. The full model is preferred if $AIC(\gamma) > AIC(\beta)$, or

$$1 < \frac{e^{AIC(\gamma)}}{e^{AIC(\beta)}} = \frac{RSS(\gamma)}{RSS(\beta)} e^{-2(s-p)/n}$$

or

$$e^{2(s-p)/n} < \frac{RSS(\gamma)}{RSS(\beta)}.$$

For a close to 0, $e^a \doteq 1 + a$, so when n is large an approximate condition for preferring the larger model is

$$\left[1 + \frac{2(s-p)}{n}\right] < \frac{RSS(\gamma)}{RSS(\beta)},$$

or

$$2 < \frac{n}{s-p}\left[\frac{RSS(\gamma)}{RSS(\beta)} - 1\right],$$

or

$$2 < \frac{n}{n-s-1}\frac{n-s-1}{s-p}\left[\frac{RSS(\gamma) - RSS(\beta)}{RSS(\beta)}\right] = \frac{n}{n-s-1}F.$$

For large n this is approximately the same condition as for C_p.

For small to moderate samples, many prefer to use a bias corrected form of AIC for evaluating candidate models,

$$AICc \equiv AIC + \frac{2(p+2)(p+3)}{n-p-3}.$$

cf. Sugiura (1978), Hurvich and Tsai (1989), Bedrick and Tsai (1994), and Cavanaugh (1997). It is commonly suggested to use AICc when any of the candidate models have $n/(p+2) < 40$.

Schwarz (1978) presented an asymptotic *Bayesian information criterion (BIC)* which is -2 times the maximum of the log-likelihood plus $\log(n)$ times the number of model parameters. For a standard normal theory regression candidate model with an intercept,

$$\begin{aligned} BIC &= -2\ell(\hat{\gamma}, \hat{\sigma}_\gamma^2) + (p+2)\log(n) \\ &= \{n\log(2\pi) - (n-2)\log(n) + n\} + n\log RSS + p\log(n), \end{aligned}$$

or effectively,

$$BIC = n\log RSS + p\log(n).$$

The derivation of BIC from Bayesian principals is described in Christensen et al. (2010).

BIC places much greater demands on variables to be included. When comparing nested models with p and s predictors, it chooses the larger model when, approximately,

$$\log(n) < F.$$

(The asymptotics kick in much slower than for AIC, which depends on $(1/n) \to 0$, because BIC relies on the much slower convergence of $\log(n)/n \to 0$.)

Exercise 14.7 Show that, for a given value of p, the R^2, Adj R^2, C_p, AIC, AICc, and BIC criteria all induce the same rankings of candidate models.

14.1.5 Cost Complexity Pruning

Cost complexity pruning is a related, but more complicated, way of determining the best model within a collection of candidate models. The collection of models can be all possible models or just the sequences of models determined by a stepwise regression method as discussed in the next section. For a given value of α, pick the model from the collection that minimizes $RSS + \alpha p$. The complexity comes because the value of α is chosen by cross-validation. Find the best model for a large number of α values and then choose a final α, and thus a final model, by cross-validation.

Specifically, randomly divide the data into K equal sized subgroups of observations, leave out one subgroup and apply the procedure to the other $K-1$. Using the $K-1$ subgroups as data, find the model within the collection that minimizes $RSS + \alpha p$, fit the model, predict the results in the omitted subgroup, and find the mean of the squared prediction errors, i.e., *MSPE*. (Mean Squared *Prediction* Error;

not the Mean Squared Pure Error discussed in Chapter 6.) In cost complexity pruning you do this for a large number of different α values to get a *MSPE* for each α, i.e., $MSPE(\alpha)$. Cycle through, leaving out a different subgroup each time, to get K different means of squared prediction errors, i.e., $MSPE_k(\alpha)$, $k = 1, \dots, K$. Pick $\hat{\alpha}$ to minimize $\sum_{k=1}^{K} MSPE_k(\alpha)$. The best model is the model that minimizes $RSS + \hat{\alpha} p$ when fitted to all the data.

James et al. (2013) discuss cost complexity pruning in the context of fitting regression trees. The term "pruning" originates from lopping off tree limbs, not from devouring desiccated plums.

14.2 Stepwise Regression

Stepwise regression methods involve adding or deleting variables one at a time.

14.2.1 Traditional Forward Selection

Example 14.2.1 Suppose we have variables y, x_1, x_2, and x_3 and the current model is

$$y_i = \beta_0 + \beta_1 x_{i1} + e_i .$$

In forward selection we must choose between adding variables x_2 and x_3. Fit the models

$$y_i = \beta_0 + \beta_1 x_{i1} + \beta_2 x_{i2} + e_i ,$$
$$y_i = \beta_0 + \beta_1 x_{i1} + \beta_3 x_{i3} + e_i .$$

Choose the model with the higher R^2. Equivalently, one could look at the t (or F) statistics for testing $H_0 : \beta_2 = 0$ and $H_0 : \beta_3 = 0$ and choose the model that gives the larger absolute value of the statistic. One could also look at $r_{y2\cdot 1}$ and $r_{y3\cdot 1}$ and pick the variable that gives the larger absolute value for the partial correlation.

Traditional forward selection sequentially adds variables to the model. Since this is a sequential procedure, the model in question is constantly changing. At any stage in the selection process, forward selection adds the variable that when added:

1. gives the largest absolute t statistic,
2. gives the largest F statistic,
3. gives the smallest P value,
4. increases R^2 the most,
5. decreases *RSS* the most,

6. gives the smallest C_p,
7. gives the smallest AIC,
8. gives the smallest BIC,
9. has the highest absolute partial correlation with y given the variables in the current model.

These criteria are equivalent because we are only considering the addition of one new variable.

Exercise 14.8 Show that the nine enumerated criteria for selecting a variable are equivalent.

Traditional forward selection stops adding variables when one of three things happens:

1. p^* variables have been added,
2. all absolute t statistics for adding variables not in the model are less than t^*,
3. the tolerance is too small for all variables not in the model.

The user (or programmer) picks the values of p^* and t^* and the tolerance limit. Tolerance was discussed in the previous chapter. No variable is ever added if its tolerance is too small, regardless of its absolute t statistic. Traditionally, one just uses the model that stopped the process. Alternatively, one could use a model selection criterion to pick the best among the models that forward selection has produced.

Although the nine criteria for adding variables are equivalent, stopping rules can get tricky. For example, a stopping rule based on having all P values above some fixed number is not equivalent to any stopping rule based on having all $|t|$ statistics below some fixed number because the P values depend on the residual degrees of freedom which keep changing. Reasonable stopping rules based on AIC or BIC might be when these no longer decrease or one can stop when Adj. R^2 fails to increase. Unfortunately, such stopping rules remove the flexibility that a self-selected stopping rule has to control the extent to which forward selection explores the set of all models. A stopping rule for forward selection based on C_p is rarely appropriate because there is rarely a largest model under consideration. (I would never use forward selection if I could fit a reasonable full model.)

The forward selection process is typically started with the initial model

$$y_i = \beta_0 + e_i.$$

Example 14.2.2 *Big Data.*

The data involve 1000 observations on a dependent variable and 100 predictor variables. By big data standards, this is a small set of big data. I performed forward selection with a stopping rule that the P value to enter the model must be below $\alpha = 0.05$. The table of parameters follows.

Table of Coefficients: Forward Selection, $\alpha = 0.05$.

Predictor	$\hat{\gamma}_i$	$SE(\hat{\gamma}_i)$	t	P
Constant	501.14	9.05	55.39	0.000
x_{24}	−21.84	9.03	−2.42	0.016
x_{56}	21.42	9.05	2.37	0.018
x_{72}	−26.34	8.91	−2.96	0.003
x_{75}	20.51	8.73	2.35	0.019
x_{82}	−18.08	8.79	−2.06	0.040

The model has a horrible R^2 of 0.0287 but it seems that there are several variables that help explain the data.

None of these variables has any actual effect! With two exceptions, all the 101,000 observations are i.i.d. standard normal. The first exception is that the y observations have a signal added to them. The other exception is that x_{49} consists of a small multiple of the signal buried within (the negative of) the white noise in x_{100}. The signal in x_{49} is almost hopelessly lost within that white noise, except for the fact that $x_{49} + x_{100}$ is precisely the small multiple of the signal. If you have both x_{49} and x_{100} in the model, you can extract the signal, but neither x_{49} nor x_{100} by itself helps. Any model that contains the two important variables will have $R^2 = 1.0000$.

I also ran the forward selection stopping when, to get a variable added, its P value must be below $\alpha = 0.25$ (the Minitab default). The idea of the higher cutoff value is to try to get important combinations of variables, like x_{49} and x_{100}, into the model so that their joint effect can be seen. It did not work here. The final model included the 5 predictors from $\alpha = 0.05$ plus 17 additional worthless predictors. Again, the model gives poor prediction with $R^2 = 0.0717$. When using a large α value, it seems inappropriate to blindly use the final model produced, since it will include a lot of variables with little predictive power. A far better procedure would be to select the best among the sequence of models produced or the best subset of the final model. While these are better procedures, in this example they are better procedures for producing garbage.

Even with $\alpha = 0.75$ and a final model that includes 79 of the 100 predictors, forward selection still did not manage to pick up both of the predictors necessary for getting a good model. Using a large α increases our chances of picking up the two good predictors, but it picks up a lot of junk predictors also. If we are lucky enough to pick up our two worthwhile predictors, clearly we would want to go back and eliminate the obvious junk. Of course $\alpha = 1$ will always find the two important variables because it will always fit the full model at the end of the sequence. But if you can fit the full model, you should be doing backward elimination or, better yet, best subset selection.

The way this example was constructed, α is pretty much the probability of finding a good prediction model using forward selection. I did not go looking for a large P value, but the P value associated with x_{100} was particularly large. (Recall that, by itself, x_{100} is unrelated to y and independent of all the predictors except x_{49}, so it's P remains pretty stable in any model that excludes x_{49}. Moreover, because the signal is *buried* in x_{49}, the same is true about x_{49} in any model that excludes x_{100}.) If the P value for x_{100} is below α, forward selection should find a good model in this example.

The problem of finding "important" effects that are actually meaningless is ubiquitous with big data. It is no accident that with $\alpha = 0.05$ we found about 5% of the meaningless predictors to be significant. It is no accident that with $\alpha = 0.25$ we found about 25% of the meaningless predictors in our model. The more tests you perform, the more meaningless things will look statistically significant. With any big data where nothing is related, you can always find something that looks related. In exploring big data, the usual standards of statistical significance do not apply.

As mentioned earlier, stopping rules based on AIC or BIC failing to decrease or Adj. R^2 failing to increase, lack the flexibility in exploring the space of possible models that one gets by selecting a P value or an absolute t statistic for inclusion.

14.2.2 Backward Elimination

Backward elimination sequentially deletes variables from the model. At any stage in the selection process, it deletes the variable with the smallest absolute t statistic or F statistic or equivalent criterion. Backward elimination often stops deleting variables when:

1. p_* variables have been eliminated,
2. the smallest absolute t statistic for eliminating a variable is greater than t_*.

The user can usually specify p_* and t_* in a computer program. Often, the process is stopped when the P value associated with $|t|$ is too small. Stopping rules are also based on AIC or BIC failing to decrease or Adj. R^2 failing to increase. Traditionally, one just uses the model that stopped the process. Alternatively, one could use a model selection criterion to pick the best among the sequence of models that backward elimination produced.

The initial model in the backward elimination procedure is the model with all of the predictor variables included,

$$y_i = \beta_0 + \beta_1 x_{i1} + \cdots + \beta_s x_{is} + e_i .$$

Backward elimination should give an adequate model. We assume that the process is started with an adequate model, and so only variables that add nothing are eliminated. The model arrived at may, however, be far from the most succinct. On the other hand, *there is no reason to believe that forward selection gives even an adequate model.*

Since we are starting with an adequate model, unlike forward selection, there is little reason to choose a stopping rule that helps us to explore the space of possible models.

Example 14.2.3 *Big Data Continued.*
I applied backward selection to the data with a P value cutoff of (the Minitab default) $\alpha = 0.10$

Table of Coefficients: Backward Elimination, $\alpha = 0.10$.

Predictor	$\hat{\gamma}_i$	SE($\hat{\gamma}_i$)	t	P
Constant	0.0458	0.0604	0.76	0.449
x_2	0.0669	0.0306	2.19	0.029
x_{34}	−0.0577	0.0302	−1.91	0.057
x_{48}	0.0632	0.0308	2.05	0.040
x_{49}	999879	104	9572.43	0.000
x_{59}	−0.0572	0.0311	−1.84	0.066
x_{66}	0.0680	0.0299	2.27	0.023
x_{95}	−0.0546	0.0293	−1.86	0.063
x_{98}	−0.0752	0.0303	−2.49	0.013
x_{100}	999879	104	9572.41	0.000

This is an extreme example so the $|t|$ statistics for the two important variables leap out. (Their P values do not!) But the point of this example is not that backward elimination found and kept the important variables. The point of this example is that backward elimination still finds 4 worthless variables that look significant by traditional standards and another 3 worthless variables that one would normally consider to be of marginal significance.

Unlike forward selection, backward elimination gives a good predictive model, one with $R^2 = 1.00$. The residual mean square is $RMS = 0.906$ which is disturbingly below the correct value 1, even with $dfE = 990$. (Based on the asymptotic normal approximation to the χ^2, RMS is more than two standard deviations below the true value.) Fitting the full model there are 899 degrees of freedom for error. From the 98 worthless predictor variables there are another 98 sums of squares with 1 degree of freedom that could all go into the error. The seven largest of those 98 sums of squares have been assigned to the model and the 91 smallest have been assigned to the error. Even when starting with 899 degrees of freedom for error from the full model that truly estimate the error, adding in the 91 smallest and leaving out the 7 biggest sums for squares biases the estimated error downward in the fitted backward elimination model.

If the final model from backward elimination is sufficiently small, one could apply best subset selection to the variables in the final model. But that is unlikely to accomplish anything that changing the α level could not accomplish and it is highly unlikely to eliminate all of the worthless predictor variables. The bigger the data set, the more stringent our requirements for significance should be.

14.2.3 Other Methods

Traditional forward selection is such an obviously faulty method that several improvements have been recommended. These consist of introducing rules for eliminating and exchanging variables. Four rules for adding, deleting, and exchanging variables follow.

1. Add the variable with the largest absolute t value if that value is greater than t^*.
2. Delete the variable with the smallest absolute t value if that value is less than t_*.
3. A variable not in the model is exchanged for a variable in the model if the exchange increases R^2.
4. The largest R^2 for each size model considered so far is saved. Delete a variable if the deletion gives a model with R^2 larger than any other model of the same size.

These rules can be used in combination. For example, 1 then 2, 1 then 2 then 3, 1 then 4, or 1 then 4 then 3. Again, no variable is ever added if its tolerance is too small.

Basically, these rules are just an attempt to get forward selection to look at a broader collection of models and it would be wise to select the best among the sequence of models generated.

Example 14.2.4 *Big Data Continued.*
Minitab's default stepwise procedure began with the intercept model, concluded with 17 variables, none of which were the important two.

14.3 Discussion of Traditional Variable Selection Techniques

Stepwise regression methods are fast, easy, cheap, and readily available. When the number of observations, n, is less than the number of variables, $s + 1$, forward selection or a modification of it is the only available method for variable selection. Backward elimination and best subset regression assume that one can fit the model that includes all the predictor variables. This is not possible when $n < s + 1$. In fact the use of t statistics, or anything equivalent to them, is probably unwise unless they are associated with a reasonable number of residual degrees of freedom.

There are serious problems with stepwise methods. They do not give the best model (based on any of the criteria we have discussed). In fact, stepwise methods can give models that contain none of the variables that are in the best regressions. That is because, as mentioned earlier, they handle variables one at a time. Another problem is nontechnical. The user of a stepwise regression program will end up with one model. The user may be inclined to think that this is *the* model. It probably is not. In fact, *the* model probably does not exist. Even though Adjusted R^2, Mallows's C_p, AIC, AICc, and BIC all define a unique best model, and could be subject to the same problem, best subset regression programs generally present several of the best models.

A problem with variable selection methods is that they tend to give models that appear to be better than they really are. For example, the Adjusted R^2 criterion chooses the model with the smallest *RMS*. Because one has selected the smallest *RMS*, the *RMS* for that model is biased toward being too small. Almost any measure of the fit of a model is related to the *RMS*, so the fit of the model will appear to be better than it is. If one could sample the data over again and fit the same model, the *RMS* would almost certainly be larger, perhaps substantially so.

When using Mallows's C_p statistic, if one wants to exploit the virtues of biased estimation, one often picks models with the smallest value of C_p. This can be justified by the fact that the model with the smallest C_p is the model with the smallest estimated expected squared error. However, as suggested by Exercise 14.5, if you are looking for a correct model the target value of C_p is the number of predictors, so it seems to make little sense to pick the model with the smallest C_p. It seems that one should pick models for which C_p is close to the number of predictors. (I pick models with small C_p.)

The result of Exercise 14.7, that for a fixed number of predictor variables the best regression criteria are equivalent, is interesting because the various criteria can be viewed as simply different methods of penalizing models that include more variables. The penalty is needed because models with more variables necessarily explain as much or more variation (have as high or higher R^2s).

14.3.1 R^2

R^2 is a good statistic for measuring the predictive ability of a model. R^2 is also a good statistic for comparing models. That is what we used it for here. But the actual value of R^2 should not be overemphasized when it is being used to identify correct models (rather than models that are merely useful for prediction). If you have data with a lot of variability, it is possible to have a very good fit to the underlying regression model without having a high R^2. For example, if the *RSS* admits a decomposition into pure error and lack of fit, it is possible to have very little lack of fit while having a substantial pure error so that R^2 is small while the fit is good.

If transformations of the dependent variable y are considered, it is inappropriate to compare R^2 for models based on different transformations. For example, it is possible for a transformation to increase R^2 without really increasing the predictive ability of the model. One way to check whether this is happening is to compare the width of confidence intervals for predicted values after transforming them to a common scale.

To compare models based on different transformations of y, say $y_1 = f_1(y)$ and $y_2 = f_2(y)$, fit models to the transformed data to obtained predicted values $\hat{y}_1$ and $\hat{y}_2$. Return these to the original scale with $\tilde{y}_1 = f_1^{-1}(\hat{y}_1)$ and $\tilde{y}_2 = f_2^{-1}(\hat{y}_2)$. Finally, define R_1^2 as the squared sample correlation between the ys and the $\tilde{y}_1$s and define

R_2^2 as the squared sample correlation between the ys and the $\tilde{y}_2$s. These R^2 values are comparable (and particularly so when the number of parameters in the two fitted models are comparable).

14.3.2 Influential Observations

Influential observations are a problem in any regression analysis. Variable selection techniques involve fitting lots of models, so the problem of influential observations is multiplied. Recall that an influential observation in one model is not necessarily influential in a different model.

Some statisticians think that the magnitude of the problem of influential observations is so great as to reject all variable selection techniques. They argue that the models arrived at from variable selection techniques depend almost exclusively on the influential observations and have little to do with any real world effects. Most statisticians, however, approve of the judicious use of variable selection techniques. (But then, by definition, everyone will approve of the *judicious* use of anything.)

14.3.3 Exploritory Data Analysis

John W. Tukey, among others, has emphasized the difference between exploratory and confirmatory data analysis. Briefly, *exploratory data analysis (EDA)* deals with situations in which you are trying to find out what is going on in a set of data. Confirmatory data analysis is for proving what you already think is going on. EDA frequently involves looking at lots of graphs. *Confirmatory data analysis* looks at things like tests and confidence intervals. Strictly speaking, you cannot do both exploratory data analysis and confirmatory data analysis on the same set of data.

Variable selection is an exploratory technique. If you know what variables are important, you do not need variable selection and should not use it. When you do use variable selection, if the model is fitted with the same set of data that determined the variable selection, then the model you eventually decide on will give biased estimates and invalid tests and confidence intervals. The biased estimates may very well be better point estimates than a full or correct model gives but tests and confidence intervals are usually over optimistic. Because you typically pick a candidate model partially because it has $RMS(\gamma) < RMS(\beta)$, confidence intervals are too narrow and tests are too significant.

If you can fit the model with all predictor variables and still have a reasonable dfE, it might be reasonable to perform tests and confidence intervals using least squares on the full model but use biased methods for point estimation and prediction. (With many predictors, you still need to use multiple comparison methods.) The alternative seems to be to use asymptotic or ad hoc methods for inference based directly on biased estimates. (Bayesians have the best of both worlds in that a proper Bayesian analysis both uses biased estimation and has exact small sample inference methods.)

One solution to this problem of selecting variables and fitting parameters with the same data is to divide the data into two parts. Do an exploratory analysis on one part and then a confirmatory analysis on the other. To do this well requires a lot of data. It also demonstrates the problem of influential observations. Depending on where the influential observations are, you can get pretty strange results. The PRESS statistic was designed to be used in procedures similar to this. However, as we have seen, the PRESS statistic is highly sensitive to influential observations.

14.3.4 Multiplicities

Methods of statistical inference were originally developed for situations where data were collected to investigate one thing. The methods work well on the one thing. In reality, even the best studies are designed to look at multiple questions, so the original methods need adjustment. Hence the need for the multiple comparisons methods of Chapter 5.

In an awful lot of studies, people collect data and muck around with it to see what they can find that is interesting. To paraphrase Seymour Geisser, they ransack the data. When mucking around with a lot of data, if you see something interesting, there is a good chance it is just random variation. Even if there is something there, the true effect is probably smaller than it looks. In this context, if you require a statistical test to show that something is important, it probably isn't important. We saw this with the Big Data examples of the previous section and those were examples with very clear structures. The general problem in mucking with the data is that to adjust for ransacking you need to keep track of *everything* you looked at that could *possibly* have been interesting. And we are just not psychologically equipped to do that.

14.3.5 Predictive Models

Predictive statistical models are based on correlation rather than causation. They work just fine as long as *nothing (important) has changed* from when the data were collected. You wake up, hear the shower on, you know your dad is making breakfast. Hearing the shower is a good predictor of Dad making breakfast. If you wake up from a nap and hear the shower at 2 in the afternoon, do you think Dad will be making breakfast?

What is the causation behind this prediction? Mom showering? It being 7am? Mom having to be to work at 8?

You cannot figure out what a change does to a system without changing the system! Yet *everybody* wants to do just that. They want to solve problems by collecting more data on present conditions. The world doesn't work that way. Without changing the system, you only have (hopefully intelligent) guesswork. But guesswork has limited value. Evaluating data from current conditions may provide ideas about what changes to try but it provides no assurance of what those changes will accomplish.

14.3.6 *Overfitting*

A big problem with having s large relative to n is the tendency to overfit. *Overfitting* is the phenomenon of fitting a model with so many parameters that the model looks like it fits the data well, e.g. has a high R^2, but does a poor job of predicting future data. Fitting any model with $r(X) = n$ gives $R^2 = 1$, so it is easy to overfit regression models just by taking an X with $r(X) \doteq n$. Our discussion of variable selection was about making X smaller (turning X into a well chosen X_0), so as not to overfit the data. If $r(X) \doteq n$, forward selection could be applied to try to avoid overfitting. When fitting regression trees and other multiple nonparametric regression models (cf. *ALM-III*, Chapter 1), the set of potential predictor variables is huge and forward selection is used to pick variables that seem appropriate.

Under normal theory for a standard linear model

$$\mathrm{E}(RMS) = \sigma^2, \qquad \mathrm{Var}(RMS) = 2\sigma^4/dfE,$$

so the coefficient of variation (CV) is $\sqrt{2/dfE}$. For *RMS* is to be a decent estimate, we need CV reasonably small. With $dfE = 2, 8, 18$, CV is 1, 1/2, 1/3, so I would like at least 18 dfE, 8 might be tolerable, and using 2 or less is fraught with danger. (I don't want to think about how often I have failed to live up to that prescription.)

Christensen (2015, Section 8.2) shows how bad predictions can be from overfitted models but it incidentally shows how bad the estimated variances are from those overfitted models. For everything except the simple linear regression, his estimated variances were well below the target value of 1.

Some rules I have seen to avoid overfitting require $n \geq 10p$, $n \geq 15p$, or $n \geq 50 + 8p$. These seem like they should work but they seem awfully stringent.

14.4 Modern Forward Selection: Boosting, Bagging, and Random Forests

For many years, forward selection was dismissed as the poor sibling of variable selection. Forward selection provides no assurance that it will find anything like the best models. Backward elimination, since it begins with a presumably reasonable full model and only does reasonable things to that model, should arrive at a decent model. Looking at the "best" subsets of variables seems like the best thing to do.

But backward elimination and best subset selection both require being able to fit a reasonable full model.

If the number of predictor variables s is big enough so that $r(X) = n$, we have a saturated full model. Least squares then gives $\hat{Y} = Y$, $SSE = 0$, $dfE = 0$, and the model will be over-fitted so that predictions of new observations typically are poor. Whenever dfE is small, we have probably over-fitted, making our full-model results dubious. In problems with $s \doteq n$ or $s > n$, forward selection, poor as it is, is about the only game in town. (Principal component regression is another.) *Boosting*, *Bagging*, and *Random Forests* are more recently developed methods of forward selection by which one can use over-fitting of models to improve predictions. Despite all of the difficulties that arose in our Big Data examples given earlier, by the standards of this section those examples have extremely well behaved data because $s << n$. Nonetheless, I do not *believe* that the improvements presented here are capable of overcoming the specific foreward selection problem built into the Big Data examples, namely that the importance of the pair of variables is not detectable from either variable separately. (Randomly picking variables for a full model could solve the problem with that example.)

Boosting is a biased estimation technique associated with forward selection. Bagging (bootstrap aggregation) involves use of the *bootstrap* to get more broad based estimates. Random forests are a modification of bagging.

Forward selection starts with some relatively small model and defines a sequence of larger and larger models. The two key features are (1) how to decide which variable gets added next and (2) when to stop adding variables. The traditional method of forward selection ranks variables based on the absolute value of the t statistic for adding them to the current model (or some equivalent statistic) and chooses the highest ranked variable.

If the predictor variables happen to have equal sample variances, forward selection could use the regression coefficients themselves to rank variables, rather than their associated t statistics. In general, using regression coefficients rather than $|t|$ statistics does not seem like a great idea, but in my *quite limited* experience, the procedure works remarkably similar to Tibshirani's (1996) lasso for standardized predictors.

My primary references for this section were James et al. (2013), Hastie et al. (2016), and Efron and Hastie (2016). This section is different from any other in the book because it contains quite a few of my speculations (clearly marked as such) about how these or related methods *might* work. (The first such speculation occurred in the previous paragraph.)

14.4.1 Boosting

The forward selection method known as *boosting* involves a sequence of model matrices X_j with ppos M_j. The procedure depends on choices for integers d and B and a scalar k.

Perform a forward selection from among the predictor vectors in X to obtain a model with d predictors,

$$Y = X_1\beta_1 + e.$$

From this obtain fitted values and residuals,

$$\hat{Y}_1 = M_1Y; \quad \hat{e}_1 = Y - \hat{Y}_1.$$

Perform another forward selection using $\hat{e}_1$ as the dependent variable to obtain another model with d predictors,

$$\hat{e}_1 = X_2\beta_2 + e.$$

From this obtain fitted values

$$\tilde{Y}_2 = M_2\hat{e}_1 = M_2(I - M_1)Y.$$

Define overall fitted values

$$\hat{Y}_2 = \hat{Y}_1 + k\tilde{Y}_2$$

and residuals,

$$\hat{e}_2 = Y - \hat{Y}_2 = (I - kM_2)(I - M_1)Y.$$

In general, given residuals $\hat{e}_j$ perform a forward selection using $\hat{e}_j$ as the dependent variable to obtain another model with d predictors,

$$\hat{e}_j = X_{j+1}\beta_{j+1} + e.$$

From this obtain fitted values

$$\tilde{Y}_{j+1} = M_{j+1}\hat{e}_j.$$

Define overall fitted values

$$\hat{Y}_{j+1} = \hat{Y}_j + k\tilde{Y}_{j+1}$$

and residuals,

$$\hat{e}_{j+1} = Y - \hat{Y}_{j+1} = (I - kM_{j+1})\cdots(I - kM_2)(I - M_1)Y.$$

The procedure stops when j reaches the predetermined value B. The accepted wisdom seems to be that picking a stopping point B that is too large can still result in overfitting the model.

If $k = 1$, so that no shrinkage of the estimates is involved, boosting seems like just a lousy way of fitting

$$Y = [X_1, \ldots, X_B]\delta + e.$$

Next we present two adjusted methods that give least squares estimates when $k = 1$.

14.4.1.1 Alternatives

Collect all of the possible predictors into the matrix X. We use ideas related to the sweep operator discussed in Chapter 9. Perform a forward selection from the columns of X to obtain a model with d predictors,

$$Y = X_1\beta_1 + e.$$

From this obtain fitted values and residuals,

$$\hat{Y}_1 = M_1 Y; \quad \hat{e}_1 = Y - \hat{Y}_1.$$

Adjust all the columns of X into $\tilde{X}_2 = (I - M_1)X$. Note that X contains the columns of X_1 but these are zeroed out in $\tilde{X}_2$ and should not be eligible for future selection. This adjustment is not a hideously expensive thing to do. The single expensive operation is computing $(X_1'X_1)^{-1}$. The other operations are numerous but individually inexpensive.

Perform a forward selection using $\hat{e}_1$ as the dependent variable and $\tilde{X}_2$ as the matrix of possible variables to obtain another model with d predictors,

$$\hat{e}_1 = X_2\beta_2 + e.$$

Notice that $C(X_1) \perp C(X_2)$ so $M_1 M_2 = 0$. From this obtain overall fitted values

$$\hat{Y}_2 = \hat{Y}_1 + kM_2\hat{e}_1 = (M_1 + kM_2)Y$$

and residuals,

$$\hat{e}_2 = Y - \hat{Y}_2 = (I - M_1 - kM_2)Y.$$

In general, given fitted values $\hat{Y}_j$, residuals $\hat{e}_j$ and the matrices X_j and $\tilde{X}_j$, construct the possible additions $\tilde{X}_{j+1} \equiv (I - M_j)\tilde{X}_j$ in which all variables that are already in the model will have been zeroed out.

Perform a forward selection using $\hat{e}_j$ as the dependent variable with the columns of $\tilde{X}_{j+1}$ as potential predictors to obtain another model with d predictors,

$$\hat{e}_j = X_{j+1}\beta_{j+1} + e.$$

Again, all of the $C(X_k)$s are orthogonal. From this model obtain overall fitted values

$$\hat{Y}_{j+1} = \hat{Y}_j + kM_{j+1}\hat{e}_j = [M_1 + k(M_2 + \cdots + M_{j+1})]Y$$

and residuals

$$\hat{e}_{j+1} = Y - \hat{Y}_{j+1} = (I - kM_{j+1})\hat{e}_j.$$

The accepted wisdom that, picking a stopping point B that is too large can still result in overfitting the model, seems related to the fact that the penalty term k remains the same for every step after the first. An alternative method, akin to exponential smoothing, might do better by defining

$$\hat{Y}_{j+1} = \hat{Y}_j + k^j \tilde{Y}_{j+1} = (M_1 + kM_2 + \cdots + k^j M_{j+1})Y.$$

The alternatives mentioned here are based on the idea that it is the shrinkage of estimates that is valuable in boosting. Although I don't see how it could be true, it is possible that the very awkwardness of adding nonorthogonalized variables could be of some benefit. Boosting was originally developed for binomial regression problems and I *suspect* behaves quite differently there.

14.4.2 Bagging

Bagging is a technique described by Hastie et al. (2016, p. 282) as "how to use the bootstrap to improve the estimate or prediction." We will see that bagging can be useful but cannot be a panacea.

The fundamental idea of *bagging* (as I see it) follows: Suppose you have an algorithm for fitting a model to a set of data with n observations. In bagging this should be an algorithm that tends to overfit the data. Take a random sample with replacement of size n from your data. Apply your algorithm on this sample of data and obtain your desired results: predictions or estimates. Do this repeatedly for many random samples and average your predictions/estimates over these samples. These averages are the result of bagging. The hope is that these averages will be better predictions and estimates than the results of the original algorithm applied just once to the original data. We will see, in a simple example, that the better the algorithm, the less likely this is to be true. But in situations where we do not know how to create a good algorithm for the particular data, bagging can provide valuable improvements.

With reasonably large collections of data, the gold standard for determining the quality of a predictive model seems to be: (1) randomly pull out a set of *test data*, (2) use the remaining *training data* to develop the predictive model, and (3) evaluate the predictive model by seeing how well it predicts the test data. This is frequently used to compare the quality of various methods of developing predictive models. In this context, overfitting consists of fitting a model that explains the training data very well but does a poor job of predicting the test data.

If we think of the collection of observed data as the entire population of possible data, and randomly select the test data, then the training data is also just a random sample from the population. It should display the same predictive relationships as the overall population. If the goal is to predict a random sample (the test sample) from the population, why not use random samples from the population to develop a predictor? Moreover, we want to use overfitting to help, rather than hinder, the predictive process.

The idea is to start with a model selection procedure that is capable of modeling the salient features in the data. In multiple nonparametric regression (*ALM-III*, Chapter 1) that involves constructing additional predictor variables that make the number of predictor variables s very large indeed. Different nonparametric regression procedures have different modeling capabilities (they create different model matrices X), so the best choice of a procedure depends on the nature of the data in ways that we will rarely understand beforehand. But merely having an X matrix with $s >> n$ is not enough. To develop a prediction model we must have some variable selection scheme, presumably a form of forward selection, that includes enough predictor variables to capture the salient features of the data, and to do this consistently seems to require some overfitting.

Take a random sample from the training data and overfit a model on it. This ensures that the salient features that are present in all samples are caught but overfitting will also include features that are unique to the particular sample being fitted. Do this for many random samples (say B) and average the results. The hope is that the salient features will appear in a similar fashion in every sample but that the unique (random) features that occur from overfitting particular samples will average themselves out over the process of repeated sampling.

Bagging is a very complicated procedure. In fact, it is notorious for providing (good) predictions that are uninterpretable. We now examine an extremely simple example of bagging to explore how it actually works.

14.4.2.1 A Simple Example

Consider a random variable y. With no potential predictor variables available, the best predictor (BP) under squared error prediction loss is $\mathrm{E}(y) = \mu_y \equiv \mu$. Now suppose we have a random sample from y, say $Y_{n\times 1}$. The best nonparametric estimate of μ is $\bar{y}_\cdot = J'Y/n$. We know it is the BLUE but for a class of distributions that includes nearly all continuous distributions that have a mean, $\bar{y}_\cdot$ is minimum variance unbiased, cf. Fraser (1957). If the distribution of y is normal, $\bar{y}_\cdot$ is again minimum variance unbiased but it is much easier to be unbiased for normal distributions than it is for all continuous distributions. For uniform distributions with unknown limits, the best unbiased estimate of the expected value is the midrange. For symmetric distributions with heavy tails, e.g. Laplace (double exponential), the median tends to be a good estimate. In this context, $\bar{y}_\cdot$ is always going to be a reasonably good estimate of the BP. Generally, for light-tailed symmetric distributions, like the uniform with unknown limits, the midrange can be a good estimate of the BP but the median will be less good. For heavy-tailed symmetric distributions, the median can be a good estimate of the BP but the midrange will be less good. Neither the median nor the midrange are linear functions of the data.

Bagging brings an element of averaging into the estimates that has virtually no effect on the linear estimate $\bar{y}_\cdot$ and cannot improve an optimal estimate, but bagging can substantially improve a poor estimate and can even improve good but suboptimal estimates.

Example 14.4.1 In the spirit of fitting a number of parameters that is a large proportion of the number of observations (and just to be able to perform the computations), suppose we have a simple random sample of size $n = 3$ with order statistics $y_{(1)} < y_{(2)} < y_{(3)}$. The best predictor is the population mean, which we want to estimate. The midrange $(y_{(1)} + y_{(3)})/2$ is optimal for uniform distributions with unknown end points. The median $y_{(2)}$ works well for heavy tailed distributions. The sample mean $(y_{(1)} + y_{(2)} + y_{(3)})/3$ is optimal for normal data or extremely broad (nonparametric) families of distributions.

Normally one would not bootstrap a sample this small but the small sample size allows us to examine what Hastie et al. (2016) refer to as the *"true" bagging estimate*. With only 3 observations, bootstrapping takes samples from a population that has 27 equally probable outcomes: Three of the outcomes are $\{y_{(j)}, y_{(j)}, y_{(j)}\}$ for $j = 1, 2, 3$. Six of the outcomes are reorderings of $\{y_{(1)}, y_{(2)}, y_{(3)}\}$. The other 18 outcomes involve the three reorderings one can get from samples of the form $\{y_{(j)}, y_{(j)}, y_{(k)}\}$, $j = 1, 2, 3$, $k \neq j$ there being six distinct outcomes of this form.

From each of the 27 outcomes in the bootstrap population we can compute the sample mean, median, and midrange statistics. The bootstrap procedure actually provides an estimate (one that we can make arbitrarily good by picking B large) of the expected value over the 27 equally probable outcomes of these statistics (sample mean, median and midrange). We want to know what function of the observed data the bootstrap is estimating, because that is the function of the data that the bootstrap uses to estimate the BP (as $B \to \infty$).

The expected value of the sample mean is easily seen to be $(9y_{(1)} + 9y_{(2)} + 9y_{(3)})/27$, so unsurprisingly the bootstrap of the sample mean is estimating the sample mean of the original data. The bootstrap expected value of the sample median is $(7y_{(1)} + 13y_{(2)} + 7y_{(3)})/27$, which is a symmetric weighted average of the original observations; one that puts more weight on the middle observation. The bootstrap expected value of the midrange is $(10y_{(1)} + 7y_{(2)} + 10y_{(3)})/27$, which is again a symmetric weighted average of the original observations but one that puts less weight on the middle observation.

Another way to think about this is that the bagged median is estimating

$$(14/27)\text{midrange} + (13/27)\text{median}$$

and the bagged midrange is estimating

$$(20/27)\text{midrange} + (7/27)\text{median}.$$

But perhaps more importantly, *both of them are closer to the sample mean than they were originally*.

For a uniform distribution, where the midrange is optimal, the bagged midrange estimate will be less good because it puts too much weight on the middle observation. However, the bagged median will be better than the median because the bagged median puts more weight on the midrange.

When the median is good, the bagged median will be less good because it puts more weight on the extreme observations. However the bagged midrange will be better than the midrange because the bagged midrange puts more weight on the median.

Bagging the sample mean is a waste of effort. Because the sample mean is the best nonparametric estimate, the sample mean is never going to be too bad.

If you don't know the distribution of the data, which you almost never do, you might as well use the sample mean and bagging is irrelevant. Bagging would be useful if for some reason you cannot use the sample mean.

The three estimates we examined were all unbiased for symmetric distributions. Lets look at a biased estimate of the mean. Consider the estimate $(y_{(2)} + y_{(3)})/2$ which is clearly biased above the mean for symmetric distributions. The bootstraped estimate has expected value $(4y_{(1)} + 10y_{(2)} + 13y_{(3)})/27$, which, while still heavily biased above the mean, is considerably less biased than the original estimate.

14.4.2.2 Discussion

The prediction problem is to estimate the best predictor, $\mathrm{E}(y|x)$. The more you know about the conditional distribution of y given x, the easier the problem becomes. By expanding our definition of x, e.g. incorporating polynomials, we can often ensure that $\mathrm{E}(y|x)$ is approximately linear in x. If we have enough data to fit a full model, we should. For homoscedastic, uncorrelated data, least squares estimates are BLUEs, so they are probably about as good as we can do to start, but then we may be able to get better point estimates by incorporating bias. If $s > n$, we cannot fit the full model in any meaningful way, so we need some way of constructing a predictive model and typically that involves some form of forward selection. We know forward selection does not work well, so it is unlikely to give good estimates of the best predictor. When you have poor estimates, bagging seems to be good at improving them by averaging them over more of the data. (My *hope* is that the bagging estimates will move them closer to appropriate estimates for the [unfitable] full model.) While it is relatively easy to see that bagging has no systematic effect on estimates that are linear functions of the data, forward selection is a profoundly nonlinear estimation process.

14.4.3 Random Forests

The *randomforest* idea modifies the forward selection procedure in conjunction with bagging. The name derives from applying the idea to regression trees.

Divide the predictor variables into G groups. The modification to forward selection is that instead of considering all of the variables as candidates for selection, one randomly chooses m of the G groups as candidates for forward selection. If you were only fitting one model, that would be a disastrous idea, but in the context of bagging, all of the important variables should show up often. Typically one takes $m \doteq G/3$ or $m \doteq \sqrt{G}$.

Example 14.4.2 *Polynomial Regression.*
Division into G groups occurs naturally in polynomial regression and many other nonparametric regression procedures that, like polynomial regression, begin with, say Q, measured predictor variables and define functions of those measured variables. A full polynomial model on Q measured variables $x_1, \ldots, x_Q$ is

$$y_i = \sum_{j_1=0}^{d_1} \cdots \sum_{j_Q=0}^{d_Q} \beta_{j_1 \cdots j_s} x_{i1}^{j_1} \cdots x_{iQ}^{j_Q} + \varepsilon_i,$$

so the total number of predictor variables is $s = \prod_{j=1}^{Q} d_j$. The G groups can conveniently be taken as $x_k^{j_k} : j_k = 1, \ldots, d_k$ for $k = 1, \ldots, Q$ which makes $G = Q$. Instead of considering all of the variables $x_k^{j_k}$, $j_k = 1, \ldots, d_k$, $k = 1, \ldots, Q$ as candidates for selection, one randomly chooses m of the Q groups as candidates for forward selection. (Forward selection typically will not result in a hierarchical polynomial that contains all the lower order terms for every term in the polynomial. Although good arguments can be made for using hierarchical polynomials, they seem inconsistent with the spirit of forward selection.)

Example 14.4.3 *Big Data.*
Divide the 100 predictors into 9 groups all but one having 11 predictor variables. Randomly pick 3 groups to be included in the variable selection. In this example, with almost everything being independent, it is hard to see how the random forest idea is going to help. Much of the time the two worthwhile predictors will not even be available for selection in the model. And even when the two good predictors are both available, nothing has happened that will increase their chances of being selected. Remember, we have to have an algorithm that randomly gets one of the two predictors into the model. Once one of them is in the model, forward selection will find the second variable (if it is available to find). So the random forest idea (or bagging alone) does not seem to help in this example, but then forward selection on these data is not a method well suited for finding the salient characteristics at the expense of some overfitting.

I would be tempted to define subgroups of variables for which a random selection would give something I consider a plausible full model and rather than averaging them all, actually look for good ones. But in the era of big data, there seems to be a premium on procedures you can run without having to supervise them.

Exercise 14.9 Compare the results of using the various model selection techniques on the data of Exercise 13.5.

References

Bedrick, E. J., & Tsai, C.-L. (1994). Model selection for multivariate regression in small samples. *Biometrics, 50*, 226–231.

Cavanaugh, J. E. (1997). Unifying the derivations of the Akaike and corrected Akaike information criteria. *Statistics and Probability Letters, 31*, 201–208.

Christensen, R. (1997). *Log-linear models and logistic regression* (2nd ed.). New York: Springer.

Christensen, R. (2015). *Analysis of variance, design, and regression: Linear modeling for unbalanced data* (2nd ed.). Boca Raton, FL: Chapman and Hall/CRC Pres.

Christensen, R., Johnson, W., Branscum, A., & Hanson, T. E. (2010). *Bayesian ideas and data analysis: An introduction for scientists and statisticians*. Boca Raton, FL: Chapman and Hall/CRC Press.

Cook, R. D., Forzani, L., & Rothman, A. J. (2013). Prediction in abundant high-dimensional linear regression. *Electronic Journal of Statistics, 7*, 3059–3088.

Cook, R. D., Forzani, L., & Rothman, A. J. (2015). Letter to the editor. *The American Statistician, 69*, 253–254.

Efron, B., & Hastie, T. (2016). *Computer age statistical inference: Algorithms, evidence, and data science*. Cambridge: Cambridge University Press.

Fraser, D. A. S. (1957). *Nonparametric methods in statistics*. New York: Wiley.

Furnival, G. M., & Wilson, R. W. (1974). Regression by leaps and bounds. *Technometrics, 16*, 499–511.

Hastie, T., Tibshirani, R., & Friedman, J. (2016). *The elements of statistical learning: Data mining, inference, and prediction* (2nd ed.). New York: Springer.

Hurvich, C. M., & Tsai, C.-L. (1989). Regression and time series model selection in small samples. *Biometrika, 76*, 297–307.

James, G., Witten, D., Hastie, T., & Tibshirani, R. (2013). *An introduction to statistical learning*. New York: Springer.

Schatzoff, M., Tsao, R., & Fienberg, S. (1968). Efficient calculations of all possible regressions. *Technometrics, 10*, 768–779.

Schwarz, G. (1978). Estimating the dimension of a model. *Annals of Statistics, 6*, 461–464.

Sugiura, N. (1978). Further analysis of the data by Akaike's information criterion and the finite corrections. *Communications in Statistics, Part A, Theory and Methods, 7*, 13–26.

Tarpey, T., Ogden, R., Petkova, E., & Christensen, R. (2015). Reply. *The American Statistician, 69*, 254–255.

Appendix A
Vector Spaces

Abstract This appendix reviews some of the basic definitions and properties of vector spaces. It presumes some basic knowledge of matrices.

Definition A.1 A set $\mathscr{M}$ is a *vector space* if, for any $x, y, z \in \mathscr{M}$ and scalars α, β, operations of vector addition and scalar multiplication are defined such that:

(1) $(x + y) + z = x + (y + z)$.
(2) $x + y = y + x$.
(3) There exists a vector $0 \in \mathscr{M}$ such that $x + 0 = x = 0 + x$ for any $x \in \mathscr{M}$.
(4) For any $x \in \mathscr{M}$, there exists $y \equiv -x$ such that $x + y = 0 = y + x$.
(5) $\alpha(x + y) = \alpha x + \alpha y$.
(6) $(\alpha + \beta)x = \alpha x + \beta x$.
(7) $(\alpha\beta)x = \alpha(\beta x)$.
(8) There exists a scalar ξ such that $\xi x = x$. (Typically, $\xi = 1$.)

We rely on context to distinguish between the vector $0 \in \mathscr{M}$ and the scalar $0 \in \mathbf{R}$. In most of our applications, we assume $\mathscr{M} \subset \mathbf{R}^n$. Vectors in $\mathbf{R}^n$ will be considered as $n \times 1$ matrices. The 0 vector referred to in Definition A.1 is just an $n \times 1$ matrix of zeros.

Definition A.2 Let $\mathscr{M}$ be a vector space, and let $\mathscr{N}$ be a set with $\mathscr{N} \subset \mathscr{M}$. $\mathscr{N}$ is a *subspace* of $\mathscr{M}$ if $\mathscr{N}$ is a vector space using the same definitions of vector addition and scalar multiplication as for $\mathscr{M}$.

Thinking of vectors in three dimensions as $(x, y, z)'$, where w' denotes the *transpose* of a matrix w. The subspace consisting of the z axis is

$$\left\{ \begin{pmatrix} 0 \\ 0 \\ z \end{pmatrix} \middle| \, z \in \mathbf{R} \right\}.$$

R. Christensen, *Plane Answers to Complex Questions*, Springer Texts in Statistics,
https://doi.org/10.1007/978-3-030-32097-3

The subspace consisting of the x, y plane is

$$\left\{ \begin{pmatrix} x \\ y \\ 0 \end{pmatrix} \middle| \, x, y \in \mathbf{R} \right\}.$$

The subspace consisting of the plane that is perpendicular to the line $x = y$ in the x, y plane is

$$\left\{ \begin{pmatrix} x \\ -x \\ z \end{pmatrix} \middle| \, x, z \in \mathbf{R} \right\}.$$

Theorem A.3 *Let $\mathcal{M}$ be a vector space, and let $\mathcal{N}$ be a nonempty subset of $\mathcal{M}$. If $\mathcal{N}$ is closed under vector addition and scalar multiplication, then $\mathcal{N}$ is a subspace of $\mathcal{M}$.*

Theorem A.4 *Let $\mathcal{M}$ be a vector space, and let $x_1, \ldots, x_r$ be in $\mathcal{M}$. The set of all linear combinations of $x_1, \ldots, x_r$, i.e., $\{v \mid v = \alpha_1 x_1 + \cdots + \alpha_r x_r, \alpha_i \in \mathbf{R}\}$, is a subspace of $\mathcal{M}$.*

Definition A.5 The set of all linear combinations of $x_1, \ldots, x_r$ is called the *space spanned by* $x_1, \ldots, x_r$. If $\mathcal{N}$ is a subspace of $\mathcal{M}$, and $\mathcal{N}$ equals the space spanned by $x_1, \ldots, x_r$, then $\{x_1, \ldots, x_r\}$ is called a *spanning set* for $\mathcal{N}$.

The space spanned by the vectors

$$x_1 = \begin{pmatrix} 1 \\ 1 \\ 1 \end{pmatrix}, \quad x_2 = \begin{pmatrix} 1 \\ 0 \\ 0 \end{pmatrix}$$

consists of all vectors of the form $(a, b, b)'$, where a and b are any real numbers.

Let A be an $n \times p$ matrix. Each column of A is a vector in $\mathbf{R}^n$. The space spanned by the columns of A is called the *column space* of A and written $C(A)$. (Some people refer to $C(A)$ as the *range space* of A and write it $R(A)$.) If B is an $n \times r$ matrix, then $C(A, B)$ is the space spanned by the $p + r$ columns of A and B.

Definition A.6 Let $x_1, \ldots, x_r$ be vectors in $\mathcal{M}$. If there exist scalars $\alpha_1, \ldots, \alpha_r$ not all zero so that $\sum \alpha_i x_i = 0$, then $x_1, \ldots, x_r$ are *linearly dependent*. If such α_is do not exist, i.e., if $\sum \alpha_i x_i = 0$ implies that $\alpha_i = 0$ for all i, then $x_1, \ldots, x_r$ are *linearly independent*.

Definition A.7 If $\mathcal{N}$ is a subspace of $\mathcal{M}$ and if $\{x_1, \ldots, x_r\}$ is a linearly independent spanning set for $\mathcal{N}$, then $\{x_1, \ldots, x_r\}$ is called a *basis* for $\mathcal{N}$.

Theorem A.8 *If $\mathcal{N}$ is a subspace of $\mathcal{M}$, all bases for $\mathcal{N}$ have the same number of vectors.*

Definition A.9 The *rank* of a subspace $\mathcal{N}$ is the number of elements in a basis for $\mathcal{N}$. The rank is written $r(\mathcal{N})$. If A is a matrix, the rank of $C(A)$ is called the rank of A and is written $r(A)$, i.e., $r(A) \equiv r[C(A)]$.

Exercise B.24 establishes that $r(A) = r(A')$.

Theorem A.10 *If $v_1, \ldots, v_r$ is a basis for $\mathcal{N}$, and $x \in \mathcal{N}$, then the characterization $x = \sum_{i=1}^{r} \alpha_i v_i$ is unique.*

Proof Suppose $x = \sum_{i=1}^{r} \alpha_i v_i$ and $x = \sum_{i=1}^{r} \beta_i v_i$. Then $0 = \sum_{i=1}^{r} (\alpha_i - \beta_i) v_i$. Since the vectors v_i are linearly independent, $\alpha_i - \beta_i = 0$ for all i. □

The vectors

$$x_1 = \begin{pmatrix} 1 \\ 1 \\ 1 \end{pmatrix}, \quad x_2 = \begin{pmatrix} 1 \\ 0 \\ 0 \end{pmatrix}, \quad x_3 = \begin{pmatrix} 2 \\ 3 \\ 3 \end{pmatrix}$$

are linearly dependent because $0 = 3x_1 - x_2 - x_3$. Any two of x_1, x_2, x_3 form a basis for the space of vectors with the form $(a, b, b)'$. This space has rank 2.

Definition A.11 Let $\mathcal{N}_1$ and $\mathcal{N}_2$ be vector subspaces. The sum of $\mathcal{N}_1$ and $\mathcal{N}_2$ is $\mathcal{N}_1 + \mathcal{N}_2 \equiv \{x | x = x_1 + x_2 \text{ for some } x_1 \in \mathcal{N}_1, x_2 \in \mathcal{N}_2\}$. This is sometimes called the *direct sum* and written $\mathcal{N}_1 \oplus \mathcal{N}_2$.

Theorem A.12 *$\mathcal{N}_1 + \mathcal{N}_2$ is a vector space and $C(A, B) = C(A) + C(B)$.*

Theorem A.13 *Let $\mathcal{N}_1$ and $\mathcal{N}_2$ be vector subspaces. If $\mathcal{N}_1 \cap \mathcal{N}_2 = \{0\}$, then any vector $x \in \mathcal{N}_1 + \mathcal{N}_2$ has a unique decomposition $x = x_1 + x_2$ with $x_1 \in \mathcal{N}_1$ and $x_2 \in \mathcal{N}_2$.*

Proof Let $x = x_1 + x_2$, $x_1 \in \mathcal{N}_1$, $x_2 \in \mathcal{N}_2$ and $x = y_1 + y_2$, $y_1 \in \mathcal{N}_1$, $y_2 \in \mathcal{N}_2$. Then $x_1 + x_2 = x = y_1 + y_2$, so $x_1 - y_1 = y_2 - x_2 \equiv v$. But $v \equiv x_1 - y_1 \in \mathcal{N}_1$ and $v = y_2 - x_2 \in \mathcal{N}_2$, so $v \in \mathcal{N}_1 \cap \mathcal{N}_2 = \{0\}$ and $x_1 = y_1$ and $y_2 = x_2$. □

Theorem A.14 *Let $\mathcal{N}_1$ and $\mathcal{N}_2$ be vector subspaces. If $\mathcal{N}_1 \cap \mathcal{N}_2 = \{0\}$, then $r(\mathcal{N}_1 + \mathcal{N}_2) = r(\mathcal{N}_1) + r(\mathcal{N}_2)$.*

Proof Let $v_1, \ldots, v_r$ be a basis for $\mathcal{N}_1$ and $w_1, \ldots, w_s$ be a basis for $\mathcal{N}_2$. It suffices to show that $v_1, \ldots, v_r, w_1, \ldots, w_s$ is a basis for $\mathcal{N}_1 + \mathcal{N}_2$. Clearly, $v_1, \ldots, v_r, w_1, \ldots, w_s$ is a spanning set for $\mathcal{N}_1 + \mathcal{N}_2$. If the vectors are linearly independent, the result is proven.

Suppose $0 = \sum_{i=1}^r \alpha_i v_i + \sum_{j=1}^s \beta_j w_j$ which implies that

$$\sum_{i=1}^r \alpha_i v_i = \sum_{j=1}^s -\beta_j w_j.$$

The left-hand side of the equation is a vector in $\mathcal{N}_1$ and the right-hand side is a vector in $\mathcal{N}_2$, so it is a vector in both $\mathcal{N}_1$ and $\mathcal{N}_2$, hence must be in the intersection which, by assumption, is the 0 vector. Since both $0 = \sum_{i=1}^r \alpha_i v_i$ and $0 = \sum_{j=1}^s \beta_j w_j$ and the v_is and w_js are each a basis, we must have $0 = \alpha_1 = \cdots = \alpha_r = \beta_1 = \cdots = \beta_s$, hence $v_1, \ldots, v_r, w_1, \ldots, w_s$ are linearly independent and a basis. It follows immediately that $r(\mathcal{N}_1 + \mathcal{N}_2) = r + s = r(\mathcal{N}_1) + r(\mathcal{N}_2)$. □

Definition A.15 The (Euclidean) *inner product* between two vectors x and y in $\mathbf{R}^n$ is $x'y$. Two vectors x and y are *orthogonal* (written $x \perp y$) if $x'y = 0$. Two subspaces $\mathcal{N}_1$ and $\mathcal{N}_2$ are orthogonal if $x \in \mathcal{N}_1$ and $y \in \mathcal{N}_2$ implies that $x'y = 0$. $\{x_1, \ldots, x_r\}$ is an *orthogonal basis* for a space $\mathcal{N}$ if $\{x_1, \ldots, x_r\}$ is a basis for $\mathcal{N}$ and for $i \neq j$, $x_i'x_j = 0$. $\{x_1, \ldots, x_r\}$ is an *orthonormal basis* for $\mathcal{N}$ if $\{x_1, \ldots, x_r\}$ is an orthogonal basis and $x_i'x_i = 1$ for $i = 1, \ldots, r$. *The terms orthogonal and perpendicular are used interchangeably.* The *length* of a vector x is $\|x\| \equiv \sqrt{x'x}$. The *distance* between two vectors x and y is the length of their difference, i.e., $\|x - y\|$.

The lengths of the vectors given earlier are

$$\|x_1\| = \sqrt{1^2 + 1^2 + 1^2} = \sqrt{3}, \quad \|x_2\| = 1, \quad \|x_3\| = \sqrt{22} \doteq 4.7.$$

If $x = (2, 1)'$, its length is $\|x\| = \sqrt{2^2 + 1^2} = \sqrt{5}$. If $y = (3, 2)'$, the distance between x and y is the length of $x - y = (2, 1)' - (3, 2)' = (-1, -1)'$, which is $\|x - y\| = \sqrt{(-1)^2 + (-1)^2} = \sqrt{2}$.

Just prior to Section B.4 and in Sections 2.7 and 6.3 we discuss more general versions of the concepts of inner product and length. In particular, a more general version of Definition A.15 is given in Subsection 6.3.5. The remaining results and definitions in this appendix are easily extended to general inner products.

Our emphasis on orthogonality and our need to find orthogonal projection matrices make both the following theorem and its proof fundamental tools in linear model theory:

Theorem A.16 The Gram–Schmidt Theorem.
Let $\mathscr{N}$ be a space with basis $\{x_1, \ldots, x_r\}$. There exists an orthonormal basis for $\mathscr{N}$, say $\{y_1, \ldots, y_r\}$, with y_s in the space spanned by $x_1, \ldots, x_s$, $s = 1, \ldots, r$.

Proof The Gram–Schmidt algorithm defines the y_is inductively:

$$y_1 = x_1 \Big/ \sqrt{x_1' x_1}\,,$$

$$w_s = x_s - \sum_{i=1}^{s-1} (x_s' y_i) y_i\,,$$

$$y_s = w_s \Big/ \sqrt{w_s' w_s}\,.$$

See Exercise A.1. □

This result is written using the Euclidean inner product, but terms like $x_s' y_i$ can be replaced with any general inner product, say, $\langle x_s, y_i \rangle$.

You can also apply the Gram–Schmidt algorithm to an arbitrary spanning set, rather than to a basis. It still gives an orthonormal basis for the space originally spanned but it involves bookkeeping issues. If the spanning set has linear dependencies, some of the vectors w_j in the algorithm will be 0 vectors, so they need to be dropped from the orthonormal basis.

The vectors

$$x_1 = \begin{pmatrix} 1 \\ 1 \\ 1 \end{pmatrix}, \quad x_2 = \begin{pmatrix} 1 \\ 0 \\ 0 \end{pmatrix}$$

are a basis for the space of vectors with the form $(a, b, b)'$. To orthonormalize this basis, take $y_1 = x_1/\sqrt{3}$. Then take

$$w_2 = \begin{pmatrix} 1 \\ 0 \\ 0 \end{pmatrix} - \frac{1}{\sqrt{3}} \begin{pmatrix} 1/\sqrt{3} \\ 1/\sqrt{3} \\ 1/\sqrt{3} \end{pmatrix} = \begin{pmatrix} 2/3 \\ -1/3 \\ -1/3 \end{pmatrix}.$$

Finally, normalize w_2 to give

$$y_2 = w_2 \Big/ \sqrt{6/9} = (2/\sqrt{6}, -1/\sqrt{6}, -1/\sqrt{6})'.$$

Note that another orthonormal basis for this space consists of the vectors

$$z_1 = \begin{pmatrix} 0 \\ 1/\sqrt{2} \\ 1/\sqrt{2} \end{pmatrix}, \quad z_2 = \begin{pmatrix} 1 \\ 0 \\ 0 \end{pmatrix}.$$

The result of Gram–Schmidt depends on the order in which you list the vectors in the basis. If you change the order, typically you get a different orthonormal basis. In fact, the z_is are the vectors you get if you change the order of the two x_is.

Definition A.17 For $\mathcal{N}$ a subspace of $\mathcal{M}$, let $\mathcal{N}_{\mathcal{M}}^{\perp} \equiv \{y \in \mathcal{M} | y \perp \mathcal{N}\}$. $\mathcal{N}_{\mathcal{M}}^{\perp}$ is called the *orthogonal complement* of $\mathcal{N}$ with respect to $\mathcal{M}$. If $\mathcal{M}$ is taken as $\mathbf{R}^n$, then $\mathcal{N}^{\perp} \equiv \mathcal{N}_{\mathbf{R}^n}^{\perp}$ is simply referred to as the orthogonal complement of $\mathcal{N}$.

Corollary A.18 *For any subspace* $\mathcal{N}$, $\mathcal{N} \cap \mathcal{N}^{\perp} = \{0\}$.

Proof For any $x \in \mathcal{N} \cap \mathcal{N}^{\perp}$, the vector must be orthogonal to itself, so $x'x = 0$ and $x = 0$. □

Theorem A.19 *Let* $\mathcal{M}$ *be a vector space, and let* $\mathcal{N}$ *be a subspace of* $\mathcal{M}$. $\mathcal{N}_{\mathcal{M}}^{\perp}$ *is a subspace of* $\mathcal{M}$ *and* $\mathcal{M} = \mathcal{N} + \mathcal{N}_{\mathcal{M}}^{\perp}$.

Before proving Theorem A.19 we state and prove the following corollary and give examples.

Corollary A.20 *Any vector* $x \in \mathcal{M}$ *can be written uniquely as* $x = x_1 + x_2$ *with* $x_1 \in \mathcal{N}$ *and* $x_2 \in \mathcal{N}_{\mathcal{M}}^{\perp}$. *Moreover,* $r(\mathcal{M}) = r(\mathcal{N}) + r\left(\mathcal{N}_{\mathcal{M}}^{\perp}\right)$.

Proof If $\mathcal{M} = \mathcal{N} + \mathcal{N}_{\mathcal{M}}^{\perp}$, the results are immediate from applying Corollary A.18 and Theorems A.13 and A.14. □

Let $\mathcal{M} = \mathbf{R}^3$ and let $\mathcal{N}$ be the space of vectors with the form $(a, b, b)'$. It is not difficult to see that the orthogonal complement of $\mathcal{N}$ consists of vectors of the form $(0, c, -c)'$. Any vector $(x, y, z)'$ can be written uniquely as

$$\begin{pmatrix} x \\ y \\ z \end{pmatrix} = \begin{pmatrix} x \\ (y+z)/2 \\ (y+z)/2 \end{pmatrix} + \begin{pmatrix} 0 \\ (y-z)/2 \\ -(y-z)/2 \end{pmatrix}.$$

The space of vectors with form $(a, b, b)'$ has rank 2, and the space $(0, c, -c)'$ has rank 1.

For additional examples, let

$$X_0 = \begin{bmatrix} 1 \\ 1 \\ 1 \end{bmatrix} \quad \text{and} \quad X = \begin{bmatrix} 1 & 1 \\ 1 & 2 \\ 1 & 3 \end{bmatrix}.$$

In this case,

$$C(X_0)^\perp = C\left(\begin{bmatrix} -1 & 1 \\ 0 & -2 \\ 1 & 1 \end{bmatrix}\right), \quad C(X_0)^\perp_{C(X)} = C\left(\begin{bmatrix} -1 \\ 0 \\ 1 \end{bmatrix}\right),$$

and

$$C(X)^\perp = C\left(\begin{bmatrix} 1 \\ -2 \\ 1 \end{bmatrix}\right).$$

Proof of Theorem A.19 It is easily seen that $\mathcal{N}^\perp_{\mathcal{M}}$ is a subspace by checking Theorem A.3. Since $\mathcal{N}$ and $\mathcal{N}^\perp_{\mathcal{M}}$ are subspaces of $\mathcal{M}$, we have $\mathcal{N} + \mathcal{N}^\perp_{\mathcal{M}} \subset \mathcal{M}$. To show equality it remains to show that $\mathcal{M} \subset \mathcal{N} + \mathcal{N}^\perp_{\mathcal{M}}$, i.e., if $x \in \mathcal{M}$, then $x \in \mathcal{N} + \mathcal{N}^\perp_{\mathcal{M}}$.

Let $r(\mathcal{M}) = n$ and $r(\mathcal{N}) = r$. Let $v_1, \ldots, v_r$ be a basis for $\mathcal{N}$ and extend this with $w_1, \ldots, w_{n-r}$ to a basis for $\mathcal{M}$. (Alternatively, take $w_1, \ldots, w_n$ a basis for $\mathcal{M}$ but define the spanning set $\{v_1, \ldots, v_r, w_1, \ldots, w_n\}$ for $\mathcal{M}$.) Apply Gram–Schmidt to get $v_1^*, \ldots, v_r^*, w_1^*, \ldots, w_{n-r}^*$ an orthonormal basis for $\mathcal{M}$ with $v_1^*, \ldots, v_r^*$ an orthonormal basis for $\mathcal{N}$. By construction $\{w_1^*, \ldots, w_{n-r}^*\} \subset \mathcal{N}^\perp_{\mathcal{M}}$.

If $x \in \mathcal{M}$, then

$$x = \sum_{i=1}^{r} \alpha_i v_i^* + \sum_{j=1}^{n-r} \beta_j w_j^*.$$

Let $x_0 \equiv \sum_{i=1}^{r} \alpha_i v_i^*$ and $x_1 \equiv \sum_{j=1}^{n-r} \beta_j w_j^*$. Then $x_0 \in \mathcal{N}$, $x_1 \in \mathcal{N}^\perp_{\mathcal{M}}$, and $x = x_0 + x_1$, so $x \in \mathcal{N} + \mathcal{N}^\perp_{\mathcal{M}}$. □

From Corollary A.20 we have $r\left(\mathcal{N}^\perp_{\mathcal{M}}\right) = n - r$, so $w_1^*, \ldots, w_{n-r}^*$ must be a basis for $\mathcal{N}^\perp_{\mathcal{M}}$. This relies on the fact that if the w_j^*s are all in $\mathcal{N}^\perp_{\mathcal{M}}$ and if they are linearly independent with the same number of vectors as in a basis for the $\mathcal{N}^\perp_{\mathcal{M}}$, then they must be a spanning set for $\mathcal{N}^\perp_{\mathcal{M}}$. But we have not shown directly that $w_1^*, \ldots, w_{n-r}^*$ is a spanning set for $\mathcal{N}^\perp_{\mathcal{M}}$. We do that now.

If $x \in \mathcal{N}^\perp_{\mathcal{M}}$, because $x \in \mathcal{M}$ write

$$x = \sum_{i=1}^{r} \alpha_i v_i^* + \sum_{j=1}^{n-r} \beta_j w_j^*.$$

Since $x \in \mathcal{N}^\perp_{\mathcal{M}}$ and $v_k^* \in \mathcal{N}$ for $k = 1, \ldots, r$,

$$\begin{aligned}
0 = x'v_k^* &= \left(\sum_{i=1}^{r} \alpha_i v_i^* + \sum_{j=1}^{n-r} \beta_j w_j^*\right)' v_k^* \\
&= \sum_{i=1}^{r} \alpha_i v_i^{*\prime} v_k^* + \sum_{j=1}^{n-r} \beta_j w_j^{*\prime} v_k^* \\
&= \alpha_k v_k^{*\prime} v_k^* = \alpha_k .
\end{aligned}$$

Thus $x = \sum_{j=1}^{n-r} \beta_j w_j^*$, implying that $\{w_1^*, \ldots, w_{n-r}^*\}$ is a spanning set and a basis for $\mathscr{N}_{\mathscr{M}}^{\perp}$.

Theorem A.21 *Let $\mathscr{N}_1$ and $\mathscr{N}_2$ be subspaces, then $(\mathscr{N}_1 \cap \mathscr{N}_2)^{\perp} = \mathscr{N}_1^{\perp} + \mathscr{N}_2^{\perp}$.*

Proof This is a restatement of Proposition 10.4.6. The proof is given there. □

Exercises

Exercise A.1 Give a detailed proof of the Gram–Schmidt theorem.

Questions A.2 through A.13 involve the following matrices:

$$A = \begin{bmatrix} 1 & 1 & 0 & 0 \\ 1 & 1 & 0 & 0 \\ 0 & 0 & 1 & 0 \\ 0 & 0 & 1 & 1 \end{bmatrix}, \quad B = \begin{bmatrix} 1 & 0 & 0 \\ 1 & 0 & 0 \\ 0 & 1 & 0 \\ 0 & 0 & 1 \end{bmatrix}, \quad D = \begin{bmatrix} 1 & 0 \\ 1 & 0 \\ 2 & 5 \\ 0 & 0 \end{bmatrix}, \quad E = \begin{bmatrix} 1 & 2 \\ 1 & 2 \\ 2 & 7 \\ 0 & 0 \end{bmatrix},$$

$$F = \begin{bmatrix} 1 & 5 & 6 \\ 1 & 5 & 6 \\ 0 & 7 & 2 \\ 0 & 0 & 9 \end{bmatrix}, \quad G = \begin{bmatrix} 1 & 0 & 5 & 2 \\ 1 & 0 & 5 & 2 \\ 2 & 5 & 7 & 9 \\ 0 & 0 & 0 & 3 \end{bmatrix}, \quad H = \begin{bmatrix} 1 & 0 & 2 & 2 & 6 \\ 1 & 0 & 2 & 2 & 6 \\ 7 & 9 & 3 & 9 & -1 \\ 0 & 0 & 0 & 3 & -7 \end{bmatrix},$$

$$K = \begin{bmatrix} 1 & 0 & 0 \\ 1 & 0 & 0 \\ 1 & 1 & 0 \\ 1 & 0 & 1 \end{bmatrix}, \quad L = \begin{bmatrix} 2 & 0 & 0 \\ 2 & 0 & 0 \\ 1 & 1 & 0 \\ 1 & 0 & 1 \end{bmatrix}, \quad N = \begin{bmatrix} 1 \\ 2 \\ 3 \\ 4 \end{bmatrix}.$$

Exercise A.2 Is the space spanned by the columns of A the same as the space spanned by the columns of B? How about the spaces spanned by the columns of K, L, F, D, and G?

Exercise A.3 Give a matrix whose column space contains $C(A)$.

Exercise A.4 Give two matrices whose column spaces contain $C(B)$.

Exercise A.5 Which of the following equalities are valid: $C(A) = C(A, D)$, $C(D) = C(A, B)$, $C(A, N) = C(A)$, $C(N) = C(A)$, $C(A) = C(F)$, $C(A) = C(G)$, $C(A) = C(H)$, $C(A) = C(D)$?

Exercise A.6 Which of the following matrices have linearly independent columns: A, B, D, N, F, H, G?

Exercise A.7 Give a basis for the space spanned by the columns of each of the following matrices: A, B, D, N, F, H, G.

Exercise A.8 Give the ranks of $A, B, D, E, F, G, H, K, L, N$.

Exercise A.9 Which of the following matrices have columns that are mutually orthogonal: B, A, D?

Exercise A.10 Give an orthogonal basis for the space spanned by the columns of each of the following matrices: A, D, N, K, H, G.

Exercise A.11 Find $C(A)^{\perp}$ and $C(B)^{\perp}$ (with respect to $\mathbf{R}^4$).

Exercise A.12 Find two linearly independent vectors in the orthogonal complement of $C(D)$ (with respect to $\mathbf{R}^4$).

Exercise A.13 Find a vector in the orthogonal complement of $C(D)$ with respect to $C(A)$.

Exercise A.14 Find an orthogonal basis for the space spanned by the columns of

$$X = \begin{bmatrix} 1 & 1 & 4 \\ 1 & 2 & 1 \\ 1 & 3 & 0 \\ 1 & 4 & 0 \\ 1 & 5 & 1 \\ 1 & 6 & 4 \end{bmatrix}.$$

Exercise A.15 For X as above, find two linearly independent vectors in the orthogonal complement of $C(X)$ (with respect to $\mathbf{R}^6$).

Exercise A.16 Let X be an $n \times p$ matrix. Prove or disprove the following statement: Every vector in $\mathbf{R}^n$ is in either $C(X)$ or $C(X)^{\perp}$ or both.

Exercise A.17 For any matrix A, prove that $C(A)$ and the null space of A' are orthogonal complements. Note: The null space is defined in Definition B.11.

Appendix B
Matrix Results

Abstract This appendix reviews standard ideas in matrix theory with emphasis given to important results that are less commonly taught in a junior/senior level linear algebra course. The appendix begins with basic definitions and results. A section devoted to eigenvalues and their applications follows. This section contains a number of standard definitions, but it also contains a number of very specific results that are unlikely to be familiar to people with only an undergraduate background in linear algebra. The third section is devoted to an intense (brief but detailed) examination of projections and their properties. The appendix closes with some miscellaneous results, some results on Kronecker products and Vec operators, and an introduction to tensors.

B.1 Basic Ideas

Definition B.1 Any matrix with the same number of rows and columns is called a *square matrix*.

Definition B.2 Let $A = [a_{ij}]$ be a matrix. The *transpose* of A, written A', is the matrix $A' = [b_{ij}]$, where $b_{ij} = a_{ji}$.

Definition B.3 If $A = A'$, then A is called *symmetric*. Note that only square matrices can be symmetric.

Definition B.4 If $A = [a_{ij}]$ is a square matrix and $a_{ij} = 0$ for $i \neq j$, then A is a *diagonal matrix*. If $\lambda_1, \ldots, \lambda_n$ are scalars, then $D(\lambda_j)$ and $\text{Diag}(\lambda_j)$ are used to indicate an $n \times n$ matrix $D = [d_{ij}]$ with $d_{ij} = 0$, $i \neq j$, and $d_{ii} = \lambda_i$. If $\lambda \equiv (\lambda_1, \ldots, \lambda_n)'$, then $D(\lambda) \equiv D(\lambda_j)$. A diagonal matrix with all 1s on the diagonal is called an *identity matrix* and is denoted I. Occasionally, I_n is used to denote an $n \times n$ identity matrix.

R. Christensen, *Plane Answers to Complex Questions*, Springer Texts in Statistics,
https://doi.org/10.1007/978-3-030-32097-3

If $A = [a_{ij}]$ is $n \times p$ and $B = [b_{ij}]$ is $n \times q$, we can write an $n \times (p+q)$ matrix $C = [A, B]$, where $c_{ij} = a_{ij}$, $i = 1, \ldots, n$, $j = 1, \ldots, p$, and $c_{ij} = b_{i,j-p}$, $i = 1, \ldots, n$, $j = p+1, \ldots, p+q$. This notation can be extended in obvious ways, e.g., $C' = \begin{bmatrix} A' \\ B' \end{bmatrix}$.

Definition B.5 Let $A = [a_{ij}]$ be an $r \times c$ matrix and $B = [b_{ij}]$ be an $s \times d$ matrix. The *Kronecker product* of A and B, written $A \otimes B$, is an $r \times c$ matrix of $s \times d$ matrices. The matrix in the ith row and jth column is $a_{ij}B$. In total, $A \otimes B$ is an $rs \times cd$ matrix.

Definition B.6 Let A be an $r \times c$ matrix. Write $A = [A_1, A_2, \ldots, A_c]$, where A_i is the ith column of A. The *Vec* operator stacks the columns of A into an $rc \times 1$ vector; thus,

$$[\text{Vec}(A)]' = [A_1', A_2', \ldots, A_c'].$$

Example B.7

$$A = \begin{bmatrix} 1 & 4 \\ 2 & 5 \end{bmatrix}, \quad B = \begin{bmatrix} 1 & 3 \\ 0 & 4 \end{bmatrix},$$

$$A \otimes B = \begin{bmatrix} 1\begin{pmatrix} 1 & 3 \\ 0 & 4 \end{pmatrix} & 4\begin{pmatrix} 1 & 3 \\ 0 & 4 \end{pmatrix} \\ 2\begin{pmatrix} 1 & 3 \\ 0 & 4 \end{pmatrix} & 5\begin{pmatrix} 1 & 3 \\ 0 & 4 \end{pmatrix} \end{bmatrix} = \begin{bmatrix} 1 & 3 & 4 & 12 \\ 0 & 4 & 0 & 16 \\ 2 & 6 & 5 & 15 \\ 0 & 8 & 0 & 20 \end{bmatrix},$$

$$\text{Vec}(A) = [1, 2, 4, 5]'.$$

Definition B.8 Let A be an $n \times n$ matrix. A is *nonsingular* if there exists a matrix A^{-1} such that $A^{-1}A = I = AA^{-1}$. If no such matrix exists, then A is singular. If A^{-1} exists, it is called the *inverse* of A.

Theorem B.9 *An $n \times n$ matrix A is nonsingular if and only if $r(A) = n$, i.e., the columns of A form a basis for $\mathbf{R}^n$.*

Corollary B.10 *$A_{n\times n}$ is singular if and only if there exists $x \neq 0$ such that $Ax = 0$.*

For any matrix A, the set of all x such that $Ax = 0$ is easily seen to be a vector space.

Definition B.11 The set of all x such that $Ax = 0$ is called the *null space* of A and written $\mathscr{N}(A)$.

Theorem B.12 *If A is $n \times n$ and $r(A) = r$, then the null space of A has rank $n - r$.*

B.2 Eigenvalues and Related Results

The material in this section deals with eigenvalues and eigenvectors either in the statements of the results or in their proofs. Again, this is meant to be a brief review of important concepts; but, in addition, there are a number of specific results that may be unfamiliar.

Definition B.13 The scalar λ is an *eigenvalue* of $A_{n \times n}$ if $A - \lambda I$ is singular. λ is an eigenvalue of *multiplicity* s if the rank of the null space of $A - \lambda I$ is s. A nonzero vector x is an *eigenvector* of A corresponding to the eigenvalue λ if x is in the null space of $A - \lambda I$, i.e., if $Ax = \lambda x$. Eigenvalues are also called *singular values* and *characteristic roots*.

For example,

$$\begin{bmatrix} 2 & 1 \\ 1 & 2 \end{bmatrix} \begin{pmatrix} 1 \\ 1 \end{pmatrix} = 3 \begin{pmatrix} 1 \\ 1 \end{pmatrix}$$

and

$$\begin{bmatrix} 2 & 1 \\ 1 & 2 \end{bmatrix} \begin{pmatrix} -1 \\ 1 \end{pmatrix} = 1 \begin{pmatrix} -1 \\ 1 \end{pmatrix}.$$

Combining the two equations gives

$$\begin{bmatrix} 2 & 1 \\ 1 & 2 \end{bmatrix} \begin{bmatrix} 1 & -1 \\ 1 & 1 \end{bmatrix} = \begin{bmatrix} 1 & -1 \\ 1 & 1 \end{bmatrix} \begin{bmatrix} 3 & 0 \\ 0 & 1 \end{bmatrix}.$$

Note that if $\lambda \neq 0$ is an eigenvalue of A, the eigenvectors corresponding to λ (along with the vector 0) form a subspace of $C(A)$. For example, if $Ax_1 = \lambda x_1$ and $Ax_2 = \lambda x_2$, then $A(x_1 + x_2) = \lambda(x_1 + x_2)$, so the set of eigenvectors is closed under vector addition. Similarly, it is closed under scalar multiplication, so it forms a subspace (except that eigenvectors cannot be 0 and every subspace contains 0). If $\lambda = 0$, the subspace is the null space of A.

If A is a symmetric matrix, and γ and λ are distinct eigenvalues, then the eigenvectors corresponding to λ and γ are orthogonal. To see this, let x be an eigenvector for λ and y an eigenvector for γ. Then $\lambda x'y = x'Ay = \gamma x'y$, which can happen only if $\lambda = \gamma$ or if $x'y = 0$. Since λ and γ are distinct, we have $x'y = 0$.

Let $\lambda_1, \ldots, \lambda_r$ be the distinct nonzero eigenvalues of a symmetric matrix A with respective multiplicities $s(1), \ldots, s(r)$. Let $v_{i1}, \ldots, v_{is(i)}$ be a basis for the space of eigenvectors of λ_i. We want to show that $v_{11}, v_{12}, \ldots, v_{rs(r)}$ is a basis for $C(A)$.

Suppose $v_{11}, v_{12}, \ldots, v_{rs(r)}$ is not a basis. Since $v_{ij} \in C(A)$ and the v_{ij}s are linearly independent, we can pick $x \in C(A)$ with $x \perp v_{ij}$ for all i and j. Note that since $Av_{ij} = \lambda_i v_{ij}$, we have $(A)^p v_{ij} = (\lambda_i)^p v_{ij}$. In particular, $x'(A)^p v_{ij} = x'(\lambda_i)^p v_{ij} = (\lambda_i)^p x' v_{ij} = 0$, so $A^p x \perp v_{ij}$ for any i, j, and p. The vectors $x, Ax, A^2x, \ldots$ cannot all be linearly independent, so there exists a smallest value $k \leq n$ such that

$$A^k x + b_{k-1} A^{k-1} x + \cdots + b_0 x = 0.$$

Since there is a solution to this, for some real number μ we can write the equation as

$$(A - \mu I)\left(A^{k-1}x + \gamma_{k-2}A^{k-2}x + \cdots + \gamma_0 x\right) = 0,$$

and μ is an eigenvalue. (See Exercise B.1.) An eigenvector for μ is $y = A^{k-1}x + \gamma_{k-2}A^{k-2}x + \cdots + \gamma_0 x$. Clearly, $y \perp v_{ij}$ for any i and j. Since k was chosen as the smallest value to get linear dependence, we have $y \neq 0$. If $\mu \neq 0$, y is an eigenvector that does not correspond to any of $\lambda_1, \ldots, \lambda_r$, a contradiction. If $\mu = 0$, we have $Ay = 0$; and since A is symmetric, y is a vector in $C(A)$ that is orthogonal to every other vector in $C(A)$, i.e., $y'y = 0$ but $y \neq 0$, a contradiction. We have proven

Theorem B.14 *If A is a symmetric matrix, then there exists a basis for $C(A)$ consisting of eigenvectors of nonzero eigenvalues. If λ is a nonzero eigenvalue of multiplicity s, then the basis will contain s eigenvectors for λ.*

If λ is an eigenvalue of A with multiplicity s, then we can think of λ as being an eigenvalue s times. With this convention, the rank of A is the number of nonzero eigenvalues. The total number of eigenvalues is n if A is an $n \times n$ matrix.

For a symmetric matrix A, if we use eigenvectors corresponding to the zero eigenvalue, we can get a basis for $\mathbf{R}^n$ consisting of eigenvectors. We already have a basis for $C(A)$, and the eigenvectors of 0 are the null space of A. For A symmetric, $C(A)$ and the null space of A are orthogonal complements. Let $\lambda_1, \ldots, \lambda_n$ be the eigenvalues of a symmetric matrix A. Let $v_1, \ldots, v_n$ denote a basis of eigenvectors for $\mathbf{R}^n$, with v_i being an eigenvector for λ_i for any i.

Theorem B.15 *If A is symmetric, there exists an orthonormal basis for $\mathbf{R}^n$ consisting of eigenvectors of A.*

Proof Assume $\lambda_{i1} = \cdots = \lambda_{ik}$ are all the λ_is equal to any particular value λ, and let $v_{i1}, \ldots, v_{ik}$ be a basis for the space of eigenvectors for λ. By Gram–Schmidt there exists an orthonormal basis $w_{i1}, \ldots, w_{ik}$ for the space of eigenvectors corresponding to λ. If we do this for each distinct eigenvalue, we get a collection of orthonormal sets that form a basis for $\mathbf{R}^n$. Since, as we have seen, for $\lambda_i \neq \lambda_j$, any eigenvector for λ_i is orthogonal to any eigenvector for λ_j, the basis is orthonormal. □

Definition B.16 A square matrix P is *orthonormal* (more often called *orthogonal*) if $P' = P^{-1}$. Note that if P is orthonormal, so is P'.

Some examples of orthonormal matrices are

$$P_1 = \frac{1}{\sqrt{6}}\begin{bmatrix} \sqrt{2} & -\sqrt{3} & 1 \\ \sqrt{2} & 0 & -2 \\ \sqrt{2} & \sqrt{3} & 1 \end{bmatrix}, \quad P_2 = \frac{1}{\sqrt{2}}\begin{bmatrix} 1 & 1 \\ 1 & -1 \end{bmatrix},$$

$$P_3 = \begin{bmatrix} 1 & 0 & 0 \\ 0 & -1 & 0 \\ 0 & 0 & 1 \end{bmatrix}.$$

Theorem B.17 $P_{n\times n}$ *is* orthonormal *if and only if the columns of P form an orthonormal basis for* $\mathbf{R}^n$.

Proof $\Leftarrow$ It is clear that if the columns of P form an orthonormal basis for $\mathbf{R}^n$, then $P'P = I$.

$\Rightarrow$ Since P is nonsingular, the columns of P form a basis for $\mathbf{R}^n$. Since $P'P = I$, the basis is orthonormal. □

Corollary B.18 $P_{n\times n}$ *is orthonormal if and only if the rows of P form an orthonormal basis for* $\mathbf{R}^n$.

Proof P is orthonormal if and only if P' is orthonormal if and only if the columns of P' are an orthonormal basis if and only if the rows of P are an orthonormal basis. □

Theorem B.19 *If A is an* $n \times n$ *symmetric matrix, then there exists an orthonormal matrix P such that* $P'AP = Diag(\lambda_i)$, *where* $\lambda_1, \lambda_2, \ldots, \lambda_n$ *are the eigenvalues of A.*

Proof Let $v_1, v_2, \ldots, v_n$ be an orthonormal set of eigenvectors of A corresponding, respectively, to $\lambda_1, \lambda_2, \ldots, \lambda_n$. Let $P = [v_1, \ldots, v_n]$. Then

$$\begin{aligned} P'AP &= \begin{bmatrix} v_1' \\ \vdots \\ v_n' \end{bmatrix} [Av_1, \ldots, Av_n] \\ &= \begin{bmatrix} v_1' \\ \vdots \\ v_n' \end{bmatrix} [\lambda_1 v_1, \ldots, \lambda_n v_n] \\ &= \begin{bmatrix} \lambda_1 v_1' v_1 & \ldots & \lambda_n v_1' v_n \\ \vdots & \ddots & \vdots \\ \lambda_1 v_n' v_1 & \ldots & \lambda_n v_n' v_n \end{bmatrix} \\ &= \text{Diag}(\lambda_i). \end{aligned}$$

□

The *singular value decomposition* for a symmetric matrix is given by the following corollary.

Corollary B.20 $A = PD(\lambda_i)P'$.

For example, using results illustrated earlier,

$$\begin{bmatrix} 2 & 1 \\ 1 & 2 \end{bmatrix} = \begin{bmatrix} 1/\sqrt{2} & -1/\sqrt{2} \\ 1/\sqrt{2} & 1/\sqrt{2} \end{bmatrix} \begin{bmatrix} 3 & 0 \\ 0 & 1 \end{bmatrix} \begin{bmatrix} 1/\sqrt{2} & 1/\sqrt{2} \\ -1/\sqrt{2} & 1/\sqrt{2} \end{bmatrix}.$$

Definition B.21 A symmetric matrix A is *positive (nonnegative) definite* if, for any nonzero vector $v \in \mathbf{R}^n$, $v'Av$ is positive (nonnegative).

Theorem B.22 *A is nonnegative definite if and only if there exists a square matrix Q such that* $A = QQ'$.

Proof $\Rightarrow$ We know that there exists P orthonormal with $P'AP = \text{Diag}(\lambda_i)$. The λ_is must all be nonnegative, because if $e_j' = (0, \ldots, 0, 1, 0, \ldots, 0)$ with the 1 in the jth place and we let $v = Pe_j$, then $0 \leq v'Av = e_j'\text{Diag}(\lambda_i)e_j = \lambda_j$. Let $Q = P\text{Diag}\left(\sqrt{\lambda_i}\right)$. Then, since $P\text{Diag}(\lambda_i)P' = A$, we have

$$QQ' = P\text{Diag}(\lambda_i)P' = A.$$

$\Leftarrow$ If $A = QQ'$, then $v'Av = (Q'v)'(Q'v) \geq 0$. □

Corollary B.23 *A is positive definite if and only if Q is nonsingular for any choice of Q.*

Proof There exists $v \neq 0$ such that $v'Av = 0$ if and only if there exists $v \neq 0$ such that $Q'v = 0$, which occurs if and only if Q' is singular. The contrapositive of this is that $v'Av > 0$ for all $v \neq 0$ if and only if Q' is nonsingular. □

In the interest of brevity, I have dropped Theorem B.24, Corollary B.25, and Corollary B.26 that appeared in earlier editions.

Definition B.27 Let $A = [a_{ij}]$ be an $n \times n$ matrix. The *trace* of A is $\text{tr}(A) = \sum_{i=1}^{n} a_{ii}$.

Theorem B.28 *For matrices* $A_{r \times s}$ *and* $B_{s \times r}$, $\text{tr}(AB) = \text{tr}(BA)$.

Proof See Exercise B.8. □

Theorem B.29 *If $A_{n\times n}$ is a symmetric matrix,* $\mathrm{tr}(A) = \sum_{i=1}^{n} \lambda_i$*, where $\lambda_1, \ldots, \lambda_n$ are the eigenvalues of A.*

Proof $A = PD(\lambda_i)P'$ with P orthonormal

$$\begin{aligned}\mathrm{tr}(A) &= \mathrm{tr}[PD(\lambda_i)P'] = \mathrm{tr}[D(\lambda_i)P'P] \\ &= \mathrm{tr}[D(\lambda_i)] = \sum_{i=1}^{n} \lambda_i.\end{aligned}$$

□

To illustrate, we saw earlier that the matrix $\begin{bmatrix} 2 & 1 \\ 1 & 2 \end{bmatrix}$ had eigenvalues of 3 and 1. In fact, a stronger result than Theorem B.29 is true. We give it without proof.

Theorem B.30 $\mathrm{tr}(A) = \sum_{i=1}^{n} \lambda_i$*, where $\lambda_1, \ldots, \lambda_n$ are the eigenvalues of A. Moreover, the determinant of A is* $\det(A) = \prod_{i=1}^{n} \lambda_i$.

B.3 Projections

This section is devoted primarily to a discussion of perpendicular projection operators. It begins with their definition, some basic properties, and two important characterizations: Theorems B.33 and B.35. A third important characterization, Theorem B.44, involves generalized inverses. Generalized inverses are defined, briefly studied, and applied to projection operators. The section continues with the examination of the relationships between two perpendicular projection operators and closes with discussions of the Gram–Schmidt theorem, eigenvalues of projection operators, and oblique (nonperpendicular) projection operators.

We begin by defining a *perpendicular projection operator (ppo)* onto an arbitrary space. To be consistent with later usage, we denote the arbitrary space $C(X)$ for some matrix X.

Definition B.31 M is a perpendicular projection operator (matrix) onto $C(X)$ if and only if

(i) $v \in C(X)$ implies $Mv = v$ (projection),
(ii) $w \perp C(X)$ implies $Mw = 0$ (perpendicularity).

For example, consider the subspace of $\mathbf{R}^2$ determined by vectors of the form $(2a, a)'$. It is not difficult to see that the orthogonal complement of this subspace consists of vectors of the form $(b, -2b)'$. The perpendicular projection operator onto the $(2a, a)'$ subspace is

$$M = \begin{bmatrix} 0.8 & 0.4 \\ 0.4 & 0.2 \end{bmatrix}.$$

To verify this note that

$$M \begin{pmatrix} 2a \\ a \end{pmatrix} = \begin{bmatrix} 0.8 & 0.4 \\ 0.4 & 0.2 \end{bmatrix} \begin{pmatrix} 2a \\ a \end{pmatrix} = \begin{pmatrix} (0.8)2a + 0.4a \\ (0.4)2a + 0.2a \end{pmatrix} = \begin{pmatrix} 2a \\ a \end{pmatrix}$$

and

$$M \begin{pmatrix} b \\ -2b \end{pmatrix} = \begin{bmatrix} 0.8 & 0.4 \\ 0.4 & 0.2 \end{bmatrix} \begin{pmatrix} b \\ -2b \end{pmatrix} = \begin{pmatrix} 0.8b + 0.4(-2b) \\ 0.4b + 0.2(-2b) \end{pmatrix} = \begin{pmatrix} 0 \\ 0 \end{pmatrix}.$$

Notationally, M is used to indicate the ppo onto $C(X)$. If A is another matrix, M_A denotes the ppo onto $C(A)$. Thus, $M \equiv M_X$. When X has a subscript we typically write the ppo onto $C(X_0)$ as $M_0 \equiv M_{X_0}$ and, similarly, $M_1 \equiv M_{X_1}$, but often $M_2 \neq M_{X_2}$.

Proposition B.32 *If M is a perpendicular projection operator onto $C(X)$, then $C(M) = C(X)$.*

Proof See Exercise B.2. □

Note that both columns of

$$M = \begin{bmatrix} 0.8 & 0.4 \\ 0.4 & 0.2 \end{bmatrix}$$

have the form $(2a, a)'$.

Theorem B.33 *M is a perpendicular projection operator on $C(M)$ if and only if $MM = M$ and $M' = M$.*

Proof $\Rightarrow$ Write $v = v_1 + v_2$, where $v_1 \in C(M)$ and $v_2 \perp C(M)$, and let $w = w_1 + w_2$ with $w_1 \in C(M)$ and $w_2 \perp C(M)$. Since $(I - M)v = (I - M)v_2 = v_2$ and $Mw = Mw_1 = w_1$, we get

$$w'M'(I - M)v = w_1'M'(I - M)v_2 = w_1'v_2 = 0.$$

This is true for any v and w, so we have $M'(I - M) = 0$ or $M' = M'M$. Since $M'M$ is symmetric, M' must also be symmetric, and this implies that $M = MM$.

$\Leftarrow$ If $M^2 = M$ and $v \in C(M)$, then since $v = Mb$ we have $Mv = MMb = Mb = v$. If $M' = M$ and $w \perp C(M)$, then $Mw = M'w = 0$ because the columns of M are in $C(M)$. □

In our example,

$$MM = \begin{bmatrix} 0.8 & 0.4 \\ 0.4 & 0.2 \end{bmatrix} \begin{bmatrix} 0.8 & 0.4 \\ 0.4 & 0.2 \end{bmatrix} = \begin{bmatrix} 0.8 & 0.4 \\ 0.4 & 0.2 \end{bmatrix} = M$$

and

$$M = \begin{bmatrix} 0.8 & 0.4 \\ 0.4 & 0.2 \end{bmatrix} = M'.$$

Proposition B.34 *Perpendicular projection operators are unique.*

Proof Let M and P be perpendicular projection operators onto some space $\mathcal{M}$. Let $v \in \mathbf{R}^n$ and write $v = v_1 + v_2$, $v_1 \in \mathcal{M}$, $v_2 \perp \mathcal{M}$. Since v is arbitrary and $Mv = v_1 = Pv$, we have $M = P$. □

For any matrix X, we will now find two ways to characterize the perpendicular projection operator onto $C(X)$. The first method depends on the Gram–Schmidt theorem; the second depends on the concept of a generalized inverse.

Theorem B.35 *Let $o_1, \ldots, o_r$ be an orthonormal basis for $C(X)$, and let $O = [o_1, \ldots, o_r]$. Then $OO' = \sum_{i=1}^{r} o_i o_i'$ is the perpendicular projection operator onto $C(X)$.*

Proof OO' is symmetric and $OO'OO' = OI_rO' = OO'$; so, by Theorem B.33, it only remains to show that $C(OO') = C(X)$. Clearly $C(OO') \subset C(O) = C(X)$. On the other hand, if $v \in C(O)$, then $v = Ob$ for some vector $b \in \mathbf{R}^r$ and $v = Ob = OI_rb = OO'Ob$; so clearly $v \in C(OO')$. □

For example, to find the perpendicular projection operator for vectors of the form $(2a, a)'$, we can find an orthonormal basis. The space has rank 1 and to normalize $(2a, a)'$, we must have

$$1 = (2a, a)\begin{pmatrix} 2a \\ a \end{pmatrix} = 4a^2 + a^2 = 5a^2;$$

so $a^2 = 1/5$ and $a = \pm 1/\sqrt{5}$. If we take $(2/\sqrt{5}, 1/\sqrt{5})'$ as our orthonormal basis, then

$$M = \begin{pmatrix} 2/\sqrt{5} \\ 1/\sqrt{5} \end{pmatrix} (2/\sqrt{5}, 1/\sqrt{5}) = \begin{bmatrix} 0.8 & 0.4 \\ 0.4 & 0.2 \end{bmatrix},$$

as was demonstrated earlier.

One use of Theorem B.35 is that, given a matrix X, one can use the Gram–Schmidt theorem to get an orthonormal basis for $C(X)$ and thus obtain the perpendicular projection operator.

We now examine properties of generalized inverses. Generalized inverses are a generalization on the concept of the inverse of a matrix. Although the most common use of generalized inverses is in solving systems of linear equations, our interest lies primarily in their relationship to projection operators. The discussion below is given for an arbitrary matrix A.

Definition B.36 A *generalized inverse* of a matrix A is any matrix G such that $AGA = A$. The notation A^- is used to indicate a generalized inverse of A.

Theorem B.37 *If A is nonsingular, the unique generalized inverse of A is A^{-1}.*

Proof $AA^{-1}A = IA = A$, so A^{-1} is a generalized inverse. If $AA^-A = A$, then $AA^- = AA^-AA^{-1} = AA^{-1} = I$; so A^- is the inverse of A. □

Theorem B.38 *For any symmetric matrix A, there exists a generalized inverse of A.*

Proof There exists P orthonormal so that $P'AP = D(\lambda_i)$ and $A = PD(\lambda_i)P'$. Let

$$\gamma_i = \begin{cases} 1/\lambda_i, & \text{if } \lambda_i \neq 0 \\ 0, & \text{if } \lambda_i = 0, \end{cases}$$

and $G = PD(\gamma_i)P'$. We now show that G is a generalized inverse of A. P is orthonormal, so $P'P = I$ and

$$\begin{aligned} AGA &= PD(\lambda_i)P'PD(\gamma_i)P'PD(\lambda_i)P' \\ &= PD(\lambda_i)D(\gamma_i)D(\lambda_i)P' \\ &= PD(\lambda_i)P' \\ &= A. \end{aligned}$$

□

Although this is the only existence result we really need, later we will show that generalized inverses exist for arbitrary matrices.

Theorem B.39 *If G_1 and G_2 are generalized inverses of A, then so is G_1AG_2.*

Proof $A(G_1AG_2)A = (AG_1A)G_2A = AG_2A = A$. □

For A symmetric, A^- need not be symmetric.

Example B.40 Consider the matrix

$$\begin{bmatrix} a & b \\ b & b^2/a \end{bmatrix}.$$

It has a generalized inverse

$$\begin{bmatrix} 1/a & -1 \\ 1 & 0 \end{bmatrix},$$

and in fact, by considering the equation

$$\begin{bmatrix} a & b \\ b & b^2/a \end{bmatrix} \begin{bmatrix} r & s \\ t & u \end{bmatrix} \begin{bmatrix} a & b \\ b & b^2/a \end{bmatrix} = \begin{bmatrix} a & b \\ b & b^2/a \end{bmatrix},$$

it can be shown that if $r = 1/a$, then any solution of $at + as + bu = 0$ gives a generalized inverse.

Corollary B.41 *For a symmetric matrix A, there exists A^- such that $A^-AA^- = A^-$ and $(A^-)' = A^-$.*

Proof Take A^- as the generalized inverse in the proof of Theorem B.38. Clearly, $A^- = PD(\gamma_i)P'$ is symmetric and

$$A^-AA^- = PD(\gamma_i)P'PD(\lambda_i)P'PD(\gamma_i)P' = PD(\gamma_i)D(\lambda_i)D(\gamma_i)P' = PD(\gamma_i)P' = A^-.$$

□

Definition B.42 A generalized inverse A^- for a matrix A that has the property $A^-AA^- = A^-$ is said to be *reflexive*.

Corollary B.41 establishes the existence of a reflexive generalized inverse for any symmetric matrix.

Generalized inverses are of interest in that they provide an alternative to the characterization of perpendicular projection matrices given in Theorem B.35. The two results immediately below characterize the perpendicular projection matrix onto $C(X)$.

Lemma B.43 *If G and H are generalized inverses of $(X'X)$, then*

(i) $XGX'X = XHX'X = X$,
(ii) $XGX' = XHX'$.

Proof For $v \in \mathbf{R}^n$, let $v = v_1 + v_2$ with $v_1 \in C(X)$ and $v_2 \perp C(X)$. Also let $v_1 = Xb$ for some vector b. Then

$$v'XGX'X = v_1'XGX'X = b'(X'X)G(X'X) = b'(X'X) = v'X.$$

Since v and G are arbitrary, we have shown (i).

To see (ii), observe that for the arbitrary vector v above,

$$XGX'v = XGX'Xb = XHX'Xb = XHX'v.$$

□

Since $X'X$ is symmetric, there exists a generalized inverse $(X'X)^-$ that is symmetric. For this generalized inverse, $X(X'X)^-X'$ is symmetric; so, by the above lemma, $X(X'X)^-X'$ must be symmetric for any choice of $(X'X)^-$.

Theorem B.44 *$X(X'X)^-X'$ is the perpendicular projection operator onto $C(X)$.*

Proof We need to establish conditions (i) and (ii) of Definition B.31. (i) For $v \in C(X)$, write $v = Xb$, so by Lemma B.43, $X(X'X)^-X'v = X(X'X)^-X'Xb = Xb = v$. (ii) If $w \perp C(X)$, $X(X'X)^-X'w = 0$. □

For example, one spanning set for the subspace of vectors with the form $(2a, a)'$ is $(2, 1)'$. It follows that

$$M = \begin{pmatrix} 2 \\ 1 \end{pmatrix} \left[(2, 1) \begin{pmatrix} 2 \\ 1 \end{pmatrix} \right]^{-1} (2, 1) = \begin{bmatrix} 0.8 & 0.4 \\ 0.4 & 0.2 \end{bmatrix},$$

as was shown earlier.

The next five results examine the relationships between two perpendicular projection matrices.

Theorem B.45 *Let M_1 and M_2 be perpendicular projection matrices on $\mathbf{R}^n$. $(M_1 + M_2)$ is the perpendicular projection matrix onto $C(M_1, M_2)$ if and only if $C(M_1) \perp C(M_2)$.*

Proof $\Leftarrow$ If $C(M_1) \perp C(M_2)$, then $M_1M_2 = M_2M_1 = 0$. Because

$$(M_1 + M_2)^2 = M_1^2 + M_2^2 + M_1M_2 + M_2M_1 = M_1^2 + M_2^2 = M_1 + M_2$$

and

$$(M_1 + M_2)' = M_1' + M_2' = M_1 + M_2,$$

$M_1 + M_2$ is the perpendicular projection matrix onto $C(M_1 + M_2)$. Clearly $C(M_1 + M_2) \subset C(M_1, M_2)$. To see that $C(M_1, M_2) \subset C(M_1 + M_2)$, write $v = M_1b_1 + M_2b_2$. Then, because $M_1M_2 = M_2M_1 = 0$, $(M_1 + M_2)v = v$. Thus, $C(M_1, M_2) = C(M_1 + M_2)$.

$\Rightarrow$ If $M_1 + M_2$ is a perpendicular projection matrix, then

$$\begin{aligned}(M_1 + M_2) &= (M_1 + M_2)^2 = M_1^2 + M_2^2 + M_1M_2 + M_2M_1 \\ &= M_1 + M_2 + M_1M_2 + M_2M_1.\end{aligned}$$

Thus, $M_1M_2 + M_2M_1 = 0$.

Multiplying by M_1 gives $0 = M_1^2M_2 + M_1M_2M_1 = M_1M_2 + M_1M_2M_1$ and thus $-M_1M_2M_1 = M_1M_2$. Since $-M_1M_2M_1$ is symmetric, so is M_1M_2. This gives $M_1M_2 = (M_1M_2)' = M_2M_1$, so the condition $M_1M_2 + M_2M_1 = 0$ becomes

$2(M_1M_2) = 0$ or $M_1M_2 = 0$. By symmetry, this says that the columns of M_1 are orthogonal to the columns of M_2. □

Theorem B.46 *If M_1 and M_2 are symmetric, $C(M_1) \perp C(M_2)$, and $(M_1 + M_2)$ is a perpendicular projection matrix, then M_1 and M_2 are perpendicular projection matrices.*

Proof

$$(M_1 + M_2) = (M_1 + M_2)^2 = M_1^2 + M_2^2 + M_1M_2 + M_2M_1.$$

Since M_1 and M_2 are symmetric with $C(M_1) \perp C(M_2)$, we have $M_1M_2 + M_2M_1 = 0$ and $M_1 + M_2 = M_1^2 + M_2^2$. Rearranging gives $M_2 - M_2^2 = M_1^2 - M_1$, so $C(M_2 - M_2^2) = C(M_1^2 - M_1)$. Now $C(M_2 - M_2^2) \subset C(M_2)$ and $C(M_1^2 - M_1) \subset C(M_1)$, so $C(M_2 - M_2^2) \perp C(M_1^2 - M_1)$. The only way a vector space can be orthogonal to itself is if it consists only of the zero vector. Thus, $M_2 - M_2^2 = M_1^2 - M_1 = 0$, and $M_2 = M_2^2$ and $M_1 = M_1^2$. □

Theorem B.47 *Let M and M_0 be perpendicular projection matrices with $C(M_0) \subset C(M)$. Then $M - M_0$ is the perpendicular projection matrix onto $C(M_0)^\perp_{C(M)}$.*

Proof Since $C(M_0) \subset C(M)$, $MM_0 = M_0$ and, by symmetry, $M_0M = M_0$. Checking the conditions of Theorem B.33, we see that $(M - M_0)^2 = M^2 - MM_0 - M_0M + M_0^2 = M - M_0 - M_0 + M_0 = M - M_0$, and $(M - M_0)' = M - M_0$, so $M - M_0$ is a ppo onto $C(M - M_0)$.

To see that $C(M - M_0) = C(M_0)^\perp_{C(M)}$ note that $C(M - M_0) \perp C(M_0)$, because $(M - M_0)M_0 = MM_0 - M_0^2 = M_0 - M_0 = 0$. Thus, $C(M - M_0) \subset C(M_0)^\perp_{C(M)}$. If $x \in C(M)$ and $x \perp C(M_0)$, then $x = Mx = (M - M_0)x + M_0x = (M - M_0)x$. Thus, $x \in C(M - M_0)$ and $C(M_0)^\perp_{C(M)} \subset C(M - M_0)$. □

Corollary B.48 $C(M - M_0) = C(M_0)^\perp_{C(M)}$.

Corollary B.49 $r(M) = r(M_0) + r(M - M_0)$.

One particular application of these results involves I, the perpendicular projection operator onto $\mathbf{R}^n$. For any other perpendicular projection operator M, $I - M$ is the perpendicular projection operator onto the orthogonal complement of $C(M)$ with respect to $\mathbf{R}^n$. For example, the subspace of vectors with the form $(2a, a)'$ has an orthogonal complement consisting of vectors with the form $(b, -2b)'$. With M as given earlier,

$$I - M = \begin{bmatrix} 1 & 0 \\ 0 & 1 \end{bmatrix} - \begin{bmatrix} 0.8 & 0.4 \\ 0.4 & 0.2 \end{bmatrix} = \begin{bmatrix} 0.2 & -0.4 \\ -0.4 & 0.8 \end{bmatrix}.$$

Note that

$$(I-M)\begin{pmatrix} b \\ -2b \end{pmatrix} = \begin{pmatrix} b \\ -2b \end{pmatrix} \quad \text{and} \quad (I-M)\begin{pmatrix} 2a \\ a \end{pmatrix} = 0;$$

so by definition $I - M$ is the perpendicular projection operator onto the space of vectors with the form $(b, -2b)'$.

At this point, we examine the relationship between perpendicular projection operations and the Gram–Schmidt theorem (Theorem A.16). Recall that in the Gram–Schmidt theorem, $x_1, \ldots, x_r$ denotes the original basis and $y_1, \ldots, y_r$ denotes the orthonormal basis. Let

$$M_s = \sum_{i=1}^{s} y_i y_i'.$$

Applying Theorem B.35, M_s is the ppo onto $C(x_1, \ldots, x_s)$. Now define

$$w_{s+1} = (I - M_s)x_{s+1}.$$

Thus, w_{s+1} is the perpendicular projection of x_{s+1} onto the orthogonal complement of $C(x_1, \ldots, x_s)$. Finally, y_{s+1} is just w_{s+1} normalized.

Consider the eigenvalues of a perpendicular projection operator M. Let $v_1, \ldots, v_r$ be a basis for $C(M)$. Then $Mv_i = v_i$, so v_i is an eigenvector of M with eigenvalue 1. In fact, 1 is an eigenvalue of M with multiplicity r. Now, let $w_1, \ldots, w_{n-r}$ be a basis for $C(M)^\perp$. $Mw_j = 0$, so 0 is an eigenvalue of M with multiplicity $n - r$. We have completely characterized the n eigenvalues of M. Since $\text{tr}(M)$ equals the sum of the eigenvalues, we have $\text{tr}(M) = r(M)$.

In fact, if A is an $n \times n$ matrix with $A^2 = A$, any basis for $C(A)$ is a basis for the space of eigenvectors for the eigenvalue 1. The null space of A is the space of eigenvectors for the eigenvalue 0. The rank of A and the rank of the null space of A add to n, and A has n eigenvalues, so all the eigenvalues are accounted for. Again, $\text{tr}(A) = r(A)$.

Definition B.50

(a) If A is a square matrix with $A^2 = A$, then A is called *idempotent*.

(b) Let $\mathcal{N}$ and $\mathcal{M}$ be two spaces with $\mathcal{N} \cap \mathcal{M} = \{0\}$ and $r(\mathcal{N}) + r(\mathcal{M}) = n$. The square matrix A is a *projection operator* onto $\mathcal{N}$ along $\mathcal{M}$ if 1) $Av = v$ for any $v \in \mathcal{N}$, and 2) $Aw = 0$ for any $w \in \mathcal{M}$.

If the square matrix A has the property that $Av = v$ for any $v \in C(A)$, then A is the projection operator (matrix) onto $C(A)$ along $C(A')^\perp$. (Note that $C(A')^\perp$ is the null space of A.) It follows immediately that if A is idempotent, then A is a projection operator onto $C(A)$ along $\mathcal{N}(A) = C(A')^\perp = C(I - A)$, see Exercise B.22.

The uniqueness of projection operators can be established like it was for perpendicular projection operators. Note that $x \in \mathbf{R}^n$ can be written uniquely as $x = v + w$

for $v \in \mathcal{N}$ and $w \in \mathcal{M}$, i.e., $\mathbf{R}^n = \mathcal{N} + \mathcal{M}$. To see this, take basis matrices for the two spaces, say N and M, respectively. The result follows from observing that $[N, M]$ is a basis matrix for $\mathbf{R}^n$. Because of the rank conditions, $[N, M]$ is an $n \times n$ matrix. It is enough to show that the columns of $[N, M]$ must be linearly independent.

$$0 = [N, M] \begin{bmatrix} b \\ c \end{bmatrix} = Nb + Mc$$

implies $Nb = M(-c)$ which, since $\mathcal{N} \cap \mathcal{M} = \{0\}$, can only happen when $Nb = 0 = M(-c)$, which, because they are basis matrices, can only happen when $b = 0 = (-c)$, which implies that $\begin{bmatrix} b \\ c \end{bmatrix} = 0$, and we are done.

Any projection operator that is not a perpendicular projection is referred to as an *oblique projection operator.*

To show that a matrix A is a projection operator onto an arbitrary space, say $C(X)$, it is necessary to show that $C(A) = C(X)$ and that for $x \in C(X)$, $Ax = x$. A typical proof runs in the following pattern. First, show that $Ax = x$ for any $x \in C(X)$. This also establishes that $C(X) \subset C(A)$. To finish the proof, it suffices to show that $Av \in C(X)$ for any $v \in \mathbf{R}^n$ because this implies that $C(A) \subset C(X)$.

In this book, our use of the word "perpendicular" is based on the standard inner product that defines Euclidean distance. In other words, for two vectors x and y, their inner product is $x'y$. By definition, the vectors x and y are orthogonal if their inner product is 0. In fact, for any two vectors x and y, let θ be the angle between x and y. Then $x'y = \sqrt{x'x}\sqrt{y'y}\cos\theta$. The length of a vector x is defined as the square root of the inner product of x with itself, i.e., $\|x\| \equiv \sqrt{x'x}$. The distance between two vectors x and y is the length of their difference, i.e., $\|x - y\|$.

These concepts can be generalized. For a positive definite matrix B, we can define an inner product between x and y as $x'By$. As before, x and y are orthogonal if their inner product is 0 and the length of x is the square root of its inner product with itself (now $\|x\|_B \equiv \sqrt{x'Bx}$). As argued above, any idempotent matrix is always a projection operator, but which one is the perpendicular projection operator depends on the inner product. As can be seen from Proposition 2.7.2 and Exercise 2.5, the matrix $X(X'BX)^-X'B$ is an oblique projection onto $C(X)$ for the standard inner product; but it is the perpendicular projection operator onto $C(X)$ with the inner product defined using the matrix B.

B.4 Miscellaneous Results

Proposition B.51 *For any matrix X, $C(XX') = C(X)$.*

Proof Clearly $C(XX') \subset C(X)$, so we need to show that $C(X) \subset C(XX')$. Let $x \in C(X)$. Then $x = Xb$ for some b. Write $b = b_0 + b_1$, where $b_0 \in C(X')$ and $b_1 \perp C(X')$. Clearly, $Xb_1 = 0$, so we have $x = Xb_0$. But $b_0 = X'd$ for some d; so $x = Xb_0 = XX'd$ and $x \in C(XX')$. □

Corollary B.52 *For any matrix X, $r(XX') = r(X)$.*

Proof See Exercise B.4. □

Corollary B.53 *If $X_{n\times p}$ has $r(X) = p$, then the $p \times p$ matrix $X'X$ is nonsingular.*

Proof See Exercise B.5. □

Proposition B.54

(a) *If $C(U_1) \subset C(U_2)$, then $C(XU_1) \subset C(XU_2)$.*
(b) *If $C(U_1) = C(U_2)$, then $C(XU_1) = C(XU_2)$.*
(c) *$C(XB) \subset C(X)$.*
(d) *If B is nonsingular, $C(XB) = C(X)$.*

Proof (a) Take $v \in C(XU_1)$. For some γ_1, $v = XU_1\gamma_1$. Because, $C(U_1) \subset C(U_2)$, there exists γ_2 so that $U_1\gamma_1 = U_2\gamma_2$. Clearly, $v = XU_1\gamma_1 = XU_2\gamma_2 \in C(XU_2)$. (b) Use (a) as is and with the roles of U_1 and U_2 reversed. (c) This is immediate from the definition of a column space but also, in a) take $U_1 = B$ and $U_2 = I$. (d) If B is nonsingular, $C(B) = C(I)$ and use (b). □

It follows immediately from Proposition B.54 that, for B nonsingular, the perpendicular projection operators onto $C(XB)$ and $C(X)$ are identical.

We now show that generalized inverses always exist.

Theorem B.55 *For any matrix X, there exists a generalized inverse X^-.*

Proof We know that $(X'X)^-$ exists. Set $X^- = (X'X)^-X'$. Then $XX^-X = X(X'X)^-X'X = X$ because $X(X'X)^-X'$ is a projection matrix onto $C(X)$. □

Note that for any X^-, the matrix $XX^-$is idempotent and hence a projection operator.

Proposition B.56 *When all inverses exist,*

$$[A + BCD]^{-1} = A^{-1} - A^{-1}B\left[C^{-1} + DA^{-1}B\right]^{-1}DA^{-1}.$$

Proof If all inverses exist

$$\begin{aligned}
&[A + BCD]\left[A^{-1} - A^{-1}B\left[C^{-1} + DA^{-1}B\right]^{-1}DA^{-1}\right]\\
&= I - B\left[C^{-1} + DA^{-1}B\right]^{-1}DA^{-1} + BCDA^{-1}\\
&\quad - BCDA^{-1}B\left[C^{-1} + DA^{-1}B\right]^{-1}DA^{-1}\\
&I - B\left[I + CDA^{-1}B\right]\left[C^{-1} + DA^{-1}B\right]^{-1}DA^{-1} + BCDA^{-1}\\
&I - BC\left[C^{-1} + DA^{-1}B\right]\left[C^{-1} + DA^{-1}B\right]^{-1}DA^{-1} + BCDA^{-1}\\
&I - BCDA^{-1} + BCDA^{-1} = I.
\end{aligned}$$

□

Proposition B.57 *Let P be a projection operator (idempotent), and let a and b be real numbers. Then*

$$[aI + bP]^{-1} = \frac{1}{a}\left[I - \frac{b}{a+b}P\right].$$

Proof

$$\frac{1}{a}\left[I - \frac{b}{a+b}P\right][aI + bP] = \frac{1}{a}\left[aI + bP - \frac{ab}{a+b}P - \frac{b^2}{a+b}P\right] = I. \quad \square$$

When we study linear models, we frequently need to refer to matrices and vectors that consist entirely of 1s. Such matrices are denoted by the letter J with various subscripts and superscripts to specify their dimensions. J_r^c is an $r \times c$ matrix of 1s. The subscript indicates the number of rows and the superscript indicates the number of columns. If there is only one column, the superscript may be suppressed, e.g., $J_r \equiv J_r^1$. In a context where we are dealing with vectors in $\mathbf{R}^n$, the subscript may also be suppressed, e.g., $J \equiv J_n \equiv J_n^1$.

A matrix of 0s is always denoted by 0.

B.5 Properties of Kronecker Products and Vec Operators

Kronecker products and Vec operators are extremely useful in multivariate analysis and some approaches to variance component estimation. (Both are discussed in *ALM-III*.) They are also often used in writing balanced ANOVA models. We now present their basic algebraic properties.

1. If the matrices are of conformable sizes, $[A \otimes (B + C)] = [A \otimes B] + [A \otimes C]$.
2. If the matrices are of conformable sizes, $[(A + B) \otimes C] = [A \otimes C] + [B \otimes C]$.
3. If a and b are scalars, $ab[A \otimes B] = [aA \otimes bB]$.
4. If the matrices are of conformable sizes, $[A \otimes B][C \otimes D] = [AC \otimes BD]$.
5. The transpose of a Kronecker product matrix is $[A \otimes B]' = [A' \otimes B']$.
6. The generalized inverse of a Kronecker product matrix is $[A \otimes B]^- = [A^- \otimes B^-]$.
7. For two vectors v and w, $\mathrm{Vec}(vw') = w \otimes v$.
8. For a matrix W and conformable matrices A and B, $\mathrm{Vec}(AWB') = [B \otimes A]\mathrm{Vec}(W)$.
9. For conformable matrices A and B, $\mathrm{Vec}(A)'\mathrm{Vec}(B) = \mathrm{tr}(A'B)$.
10. The Vec operator commutes with any matrix operation that is performed elementwise. For example, $\mathrm{E}\{\mathrm{Vec}(W)\} = \mathrm{Vec}\{\mathrm{E}(W)\}$ when W is a random matrix. Similarly, for conformable matrices A and B and scalar ϕ, $\mathrm{Vec}(A + B) = \mathrm{Vec}(A) + \mathrm{Vec}(B)$ and $\mathrm{Vec}(\phi A) = \phi\mathrm{Vec}(A)$.
11. If A and B are positive definite, then $A \otimes B$ is positive definite.

Most of these are well-known facts and easy to establish. Two of them are somewhat more unusual. Proofs for Items 8 and 11 are given in *ALM-III*, Appendix A.2 and in earlier editions of this book.

B.6 Tensors

Tensors are simply an alternative notation for writing vectors. This notation has substantial advantages when dealing with quadratic forms and when dealing with more general concepts than quadratic forms. Our main purpose in discussing them here is simply to illustrate how flexibly subscripts can be used in writing vectors.

Consider a vector $Y = (y_1, \ldots, y_n)'$. The tensor notation for this is simply y_i. We can write another vector $a = (a_1, \ldots, a_n)'$ as a_i. When written individually, the subscript is not important. In other words, a_i is the same vector as a_j. Note that the length of these vectors needs to be understood from the context. Just as when we write Y and a in conventional vector notation, there is nothing in the notation y_i or a_i to tell us how many elements are in the vector.

If we want the inner product $a'Y$, in tensor notation we write $a_i y_i$. Here we are using something called the *summation convention*. Because the subscripts on a_i and y_i are the same, $a_i y_i$ is taken to mean $\sum_{i=1}^n a_i y_i$. If, on the other hand, we wrote $a_i y_j$, this means something completely different. $a_i y_j$ is an alternative notation for the Kronecker product $[a \otimes Y] = (a_1 y_1, \ldots, a_1 y_n, a_2 y_1, \ldots, a_n y_n)'$. In $[a \otimes Y] \equiv a_i y_j$, we have two subscripts identifying the rows of the vector.

Now, suppose we want to look at a quadratic form $Y'AY$, where Y is an n vector and A is $n \times n$. One way to rewrite this is

$$Y'AY = \sum_{i=1}^{n} \sum_{j=1}^{n} y_i a_{ij} y_j = \sum_{i=1}^{n} \sum_{j=1}^{n} a_{ij} y_i y_j = \text{Vec}(A)'[Y \otimes Y].$$

(From Property B.5.8 we also have $Y'AY = [Y' \otimes Y']\text{Vec}(A)$.) Here we have rewritten the quadratic form as a linear combination of the elements in the vector $[Y \otimes Y]$. The linear combination is determined by the elements of the vector $\text{Vec}(A)$. In tensor notation, this becomes quite simple. Using the summation convention in which objects with the same subscript are summed over,

$$Y'AY = y_i a_{ij} y_j = a_{ij} y_i y_j.$$

The second term just has the summation signs removed, but the third term, which obviously gives the same sum as the second, is actually the tensor notation for $\text{Vec}(A)'[Y \otimes Y]$. Again, $\text{Vec}(A) = (a_{11}, a_{21}, a_{31}, \ldots, a_{nn})'$ uses two subscripts to identify rows of the vector. Obviously, if you had a need to consider things like

$$\sum_{i=1}^{n}\sum_{j=1}^{n}\sum_{k=1}^{n} a_{ijk}y_i y_j y_k \equiv a_{ijk}y_i y_j y_k,$$

the tensor version $a_{ijk}y_i y_j y_k$ saves some work.

There is one slight complication in how we have been writing things. Suppose A is not symmetric and we have another n vector W. Then we might want to consider

$$W'AY = \sum_{i=1}^{n}\sum_{j=1}^{n} w_i a_{ij} y_j.$$

From item 8 in the previous subsection,

$$W'AY = \text{Vec}(W'AY) = [Y' \otimes W']\text{Vec}(A).$$

Alternatively,

$$W'AY = \sum_{i=1}^{n}\sum_{j=1}^{n} w_i a_{ij} y_j = \sum_{i=1}^{n}\sum_{j=1}^{n} a_{ij} y_j w_i = \text{Vec}(A)'[Y \otimes W]$$

or $W'AY = Y'A'W = \text{Vec}(A')'[W \otimes Y]$. However, with A nonsymmetric, $W'A'Y = \text{Vec}(A')'[Y \otimes W]$ is typically different from $W'AY$. The Kronecker notation requires that care be taken in specifying the order of the vectors in the Kronecker product, and whether or not to transpose A before using the Vec operator. In tensor notation, $W'AY$ is simply $w_i a_{ij} y_j$. In fact, the orders of the vectors can be permuted in any way; so, for example, $a_{ij} y_j w_i$ means the same thing. $W'A'Y$ is simply $w_i a_{ji} y_j$. The tensor notation and the matrix notation require less effort than the Kronecker notation.

For our purposes, the real moral here is simply that the subscripting of an individual vector does not matter. We can write a vector $Y = (y_1, \ldots, y_n)'$ as $Y = [y_k]$ (in tensor notation as simply y_k), or we can write the same n vector as $Y = [y_{ij}]$ (in tensor notation, simply y_{ij}), where, as long as we know the possible values that i and j can take on, the actual order in which we list the elements is not of much importance. Thus, if $i = 1, \ldots, t$ and $j = 1, \ldots, N_i$, with $n = \sum_{i=1}^{t} N_i$, it really does not matter if we write a vector Y as $(y_1, \ldots, y_n)$, or $(y_{11}, \ldots, y_{1N_1}, y_{21}, \ldots, y_{tN_t})'$ or $(y_{t1}, \ldots, y_{tN_t}, y_{t-1,1}, \ldots, y_{1N_1})'$ or in any other fashion we may choose, as long as we keep straight which row of the vector is which. Thus, a linear combination $a'Y$ can be written $\sum_{k=1}^{n} a_k y_k$ or $\sum_{i=1}^{t}\sum_{j=1}^{N_i} a_{ij} y_{ij}$. In tensor notation, the first of these is simply $a_k y_k$ and the second is $a_{ij} y_{ij}$. These ideas become very handy in examining analysis of variance models, where the standard approach is to use multiple subscripts to identify the various observations. The subscripting has no intrinsic importance; the only thing that matters is knowing which row is which in the vectors. The subscripts are an aid in this identification, but they do not create any problems. We can still put all of the observations into a vector and use standard operations on them.

B.7 Exercises

Exercise B.0

(a) Let $X = [X_0, X_1]$ with M and M_0 the ppos onto $C(X)$ and $C(X_0)$, respectively. Show that $(I - M_0)X_1[X_0'(I - M_0)X_1]^- X_1'(I - M_0)$ is the ppo onto $C(X_0)^\perp_{C(X)}$.

(b) Let r and s be two n vectors. Let M_r be the ppo onto $C(r)$, then $s'(I - M_r)s \geq 0$. Use this fact to prove the Cauchy–Schwarz inequality,

$$(s'r)^2 \leq s's\, r'r.$$

Exercise B.1

(a) Show that

$$A^k x + b_{k-1}A^{k-1}x + \cdots + b_0 x = (A - \mu I)\left(A^{k-1}x + \tau_{k-2}A^{k-2}x + \cdots + \tau_0 x\right) = 0,$$

where μ is any nonzero solution of $b_0 + b_1 w + \cdots + b_k w^k = 0$ with $b_k = 1$ and $\tau_j = -(b_0 + b_1\mu + \cdots + b_j\mu^j)/\mu^{j+1}$, $j = 0, \ldots, k$.

(b) Show that if the only root of $b_0 + b_1 w + \cdots + b_k w^k$ is zero, then the factorization in (a) still holds.

(c) The solution μ used in (a) need not be a real number, in which case μ is a complex eigenvalue and the τ_is are complex; so the eigenvector is complex. Show that with A symmetric, μ must be real because the eigenvalues of A must be real. In particular, assume that

$$A(y + iz) = (\lambda + i\gamma)(y + iz),$$

for y, z, λ, and γ real vectors and scalars, respectively, set $Ay = \lambda y - \gamma z$, $Az = \gamma y + \lambda z$, and examine $z'Ay = y'Az$.

Exercise B.2 Prove Proposition B.32.

Exercise B.3 Show that any nonzero symmetric matrix A can be written as $A = PDP'$, where $C(A) = C(P)$, $P'P = I$, and D is nonsingular.

Exercise B.4 Prove Corollary B.52.

Exercise B.5 Prove Corollary B.53.

Exercise B.6 Show $\text{tr}(cI_n) = nc$.

Exercise B.7 Let a, b, c, and d be real numbers. If $ad - bc \neq 0$, find the inverse of

$$\begin{bmatrix} a & b \\ c & d \end{bmatrix}.$$

Exercise B.8 Prove Theorem B.28, i.e., let A be an $r \times s$ matrix, let B be an $s \times r$ matrix, and show that $\text{tr}(AB) = \text{tr}(BA)$.

Exercise B.9 Determine whether the matrices given below are positive definite, nonnegative definite, or neither.

$$\begin{bmatrix} 3 & 2 & -2 \\ 2 & 2 & -2 \\ -2 & -2 & 10 \end{bmatrix}, \begin{bmatrix} 26 & -2 & -7 \\ -2 & 4 & -6 \\ -7 & -6 & 13 \end{bmatrix}, \begin{bmatrix} 26 & 2 & 13 \\ 2 & 4 & 6 \\ 13 & 6 & 13 \end{bmatrix}, \begin{bmatrix} 3 & 2 & -2 \\ 2 & -2 & -2 \\ -2 & -2 & 10 \end{bmatrix}.$$

Exercise B.10 Show that the matrix B given below is positive definite, and find a matrix Q such that $B = QQ'$. (Hint: The first row of Q can be taken as $(1, -1, 0)$.)

$$B = \begin{bmatrix} 2 & -1 & 1 \\ -1 & 1 & 0 \\ 1 & 0 & 2 \end{bmatrix}.$$

Exercise B.11 Let

$$A = \begin{bmatrix} 2 & 0 & 4 \\ 1 & 5 & 7 \\ 1 & -5 & -3 \end{bmatrix}, \quad B = \begin{bmatrix} 1 & 0 & 0 \\ 0 & 0 & 1 \\ 0 & 1 & 0 \end{bmatrix}, \quad C = \begin{bmatrix} 1 & 4 & 1 \\ 2 & 5 & 1 \\ -3 & 0 & 1 \end{bmatrix}.$$

Use Theorem B.35 to find the perpendicular projection operator onto the column space of each matrix.

Exercise B.12 Show that for a perpendicular projection matrix M,

$$\sum_i \sum_j m_{ij}^2 = r(M).$$

Exercise B.13 Prove that if $M = M'M$, then $M = M'$ and $M = M^2$.

Exercise B.14 Let M_1 and M_2 be perpendicular projection matrices, and let M_0 be a perpendicular projection operator onto $C(M_1) \cap C(M_2)$. Show that the following are equivalent:

(a) $M_1 M_2 = M_2 M_1$.

(b) $M_1 M_2 = M_0$.

(c) $\{C(M_1) \cap [C(M_1) \cap C(M_2)]^\perp\} \perp \{C(M_2) \cap [C(M_1) \cap C(M_2)]^\perp\}$.

Hints: (i) Show that $M_1 M_2$ is a projection operator. (ii) Show that $M_1 M_2$ is symmetric. (iii) Note that $C(M_1) \cap [C(M_1) \cap C(M_2)]^\perp = C(M_1 - M_0)$.

Exercise B.15 Let M_1 and M_2 be perpendicular projection matrices. Show that

(a) the eigenvalues of $M_1 M_2$ are no greater than 1 in absolute value (they may be complex);

(b) $\mathrm{tr}(M_1 M_2) \leq r(M_1 M_2)$.

Hints: For part (a) show that with $x'Mx \equiv \|Mx\|^2$, $\|Mx\| \leq \|x\|$ for any perpendicular projection operator M. Use this to show that if $M_1 M_2 x = \lambda x$, then $\|M_1 M_2 x\| \geq |\lambda| \, \|M_1 M_2 x\|$.

Exercise B.16 For vectors x and y, let $M_x = x(x'x)^{-1}x'$ and $M_y = y(y'y)^{-1}y'$. Show that $M_x M_y = M_y M_x$ if and only if $C(x) = C(y)$ or $x \perp y$.

Exercise B.17 Consider the matrix

$$A = \begin{bmatrix} 0 & 1 \\ 0 & 1 \end{bmatrix}.$$

(a) Show that A is a projection matrix.

(b) Is A a perpendicular projection matrix? Why or why not?

(c) Describe the space that A projects onto and the space that A projects along. Sketch these spaces.

(d) Find another projection operator onto the space that A projects onto.

Exercise B.18 Let A be an arbitrary projection matrix. Show that $C(I - A) = C(A')^\perp$.

Hints: Recall that $C(A')^\perp$ is the null space of A. Show that $(I - A)$ is a projection matrix.

Exercise B.19 Show that if A^- is a generalized inverse of A, then so is

$$G = A^- A A^- + (I - A^- A)B_1 + B_2(I - AA^-)$$

for any choices of B_1 and B_2 with conformable dimensions.

Exercise B.20 Let A be positive definite with eigenvalues $\lambda_1, \ldots, \lambda_n$. Show that A^{-1} has eigenvalues $1/\lambda_1, \ldots, 1/\lambda_n$ and the same eigenvectors as A.

Exercise B.21 For A nonsingular, let

$$A = \begin{bmatrix} A_{11} & A_{12} \\ A_{21} & A_{22} \end{bmatrix},$$

and let $A_{1\cdot 2} = A_{11} - A_{12}A_{22}^{-1}A_{21}$. Show that if all inverses exist,

$$A^{-1} = \begin{bmatrix} A_{1\cdot 2}^{-1} & -A_{1\cdot 2}^{-1}A_{12}A_{22}^{-1} \\ -A_{22}^{-1}A_{21}A_{1\cdot 2}^{-1} & A_{22}^{-1} + A_{22}^{-1}A_{21}A_{1\cdot 2}^{-1}A_{12}A_{22}^{-1} \end{bmatrix}$$

and that

$$A_{22}^{-1} + A_{22}^{-1}A_{21}A_{1\cdot 2}^{-1}A_{12}A_{22}^{-1} = \left[A_{22} - A_{21}A_{11}^{-1}A_{12}\right]^{-1}.$$

Exercise B.22 Show that if A is idempotent, then $\mathcal{N}(A) = C(I - A)$. Hint: Show that each set is contained in the other.

Exercise B.23 Consider the vectors that are the columns of a matrix $X_{n\times p}$ with $r(X) = r$. A *rotation* of these vectors keeps their lengths the same and the angles between them the same. In other words, it keeps all of the inner products between them the same. If P is an orthonormal matrix, then $Z = PX$ is a rotation of the columns of X because $Z'Z = X'P'PX = X'X$.

Rotating vectors within a subspace is more difficult. Let the $n \times r$ matrix Q have columns that are an orthonormal basis for $C(X)$. Write $X = QB$. Show that a rotation of the columns of X that remains within $C(X)$ is obtained from $Z = QPB$ where P is an $r \times r$ orthonormal matrix.

Exercise B.24 Show that $r(X) = r(X')$. Hints: Let X be $n \times p$ with $r(X') = r$. Let $\tilde{X}$ be an $r \times p$ matrix with the rows of $\tilde{X}$ forming a basis for $C(X')$. Write $X = B\tilde{X}$ and argue that $r(X) \leq r = r(X')$. Reverse the roles of X and X'.

Appendix C
Some Univariate Distributions

Abstract The tests and confidence intervals presented in this book rely almost exclusively on the χ^2, t, and F distributions. This appendix defines each of the distributions.

Definition C.1 Let $Z_1, \ldots, Z_n$ be independent with $Z_i \sim N(\mu_i, 1)$. Then if

$$W \sim \sum_{i=1}^{n} Z_i^2$$

we say W has a *noncentral chi-squared distribution* with n degrees of freedom and *noncentrality parameter* $\gamma = \sum_{i=1}^{n} \mu_i^2/2$. Write $W \sim \chi^2(n, \gamma)$.

See Rao (1973, Section 3b.2) for a proof that the distribution of W depends only on n and γ.

It is evident from the definition that if $X \sim \chi^2(r, \gamma)$ and $Y \sim \chi^2(s, \delta)$ with X and Y independent, then $(X + Y) \sim \chi^2(r + s, \gamma + \delta)$. A central χ^2 distribution is a distribution with a noncentrality parameter of zero, i.e., $\chi^2(r, 0)$. We will use $\chi^2(r)$ to denote a $\chi^2(r, 0)$ distribution. The 100αth percentile of a $\chi^2(r)$ distribution is the point $\chi^2(\alpha, r)$ that satisfies the equation

$$\Pr\left[\chi^2(r) \leq \chi^2(\alpha, r)\right] = \alpha.$$

Note that if $0 \leq a < 1$, the $100a$ percentile of a central $\chi^2(b)$ is denoted $\chi^2(a, b)$. However, if a is a positive integer, $\chi^2(a, b)$ denotes a noncentral chi-squared distribution.

Definition C.2 Let $X \sim N(\mu, 1)$ and $Y \sim \chi^2(n)$ with X and Y independent. Then

R. Christensen, *Plane Answers to Complex Questions*, Springer Texts in Statistics,
https://doi.org/10.1007/978-3-030-32097-3

$$W = \frac{X}{\sqrt{Y/n}}$$

has a *noncentral t distribution* with n degrees of freedom and noncentrality parameter μ. Write $W \sim t(n, \mu)$. If $\mu = 0$, we say that the distribution is a central t distribution and write $W \sim t(n)$. The 100αth percentile of a $t(n)$ distribution is denoted $t(\alpha, n)$.

Definition C.3 Let $X \sim \chi^2(r, \gamma)$ and $Y \sim \chi^2(s, 0)$ with X and Y independent. Then

$$W = \frac{X/r}{Y/s}$$

has a *noncentral F distribution* with r numerator and s denominator degrees of freedom and noncentrality parameter γ. Write $W \sim F(r, s, \gamma)$. If $\gamma = 0$, write $W \sim F(r, s)$ for the central F distribution. The 100αth percentile of $F(r, s)$ is denoted $F(\alpha, r, s)$.

As indicated, if the noncentrality parameter of any of these distributions is zero, the distribution is referred to as a *central distribution* (e.g., central F distribution). The central distributions are those commonly used in statistical methods courses. If any of these distributions is not specifically identified as a noncentral distribution, it should be assumed to be a central distribution.

It is easily seen from Definition C.1 that any noncentral chi-squared distribution *tends* to be larger than the central chi-squared distribution with the same number of degrees of freedom. Similarly, from Definition C.3, a noncentral F tends to be larger than the corresponding central F distribution. (These ideas are made rigorous in Exercise C.1.) The fact that the noncentral F distribution tends to be larger than the corresponding central F distribution is the basis for many of the tests used in linear models. Typically, test statistics are used that have a central F distribution if the reduced (null) model is true and a noncentral F distribution if the full model is true but the null model is not. Since the noncentral F distribution tends to be larger, large values of the test statistic are more consistent with the full model than with the null. Thus, the form of an appropriate rejection region when the full model is true is to reject the null hypothesis for large values of the test statistic.

The power of these F tests is simply a function of the noncentrality parameter. Given a value for the noncentrality parameter, there is no theoretical difficulty in finding the power of an F test. The power simply involves computing the probability of the rejection region when the probability distribution is a noncentral F. Davies (1980) gives an algorithm for making these and more general computations.

We now prove a theorem about central F distributions that will be useful in Chapter 5.

Theorem C.4 *If $s > t$, then $sF(1 - \alpha, s, v) \geq tF(1 - \alpha, t, v)$.*

Proof Let $X \sim \chi^2(s)$, $Y \sim \chi^2(t)$, and $Z \sim \chi^2(v)$. Let Z be independent of X and Y. Note that $(X/s)\big/(Z/v)$ has an $F(s, v)$ distribution; so $sF(1-\alpha, s, v)$ is the $100(1-\alpha)$ percentile of the distribution of $X\big/(Z/v)$. Similarly, $tF(1-\alpha, t, v)$ is the $100(1-\alpha)$ percentile of the distribution of $Y\big/(Z/v)$.

We will first argue that to prove the theorem it is enough to show that

$$\Pr[X \le d] \le \Pr[Y \le d] \tag{1}$$

for all real numbers d. We will then show that (1) is true.

If (1) is true, if c is any real number, and if $Z = z$, by independence we have

$$\Pr[X \le cz/v] = \Pr[X \le cz/v|Z = z] \le \Pr[Y \le cz/v|Z = z] = \Pr[Y \le cz/v].$$

Taking expectations with respect to Z,

$$\begin{aligned}\Pr\left[X\big/(Z/v) \le c\right] &= \mathrm{E}\left(\Pr[X \le cz/v|Z = z]\right) \\ &\le \mathrm{E}\left(\Pr[Y \le cz/v|Z = z]\right) \\ &= \Pr\left[Y\big/(Z/v) \le c\right].\end{aligned}$$

Since the cumulative distribution function (cdf) for $X\big/(Z/v)$ is always no greater than the cdf for $Y\big/(Z/v)$, the point at which a probability of $1-\alpha$ is attained for $X\big/(Z/v)$ must be no less than the similar point for $Y\big/(Z/v)$. Therefore,

$$sF(1-\alpha, s, v) \ge tF(1-\alpha, t, v).$$

To see that (1) holds, let Q be independent of Y and $Q \sim \chi^2(s-t)$. Then, because Q is nonnegative,

$$\Pr[X \le d] = \Pr[Y + Q \le d] \le \Pr[Y \le d].$$

□

Exercise

Definition C.5 Consider two random variables W_1 and W_2. W_2 is said to be *stochastically larger* than W_1 if for every real number w

$$\Pr[W_1 > w] \le \Pr[W_2 > w].$$

If for some random variables W_1 and W_2, W_2 is stochastically larger than W_1, then we also say that the distribution of W_2 is stochastically larger than the distribution of W_1.

Exercise C.1 Show that a noncentral chi-squared distribution is stochastically larger than the central chi-squared distribution with the same degrees of freedom. Show that a noncentral F distribution is stochastically larger than the corresponding central F distribution.

Appendix D
Multivariate Distributions

Abstract This appendix reviews properties of multivariate distributions. It also examines the concept of identifiable parameters.

Let $(x_1, \ldots, x_n)'$ be a random vector. The joint cumulative distribution function (cdf) of $(x_1, \ldots, x_n)'$ is

$$F(u_1, \ldots, u_n) \equiv \Pr[x_1 \le u_1, \ldots, x_n \le u_n].$$

If $F(u_1, \ldots, u_n)$ is the cdf of a discrete random variable, we can define a (joint) probability mass function

$$f(u_1, \ldots, u_n) \equiv \Pr[x_1 = u_1, \ldots, x_n = u_n].$$

If $F(u_1, \ldots, u_n)$ admits the nth order mixed partial derivative, then we can define a (joint) density function

$$f(u_1, \ldots, u_n) \equiv \frac{\partial^n}{\partial u_1 \cdots \partial u_n} F(u_1, \ldots, u_n).$$

The cdf can be recovered from the density as

$$F(u_1, \ldots, u_n) = \int_{-\infty}^{u_1} \cdots \int_{-\infty}^{u_n} f(w_1, \ldots, w_n) dw_1 \cdots dw_n.$$

For a function $g(\cdot)$ of $(x_1, \ldots, x_n)'$ into $\mathbf{R}$, the expected value is defined as

$$\mathrm{E}[g(x_1, \ldots, x_n)] = \int_{-\infty}^{\infty} \cdots \int_{-\infty}^{\infty} g(u_1, \ldots, u_n) f(u_1, \ldots, u_n) du_1 \cdots du_n.$$

We might also write this as $\mathrm{E}_x[g(x)]$.

R. Christensen, *Plane Answers to Complex Questions*, Springer Texts in Statistics,
https://doi.org/10.1007/978-3-030-32097-3

We now consider relationships between two random vectors, say $x = (x_1, \ldots, x_n)'$ and $y = (y_1, \ldots, y_m)'$. Assume that the joint vector $(x', y')' = (x_1, \ldots, x_n, y_1, \ldots, y_m)'$ has a density function

$$f_{x,y}(u, v) \equiv f_{x,y}(u_1, \ldots, u_n, v_1, \ldots, v_m).$$

Similar definitions and results hold if $(x', y')'$ has a probability mass function.

The distribution of one random vector, say x, ignoring the other vector, y, is called the *marginal distribution* of x. The marginal cdf of x can be obtained by substituting the value $+\infty$ into the joint cdf for all of the y variables:

$$F_x(u) = F_{x,y}(u_1, \ldots, u_n, +\infty, \ldots, +\infty).$$

The marginal density can be obtained either by partial differentiation of $F_x(u)$ or by integrating the joint density over the y variables:

$$f_x(u) = \int_{-\infty}^{\infty} \cdots \int_{-\infty}^{\infty} f_{x,y}(u_1, \ldots, u_n, v_1, \ldots, v_m) dv_1 \cdots dv_m.$$

The conditional density of a vector, say x, given the value of the other vector, say $y = v$, is obtained by dividing the density of $(x', y')'$ by the density of y evaluated at v, i.e.,

$$f_{x|y}(u|v) \equiv f_{x,y}(u, v) \big/ f_y(v).$$

The conditional density is a well-defined density, so expectations with respect to it are well defined. Let g be a function from $\mathbf{R}^n$ into $\mathbf{R}$,

$$\mathrm{E}[g(x)|y = v] = \int_{-\infty}^{\infty} \cdots \int_{-\infty}^{\infty} g(u) f_{x|y}(u|v) du,$$

where $du \equiv du_1 du_2 \cdots du_n$. Sometimes we write

$$\mathrm{E}_{x|y=v}[g(x)] \equiv \mathrm{E}[g(x)|y = v].$$

The standard properties of expectations hold for conditional expectations. For example, with a and b real,

$$\mathrm{E}[a g_1(x) + b g_2(x)|y = v] = a\mathrm{E}[g_1(x)|y = v] + b\mathrm{E}[g_2(x)|y = v].$$

The conditional expectation of $\mathrm{E}[g(x)|y = v]$ is a function of the value v. Since y is random, we can consider $\mathrm{E}[g(x)|y = v]$ as a random variable. In this context we write $\mathrm{E}[g(x)|y]$ or $\mathrm{E}_{x|y}[g(x)]$. An important property of conditional expectations is

$$\mathrm{E}[g(x)] = \mathrm{E}[\,\mathrm{E}[g(x)|y]\,].$$

To see this, note that $f_{x|y}(u|v) f_y(v) = f_{x,y}(u, v)$ and

$$
\begin{aligned}
\mathrm{E}[\,\mathrm{E}[g(x)|y]\,] &= \int_{-\infty}^{\infty}\cdots\int_{-\infty}^{\infty} \mathrm{E}[g(x)|y=v]\,f_y(v)dv \\
&= \int_{-\infty}^{\infty}\cdots\int_{-\infty}^{\infty}\left[\int_{-\infty}^{\infty}\cdots\int_{-\infty}^{\infty} g(u)f_{x|y}(u|v)du\right] f_y(v)dv \\
&= \int_{-\infty}^{\infty}\cdots\int_{-\infty}^{\infty} g(u)f_{x|y}(u|v)f_y(v)du\,dv \\
&= \int_{-\infty}^{\infty}\cdots\int_{-\infty}^{\infty} g(u)f_{x,y}(u,v)du\,dv \\
&= \mathrm{E}[g(x)]\,.
\end{aligned}
$$

In fact, both the notion of conditional expectation and this result can be generalized. Consider a function $g(x, y)$ from $\mathbf{R}^{n+m}$ into $\mathbf{R}$. If $y = v$, we can define $\mathrm{E}[g(x, y)|y = v]$ in a natural manner. If we consider y as random, we write $\mathrm{E}[g(x, y)|y]$. It can be easily shown that

$$
\mathrm{E}[g(x, y)] = \mathrm{E}[\mathrm{E}[g(x, y)|y]]\,.
$$

A function of x or y alone can also be considered as a function from $\mathbf{R}^{n+m}$ into $\mathbf{R}$.

A second important property of conditional expectations is that if $h(y)$ is a function from $\mathbf{R}^m$ into $\mathbf{R}$, we have

$$
\mathrm{E}[h(y)g(x, y)|y] = h(y)\mathrm{E}[g(x, y)|y]\,. \tag{1}
$$

This follows because if $y = v$,

$$
\begin{aligned}
\mathrm{E}[h(y)g(x, y)|y = v] &= \int_{-\infty}^{\infty}\cdots\int_{-\infty}^{\infty} h(v)g(u, v)f_{x|y}(u|v)du \\
&= h(v)\int_{-\infty}^{\infty}\cdots\int_{-\infty}^{\infty} g(u, v)f_{x|y}(u|v)du \\
&= h(v)\mathrm{E}[g(x, y)|y = v]\,.
\end{aligned}
$$

This is true for all v, so (1) holds. In particular, if $g(x, y) \equiv 1$, we get

$$
\mathrm{E}[h(y)|y] = h(y).
$$

Finally, we can extend the idea of conditional expectation to a function $g(x, y)$ from $\mathbf{R}^{n+m}$ into $\mathbf{R}^s$. Write $g(x, y) = [g_1(x, y), \ldots, g_s(x, y)]'$. Then define

$$
\mathrm{E}[g(x, y)|y] = (\mathrm{E}[g_1(x, y)|y]\,, \ldots, \mathrm{E}[g_s(x, y)|y])'\,.
$$

If their densities exist, two random vectors are *independent* if and only if their joint density is equal to the product of their marginal densities, i.e., x and y are independent if and only if

$$f_{x,y}(u, v) = f_x(u) f_y(v).$$

Note that if x and y are independent,

$$f_{x|y}(u|v) = f_x(u).$$

If the random vectors x and y are independent, then any (reasonable) vector-valued functions of them, say $g(x)$ and $h(y)$, are also independent. This follows easily from a more general definition of the independence of two random vectors: The random vectors x and y are independent if for any two (reasonable) sets A and B,

$$\Pr[x \in A, y \in B] = \Pr[x \in A]\Pr[y \in B].$$

To prove that functions of random variables are independent, recall that the set inverse of a function $g(u)$ on a set A_0 is $g^{-1}(A_0) \equiv \{u|g(u) \in A_0\}$. That $g(x)$ and $h(y)$ are independent follows from the fact that for any (reasonable) sets A_0 and B_0,

$$\begin{aligned}
\Pr[g(x) \in A_0, h(y) \in B_0] &= \Pr[x \in g^{-1}(A_0), y \in h^{-1}(B_0)] \\
&= \Pr[x \in g^{-1}(A_0)] \Pr[y \in h^{-1}(B_0)] \\
&= \Pr[g(x) \in A_0] \Pr[h(y) \in B_0].
\end{aligned}$$

By "reasonable" I mean things that satisfy the mathematical definitions of being measurable.

The *characteristic function* of a random vector $x = (x_1, \dots, x_n)'$ is a function from $\mathbf{R}^n$ to $\mathbf{C}$, the complex numbers. It is defined by

$$\varphi_x(t_1, \dots, t_n) = \int_{-\infty}^{\infty} \cdots \int_{-\infty}^{\infty} \exp\left[i \sum_{j=1}^{n} t_j u_j \right] f_x(u_1, \dots, u_n) du_1 \cdots du_n.$$

We are interested in characteristic functions because if $x = (x_1, \dots, x_n)'$ and $y = (y_1, \dots, y_n)'$ are random vectors and if

$$\varphi_x(t_1, \dots, t_n) = \varphi_y(t_1, \dots, t_n)$$

for all $(t_1, \dots, t_n)$, then x and y have the same distribution.

For convenience, we have assumed the existence of densities. With minor modifications, the definitions and results of this appendix hold for any probability defined on $\mathbf{R}^n$.

D.1 Identifiability

For better or worse (usually worse) much of statistical practice focuses on estimating and testing parameters. Identifiability is a property that ensures that this process is a sensible one.

Consider a collection of probability distributions $Y \sim P_\theta, \theta \in \Theta$. The parameter θ merely provides the name (index) for each distribution in the collection. Identifiability ensures that each distribution has a unique name/index.

Definition D.1 The parameterization $\theta \in \Theta$ is *identifiable* if $Y_1 \sim P_{\theta_1}, Y_2 \sim P_{\theta_2}$, and $Y_1 \sim Y_2$ imply that $\theta_1 = \theta_2$.

Being identifiable is easily confused with the concept of being well defined.

Definition D.2 The parameterization $\theta \in \Theta$ is *well defined* if $Y_1 \sim P_{\theta_1}, Y_2 \sim P_{\theta_2}$, and $\theta_1 = \theta_2$ imply that $Y_1 \sim Y_2$.

The problem with not being identifiable is that some distributions have more than one name. Observed data give you information about the correct distribution and thus about the correct name. Typically, the more data you have, the more information you have about the correct name. Estimation is about getting close to the correct name and testing hypotheses is about deciding which of two lists contains the correct name. If a distribution has more than one name, it could be in both lists. (Significance testing is about whether it seems plausible that a name is on a list, so identifiability seems less of an issue.) If a distribution has more than one name, does getting close to one of those names really help? In applications to linear models, typically distributions have only one name or they have an infinite number of names.

The ideas are roughly this. If the distributions are well defined and I know that Wesley O. Johnson (θ_1),and O. Wesley Johnson (θ_2) are the same person ($\theta_1 = \theta_2$), then, say, any collection of blood pressure readings on Wesley O. should look pretty much the same as comparable readings on O. Wesley. They would be two samples from the same distribution. Identifiability is the following: if all the samples I have taken or ever could take on Wesley O. look pretty much the same as samples on O. Wesley, then Wesley O. would have to be the same person as O. Wesley. (The reader might consider whether personhood is actually an identifiable parameter for blood pressure.)

For the multivariate normal distributions of Section 1.2, being well defined is the requirement that if $Y_1 \sim N(\mu_1, V_1)$, $Y_2 \sim N(\mu_2, V_2)$, and $\mu_1 = \mu_2$ and $V_1 = V_2$, then $Y_1 \sim Y_2$. Theorem 1.2.2 establishes that the mean and covariance of a multivariate normal determine the distribution. Being identifiable is that if $Y_1 \sim N(\mu_1, V_1)$, $Y_2 \sim N(\mu_2, V_2)$, and $Y_1 \sim Y_2$, then $\mu_1 = \mu_2$ and $V_1 = V_2$. Obviously, two random vectors with the same distribution have to have the same mean vector and covariance matrix. But life gets more complicated.

The more interesting problem for multivariate normality is a model

$$Y \sim N[F(\beta), V(\phi)]$$

where F and V are known functions of parameter vectors β and ϕ. To show that β and ϕ are identifiable we need to consider

$$Y_1 \sim N[F(\beta_1), V(\phi_1)], \qquad Y_2 \sim N[F(\beta_2), V(\phi_2)]$$

and show that if $Y_1 \sim Y_2$ then $\beta_1 = \beta_2$ and $\phi_1 = \phi_2$. From our earlier discussion, if $Y_1 \sim Y_2$ then $F(\beta_1) = F(\beta_2)$ and $V(\phi_1) = V(\phi_2)$. We need to check that $F(\beta_1) = F(\beta_2)$ implies $\beta_1 = \beta_2$ and that $V(\phi_1) = V(\phi_2)$ implies $\phi_1 = \phi_2$.

Section 2.1 gives an extensive discussion of when the mean parameterization is identifiable, i.e., when $F(\beta_1) = F(\beta_2)$ implies $\beta_1 = \beta_2$. There we defined identifiable functions of β as those that are functions of $F(\beta)$.

In this book, we mostly consider simple models for the covariance parameterization $V(\phi)$; models that are clearly identifiable because they involve at most one scalar parameter. We consider $V(\phi) \equiv \sigma^2 I$, and $V(\phi) \equiv \sigma^2 V$ where V is a known nonnegative definite matrix, and, again with known V, $V(\phi) \equiv V$, which involves no parameterization. For example, if $V(\phi_1) \equiv \sigma_1^2 I = V(\phi_2) \equiv \sigma_2^2 I$, we must have $\phi_1 \equiv \sigma_1^2 = \phi_2 \equiv \sigma_2^2$. As long as V is not the zero matrix, the covariance parameterizations in this book are identifiable.

ALM-III examines many commonly used models for $V(\phi)$ using similar notation to that used here. In particular, linear covariance parameterizations of the form $\sum_{r=0}^{s} \phi_r V_r$ for nonnegative ϕ_rs and nonnegative definite known V_rs are identifiable if and only if the V_rs are linearly independent, i.e., the $\text{Vec}(V_r)$s are linearly independent. The covariance matrices of Chapter 11 fall into this category.

Exercise

Exercise D.1 Let x and y be independent. Show that

(a) $\text{E}[g(x)|y] = \text{E}[g(x)]$;

(b) $\text{E}[g(x)h(y)] = \text{E}[g(x)]\,\text{E}[h(y)]$.

Appendix E
Inference for One Parameter

Abstract Since the third edition of this book, I have thought hard about the philosophy of testing as a basis for non-Bayesian statistical inference, cf. Christensen (2005, 2008). This appendix has been modified accordingly. The approach taken is one I call Fisherian, as opposed to the Neyman–Pearson approach. The theory presented here has no formal role for alternative hypotheses. A more extensive discussion of these ideas appears in Chapter 3 of Christensen (2015).

A significance testing problem is essentially a form of proof by contradiction. We have a *null model* for the data and we determine whether the observed data seem to contradict that null model or whether they are consistent with it. If the data contradict the null model, something must be wrong with the null model. Having data consistent with the null model certainly does not suggest that the null model is correct but may suggest that the model is tentatively adequate. The catch is that we rarely get an absolute contradiction to the null model, so we use probability to determine the extent to which the data seem inconsistent with the null model.

In the current discussion, *it is convenient to break the null model into two parts: a general model for the data and a particular statement about a single parameter of interest, called the null hypothesis* (H_0).

Many statistical tests and confidence intervals for a single parameter are applications of the same theory. (Tests and confidence intervals for variances are an exception.) To use this theory we need to know four things: [1] The unobservable *parameter* of interest (*Par*). [2] The *estimate* of the parameter (*Est*). [3] The *standard error* of the estimate (SE(*Est*)), wherein SE(*Est*) is typically an estimate of the standard deviation of *Est*, but if we happened to know the actual standard deviation, we would be happy to use it. And [4] an appropriate *reference distribution*. Specifically, we need the distribution of

$$\frac{Est - Par}{\mathrm{SE}(Est)}.$$

If the SE(*Est*) is estimated, the reference distribution is usually the t distribution with

R. Christensen, *Plane Answers to Complex Questions*, Springer Texts in Statistics,
https://doi.org/10.1007/978-3-030-32097-3

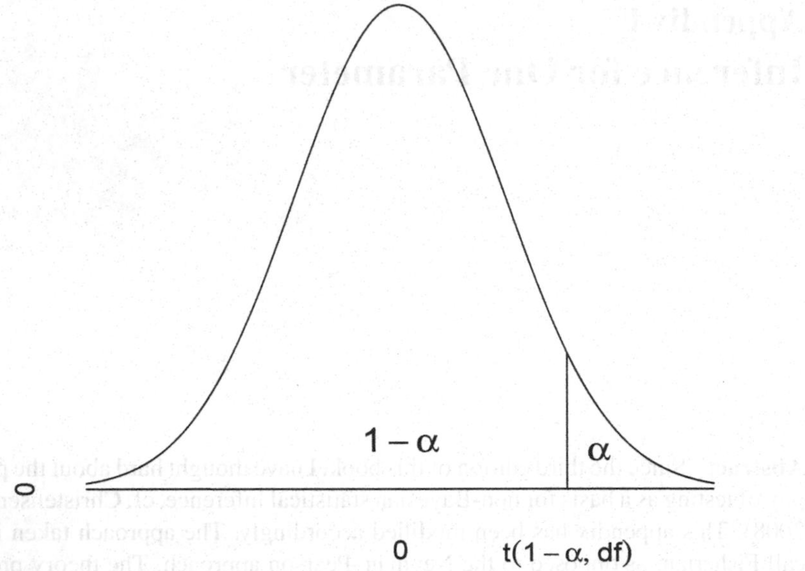

Fig. E.1 Percentiles of $t(df)$ distributions

some known number of degrees of freedom df, say, $t(df)$. If the SE(*Est*) is known, then the distribution is usually the standard normal distribution, i.e., a $t(\infty)$. In some problems (e.g., problems involving the binomial distribution) large sample results are used to get an approximate distribution and then the technique proceeds as if the approximate distribution were correct. When appealing to large sample results, the known distribution of part [4] is the standard normal (although I suspect that a $t(df)$ distribution with a reasonable, finite number of degrees of freedom would give more realistic results).

These four required items are derived from the model for the data (although sometimes the standard error incorporates the null hypothesis). For convenience, we may refer to these four items as "the model."

The $1 - \alpha$ percentile of a distribution is the point that cuts off the top α of the distribution. For a t distribution, denote this $t(1 - \alpha, df)$ as seen in Figure E.1. Formally, we can write

$$\Pr\left[\frac{Est - Par}{\mathrm{SE}(Est)} \geq t(1 - \alpha, df)\right] = \alpha.$$

By symmetry about zero, we also have

$$\Pr\left[\frac{Est - Par}{\mathrm{SE}(Est)} \leq -t(1 - \alpha, df)\right] = \alpha.$$

To keep the discussion as simple as possible, numerical examples have been restricted to one-sample normal theory. However, the results also apply to inferences on each individual mean and the difference between the means in two-sample problems, contrasts in analysis of variance, coefficients in regression, and, in general, to one-dimension estimable parametric functions in arbitrary linear models.

E.1 Testing

We want to test the null hypothesis

$$H_0 : Par = m,$$

where m is some known number. In *significance (Fisherian) testing*, we cannot do that. *What we can do* is test the null model, which is the combination of the model and the null hypothesis. The test is based on the assumption that both the model and H_0 are true. As mentioned earlier, it is rare that data contradict the null model absolutely, so we check to see if the data seem inconsistent with the null model.

What kind of data are inconsistent with the null model? Consider the *test statistic*

$$\frac{Est - m}{\mathrm{SE}(Est)}.$$

With m known, the test statistic is an observable random variable. If the null model is true, the test statistic has a known $t(df)$ distribution as illustrated in Figure E.1. The $t(df)$ distribution is likely to give values near 0 and is increasingly less likely to give values far from 0. Therefore, weird data, i.e., those that are most inconsistent with the null model, are large positive and large negative values of $[Est - m]/\mathrm{SE}(Est)$. The density (shape) of the $t(df)$ distribution allows us to order the possible values of the test statistic in terms of how weird they are relative to the null model.

To decide on a formal test, we need to decide which values of the test statistic will cause us to reject the null model and which will not. In other words, "How weird must data be before we question the null model?" We solve this problem by picking a small probability α that determines a *rejection region*, somtimes called a *critical region*. The rejection region consists of the weirdest test statistic values under the null model, but is restricted to have a probability of only α under the null model. Since a $t(df)$ distribution is symmetric about 0 and the density decreases as we go away from 0, the α critical region consists of points less than $-t(1-\alpha/2, df)$ and points larger than $t(1-\alpha/2, df)$. In other words, the α level test for the model with $H_0 : Par = m$ is to reject the null model if

$$\frac{Est - m}{\mathrm{SE}(Est)} \geq t\left(1 - \frac{\alpha}{2}, df\right)$$

or if

$$\frac{Est - m}{\mathrm{SE}(Est)} \leq -t\left(1 - \frac{\alpha}{2}, df\right).$$

This is equivalent to rejecting the null model if

$$\frac{|Est - m|}{\mathrm{SE}(Est)} \geq t\left(1 - \frac{\alpha}{2}, df\right).$$

What causes us to reject the null model? Either having a true model that is so different from the null that the data look "weird," or having the null model true and getting unlucky with the data.

Observing weird data, i.e., data that are inconsistent with the null model, gives us cause to question the validity of the null model. Specifying a small α level merely ensures that everything in the rejection region really constitutes weird data. More properly, specifying a small α level is our means of determining what constitutes weird data. Although α can be viewed as a probability, it is better viewed as a measure of how weird the data must be relative to the null model before we will reject. We want α small so that we only reject the null model for data that are truly weird, but we do not want α so small that we fail to reject the null model even when very strange data occur.

Rejecting the null model means that *either* the null hypothesis *or* the model is deemed incorrect. Only if we are confident that the model is correct can we conclude that the null hypothesis is wrong. If we want to make conclusions about the null hypothesis, it is important to do everything possible to assure ourselves that the model is reasonable.

If we do not reject the null model, we merely have data that are consistent with the null model. That in no way implies that the null model is true. Many other models will also be consistent with the data. Typically, $Par = m + 0.00001$ fits the data about as well as the null model. Not rejecting the test does not imply that the null model is true any more than rejecting the null model implies that the underlying model is true.

Example E.1 Suppose that 16 independent observations are taken from a normal population. Test $H_0 : \mu = 20$ with α level 0.01. The observed values of $\bar{y}_\cdot$ and s^2 were 19.78 and 0.25, respectively.

[1] $Par = \mu$,
[2] $Est = \bar{y}_\cdot$,
[3] $\mathrm{SE}(Est) = \sqrt{s^2/16}$. In this case, the $\mathrm{SE}(Est)$ is estimated.
[4] $[Est - Par]/\mathrm{SE}(Est) = [\bar{y}_\cdot - \mu]\big/\sqrt{s^2/16}$ has a $t(15)$ distribution.

With $m = 20$, the $\alpha = 0.01$ test is to reject the H_0 model if

$$|\bar{y}_\cdot - 20|\big/[s/4] \geq 2.947 = t(0.995, 15).$$

Having $\bar{y}_\cdot = 19.78$ and $s^2 = 0.25$, we reject if

$$\frac{|19.78 - 20|}{\sqrt{.25/16}} \geq 2.947.$$

Since $|19.78 - 20|/\sqrt{.25/16} = |-1.76|$ is less than 2.947, we do not reject the null model at the $\alpha = 0.01$ level.

Nobody actually does this! Or at least, nobody should do it. Although this procedure provides a philosophical basis for our statistical inferences, there are two other procedures, both based on this, that give uniformly more information. This procedure requires us to specify the model, the null hypothesis parameter value m, and the α level. For a fixed model and a fixed null parameter m, P values are more informative because they allow us to report test results for all α levels. Alternatively, for a fixed model and a fixed α level, confidence intervals report the values of all parameters that are consistent with the model and the data. (Parameter values that are inconsistent with the model and the data are those that would be rejected, assuming the model is true.) We now discuss these other procedures.

E.2 *P* Values

The P value of a test is the probability under the null model of seeing data as weird or weirder than we actually saw. Weirdness is determined by the distribution of the test statistic. If the observed value of the test statistic from Section 1 is t_{obs}, then the P value is the probability of seeing data as far or farther from 0 than t_{obs}. In general, we do not know if t_{obs} will be positive or negative, but its distance from 0 is $|t_{obs}|$. The P value is the probability that a $t(df)$ distribution is less than or equal to $-|t_{obs}|$ or greater than or equal to $|t_{obs}|$.

In Example E.1, the value of the test statistic is -1.76. Since $t(0.95, 15) = 1.75$, the P value of the test is approximately (just smaller than) 0.10. An $\alpha = 0.10$ test would use the $t(0.95, 15)$ value.

It is not difficult to see that the P value is the α level at which the test would just barely be rejected. So if $P \leq \alpha$, the null model is rejected, and if $P > \alpha$, the data are deemed consistent with the null model. Knowing the P value lets us do all α level tests of the null model. In fact, historically and philosophically, P values come before α level tests. Rather than noticing that the α level test has this relationship with P values, it is more general to define the α level test as rejecting precisely when $P \leq \alpha$. We can then observe that, for our setup, the α level test has the form given in Section 1.

While an α level constitutes a particular choice about how weird the data must be before we decide to reject the null model, the P value measures the evidence against the null hypothesis. The smaller the P value, the more evidence against the null model.

E.3 Confidence Intervals

A $(1-\alpha)100\%$ *confidence interval (CI)* for *Par* is defined to be the set of all parameter values m that would not be rejected by an α level test. In Section 1 we gave the rule for when an α level test of $H_0 : Par = m$ rejects. Conversely, the null model will not be rejected if

$$-t\left(1-\frac{\alpha}{2}, df\right) < \frac{Est - m}{\text{SE}(Est)} < t\left(1-\frac{\alpha}{2}, df\right). \tag{1}$$

Some algebra, given later, establishes that we do not reject the null model if and only if

$$Est - t\left(1-\frac{\alpha}{2}, df\right)\text{SE}(Est) < m < Est + t\left(1-\frac{\alpha}{2}, df\right)\text{SE}(Est). \tag{2}$$

This interval consists of all the parameter values m that are consistent with the data and the model as determined by an α *level test.* The endpoints of the CI can be written

$$Est \pm t\left(1-\frac{\alpha}{2}, df\right)\text{SE}(Est).$$

On occasion (as with binomial data), when doing an α level test or a P value, we may let the standard error depend on the null hypothesis. To obtain a confidence interval using this approach, we need a standard error that does not depend on m.

Example E.2 We have 10 independent observations from a normal population with variance 6. $\bar{y}_.$ is observed to be 17. We find a 95% CI for μ, the mean of the population.

[1] $Par = \mu$,
[3] $Est = \bar{y}_.$,
[3] $\text{SE}(Est) = \sqrt{6/10}$. In this case, SE(*Est*) is known and not estimated.
[4] $[Est - Par]/\text{SE}(Est) = [\bar{y}_. - \mu]\big/\sqrt{6/10} \sim N(0,1) = t(\infty)$.

The confidence coefficient is $95\% = (1-\alpha)100\%$, so $1-\alpha = 0.95$ and $\alpha = 0.05$. The percentage point from the normal distribution that we require is $t\left(1-\frac{\alpha}{2}, \infty\right) = t(0.975, \infty) = 1.96$. The limits of the 95% CI are, in general,

$$\bar{y}_. \pm 1.96\sqrt{6/10}$$

or, since $\bar{y}_. = 17$,

$$17 \pm 1.96\sqrt{6/10}.$$

The μ values in the interval (15.48, 18.52) are consistent with the data and the normal random sampling model as determined by an $\alpha = 0.05$ test.

To see that statements (1) and (2) are algebraically equivalent, the argument runs as follows:

$$-t\left(1-\frac{\alpha}{2}, df\right) < \frac{Est - m}{\mathrm{SE}(Est)} < t\left(1-\frac{\alpha}{2}, df\right)$$

if and only if $-t\left(1-\frac{\alpha}{2}, df\right)\mathrm{SE}(Est) < Est - m < t\left(1-\frac{\alpha}{2}, df\right)\mathrm{SE}(Est)$;
if and only if $t\left(1-\frac{\alpha}{2}, df\right)\mathrm{SE}(Est) > -Est + m > -t\left(1-\frac{\alpha}{2}, df\right)\mathrm{SE}(Est)$;
if and only if $Est + t\left(1-\frac{\alpha}{2}, df\right)\mathrm{SE}(Est) > m > Est - t\left(1-\frac{\alpha}{2}, df\right)\mathrm{SE}(Est)$;
if and only if $Est - t\left(1-\frac{\alpha}{2}, df\right)\mathrm{SE}(Est) < m < Est + t\left(1-\frac{\alpha}{2}, df\right)\mathrm{SE}(Est)$.

E.4 Final Comments on Significance Testing

The most arbitrary element in Fisherian testing is the choice of a test statistic. Although alternative hypotheses do not play a formal role in significance testing, interesting possible alternative hypotheses do inform the choice of test statistic.

For example, in linear models we often test a full model $Y = X\beta + e$ against a reduced model $Y = X_0\gamma + e$, with $e \sim N(0, \sigma^2 I)$ and $C(X_0) \subset C(X)$. Although we choose a test statistic based on comparing these models, the significance test is only a test of whether the data are consistent with the reduced model (and is a two-sided F test when $r(X) - r(X_0) \geq 3$). Rejecting the F test does not suggest that the full model is correct, it only suggests that the reduced model is wrong. Nonetheless, it is of interest to see how the test behaves if the full model is correct. But models other than the full model can also cause the test to reject, see Appendix F, especially Section F.2. For example, it is of interest to examine the *power* of a test. The power of an α level test at some alternative model is the probability of rejecting the null model when the alternative model is true. But in significance testing, there is no thought of accepting any alternative model. Any number of things can cause the rejection of the null model. Similar comments hold for testing generalized linear models.

When testing a null model based on a single parameter hypothesis $H_0 : Par = m$, interesting possible alternatives include $Par \neq m$. Our test statistic is designed to be sensitive to these alternatives, but problems with the null model other than $Par \neq m$ can cause us to reject the null model.

In general, a test statistic can be any function of the data for which the distribution under the null model is known (or can be approximated). But finding a usable test statistic can be difficult. Having to choose between alternative test statistics for the same null model is something of a luxury. For example, to test the null model with equal means in a balanced one-way ANOVA, we can use either the F test of Chapter 4 or the Studentized range test of Section 5.5.

Appendix F
Significantly Insignificant Tests

Abstract Computer programs for fitting linear models typical focus on the significance of large F statistics. This appendix discusses why one should always be concerned about observing F statistics very near 0 (when the numerator degrees of freedom are 3 or more).

Philosophically, the test of a null model occurs almost in a vacuum. Either the data contradict the null model or they are consistent with it. The discussion of model testing in Section 3.2 largely assumes that the full model is true. While it is interesting to explore the behavior of the F test statistic when the full model is true, and indeed it is reasonable and appropriate to choose a test statistic that will work well when the full model is true, the act of rejecting the null model in no way implies that the full model is true. It is perfectly reasonable that the null (reduced) model can be rejected when the full model is false.

Throughout this book we have examined standard approaches to testing in which F tests are rejected only for large values. The rationale for this is based on the full model being true. We now examine the significance of small F statistics. Small F statistics can be caused by an unsuspected lack of fit or, when the mean structure of the reduced model is correct, they can be caused by not accounting for negatively correlated data or not accounting for heteroscedasticity. We also demonstrate that large F statistics can be generated by not accounting for positively correlated data or heteroscedasticity, even when the mean structure of the reduced model is correct.

Christensen (1995, 2005, 2008) argues that (non-Bayesian) testing should be viewed as an exercise in examining whether or not the data are consistent with a particular (predictive) model. While possible alternative hypotheses may drive the choice of a test statistic, any unusual values of the test statistic should be considered important. By this standard, perhaps the only general way to decide which values of the test statistic are unusual is to identify as unusual those values that have small probabilities or small densities under the model being tested.

The F test statistic is driven by the idea of testing the reduced model against the full model. However, given the test statistic, any unusual values of that statistic

R. Christensen, *Plane Answers to Complex Questions*, Springer Texts in Statistics,
https://doi.org/10.1007/978-3-030-32097-3

should be recognized as indicating data that are inconsistent with the model being tested. If the full model is true, values of F much larger than 1 are inconsistent with the reduced model. Values of F much larger than 1 are consistent with the full model but, as we shall see, they are consistent with other models as well. Similarly, (when the numerator degrees of freedom are 3 or more) values of F much smaller than 1 are also inconsistent with the reduced model and we will examine models that can generate small F statistics.

I have been hesitant to discuss what I think of as a Fisherian F test, since nobody actually performs them. (That includes me, because it is so much easier to use the reported P values provided by standard computer programs.) Although the test statistic comes from considering both the reduced (null) model and the full model, once the test statistic is chosen, the full model no longer plays a role. From Theorem 3.2.1(ii), if the reduced model is true,

$$F \equiv \frac{Y'(M-M_0)Y/r(M-M_0)}{Y'(I-M)Y/r(I-M)} \sim F(r(M-M_0), r(I-M), 0).$$

We use the density to define "weird" values of the F distribution. The smaller the density, the weirder the observation. Write $r_1 \equiv r(M-M_0)$ and $r_2 \equiv r(I-M)$, denote the density $g(f|r_1,r_2)$, and let F_{obs} denote the observed value of the F statistic. Since the P value of a test is the probability under the null model of seeing data as weird or weirder than we actually saw, and weirdness is defined by the density, the P value of the test is

$$P = \Pr[g(F|r_1,r_2) \leq g(F_{obs}|r_1,r_2)],$$

wherein F_{obs} is treated as fixed and known. This is computed under the only distribution we have, the $F(r_1,r_2)$ distribution. An α level test is defined as rejecting the null model precisely when $P \leq \alpha$.

If $r_1 > 2$, the $F(r_1,r_2)$ density has the familiar shape that starts at 0, rises to a maximum in the vicinity of 1, and drops back down to zero for large values. Unless F_{obs} happens to be the mode, there are two values $f_1 < f_2$ that have

$$g(F_{obs}|r_1,r_2) = g(f_1|r_1,r_2) = g(f_2|r_1,r_2).$$

(One of f_1 and f_2 will be F_{obs}.) In this case, the P value reduces to

$$P = \Pr[F \leq f_1] + \Pr[F \geq f_2].$$

In other words, the Fisherian F test is a two-sided F test, rejecting both for very small and very large values of F_{obs}. For $r_1 = 1, 2$, the Fisherian test agrees with the usual test because then the $F(r_1,r_2)$ density starts high and decreases as f gets larger.

I should also admit that there remain open questions about the appropriateness of using densities, rather than actual probabilities, to define the weirdness of observations. In fact, I have long speculated whether *Fisher's "z" distribution*, i.e., $z \equiv \log(F)/2$ might not provide a more appropriate density for significance testing than the F density. The remainder of this appendix is closely related to Christensen (2003). See also Högfeldt (1979).

F.1 Lack of Fit and Small F Statistics

The standard assumption in testing models is that there is a full model $Y = X\beta + e$, $\mathrm{E}(e) = 0$, $\mathrm{Cov}(e) = \sigma^2 I$ that fits the data. We then test the adequacy of a reduced model $Y = X_0\gamma + e$, $\mathrm{E}(e) = 0$, $\mathrm{Cov}(e) = \sigma^2 I$ in which $C(X_0) \subset C(X)$, cf. Section 3.2. Based on second moment arguments, the test statistic is a ratio of variance estimates. We construct an unbiased estimate of σ^2, $Y'(I-M)Y/r(I-M)$, and another statistic $Y'(M-M_0)Y/r(M-M_0)$ that has $\mathrm{E}[Y'(M-M_0)Y/r(M-M_0)] = \sigma^2 + \beta'X'(M-M_0)X\beta/r(M-M_0)$. Under the assumed covariance structure, this second statistic is an unbiased estimate of σ^2 if and only if the reduced model is correct. The test statistic

$$F = \frac{Y'(M-M_0)Y/r(M-M_0)}{Y'(I-M)Y/r(I-M)}$$

is a (biased) estimate of

$$\frac{\sigma^2 + \beta'X'(M-M_0)X\beta/r(M-M_0)}{\sigma^2} = 1 + \frac{\beta'X'(M-M_0)X\beta}{\sigma^2\, r(M-M_0)}.$$

Under the null model, F is an estimate of the number 1. When the full model is true, values of F much larger than 1 suggest that F is estimating something larger than 1, which suggests that $\beta'X'(M-M_0)X\beta/\sigma^2\, r(M-M_0) > 0$, something that occurs if and only if the reduced model is false. The standard normality assumption leads to an exact central F distribution for the test statistic under the null model, so we are able to quantify how unusual it is to observe any F statistic greater than 1. Although the test is based on second moment considerations, under the normality assumption it is also the generalized likelihood ratio test, see Exercise 3.1, and a uniformly most powerful invariant test, see Lehmann (1986, Section 7.1).

In testing lack of fit, the same basic ideas apply except that we start with the (reduced) model $Y = X\beta + e$. The ideal situation would be to know that if $Y = X\beta + e$ has the wrong mean structure, then a model of the form

$$Y = X\beta + W\delta + e, \quad C(W) \perp C(X) \tag{1}$$

fits the data where assuming $C(W) \perp C(X)$ creates no loss of generality. Unfortunately, there is rarely anyone to tell us the true matrix W. Lack-or-fit testing is largely about constructing a full model, say, $Y = X_*\beta_* + e$ with $C(X) \subset C(X_*)$ based on reasonable assumptions about the nature of any lack of fit. The test for lack of fit is simply the test of $Y = X\beta + e$ against the constructed model $Y = X_*\beta_* + e$. Typically, the constructed full model involves somehow generalizing the structure already observed in $Y = X\beta + e$. Section 6.7 discusses the rationale for several choices of constructed full models. For example, the traditional lack-or-fit test for simple linear regression begins with the replication model $y_{ij} = \beta_0 + \beta_1 x_i + e_{ij}$, $i = 1, \ldots, a$, $j = 1, \ldots, N_i$. It then assumes $\mathrm{E}(y_{ij}) = f(x_i)$ for some function $f(\cdot)$, in other words, it assumes that the several observations associated with x_i have the same expected value. Making no additional assumptions leads to fitting the full model $y_{ij} = \mu_i + e_{ij}$ and the traditional lack-or-fit test. Another way to think of this traditional test views the reduced model relative to the one-way ANOVA as having only the linear contrast important. The traditional lack-or-fit test statistic becomes

$$F = \frac{SSTrts - SS(lin)}{a-2} \Big/ MSE, \tag{2}$$

where $SS(lin)$ is the sum of squares for the linear contrast. If there is no lack of fit in the reduced model, F should be near 1. If lack of fit exists because the more general mean structure of the one-way ANOVA fits the data better than the simple linear regression model, the F statistic tends to be larger than 1.

Unfortunately, if the lack of fit exists because of features that are not part of the original model, generalizing the structure observed in $Y = X\beta + e$ is often inappropriate. Suppose that the simple linear regression model is balanced, i.e., all $N_i = N$, that for each i the data are taken in time order $t_1 < t_2 < \cdots < t_N$, and that the lack of fit is due to the true model being

$$y_{ij} = \beta_0 + \beta_1 x_i + \delta t_j + e_{ij}, \quad \delta \neq 0. \tag{3}$$

Thus, depending on the sign of δ, the observations within each group are subject to an increasing or decreasing trend. Note that in this model, for fixed i, the $\mathrm{E}(y_{ij})$s are *not* the same for all j, thus invalidating the assumption of the traditional test. In fact, this causes the traditional lack of fit test to have a *small* F statistic. One way to see this is to view the problem in terms of a balanced two-way ANOVA. The true model (3) is a special case of the two-way ANOVA model $y_{ij} = \mu + \alpha_i + \eta_j + e_{ij}$ in which the only nonzero terms are the linear contrast in the α_is and the linear contrast in the η_js. Under model (3), the numerator of the statistic (2) gives an unbiased estimate of σ^2 because $SSTrts$ in (2) is $SS(\alpha)$ for the two-way model and the only nonzero α effect is being eliminated from the treatments. However, the mean squared error in the denominator of (2) is a weighted average of the error mean square from the two-way model and the mean square for the η_js in the two-way model. The sum of squares for the significant linear contrast in the η_js from model (3) is included in the error term of the lack-or-fit test (2), thus biasing the error term to estimate something larger than σ^2. In particular, the denominator has an expected value of $\sigma^2 + \delta^2 a \sum_{j=1}^{N}(t_j - \bar{t}_\cdot)^2/a(N-1)$. Thus, if the appropriate model is (3), the statistic in (2) estimates $\sigma^2/[\sigma^2 + \delta^2 a \sum_{j=1}^{N}(t_j - \bar{t}_\cdot)^2/a(N-1)]$ which is a number that is less than 1. Values of F much smaller than 1, i.e., very near 0, are consistent with a lack of fit that exists within the groups of the one-way ANOVA. Note that in this balanced case, true models involving interaction terms, e.g., models like

$$y_{ij} = \beta_0 + \beta_1 x_i + \delta t_j + \gamma x_i t_j + e_{ij},$$

also tend to make the F statistic small if either $\delta \neq 0$ or $\gamma \neq 0$. Finally, if there exists lack of fit both between the groups of observations and within the groups, if can be very difficult to identify. For example, if $\beta_2 \neq 0$ and either $\delta \neq 0$ or $\gamma \neq 0$ in the true model

$$y_{ij} = \beta_0 + \beta_1 x_i + \beta_2 x_i^2 + \delta t_j + \gamma x_i t_j + e_{ij},$$

there is both a traditional lack of fit between the groups (the significant $\beta_2 x_i^2$ term) and lack of fit within the groups ($\delta t_j + \gamma x_i t_j$). In this case, neither the numerator nor the denominator in (2) is an estimate of σ^2.

More generally, start with a model $Y = X\beta + e$. This is tested against a larger model $Y = X_*\beta_* + e$ with $C(X) \subset C(X_*)$, regardless of where the larger model comes from. The F statistic is

$$F = \frac{Y'(M_* - M)Y/r(M_* - M)}{Y'(I - M_*)Y/r(I - M_*)}.$$

We assume that the true model is (1). The F statistic estimates 1 if the original model $Y = X\beta + e$ is correct. It estimates something greater than 1 if the larger model $Y = X_*\beta_* + e$ is correct, i.e., if $W\delta \in C(X)^{\perp}_{C(X_*)}$. F estimates something less than 1 if $W\delta \in C(X_*)^{\perp}$, i.e., if $W\delta$ is actually in the error space of the larger model, because then the numerator estimates σ^2 but the denominator estimates

$$\sigma^2 + \delta' W'(I - M_*)W\delta/r(I - M_*) = \sigma^2 + \delta' W' W\delta/r(I - M_*).$$

If $W\delta$ is in neither of $C(X)^{\perp}_{C(X_*)}$ nor $C(X_*)^{\perp}$, it is not clear how the test will behave because neither the numerator nor the denominator estimates σ^2. Christensen (1989, 1991) contains related discussion of these concepts.

The main point is that, when testing a full model $Y = X\beta + e, \mathrm{E}(e) = 0, \mathrm{Cov}(e) = \sigma^2 I$ against a reduced model $Y = X_0\gamma + e$, $C(X_0) \subset C(X)$, if the F statistic is small, it suggests that $Y = X_0\gamma + e$ may suffer from lack of fit in which the lack of fit exists in the error space of $Y = X\beta + e$. We will see in the next section that other possible explanations for a small F statistic are the existence of "negative correlation" in the data or heteroscedasticity.

F.2 The Effect of Correlation and Heteroscedasticity on F Statistics

The test of a reduced model assumes that the full model $Y = X\beta + e$, $\mathrm{E}(e) = 0$, $\mathrm{Cov}(e) = \sigma^2 I$ holds and tests the adequacy of a reduced model $Y = X_0\gamma + e$, $\mathrm{E}(e) = 0$, $\mathrm{Cov}(e) = \sigma^2 I$, $C(X_0) \subset C(X)$. Rejecting the reduced model does not imply that the full model is correct. The mean structure of the reduced model may be perfectly valid, but the F statistic can become large or small because the assumed covariance structure is incorrect.

We begin with a concrete example, one-way ANOVA. Let $i = 1, \ldots, a$, $j = 1, \ldots, N$, and $n \equiv aN$. Consider a reduced model $y_{ij} = \mu + e_{ij}$ which in matrix terms we write $Y = J\mu + e$, and a full model $y_{ij} = \mu_i + e_{ij}$, which we write $Y = Z\gamma + e$. In matrix terms the usual one-way ANOVA F statistic is

$$F = \frac{Y'[M_Z - (1/n)J_n^n]Y/(a-1)}{Y'(I - M_Z)Y/a(N-1)}. \tag{1}$$

We now assume that the true model is $Y = J\mu + e$, $\mathrm{E}(e) = 0$, $\mathrm{Cov}(e) = \sigma^2 V$ and examine the behavior of the F statistic (1).

For a homoscedastic balanced one-way ANOVA we want to characterize the concepts of overall positive correlation, positive correlation within groups, and positive correlation for evaluating differences between groups. Consider first a simple example with $a = 2$, $N = 2$. The first two observations are a group and the last two are a group. Consider a covariance structure

$$V_1 = \begin{bmatrix} 1 & 0.9 & 0.1 & 0.09 \\ 0.9 & 1 & 0.09 & 0.1 \\ 0.1 & 0.09 & 1 & 0.9 \\ 0.09 & 0.1 & 0.9 & 1 \end{bmatrix}.$$

There is an overall positive correlation, high positive correlation between the two observations in each group, and weak positive correlation between the groups. A second example,

$$V_2 = \begin{bmatrix} 1 & 0.1 & 0.9 & 0.09 \\ 0.1 & 1 & 0.09 & 0.9 \\ 0.9 & 0.09 & 1 & 0.1 \\ 0.09 & 0.9 & 0.1 & 1 \end{bmatrix},$$

has an overall positive correlation but weak positive correlation between the two observations in each group, with high positive correlation between some observations in different groups.

We now make a series of definitions for homoscedastic balanced one-way ANOVA based on the projection operators in (1) and V. Overall positive correlation is characterized by $\mathrm{Var}(\bar{y}_{\cdot\cdot}) > \sigma^2/n$, which in matrix terms is written

$$n\frac{\mathrm{Var}(\bar{y}_{\cdot\cdot})}{\sigma^2} = \mathrm{tr}[(1/n)JJ'V] > \frac{1}{n}\mathrm{tr}(V)\mathrm{tr}[(1/n)JJ'] = \frac{1}{n}\mathrm{tr}(V). \tag{2}$$

Overall negative correlation is characterized by the reverse inequality. For homoscedastic models the term $\mathrm{tr}(V)/n$ is 1. For heteroscedastic models the term on the right is the average variance of the observations divided by σ^2.

Positive correlation within groups is characterized by $\sum_{i=1}^{a} \mathrm{Var}(\bar{y}_{i\cdot})/a > \sigma^2/N$, which in matrix terms is written

$$\sum_{i=1}^{a} N \frac{\mathrm{Var}(\bar{y}_{i\cdot})}{\sigma^2} = \mathrm{tr}[M_Z V] > \frac{1}{n}\mathrm{tr}(V)\mathrm{tr}[M_Z] = \frac{a}{n}\mathrm{tr}(V). \tag{3}$$

Negative correlation within groups is characterized by the reverse inequality.

Positive correlation for evaluating differences between groups is characterized by

$$\frac{\sum_{i=1}^{a} \mathrm{Var}(\bar{y}_{i\cdot} - \bar{y}_{\cdot\cdot})}{a} > \frac{a-1}{a}\frac{\sigma^2}{N}.$$

Note that equality obtains if $V = I$. In matrix terms, this is written

$$\frac{N}{\sigma^2}\sum_{i=1}^{a} \mathrm{Var}(\bar{y}_{i\cdot} - \bar{y}_{\cdot\cdot}) = \mathrm{tr}([M_Z - (1/n)JJ']V)$$

$$> \frac{1}{n}\mathrm{tr}(V)\mathrm{tr}[M_Z - (1/n)JJ'] = \frac{a-1}{n}\mathrm{tr}(V) \tag{4}$$

and negative correlation for evaluating differences between groups is characterized by the reverse inequality. If all the observations in different groups are uncorrelated, there will be positive correlation for evaluating differences between groups if and only if there is positive correlation within groups. This follows because having a block diagonal covariance matrix $\sigma^2 V$ implies that $\mathrm{tr}(M_Z V) = \mathrm{tr}[(1/N)Z'VZ] = a\mathrm{tr}[(1/n)J'VJ] = a\mathrm{tr}[(1/n)JJ'V]$.

For our example V_1,

$$2.09 = (1/4)[4(2.09)] = \mathrm{tr}[(1/n)J_n^n V_1] > \frac{1}{n}\mathrm{tr}(V_1) = 4/4 = 1,$$

so there is an overall positive correlation,

$$3.8 = 2(1/2)[3.8] = \mathrm{tr}[M_Z V_1] > \frac{a}{n}\mathrm{tr}(V_1) = (2/4)4 = 2,$$

so there is positive correlation within groups, and

$$1.71 = 3.8 - 2.09 = \mathrm{tr}([M_Z - (1/n)J_n^n]V_1) > \frac{a-1}{n}\mathrm{tr}(V_1) = (1/4)4 = 1,$$

so there is positive correlation for evaluating differences between groups.

For the second example V_2,

$$2.09 = (1/4)[4(2.09)] = \mathrm{tr}[(1/n)J_n^n V_2] > \frac{1}{n}\mathrm{tr}(V_2) = 4/4 = 1,$$

so there is an overall positive correlation,

$$2.2 = 2(1/2)[2.2] = \text{tr}[M_Z V_2] > \frac{a}{n}\text{tr}(V_2) = (2/4)4 = 2,$$

so there is positive correlation within groups, but

$$0.11 = 2.2 - 2.09 = \text{tr}([M_Z - (1/n)J_n^n]V_2) < \frac{a-1}{n}\text{tr}(V_2) = (1/4)4 = 1,$$

so positive correlation for evaluating differences between groups does not exist.

The existence of positive correlation within groups and positive correlation for evaluating differences between groups causes the one-way ANOVA F statistic in (1) to get large even when there are no differences in the group means. Assuming that the correct model is $Y = J\mu + e$, $\text{E}(e) = 0$, $\text{Cov}(e) = \sigma^2 V$, by Theorem 1.3.1, the numerator of the F statistic estimates

$$\begin{aligned} \text{E}\{Y'[M_Z - (1/n)J_n^n]Y/(a-1)\} &= \text{tr}\{[M_Z - (1/n)J_n^n]V\}/(a-1) \\ &> \frac{a-1}{n}\text{tr}(V)/(a-1) = \text{tr}(V)/n \end{aligned}$$

and the denominator of the F statistic estimates

$$\begin{aligned} \text{E}\{Y'(I - M_Z)Y/a(N-1)\} &= \text{tr}\{[I - M_Z]V\}/a(N-1) \\ &= (\text{tr}\{V\} - \text{tr}\{[M_Z]V\})/a(N-1) \\ &< \left(\text{tr}\{V\} - \frac{a}{n}\text{tr}(V)\right)/a(N-1) \\ &= \frac{n-a}{n}\text{tr}(V)/a(N-1) = \text{tr}(V)/n. \end{aligned}$$

In (1), F is an estimate of

$$\begin{aligned} \frac{\text{E}\{Y'[M_Z - (1/n)J_n^n]Y/(a-1)\}}{\text{E}\{Y'(I - M_Z)Y/a(N-1)\}} &= \frac{\text{tr}\{[M_Z - (1/n)J_n^n]V\}/(a-1)}{\text{tr}\{[I - M_Z]V\}/a(N-1)} \\ &> \frac{\text{tr}(V)/n}{\text{tr}(V)/n} = 1, \end{aligned}$$

so having both positive correlation within groups and positive correlation for evaluating differences between groups tends to make F statistics large. Exactly analogous computations show that having both negative correlation within groups and negative correlation for evaluating differences between groups tends to make F statistics less than 1.

Another example elucidates some additional points. Suppose the observations have the AR(1) correlation structure discussed in Subsection 12.3.1:

$$V_3 = \begin{bmatrix} 1 & \rho & \rho^2 & \rho^3 \\ \rho & 1 & \rho & \rho^2 \\ \rho^2 & \rho & 1 & \rho \\ \rho^3 & \rho^2 & \rho & 1 \end{bmatrix}.$$

Using the same grouping structure as before, when $0 < \rho < 1$, we have overall positive correlation because

$$1 + \frac{\rho}{2}(3 + 2\rho + \rho^2) = \mathrm{tr}[(1/n)JJ'V_3] > 1,$$

and we have positive correlation within groups because

$$2(1+\rho) = \mathrm{tr}[M_Z V_3] > 2.$$

If $-1 < \rho < 0$, the inequalities are reversed. Similarly, for $-1 < \rho < 0$ we have negative correlation for evaluating differences between groups because

$$1 + \frac{\rho}{2}(1 - 2\rho - \rho^2)^2 = \mathrm{tr}([M_Z - (1/n)JJ']V_3) < 1.$$

However, we only get positive correlation for evaluating differences between groups when $0 < \rho < \sqrt{2} - 1$. Thus, for negative ρ we tend to get small F statistics, for $0 < \rho < \sqrt{2} - 1$ we tend to get large F statistics, and for $\sqrt{2} - 1 < \rho < 1$ the result is not clear.

To illustrate, suppose $\rho = 1$ and the observations all have the same mean, then with probability 1, all the observations are equal and, in particular, $\bar{y}_{i\cdot} = \bar{y}_{\cdot\cdot}$ with probability 1. It follows that

$$0 = \frac{\sum_{i=1}^{a} \mathrm{Var}(\bar{y}_{i\cdot} - \bar{y}_{\cdot\cdot})}{a} < \frac{a-1}{a}\frac{\sigma^2}{N}$$

and no positive correlation exists for evaluating differences between groups. More generally, for very strong positive correlations, both the numerator and the denominator of the F statistic estimate numbers close to 0 and both are smaller than they would be under $V = I$. On the other hand, it is not difficult to see that, for $\rho = -1$, the F statistic is 0.

In the balanced heteroscedastic one-way ANOVA, V is diagonal. This generates equality between the left sides and right sides of (2), (3), and (4), so under heteroscedasticity F still estimates the number 1. We now generalize the ideas of within group correlation and correlation for evaluating differences between groups, and see that heteroscedasticity can affect unbalanced one-way ANOVA.

In general, we test a full model $Y = X\beta + e$, $\mathrm{E}(e) = 0$, $\mathrm{Cov}(e) = \sigma^2 I$ against a reduced model $Y = X_0\gamma + e$, in which $C(X_0) \subset C(X)$. We examine the F statistic when the true model is $Y = X_0\gamma + e$, $\mathrm{E}(e) = 0$, $\mathrm{Cov}(e) = \sigma^2 V$. Using arguments similar to those for balanced one-way ANOVA, having

$$\text{tr}[MV] > \frac{1}{n}\text{tr}(V)\text{tr}[M] = \frac{r(X)}{n}\text{tr}(V)$$

and

$$\text{tr}([M - M_0]V) > \frac{1}{n}\text{tr}(V)\text{tr}[M - M_0] = \frac{r(X) - r(X_0)}{n}\text{tr}(V)$$

causes large F statistics even when the mean structure of the reduced model is true, and reversing the inequalities causes small F statistics. These are merely sufficient conditions so that the tests intuitively behave certain ways. The actual behavior of the tests under normal distributions can be determined numerically, cf. Christensen and Bedrick (1997).

These covariance conditions can be caused by patterns of positive and negative correlations as discussed earlier, but they can also be caused by heteroscedasticity. For example, consider the behavior of the unbalanced one-way ANOVA F test when the observations are uncorrelated but heteroscedastic. For concreteness, assume that $\text{Var}(y_{ij}) = \sigma_i^2$. Because the observations are uncorrelated, we need only check the condition

$$\text{tr}[MV] \equiv \text{tr}[M_Z V] > \frac{1}{n}\text{tr}(V)\text{tr}[M_Z] = \frac{a}{n}\text{tr}(V),$$

which amounts to

$$\sum_{i=1}^{a} \sigma_i^2/a > \sum_{i=1}^{a} \frac{N_i}{n}\sigma_i^2.$$

Thus, when the groups' means are equal, F statistics will get large if many observations are taken in groups with small variances and few observations are taken on groups with large variances. F statistics will get small if the reverse relationship holds.

The general condition

$$\text{tr}[MV] > \frac{1}{n}\text{tr}(V)\text{tr}[M] = \frac{r(X)}{n}\text{tr}(V)$$

is equivalent to

$$\frac{\sum_{i=1}^{n} \text{Var}(x_i'\hat{\beta})}{r(X)} > \frac{\sum_{i=1}^{n} \text{Var}(y_i)}{n}.$$

So, under homoscedasticity, positive correlation in the full model amounts to having an average variance for the predicted values (averaging over the rank of the covariance matrix of the predicted values) that is larger than the common variance of the observations. Negative correlation in the full model involves reversing the inequality. Similarly, having positive correlation for distinguishing the full model from the reduced model means

$$\frac{\sum_{i=1}^{n} \text{Var}(x_i'\hat{\beta} - x_{0i}'\hat{\gamma})}{r(X) - r(X_0)} = \frac{\text{tr}[(M - M_0)V]}{r(M - M_0)} > \frac{\text{tr}(V)}{n} = \frac{\sum_{i=1}^{n} \text{Var}(y_i)}{n}.$$

Appendix G
Randomization Theory Models

Abstract This appendix introduces randomization theory models in which, rather than assuming the existence of an error term with certain properties, the random variability in the data is constructed either by random sampling from a population or by randomly assigning treatments to experimental units.

The division of labor in statistics has traditionally designated randomization theory as an area of nonparametric statistics. Randomization theory is also of special interest in the theory of experimental design because randomization has been used to justify the analysis of designed experiments.

It can be argued that the linear models given in Chapter 8 are merely good approximations to more appropriate models based on randomization theory. One aspect of this argument is that the F tests based on the theory of normal errors are a good approximation to randomization (permutation) tests. Investigating this is beyond the scope of a linear models book, cf. Hinkelmann and Kempthorne (2005) and Puri and Sen (1971). Another aspect of the approximation argument is that the BLUEs under randomization theory are precisely the least squares estimates. By Theorem 10.4.5, to establish this we need to show that $C(VX) \subset C(X)$ for the model

$$Y = X\beta + e, \quad \mathrm{E}(e) = 0, \quad \mathrm{Cov}(e) = V,$$

where V is the covariance matrix under randomization theory. This argument will be examined here for two experimental design models: the model for a completely randomized design and the model for a randomized complete block design. First, we introduce the subject with a discussion of simple random sampling.

G.1 Simple Random Sampling

Randomization theory for a simple random sample assumes that observations y_i are picked at random (without replacement) from a larger finite population. Suppose the elements of the population are $s_1, s_2, \ldots, s_N$. We can define elementary sampling

R. Christensen, *Plane Answers to Complex Questions*, Springer Texts in Statistics,
https://doi.org/10.1007/978-3-030-32097-3

random variables for $i = 1, \ldots, n$ and $j = 1, \ldots, N$,

$$\delta_j^i = \begin{cases} 1, & \text{if } y_i = s_j \\ 0, & \text{otherwise.} \end{cases}$$

Under simple random sampling without replacement

$$\mathrm{E}[\delta_j^i] = \Pr[\delta_j^i = 1] = \frac{1}{N}.$$

$$\mathrm{E}[\delta_j^i \delta_{j'}^{i'}] = \Pr[\delta_j^i \delta_{j'}^{i'} = 1] = \begin{cases} 1/N, & \text{if } (i, j) = (i', j') \\ 1/N(N-1), & \text{if } i \neq i' \text{ and } j \neq j' \\ 0, & \text{otherwise.} \end{cases}$$

If we write $\mu = \sum_{j=1}^N s_j/N$ and $\sigma^2 = \sum_{j=1}^N (s_j - \mu)^2/N$, then

$$y_i = \sum_{j=1}^N \delta_j^i s_j = \mu + \sum_{j=1}^N \delta_j^i (s_j - \mu).$$

Letting $e_i = \sum_{j=1}^N \delta_j^i (s_j - \mu)$ gives the linear model

$$y_i = \mu + e_i.$$

The population mean μ is a fixed unknown constant. The e_is have the properties

$$\mathrm{E}[e_i] = \mathrm{E}\left[\sum_{j=1}^N \delta_j^i (s_j - \mu)\right] = \sum_{j=1}^N \mathrm{E}[\delta_j^i](s_j - \mu) = \sum_{j=1}^N (s_j - \mu)/N = 0,$$

$$\mathrm{Var}(e_i) = \mathrm{E}[e_i^2] = \sum_{j=1}^N \sum_{j'=1}^N (s_j - \mu)(s_{j'} - \mu)\mathrm{E}[\delta_j^i \delta_{j'}^i] = \sum_{j=1}^N (s_j - \mu)^2/N = \sigma^2.$$

For $i \neq i'$,

$$\begin{aligned} \mathrm{Cov}(e_i, e_{i'}) = \mathrm{E}[e_i e_{i'}] &= \sum_{j=1}^N \sum_{j'=1}^N (s_j - \mu)(s_{j'} - \mu)\mathrm{E}[\delta_j^i \delta_{j'}^{i'}] \\ &= [N(N-1)]^{-1} \sum_{j \neq j'} (s_j - \mu)(s_{j'} - \mu) \\ &= [N(N-1)]^{-1} \left(\left[\sum_{j=1}^N (s_j - \mu)\right]^2 - \sum_{j=1}^N (s_j - \mu)^2 \right) \\ &= -\sigma^2/(N-1). \end{aligned}$$

In matrix terms, the linear model can be written

$$Y = J\mu + e, \quad \mathrm{E}(e) = 0, \quad \mathrm{Cov}(e) = \sigma^2 V,$$

where

$$V = \begin{bmatrix} 1 & -(N-1)^{-1} & -(N-1)^{-1} & \cdots & -(N-1)^{-1} \\ -(N-1)^{-1} & 1 & -(N-1)^{-1} & \cdots & -(N-1)^{-1} \\ -(N-1)^{-1} & -(N-1)^{-1} & 1 & \cdots & -(N-1)^{-1} \\ \vdots & \vdots & \vdots & \ddots & \vdots \\ -(N-1)^{-1} & -(N-1)^{-1} & -(N-1)^{-1} & \cdots & 1 \end{bmatrix}.$$

Clearly $VJ = [(N-n)/(N-1)]\,J$, so the BLUE of μ is $\bar{y}_{\cdot\cdot}$.

G.2 Completely Randomized Designs

Suppose that there are t treatments, each to be randomly assigned to N units out of a collection of $n = tN$ experimental units. A one-way ANOVA model for this design is

$$y_{ij} = \mu_i + e_{ij}, \tag{1}$$

$i = 1, \ldots, t$, $j = 1, \ldots, N$. Suppose further that the ith treatment has an effect τ_i and that the experimental units without treatment effects would have readings $s_1, \ldots, s_n$. The elementary sampling random variables are

$$\delta_k^{ij} = \begin{cases} 1, & \text{if replication } j \text{ of treatment } i \text{ is assigned to unit } k \\ 0, & \text{otherwise.} \end{cases}$$

With this restricted random sampling,

$$\mathrm{E}[\delta_k^{ij}] = \Pr[\delta_k^{ij} = 1] = \frac{1}{n}$$

$$\mathrm{E}[\delta_k^{ij}\delta_{k'}^{i'j'}] = \Pr[\delta_k^{ij}\delta_{k'}^{i'j'} = 1] = \begin{cases} 1/n, & \text{if } (i, j, k) = (i', j', k') \\ 1/n(n-1), & \text{if } k \neq k' \text{ and } (i, j) \neq (i', j)' \\ 0, & \text{otherwise.} \end{cases}$$

We can write

$$y_{ij} = \tau_i + \sum_{k=1}^{n} \delta_k^{ij} s_k.$$

Taking $\mu = \sum_{k=1}^{n} s_k/n$ and $\mu_i = \mu + \tau_i$ gives

$$y_{ij} = \mu_i + \sum_{k=1}^{n} \delta_k^{ij}(s_k - \mu).$$

To obtain the linear model (1), let sp $e_{ij} = \sum_{k=1}^{n} \delta_k^{ij}(s_k - \mu)$. Write $\sigma^2 = \sum_{k=1}^{n}(s_k - \mu)^2/n$. Then

$$\mathrm{E}[e_{ij}] = \mathrm{E}\left[\sum_{k=1}^{n} \delta_k^{ij}(s_k - \mu)\right] = \sum_{k=1}^{n} \mathrm{E}\left[\delta_k^{ij}\right](s_k - \mu) = \sum_{k=1}^{n}(s_k - \mu)/n = 0,$$

$$\mathrm{Var}(e_{ij}) = \mathrm{E}[e_{ij}^2] = \sum_{k=1}^{n}\sum_{k'=1}^{n}(s_k - \mu)(s_{k'} - \mu)\mathrm{E}[\delta_k^{ij}\delta_{k'}^{ij}] = \sum_{k=1}^{n}(s_k - \mu)^2/n = \sigma^2.$$

For $(i, j) \neq (i', j')$,

$$\begin{aligned}
\mathrm{Cov}(e_{ij}, e_{i'j'}) &= \mathrm{E}[e_{ij}e_{i'j'}] = \sum_{k=1}^{n}\sum_{k'=1}^{n}(s_k - \mu)(s_{k'} - \mu)\mathrm{E}[\delta_k^{ij}\delta_{k'}^{i'j'}] \\
&= [n(n-1)]^{-1}\sum_{k \neq k'}(s_k - \mu)(s_{k'} - \mu) \\
&= [n(n-1)]^{-1}\left(\left[\sum_{k=1}^{n}(s_k - \mu)\right]^2 - \sum_{k=1}^{n}(s_k - \mu)^2\right) \\
&= -\sigma^2/(n-1).
\end{aligned}$$

In matrix terms, writing $Y = (y_{11}, y_{12}, \ldots, y_{tN})'$, we get

$$Y = X\begin{bmatrix}\mu_1 \\ \vdots \\ \mu_t\end{bmatrix} + e, \quad \mathrm{E}(e) = 0, \quad \mathrm{Cov}(e) = \sigma^2 V,$$

where

$$\begin{aligned}
V &= \begin{bmatrix}
1 & -1/(n-1) & -1/(n-1) & \cdots & -1/(n-1) \\
-1/(n-1) & 1 & -1/(n-1) & \cdots & -1/(n-1) \\
-1/(n-1) & -1/(n-1) & 1 & \cdots & -1/(n-1) \\
\vdots & \vdots & \vdots & \ddots & \vdots \\
-1/(n-1) & -1/(n-1) & -1/(n-1) & \cdots & 1
\end{bmatrix} \\
&= \frac{n}{n-1}I - \frac{1}{n-1}J_n^n.
\end{aligned}$$

It follows that

$$VX = \frac{n}{n-1}X - \frac{1}{n-1}J_n^n X.$$

Since $J \in C(X)$, $C(VX) \subset C(X)$, and least squares estimates are BLUEs. Standard errors for estimable functions can be found as in Section 11.1 using the fact that this model involves only one cluster.

Exercise G.1 Establish whether least squares estimates are BLUEs in a completely randomized design with unequal numbers of observations on the treatments.

G.3 Randomized Complete Block Designs

Suppose there are a treatments and b blocks. The experimental units must be grouped into b blocks, each of a units. Let the experimental unit effects be s_{kj}, $k = 1, \ldots, a$, $j = 1, \ldots, b$. Treatments are assigned at random to the a units in each block. The elementary sampling random variables are

$$\delta_{kj}^i = \begin{cases} 1, & \text{if treatment } i \text{ is assigned to unit } k \text{ in block } j \\ 0, & \text{otherwise.} \end{cases}$$

$$\mathrm{E}[\delta_{kj}^i] = \Pr[\delta_{kj}^i = 1] = \frac{1}{a}.$$

$$\mathrm{E}[\delta_{kj}^i \delta_{k'j'}^{i'}] = \Pr[\delta_{kj}^i \delta_{k'j'}^{i'} = 1] = \begin{cases} 1/a, & \text{if } (i, j, k) = (i', j', k') \\ 1/a^2, & \text{if } j \neq j' \\ 1/a(a-1), & \text{if } j = j', k \neq k', i \neq i' \\ 0, & \text{otherwise.} \end{cases}$$

If α_i is the additive effect of the ith treatment and $\beta_j \equiv \bar{s}_{\cdot j}$, then

$$y_{ij} = \alpha_i + \beta_j + \sum_{k=1}^{a} \delta_{kj}^i (s_{kj} - \beta_j).$$

Letting $e_{ij} = \sum_{k=1}^{a} \delta_{kj}^i (s_{kj} - \beta_j)$ gives the linear model

$$y_{ij} = \alpha_i + \beta_j + e_{ij}. \tag{1}$$

The column space of the design matrix for this model is precisely that of the model considered in Section 8.3. Let $\sigma_j^2 = \sum_{k=1}^{a} (s_{kj} - \beta_j)^2 / a$. Then

$$\mathrm{E}[e_{ij}] = \sum_{k=1}^{a}(s_{kj} - \beta_j)/a = 0,$$

$$\begin{aligned}\mathrm{Var}(e_{ij}) &= \sum_{k=1}^{a}\sum_{k'=1}^{a}(s_{kj} - \beta_j)(s_{k'j} - \beta_j)\mathrm{E}[\delta_{kj}^{i}\delta_{k'j}^{i}] \\ &= \sum_{k=1}^{a}(s_{kj} - \beta_j)^2/a = \sigma_j^2.\end{aligned}$$

For $j \neq j'$,

$$\begin{aligned}\mathrm{Cov}(e_{ij}, e_{i'j'}) &= \sum_{k=1}^{a}\sum_{k'=1}^{a}(s_{kj} - \beta_j)(s_{k'j'} - \beta_{j'})\mathrm{E}[\delta_{kj}^{i}\delta_{k'j'}^{i'}] \\ &= a^{-2}\sum_{k=1}^{a}(s_{kj} - \beta_j)\sum_{k'=1}^{a}(s_{k'j'} - \beta_{j'}) \\ &= 0.\end{aligned}$$

For $j = j', i \neq i'$,

$$\begin{aligned}\mathrm{Cov}(e_{ij}, e_{i'j'}) &= \sum_{k=1}^{a}\sum_{k'=1}^{a}(s_{kj} - \beta_j)(s_{k'j} - \beta_j)\mathrm{E}[\delta_{kj}^{i}\delta_{k'j}^{i'}] \\ &= \sum_{k\neq k'}(s_{kj} - \beta_j)(s_{k'j} - \beta_j)/a(a-1) \\ &= [a(a-1)]^{-1}\left(\left[\sum_{k=1}^{a}(s_{kj} - \beta_j)\right]^2 - \sum_{k=1}^{a}(s_{kj} - \beta_j)^2\right) \\ &= -\sigma_j^2/(a-1).\end{aligned}$$

Before proceeding, we show that although the terms β_j are not known, the differences among these are known constants under randomization theory. For any unit k in block j, some treatment is assigned, so $\sum_{i=1}^{a}\delta_{kj}^{i} = 1$.

$$\begin{aligned}\bar{y}_{\cdot j} &= \frac{1}{a}\left[\sum_{i=1}^{a}\left(\alpha_i + \beta_j + \sum_{k=1}^{a}\delta_{kj}^{i}(s_{kj} - \beta_j)\right)\right] \\ &= \frac{1}{a}\left[\sum_{i=1}^{a}\alpha_i + a\beta_j + \sum_{k=1}^{a}(s_{kj} - \beta_j)\sum_{i=1}^{a}\delta_{kj}^{i}\right] \\ &= \bar{\alpha}_{\cdot} + \beta_j + \sum_{k=1}^{a}(s_{kj} - \beta_j) \\ &= \bar{\alpha}_{\cdot} + \beta_j\,.\end{aligned}$$

Therefore, $\bar{y}_{\cdot j} - \bar{y}_{\cdot j'} = \beta_j - \beta_{j'} = \bar{s}_{\cdot j} - \bar{s}_{\cdot j'}$. Since these differences are fixed and known, there is no basis for a test of $H_0 : \beta_1 = \cdots = \beta_b$. In fact, the linear model is not just model (1) but model (1) subject to these estimable constraints on the βs.

To get best linear unbiased estimates we need to assume that $\sigma_1^2 = \sigma_2^2 = \cdots = \sigma_b^2 = \sigma^2$. We can now write the linear model in matrix form and establish that least squares estimates of treatment means and contrasts in the α_is are BLUEs. In the discussion that follows, we use notation from Section 7.1. Model (1) can be rewritten

$$Y = X\eta + e, \quad \mathrm{E}(e) = 0, \quad \mathrm{Cov}(e) = V, \tag{2}$$

where $\eta = [\mu, \alpha_1, \ldots, \alpha_a, \beta_1, \ldots, \beta_b]'$. If we let X_2 be the columns of X corresponding to $\beta_1, \ldots, \beta_b$, then (cf. Section 11.1)

$$V = \sigma^2[a/(a-1)]\left[I - (1/a)X_2X_2'\right] = \sigma^2[a/(a-1)]\left[I - M_\mu - M_\beta\right].$$

If model (2) were the appropriate model, checking that $C(VX) \subset C(X)$ would be trivial based on the fact that $C(X_2) \subset C(X)$. However, we must account for the estimable constraints on the model discussed above. In particular, consider

$$M_\beta X\eta = [t_{ij}],$$

where

$$t_{ij} = \beta_j - \bar{\beta}_\cdot = \bar{y}_{\cdot j} - \bar{y}_{\cdot\cdot} = \bar{s}_{\cdot j} - \bar{s}_{\cdot\cdot}\,.$$

This is a fixed known quantity. Proceeding as in Section 3.3, the model is subject to the estimable constraint

$$M_\beta X\eta = M_\beta Y.$$

Normally a constraint has the form $\Lambda'\beta = d$, where d is known. Here $d = M_\beta Y$, which appears to be random but, as discussed, $M_\beta Y$ is not random; it is fixed and upon observing Y it is known.

The equivalent reduced model involves $X_0 = (I - M_{MP})X = (I - M_\beta)X$ and a known vector $Xb = M_\beta Y$. Thus, the constrained model is equivalent to

$$(Y - M_\beta Y) = (I - M_\beta)X\gamma + e. \tag{3}$$

We want to show that least squares estimates of contrasts in the αs based on Y are BLUEs with respect to this model. First we show that least squares estimates from model (3) based on $(Y - M_\beta Y) = (I - M_\beta)Y$ are BLUEs. We need to show that

$$C(V(I - M_\beta)X) = C[(I - M_\mu - M_\beta)(I - M_\beta)X] \subset C[(I - M_\beta)X].$$

Because $(I - M_\mu - M_\beta)(I - M_\beta) = (I - M_\mu - M_\beta)$, we have

$$C(V(I - M_\beta)X) = C[(I - M_\mu - M_\beta)X],$$

and because $C(I - M_\mu - M_\beta) \subset C(I - M_\beta)$ we have

$$C[(I - M_\mu - M_\beta)X] \subset C[(I - M_\beta)X].$$

To finish the proof that least squares estimates based on Y are BLUEs, note that the estimation space for model (3) is $C[(I - M_\beta)X] = C(M_\mu + M_\alpha)$. BLUEs are based on

$$(M_\mu + M_\alpha)(I - M_\beta)Y = (M_\mu + M_\alpha)Y.$$

Thus, any linear parametric function in model (2) that generates a constraint on $C(M_\mu + M_\alpha)$ has a BLUE based on $(M_\mu + M_\alpha)Y$ (cf. Exercise 3.9.5). In particular, this is true for contrasts in the αs. Standard errors for estimable functions are found in a manner analogous to Section 11.1. This is true even though model (3) is not the form considered in Section 11.1 and is a result of the orthogonality relationships that are present.

The assumption that $\sigma_1^2 = \sigma_2^2 = \cdots = \sigma_b^2$ is a substantial one. Least squares estimates without this assumption are unbiased, but may be far from optimal. It is important to choose blocks so that their variances are approximately equal.

Exercise G.2 Find the standard error for a contrast in the α_is of model (1).

References

Christensen, R. (1989). Lack of fit tests based on near or exact replicates. *The Annals of Statistics, 17*, 673–683.

Christensen, R. (1991). Small sample characterizations of near replicate lack of fit tests. *Journal of the American Statistical Association, 86*, 752–756.

Christensen, R. (1995). Comment on Inman (1994). *The American Statistician, 49*, 400.

Christensen, R. (2003). Significantly insignificant F tests. *The American Statistician, 57*, 27–32.

Christensen, R. (2005). Testing Fisher, Neyman, Pearson, and Bayes. *The American Statistician, 59*, 121–126.

Christensen, R. (2008). Review of *Principals of statistical inference* by D. R. Cox. *Journal of the American Statistical Association, 103*, 1719–1723.

Christensen, R. (2015). *Analysis of variance, design, and regression: Linear modeling for unbalanced data* (2nd ed.). Boca Raton, FL: Chapman and Hall/CRC Pres.

Christensen, R., & Bedrick, E. J. (1997). Testing the independence assumption in linear models. *Journal of the American Statistical Association, 92*, 1006–1016.

Davies, R. B. (1980). The distribution of linear combinations of χ^2 random variables. *Applied Statistics, 29*, 323–333.

Hinkelmann, K., & Kempthorne, O. (2005). *Design and analysis of experiments: Volume 2, Advanced experimental design*. Hoboken, NJ: Wiley.

Högfeldt, P. (1979). On low F-test values in linearmodels. *Scandinavian Journal of Statistics, 6*, 175–178.

Lehmann, E. L. (1986). *Testing statistical hypotheses* (2nd ed.). New York: Wiley.

Puri, M. L., & Sen, P. K. (1971). *Nonparametric methods in multivariate analysis*. New York: Wiley.

Rao, C. R. (1973). *Linear statistical inference and its applications* (2nd ed.). New York: Wiley.

Author Index

R. Christensen, *Plane Answers to Complex Questions*, Springer Texts in Statistics,
https://doi.org/10.1007/978-3-030-32097-3

Subject Index

R. Christensen, *Plane Answers to Complex Questions*, Springer Texts in Statistics,
https://doi.org/10.1007/978-3-030-32097-3

Y

Z

Printed in the United States
by Baker & Taylor Publisher Services